The History of Negation in the Languages of Europe and the Mediterranean

Volume I: *Case Studies*

OXFORD STUDIES IN DIACHRONIC AND HISTORICAL LINGUISTICS

RECENTLY PUBLISHED IN THE SERIES

3
Case in Semitic
Roles, Relations, and Reconstruction
Rebecca Hasselbach

4
The Boundaries of Pure Morphology
Diachronic and Synchronic Perspectives
Edited by Silvio Cruschina, Martin Maiden, and John Charles Smith

5
The History of Negation in the Languages of Europe and the Mediterranean
Volume I: *Case Studies*
Edited by David Willis, Christopher Lucas, and Anne Breitbarth

6
Constructionalization and Constructional Changes
Elizabeth Traugott and Graeme Trousdale

FORTHCOMING

The History of Negation in the Languages of Europe and the Mediterranean
Volume II: *Patterns and Processes*
David Willis, Christopher Lucas, and Anne Breitbarth

[For a complete list of books published and in preparation for the series see pp. 534–5.]

The History of Negation in the Languages of Europe and the Mediterranean

Volume I: *Case Studies*

Edited by
DAVID WILLIS, CHRISTOPHER LUCAS,
AND ANNE BREITBARTH

OXFORD
UNIVERSITY PRESS

OXFORD
UNIVERSITY PRESS

Great Clarendon Street, Oxford, OX2 6DP,
United Kingdom

Oxford University Press is a department of the University of Oxford.
It furthers the University's objective of excellence in research, scholarship,
and education by publishing worldwide. Oxford is a registered trade mark of
Oxford University Press in the UK and in certain other countries.

First published in 2013

Impression: 1

Published in the United States of America by Oxford University Press
198 Madison Avenue, New York, NY 10016, United States of America

British Library Cataloguing in Publication Data
Data available

ISBN 978-0-19-960253-7

As printed and bound by
CPI Group (UK) Ltd, Croydon, CRO 4YY

Contents

Series preface

Modern diachronic linguistics has important contacts with other subdisciplines, notably first-language acquisition, learnability theory, computational linguistics, sociolinguistics, and the traditional philological study of texts. It is now recognized in the wider field that diachronic linguistics can make a novel contribution to linguistic theory, to historical linguistics, and arguably to cognitive science more widely.

This series provides a forum for work in both diachronic and historical linguistics, including work on change in grammar, sound, and meaning within and across languages; synchronic studies of languages in the past; and descriptive histories of one or more languages. It is intended to reflect and encourage the links between these subjects and fields such as those mentioned above.

The goal of the series is to publish high-quality monographs and collections of papers in diachronic linguistics generally, i.e. studies focussing on change in linguistic structure, and/or change in grammars, which are also intended to make a contribution to linguistic theory, by developing and adopting a current theoretical model, by raising wider questions concerning the nature of language change, or by developing theoretical connections with other areas of linguistics and cognitive science as listed above. There is no bias towards a particular language or language family, or towards a particular theoretical framework; work in all theoretical frameworks, and work based on the descriptive tradition of language typology, as well as quantitatively based work using theoretical ideas, also feature in the series.

Adam Ledgeway and Ian Roberts
University of Cambridge

Preface

This is the first of two volumes setting out the main findings of the project *The development of negation in the languages of Europe*. The second volume will examine the processes and mechanisms by which diachronic developments in negation take place and how these can inform more general theories of linguistic change. This first volume is more narrowly empirical in scope, aiming to document in detail the recurring patterns of development in the expression of negation and indefinites as they are found in the languages of Europe and the Mediterranean. The first chapter provides an overview of the most salient developments in these languages, while the remaining ten chapters present case studies of the history of negation in French, Italo-Romance, English, High German, Low German and Dutch, Brythonic Celtic, Greek, Slavonic, Arabic and Afro-Asiatic, and Mordvin. These case-study chapters have been contributed by nine leading experts on the study of negation in each of these languages or language groups.

This volume and the project from which it arose would not have been possible without the generous funding of the Arts and Humanities Research Council, whose support is hereby gratefully acknowledged.

We would also like to thank the considerable number of friends and colleagues whose help, input, and criticism have undoubtedly improved the content of this volume and the next. While individual contributors' acknowledgements are contained within each chapter, we would like to convey our special thanks to Johan van der Auwera, Liliane Haegeman, and Sten Vikner; to Ian Roberts and Adam Ledgeway for inviting us to include this volume as part of the series *Studies in Diachronic and Historical Linguistics*, as well as to John Davey as linguistics editor at Oxford University Press, for his help and encouragement throughout the project.

List of tables

List of figures

List of grammatical glosses and abbreviations

Grammatical glosses

1SG etc.	first person singular etc.
3FSG, 3MSG, etc.	third person feminine singular, third person masculine singular, etc.
1.SBJM etc.	class 1 etc. subject marker
ACC	accusative (case)
ACTP	active participle
ADV	adverb
ALL	allative (case)
AOR	aorist
AUX	auxiliary
CL	clitic
CNG	connegative (verb form)
COIMP	compound imperfect
COMP	complementizer
COMPAR	comparative
COND	conditional
CONDCONJ	conditional–conjunctive mood
CONJ	conjunctive pronoun
CONSTR	construct-state marker
DAT	dative
DEF	definite
DEM	demonstrative
DEP	dependent verb form
DET	determiner
DIM	diminutive
DO	direct object
ELA	elative (case)
EMPH	emphatic particle

EXPL	expletive
EX	existential (verb)
F	feminine
FOC	focus
FPL	feminine plural
FUT	future
GEN	genitive (case)
GER	gerund
HAB	habitual
ILL	illative (case)
IMP	imperative
IMPERS	impersonal
IND	indicative
INDEF	indefinite
INE	inessive (case)
INF	infinitive or infinitive marker
INTERROG	interrogative
IPFV	imperfective
IRR	irrealis
JUSS	jussive
LAT	lative (case)
LOC	locative (case)
M	masculine
N	neuter
NEG	sentential negative marker
NEG-R	negative reinforcer
NOM	nominative (case)
NONPAST	non-past
NPA	negative polarity adverb
NPI	negative polarity item
OBJ	object
OBL	oblique case
OCL	object clitic
OPT	optative particle
PAST	past tense

PFV	perfective
PFX	prefix
PL	plural
PLUPRF	pluperfect
POSS	possessive
PP	past participle
PRED	predicate marker
PRES	present tense
PRESP	present participle
PRF	perfect
PROG	progressive
PTCL	particle
PTCP	participle
QU	question particle
RECIP	reciprocal
REFL	reflexive
REL	relative marker
SBJ	subject
SCL	subject clitic
SG	singular
SM	soft mutation
SUBJUNC	subjunctive
TAM	tense–aspect–mood marker
VOC	vocative

List of abbreviations

CLEC	Common Lowland East Cushitic
EME	Early Middle English
Eng.	English
ENHG	Early New High German
ex.	example
Fr.	French
Gmc.	Germanic
It.	Italian
l.	line

Lat.	Latin
LF	Logical Form
LME	Late Middle English
Lomb.	Lombard
MHG	Middle High German
MLG	Middle Low German
Mil.	Milanese
Mod. It.	Modern Italian
MSA	Modern Standard Arabic
Neap.	Neapolitan
NC	negative concord
NPA	negative polarity adverb
NPI	negative polarity item
OE	Old English
OHG	Old High German
OLG	Old Low German
p.	page
Pied.	Piedmontese
Sic.	Sicilian
SMG	Standard Modern Greek
TNI	true negative imperative
V	verb
V2	verb second
v.	verse
#	pragmatically infelicitous
*	(in discussion of syntax) ungrammatical (syntactically ill-formed) (in discussion of reconstruction) reconstructed, unattested form
=	clitic boundary
-	morpheme boundary

List of contributors

ANNE BREITBARTH Vakgroep Taalkunde, University of Ghent

ARJA HAMARI Department of Finnish, Finno-Ugrian and Scandinavian Studies, University of Helsinki

MAJ-BRITT MOSEGAARD HANSEN School of Arts, Languages and Cultures, University of Manchester

RICHARD INGHAM School of English, Birmingham City University

AGNES JÄGER Institut für deutsche Sprache und Literatur I, University of Cologne

CHRISTOPHER LUCAS Department of Linguistics, SOAS, University of London

MAIR PARRY School of Modern Languages, University of Bristol

DAVID WILLIS Department of Theoretical and Applied Linguistics, University of Cambridge

JO WILLMOTT Faculty of Classics, University of Cambridge

1

Comparing diachronies of negation

DAVID WILLIS, CHRISTOPHER LUCAS,
AND ANNE BREITBARTH

Negation is one of the few truly universal grammatical categories: every language seems to have some grammaticalized means to deny the truth of an ordinary declarative sentence. Yet the expression of this category varies significantly both from language to language and historically within the same language. For the historical linguist, changes in the way that negation is expressed are therefore an ideal testing ground for theories of change, with every language having the potential to provide important data. Core phenomena in language change are amply exemplified in common developments in negation. In the emergence of new negative markers, we find grammaticalization of lexical items as new grammatical markers of negation: a noun like French *pas* 'step' comes to be reinterpreted as marking negation. We find reanalysis of syntactic structure and of syntactic category: the ancestor of English *not* was once an indefinite pronoun meaning 'nothing' and the object of the verb to which it related, while today it is a specialized negative adverb. We find interactions between syntax and semantics in the form of cyclic developments, as markers of negation are constantly being renewed and replaced by newly emerging ones. We see the results of language contact in the replication of patterns across language boundaries and the emergence of areal preferences for certain patterns of marking negation. Negation provides a wealth of material through which the mechanisms of these core phenomena may be investigated.

The two-volume publication of which this book forms the first part aims to document these patterns of change in negation as they are found repeatedly across the different languages of Europe and the Mediterranean, examining how they can be accounted for, integrating perspectives from formal models of change grounded in language acquisition, from the study of grammaticalization, and from models of language contact. This volume constitutes the first step towards that goal, concentrating on documenting in some detail the patterns of change attested in the histories of individual languages or language groups. This chapter provides an overview of the developments that we find. Thereafter, we present ten case studies that we have commissioned across seven different language groups: five Indo-European (Romance, Germanic, Celtic, Greek, Slavonic), as well as Afro-Asiatic and Uralic. The selection of languages includes a mix of those, such as French and English, with negative systems that have already been the subject of extensive research and which have served as paradigm examples in the general historical linguistic literature, and those, such as Low German or Brythonic Celtic, where research is still very much in its infancy, but whose histories may nevertheless make a significant contribution to our understanding of how negative systems change.

While the second volume will discuss the processes involved in the diachronic development of the expression of negation in more general and more theoretical terms, the present volume aims to showcase the developments in selected languages of Europe and the Mediterranean in order to bring out the common features of these developments, as well as to highlight their differences. At the same time, it presents the state of the art of research into the development of negation in the languages concerned. To this end, leading researchers with expertise in each of the surveyed languages or language groups have been invited to contribute and have been encouraged to integrate their own ongoing research into a presentation of our current understanding of the development of negation in the various language groups.

The second volume is more theoretically oriented and theoretically homogeneous, and aims to generalize across the patterns of diachronic development in negation, bringing out the role of language acquisition and language contact in these developments, and assessing the contributions that formal and functional models of change can make to explaining them. However, the focus of the present volume being on the empirical phenomena in the individual languages, no uniform theoretical approach has been imposed on the individual authors. Nevertheless, a broadly unified format was chosen for the presentation of the material, assuring comparability between the chapters. All chapters begin with a discussion of the changes, where applicable, affecting the expression of sentential negation in ordinary contexts not involving indefinites in the relevant language or language group, and subsequently address any changes affecting indefinites in the scope of negation, as well as possible interactions of these with the expression of negation, including the phenomenon of negative concord. The chapters then move on to discuss other relevant phenomena such as the negation of infinitives, negative imperatives, negative complementizers, and constituent negation wherever these show distinctive developments of their own.

Negation has been the subject of a vast body of scientific literature. Typological work, building on the pioneering surveys of Dahl (1979), Dryer (1988), and more recently Miestamo (2005) and van der Auwera (2010a), has revealed the diverse range of ways in which negation is expressed in the world's languages. Generative work (for instance, Déprez 2000, Giannakidou 2000, 2006, Haegeman 1995, Rowlett 1998, Zanuttini 1997, among many others) has attempted to account for the distribution of the various elements found in negative contexts and to derive their semantics. There is also a long tradition of work in truth-conditional semantics on negation (Horn 1989 and the references cited therein), as well as work from a functionalist perspective (Givón 1978, 2001: ch. 8).

The approach in this volume is not specifically typological, although it is informed by much groundbreaking work in the typological tradition. Rather, it locates itself within the research paradigm of 'comparing diachronies' (cf. the title of a forthcoming collection edited by Jürg Fleischer and Horst Simon, based on a workshop of the same name at the 29th Annual Meeting of the Deutsche Gesellschaft für Sprachwissenschaft in 2008). The result is a narrower focus on a specific subset of the world's languages where detailed historical trends can be established. This commitment to in-depth yet comparative historical analysis leads us to focus in the first instance on the languages of Europe and the Mediterranean. This choice is motivated by the need to take the bulk of our data from languages whose diachronies have been or can be established on the basis of written textual sources, rather than through comparative or internal reconstruction. Such languages allow their historical developments to be established in some detail, including data on the internal progress of change, namely the rate at which changes were implemented and their relative order—facts that can never be established by reconstruction. Most Indo-European languages fall into this category, and we have chosen a selection for inclusion as separate chapters in this volume. Afro-Asiatic was included, even though parts of it are not well documented historically, since North Africa (in Arabic, Berber, and Coptic) manifests one of the main historical developments characteristic of European languages, namely Jespersen's cycle, providing a detailed point of comparison. We have also included a chapter on the Mordvin languages, as representative of the sorts of developments typical in the Uralic languages, partly for geographical completeness and partly to show the contrast with the relatively uniform synchronic and diachronic patterns found in Western Europe. Inevitably, this means that some elements of syntactic reconstruction have been included in the volume, but not to the extent that generalizations about patterns of development will be overly dependent on the use of reconstruction. Varieties of European languages spoken outside of Europe, descendants of European languages, and creoles derived historically from European languages are not specifically included in the scope of this work, but we have adopted a pragmatic approach, and they will be referred to when they instantiate important developments. Naturally, coverage in this volume is better for languages to which a full chapter is dedicated. However, where important or unique developments have occurred in languages not the focus of a dedicated chapter, we have attempted to highlight them in the general survey that follows.

1.1 Sentential negation

Much typological work has focused on the expression of standard negation in the world's languages, defined as 'that type of negation that can apply to the most minimal and basic sentences' (Payne 1985: 198). This is essentially a morphosyntactic notion, the motivation for which is to allow the linguist to establish the main strategy for expressing negation in order to aid cross-linguistic comparison. Standard negation is typically the form of negation found in ordinary main clauses. Only negative constructions that are a productive means of reversing the truth value of a proposition can be considered expressions of standard negation (Miestamo 2005: 42). Hence, for example, the present-day French particle *ne*, which may appear in certain embedded clauses interpreted affirmatively, is not, on its own, considered to be a marker of standard negation. Standard negation is adopted as the basis for discussion in chapters 2 (French), 8 (Greek), and 11 (Mordvin).

A related, but distinct notion is sentential negation, a semantic concept that refers to any instance where an entire proposition, not just some subpart, is negated. This is used as the starting point in the remaining chapters. Sentential negation in English is identified by reference to the classic diagnostic tests of Klima (1964). Negative clauses allow neutral tags without *not*, as in (1); they allow an appositive tag beginning with *not even*, as in (2); and they may be conjoined with a clause of the form *and... either*, as in (3).

(1) Dogs don't like rain, do they?

(2) Dogs don't like rain, not even in summer.

(3) Dogs don't like rain, and cats don't either.

Conversely, a pragmatically neutral tag question attached to an affirmative (or, equivalently, 'positive') clause will contain *not/-n't*, as in (4);[1] it will take an appositive tag with *even* rather than *not even*, as in (5); and, rather than a continuation with *and... either*, only a continuation with *and... too* is possible, as in (6).

(4) Dogs like biscuits, don't they?

(5) Dogs like biscuits, (*not) even in summer.

(6) Dogs like biscuits, and cats do too/*either.

While these tests reliably identify core cases of sentential negation in English, analogues will not necessarily be available in all languages, and, even in English, there are sentences containing negation for which the tests produce unclear results.

[1] The alternative without *-n't*, namely *Dogs like biscuits, do they?* is of course possible, but is not pragmatically neutral: it presupposes the truth of the proposition, while treating it as new, possibly surprising, information.

A means of identifying sentential negation which largely overcomes these difficulties is the 'performative paraphrase' suggested by Payne (1985: 200): if a clause contains sentential negation then it will be paraphrasable in the form *I say (of X) that it is not true that Y*. If we apply this to (1)–(3), for example, we get: *I say of dogs that it is not true that they like rain*. Clearly, affirmative clauses as in (4)–(6) will not be paraphrasable in this way. Here an appropriate paraphrase would be: *I say of dogs that it is true that they like biscuits*.

In Klima's (1964) original discussion of these issues (as also in Payne 1985), the label given to any instance of negation in a clause which does not pass these diagnostics is 'constituent negation'. This therefore covers both instances of morphological negation with, for example, prefixes such as *un-* and *non-* in English *unhappy* or *non-toxic*, as well as negation of some constituent in an otherwise affirmative clause, as in (7).

(7) Not long after, I decided to go home.

It seems intuitively clear that (7) is an affirmative sentence, and this is confirmed by application of Klima's tests and the performative paraphrase.[2]

As Jäger (2008: 21) points out (see also section 5.3.1), it is rarely the case that all that a negative clause conveys is that the proposition expressed is false. Often, contrastive focus is also placed on some element within the clause, inviting the inference that replacing this element with some other member from a set of alternatives would render the proposition expressed no longer false. This is called 'focus of negation'. In the sentence in (8), for example, focus of negation may be placed variously on *burglar*, *window*, *cricket bat*, or the entire verb phrase, but, in each case, we have sentential, not constituent negation, as application of Klima's tests and the performative paraphrase will attest.

(8) The burglar didn't break the window with the cricket bat.

The above outlines the difference between sentential and constituent negation as Klima originally intended it. It is important to note, however, that in practice the term 'constituent negation' is often used more widely (and 'sentential negation' more narrowly) than this. In particular, narrow focus of sentential negation on some lexical phrasal element is often referred to as constituent negation; in English and some other languages, this is especially the case if the negator immediately precedes that element, as in (9), or if the rest of the negated clause has undergone ellipsis, as in (10).

(9) Not everyone likes chips.

(10) I like mashed potatoes, not chips.

[2] Note, however, that appositive tags with (*not*) *even* are only grammatical with generic statements as in (1)–(6), not with episodic clauses as in (7).

Although this use blurs the otherwise clear semantic distinction between sentential and constituent negation, a powerful justification for it is that a number of languages have negators which are specialized not just for true constituent negation, but also narrow-focus sentential negation. Not unreasonably, these items are usually called simply 'constituent negators', although according to standard definitions they mark both constituent negation and focus negation. One such item is Old High German *nalles*: it is used both as a true constituent negator and to mark narrow focus of sentential negation, but not with wide focus of sentential negation (section 5.3.1). The equivalent item *nalles* in Old English has a similar distribution (section 4.2.4), as do Latin *haud* and Welsh *nid* (section 7.11).

Expression of constituent and narrow-focus negation seems to be fairly stable diachronically, though changes in the inventory of constituent negators are occasionally observed. Since changes in standard and (wide-focus) sentential negation are probably more common and certainly more salient, the bulk of the focus in the remainder of this section, as in the book as a whole, is on these.

1.2 Jespersen's cycle

In a now much-cited quotation from his *Negation in English and other languages*, Otto Jespersen observed a development, repeated across various languages, in which a new negative marker emerges and replaces an existing negative marker:

> The history of negative expressions in various languages makes us witness the following curious fluctuation: the original negative adverb is first weakened, then found insufficient and therefore strengthened, generally through some additional word, and this in its turn may be felt as the negative proper and may then in course of time be subject to the same development as the original word. (Jespersen 1917: 4)

This process can recur, leading to a cyclic process of constant renewal, such as that observed in the double cycle from early Latin through to Modern French. Thus, the original negator *ne* was reinforced by *oenum* '(not) one (thing)', the two elements merging as Latin *non*; *non* was reduced phonologically to *ne*, which was in turn reinforced by *pas* '(not one) step' in French (see further below) (Schwegler 1983, 1988, 1990: 151–74). Although the same basic idea had been noted earlier by Gardiner (1904), the phenomenon has generally been associated with Jespersen and was dubbed 'Jespersen's cycle' by Dahl (1979: 88). As Jespersen had already noted, such renewals are observed in other domains too, particularly demonstratives and pronouns (see van Gelderen 2009, 2011). Much linguistic change then consists of the erosion, both phonological and semantic, of existing means of expression, and the countervailing processes by which new forms of expression are created.

The cycle is presented schematically in Table 1.1 for the two most commonly cited languages, English and French. An early preverbal negator, in both cases *ne*, was

TABLE 1.1 **Schematic representation of Jespersen's cycle**

	stage I	stage II	stage III	stage III′
English	ic ne secge	I ne seye not	I say not	I don't say
	(Old English)	(Middle English)	(Early Modern English)	(Present-day English)
French	jeo ne dis	je ne dis pas	je dis pas	
	(Old French)	(Middle and Modern written French)	(Colloquial French)	

reinforced by a newly created emphatic element. In English the source was an indefinite pronoun *not* < *nāwiht* 'nothing', while in French it was a generic noun *pas* 'step' used as a minimizer in expressions such as 'I did not go a step'. Once this element has been bleached, integrated into the negative system, and has become compulsory, the language develops bipartite expression of negation. Finally, the original element is lost, with the language reverting to single marking of negation, but now with a postverbal marker distinct from the original marking. Developments specific to English, involving the emergence of *do*-support, have led to the partial re-establishment of the original situation, with the negative marker *not/-n't* preceding the lexical verb today.

In formal syntactic approaches, Jespersen's cycle has been interpreted as the creation of a new negative specifier, which at first coexists with the former negative head, and eventually replaces it (Rowlett 1998: 92–7, Roberts and Roussou 2003: 154–61, van Gelderen 2008). Following a tradition that goes back to Laka (1990) and Haegeman (1995), and ultimately to Pollock's (1989) split-Infl hypothesis, the syntactic locus of negation is often assumed to be a dedicated negative projection, NegP. A language may vary as to whether it has a marker to fill the head position of NegP, its specifier, or both. It has also been suggested (Zanuttini 1997) that there are several different syntactic positions for NegP, higher and lower in the clause. On this approach, Old English *ne* would be a Neg-head, while *not*, once it entered the negative system, would be a specifier of NegP. At stage I, in (11), *ne* is in Neg^0, and the verb undergoes successive head movement from V to Neg to T (shown by strike-through in the tree), yielding subject–NEG–verb order.[3]

[3] For simplicity, these structures abstract away entirely from the verb-second system of earlier forms of English.

(11)

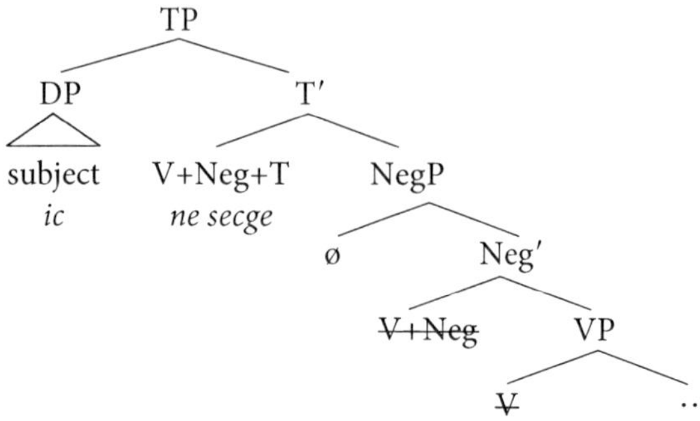

At stage II, *not* comes to express negation, as the specifier of NegP, while all other operations remain the same:

(12)

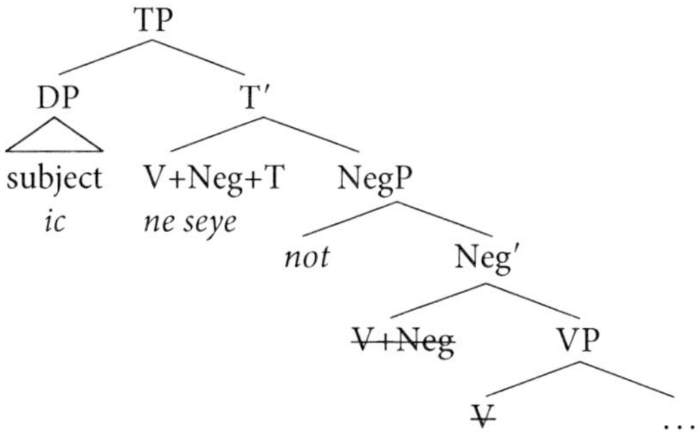

At stage III, *ne* ceases to express the Neg^0-head, which now has zero phonological realization:

(13)

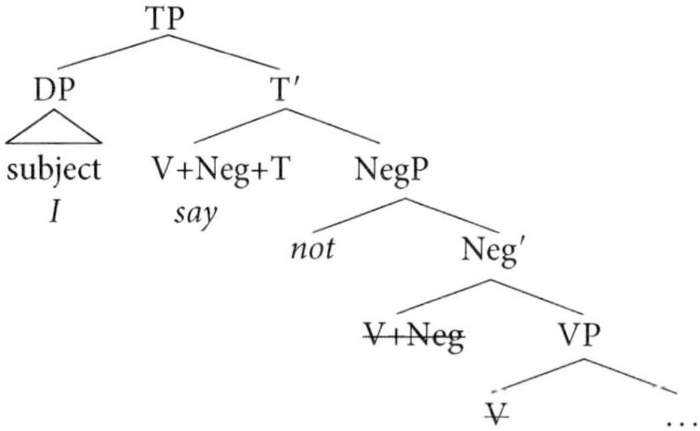

Stage III′, shown in (14), involves the loss of verb movement, unique to English, and its replacement by *do*-insertion. At this stage, *-n't* is still the realization of the specifier of NegP, and the head is empty, but the way is open to a return to stage I, via reanalysis of *-n't* as the head of Neg.

(14)

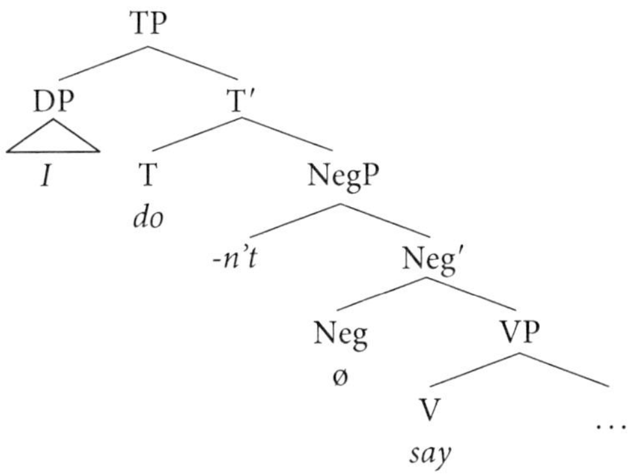

This presentation is of course highly schematic, and omits some important details and complications that will be discussed below. The situations are referred to as distinct stages in Table 1.1, with stage I representing single marking of negation in preverbal position, stage II representing bipartite negation, and stage III representing single marking of negation in postverbal position. However, more stages could be postulated (see section 1.5 below), and the stages do not necessarily reflect discrete time periods in a language's development. They might in fact be better thought of as 'syntactic structures' or 'constructions' rather than 'stages', since more than one (and indeed all three) may be found in complex patterns of variation at any given time. Furthermore, the idealized stages gloss over a number of other important issues, in particular, the syntactic and semantico-pragmatic status of the new element—is it a noun (phrase), an adverb (phrase) or a negator? Is it emphatic or neutral? The speed with which a language progresses through the stages is also highly variable, with some languages showing stable, long-term variation between two or more patterns, while others shift from stage I to stage III with stage II more closely resembling a historical linguist's idealization than a distinct historical period.

1.3 The cross-linguistic distribution of Jespersen's cycle

Jespersen's cycle has occurred extensively in the languages of Western Europe and North Africa and thus receives a prominent place in this volume. We find a full shift to stage III within Germanic, in English (section 4.2.1), Dutch and Low German (section 6.2), High German (section 5.1), and Scandinavian (Old Norse) (see below in this section); in Italic/Romance, in early Latin, colloquial French (section 2.2), some northern Italian dialects (section 3.2), and most Romansh varieties (Krefeld 1997); in Celtic, in Welsh (section 7.3); and perhaps in the history of Greek (section 8.2) and Hungarian (section 1.4). There is a partial cycle, with a shift to (broadly) obligatory stage II in Afrikaans, standard French (section 2.2) and Breton (section 7.4). Coexistence of stage I and stage II constructions is probably in fact very widespread (depending on how strictly stage II is defined, see below), but well-established stage II constructions are found alongside stage I in Catalan, standard Italian and some northern Italian dialects (section 3.2), and Estonian (section 1.4). Brazilian Portuguese allows all stages in some variety (Schwenter 2005: 1428–9). Finally, we find various stages of Jespersen's cycle in numerous Afro-Asiatic languages, such as Arabic (stage II throughout North African dialects, stage III in Palestinian), Berber (stage II in Morocco and Algeria, stage III in parts of Libya), Coptic (stage III), and others besides (section 10.2). Jespersen's cycle is often thought of as mediating a shift from preverbal to postverbal negation, and indeed the idea that the cycle allowed a language to realign its word order along more harmonic lines was crucial to some typological approaches in the 1970s (Vennemann 1974, Harris 1978). However, we take the concept to be broader than this, to include the emergence and spread of any new pragmatically marked negator along a trajectory towards possible eventual replacement of an earlier negator (cf. Schwegler 1983: 301). The defining property of Jespersen's cycle is thus the cyclic strengthening and weakening of negative markers, rather than the concomitant changes in word order that are often present.

As a further example, let us consider the development of negation in historical Scandinavian, which underwent Jespersen's cycle not just once, but twice in its early history (Eythórsson 1995: ch. 2, 2002). In early Old Norse (prior to the 7th century), the inherited Germanic preverbal negator *ni/ne*, illustrated in (15), could already be amplified by a postverbal clitic *-(a)t/-a*. This is illustrated in (16).

(15) **ni** s solu sot uk **ni** sakse stain skorin
NEG is sun.DAT hit.PP and NEG knife.DAT stone cut.PP
'It is not hit by the sun and a stone is not cut with a knife.' (Eggja inscription, Eythórsson 2002: 196)

(16) ef Gunnarr **ne** kømr=**að**
if Gunnarr NEG come.PRES.3SG=NEG
'if Gunnarr does not come' (Akv 11, Eythórsson 2002: 194)

These items may have arisen historically from combinations containing an adverb 'ever', for instance *-a* from *aiwa- 'ever' and *-at* from *aiwa-wehti- 'ever anything' (Eythórsson 2002: 194) (cf. the use of Old English *nā* 'never' as a sentential negator and *nān wiht* 'no thing, nothing' as the ancestor of Modern English *not*) or may possibly derive from the numeral 'one' (*ainata one.NEUT / *aina one.MASC) (Eythórsson 2002: 219).

In Old Norse (Old Icelandic), between the seventh and ninth centuries (between the Eggja inscription and the earliest eddic and skaldic poems), preverbal *ne* became optional and disappeared, as in (17).

(17) gaft=**at**=tu ástgiafar
give.PAST=NEG=you.SG love.presents
'You did not give love presents.' (Rm 7, Eythórsson 2002: 199)

In negative main clauses, the finite verb would front before the subject, leaving *ne* in sentence-initial position. Eythórsson suggests that its disappearance was due to phonological deletion that affected other unstressed syllables in this position as well, cf. Jespersen's (1917: 5) suggestion that negative elements in sentence-initial position are uniquely vulnerable to loss.

Old Norse *-at/-a* was limited to finite verbs and imperatives, and was complemented by *eigi* 'not' < (*ne*) *eitt=gi* 'not one.NEUT=at.all' (Christensen 2003) between the ninth and eleventh centuries, first in non-finite contexts (18), then generally (19).[4] The suffix *-at/-a* disappeared by the fourteenth century (Eythórsson 2002: 217).

(18) Enn Atli qvað=z **eigi** vilia.
but Atli say.PAST.3SG=REFL NEG want.INF
'But Atli said that he did not want.' (Od 22, Eythórsson 2002: 195)

(19) **Eigi** má ek þat vita.
NEG can.PRES.1SG I that know.INF
'I cannot know that.' (Christensen 2003: 24)

The history of the Scandinavian languages has thus witnessed two renewals of the expression of negation along Jespersen's cycle: (i) reinforcement of *ne* with *-a/-at*; and (ii) replacement of *-a/-at* by *eigi*. The latter element is the ancestor of all modern Scandinavian negators: Swedish *inte/icke*, Danish and Norwegian (Bokmål) *ikke*, Neo-Norwegian (Nynorsk) *ikkje*, Icelandic *ekki*, and Faroese *ikki* (Eythórsson 2002: 190).

Outside of Europe, Jespersen's cycle has been identified less frequently and may be less common. Synchronically, languages with bipartite expression of negation reminiscent of stage II of Jespersen's cycle are found sporadically across Niger–Congo languages in Central and West Africa. These, along with other Niger–Congo languages which express negation using a postverbal particle, seem to have undergone Jespersen's cycle developments against the background of a protolanguage with prefixal negation (Güldemann 2011: 117). Devos, Tshibanda, and van der Auwera (2010) discuss in some detail the (reconstructed) history of negation in the Bantu language, Kanincin, arguing for multiple, overlaid instances of Jespersen's cycle in its development. Kanincin negation may be expressed in a number of ways, including bipartite *ka-* ... *-p(a)* or tripartite *ka-* ... *-p(a)* ... *kwend* etc., as illustrated in (20) and (21). For phonological reasons, *-p(a)* mostly appears as *-ap* in the examples.

(20) (mvûl) wù-nàk-áán
(rain) 1.SBJM-rain-TAM
'It's raining.' (Kanincin) (Devos, Tshibanda, and van der Auwera 2010: 6)

[4] There may also be some dialect and stylistic variation, with *-at/-a* characteristic of West Norse (Icelandic and Norwegian) poetry, while *ekki* is characteristic of East Norse and prose (Eythórsson 2002: 195–6).

(21) (mvûl) **kà**-nák-ááŋ-**áp** **(kwénd)**
(rain) NEG.1.SBJM-rain-TAM-NEG NEG
'It's not raining.' (Kanincin) (Devos, Tshibanda, and van der Auwera 2010: 7)

The original (stage I) negator *ka-* was reinforced by *-pa*, originally a locative (noun-class-16) pronoun, but also used to mean 'a little':

(22) táámbúl-**àp**
take.IMP-16.LOC
'Take a little!' (Devos, Tshibanda, and van der Auwera 2010: 14)

In the scope of negation, then, this would once have acted as a minimizer 'not (even) a little, not at all', from where an emphatic negative meaning developed. The second element in (21), *kwend*, is a possessive adjective with a (class-17) pronominal prefix ('at his' hence 'as for him'), used to mark contrastive focus in affirmative sentences:

(23) à-y-á **kw-énd**
1.SBJM-go-SUBJUNC 17.at-1.POSS
'He should just go.' (i.e. he should not do anything else) (Devos, Tshibanda, and van der Auwera 2010: 20)

In negative clauses, however, as in (21), it has become part of the negation and is no longer felt to convey emphatic negation (see also Devos and van der Auwera forthcoming). This entire historical scenario parallels the Western European Jespersen cycles remarkably closely.

Similarly, the Australian (Western Daly) language Maranungku marks negation using any of the following patterns (Miestamo 2005: 62–3, Tryon 1970):

(24) *way piya* Aux V
piya Aux V *way* (emphatic)
V *way piya* Aux
piya V Aux

Both bipartite marking with *way piya* and single marking with *piya* alone are possible. *Piya* originally meant 'head' (Miestamo 2005: 63), hence has a plausible origin as a minimizer, suggesting that the language, having innovated a new negator *piya* from a minimizer source, has been generalizing it, and is currently in the transition between stages II and III of Jespersen's cycle.

Nevertheless, the synchronic frequency of bipartite negation is not a reliable guide to the frequency of Jespersen's cycle, particularly if stage II of the cycle is inherently unstable. A language that has progressed from stage II to stage III of Jespersen's cycle ends up with a single negative marker. It may therefore be difficult to establish if a language with a poorly documented textual record has undergone Jespersen's cycle unless comparative reconstruction or internal alternations imply an appropriate etymology for the negative marker.

Jespersen's cycle raises a number of important issues that can only seriously be addressed on a comparative basis. Many of these questions concern the internal structure of the cycle, what elements are recruited as new negators, and how the cycle moves forward. It is to these that we now turn.

1.4 Reanalysis, bridging contexts, and incipient markers of negation

Consider first the initial phase of the cycle, which we refer to as 'incipient Jespersen's cycle' (following Breitbarth, Lucas, and Willis 2013). The sources of new negators seem to be relatively constrained. New negators are overwhelmingly derived from nominal minimizers such as (*not*) *a drop* or generalizers such as (*not*) *anything* (*at all*). Noun phrases are used as minimizers when they are used to express the idea that a proposition does not hold even at the lowest point on some relevant scale (cf. Eckardt 2006). Generalizers, such as free-choice items, invite the hearer to expand the set of situations under consideration to include all possible worlds, expressing the idea that the current proposition does not hold in any of them (cf. Giannakidou 2001). Nouns denoting small quantities are typical minimizers; those with a fairly generic meaning are more likely to be frequent enough to become new markers of negation. Indefinite pronouns are also plausible candidates for participation in Jespersen's cycle because they may act both as minimizers ('not even a single thing') or as generalizers when they give rise to free-choice interpretations ('not any of the things that could possibly be imagined'):

(25) Mary didn't eat anything at all (not even a crumb) (i.e. not even the smallest thing)

(26) Mary didn't notice anything at all (anything you could conceive of her noticing) (i.e. for any arbitrarily chosen thing she might have noticed, Mary didn't notice it)

Nouns used as minimizers are the basis for a full, partial, or incipient Jespersen cycle in many Romance varieties (Schwegler 1988, Muller 1991: 211): French (*pas* 'step', *mie* 'crumb', *point* 'point', *goutte* 'drop') (section 2.2.1), Occitan (*pas* 'step') (Granda 1999), Catalan, some northern Italian dialects (Italian *mica* etymologically 'crumb', Modenese *brisa* 'crumb') (section 3.2), and Romansh (Engadine *brich*(*a*) < 'crumb', cf.Italian *bricia* 'crumb', Romagnolo *briʃa* 'not'; Central Romansh *betg*(*a*) perhaps < Latin *bāca* (*m*) 'berry' or a variant of *brich*(*a*); Sursilvan Romansh *buc*(*a*) perhaps < Latin *bucca* (*m*) 'mouth(ful)' or a variant of *betg*(*a*)) (Planta, Meicher, and Pult 1939: ii. 499–507, Avery 1978, Krefeld 1997, Liver 1999: 147–8, Detges 2003). Incipient developments of this type are also found elsewhere (see below).

Indefinite pronouns are the basis for Jespersen's cycle in some Italo-Romance varieties, such as Piedmontese *nen* < *nent* 'nothing' < *ne*(*c*) *gente*(*m*) 'no people' (section 3.2), English *not* < *nāwiht* 'not anything, nothing' (section 4.2.1), German *nicht* (section 5.1.1.2), and Dutch *niet* (section 6.2) (both from *ni-eo-wiht* 'not ever a thing, nothing'), Old Norse *-at* and *eigi* (see above), Welsh *ddim* < 'anything, thing' (section 7.3.2), Greek *δen* < Classical Greek *ouden* 'nothing' (section 8.2.2), North African Arabic dialects *-š* < Classical Arabic *šayʔ* 'thing, anything' (section 10.2.1.1), Central Atlas Tamazight Berber *ša* < *kra* 'thing' (section 10.2.1.3), and Hungarian *nem* < *n- + pronoun *mi* 'what; thing', cf. Komi-Zyrian, *ńi-nẹm* 'nothing' (Rédei 1970, Honti 1997: 164). Incipient developments along these lines are also common elsewhere (see below).

Languages usually have already adverbial means of strengthening the expression of negation, such as English *at all* or *for the life of me*, or Old Low German (Old Saxon) *an thesaru uueroldi* 'in this world'. Such expressions seem to form the basis of a successful Jespersen cycle only rarely, if at all. They may, however, grammaticalize to reinforce the expression of negation or negative polarity in the system of indefinites, which, like the expression of sentential negation, seems to be subject to cyclic renewal (see section 1.9 below). This has happened extensively in the Celtic languages; for instance, Middle Welsh *dim yn y byd* 'anything in the world' was contracted to Modern Welsh *dim byd* 'anything, nothing' (section 7.6.2), and similar developments are evident in the emergence of other elements that form negative indefinites, such as Breton *ebet* (< Middle Breton *en bet* 'in the world') (section 7.7), or form negative polarity indefinites, such as Irish *ar bith* and Scottish Gaelic *sam bith* (< Old Irish *for bith* 'on the world', *isin bith* 'in the world') (section 7.9.3).

A more successful adverbial source for Jespersen's cycle involves temporal adverbs. Use of temporal adverbs in generalizing senses, 'never' (> 'in no possible world') > 'not', gives rise to new emphatic negators in Jespersen-type developments in various languages, but does not often lead to full Jespersen's cycle. Lucas and Willis (2012) examine how *never* in Modern English has come to be used to express sentential negation via this development in some varieties, as in:

(27) You sure you **never** nicked it? (Cheshire 1998: 47)

A full (accelerated) development of this type is attested in Cape Verdean Portuguese Creole, where the preverbal negator *ka* derives from Portuguese *nunca* 'never' (Naro 1978: 330–3, Teyssier 1986):

(28) Bu **ka** paga kel renda.
you NEG pay that rent
'You didn't pay rent.' (Cape Verdean Portuguese Creole) (Baptista 2002: 116)

Abortive cases of this development (where the item was used as an emphatic negator for some time, but is not the majority development subsequently) include Old

English *nā* 'never, not at all' (section 4.2.1) and its Old High German cognate *nio* 'never, not at all' (section 5.1.1.2).

A third option is rather different but also involves a structural reorganization, this time of a clause-final anaphoric negator 'no'—'I didn't do it, no'—or resumption of the ordinary sentential negator—'I didn't do it, not at all'. This seems to form the basis for something akin to Jespersen's cycle in Afrikaans (Biberauer 2009, 2012) and some varieties of Dutch (e.g. Brabantic, van der Auwera 2009: 49–53), Brazilian Portuguese (Schwegler 1991b, Schwenter 2005), some northern Italian dialects, notably Milanese (section 3.2), several Modern South Arabian languages (section 10.2.1.2) and perhaps Hausa (section 10.1.2), along with various creoles, such as Palenquero (Schwegler 1991a). This development is illustrated in (29) for Afrikaans, where bipartite negation is now usual, and in (30) for Brazilian Portuguese, where it remains pragmatically marked.

(29) Ek is **nie** ryk **nie**.
I be.PRES NEG rich NEG
'I'm not rich.' (Afrikaans) (Biberauer 2012: 6)

(30) A Cláudia **não** veio à festa **não**.
the Cláudia NEG come.PAST.3SG to party NEG
'Cláudia didn't come to the party.' (Brazilian Portuguese) (Schwenter 2005: 1429)

In Italo-Romance varieties and perhaps also in Palenquero, the anaphoric or resumptive negator comes to be increasingly integrated into the clause, coming with time to occupy an earlier syntactic position. It is unclear though if these form a unified class, since the Dutch–Afrikaans case has been analysed as arising historically from extension of negative concord from indefinites to the sentential negator (van der Auwera 2009: 49–53), and this seems at least as plausible as reanalysis of the anaphoric negator. Furthermore, the possibility that these patterns arise in some cases from replication of syntactic patterns via contact (cf. Bernini and Ramat 1996: 64–81) cannot be excluded.

In addition to identifying the possible sources of new markers, we need to consider the syntactic, semantic, and pragmatic changes that they undergo and compare these from language to language. An item in a language may be said to be involved in incipient Jespersen's cycle when it first begins to stray beyond its historical semantic and syntactic environment, becoming a plausible candidate for subsequent recruitment as a new marker of negation. Entry into Jespersen's cycle involves several steps. The process may stop at any one of these steps, and, in fact, often seems to stop early, leaving the new item as a marker of emphatic negation but nothing more. Such elements are often stable, and show no signs of replacing the previous exponent of ordinary sentential negation. The processes involved at the early stages of Jespersen's cycle are the same, however, regardless of whether the item in question goes on to become the main exponent of ordinary sentential negation in the language or remains a minor adverbial means of emphasizing polarity.

Many of the typical sources for Jespersen's cycle are initially noun phrases, and hence start out as arguments of whatever class of verb is appropriate to their lexical meaning. For instance, a noun phrase meaning 'drop' is naturally restricted to acting as a minimizer as the direct object of a verb involving the manipulation of fluids, such as 'drink' or 'spill', but not verbs such as 'see' or 'read'; even a pronoun such as 'anything' begins life as an inanimate direct object and must be reanalysed as an adverb and lose restrictions on animacy. All these items must be acquired successfully, but, in many contexts, they may be acquisitionally ambiguous between an analysis as a noun-phrase object or as an adverbial, thereby providing a syntactic 'bridging context' for change. This is the case for an optionally transitive verb such as *drink* in (31) and (32).

(31) I didn't drink a drop.

(32) I didn't drink anything.

In both of these cases, there is acquisitional ambiguity, and an acquirer must consider both the possibility that *a drop* and *anything* are noun phrase objects and that they are adverbials. If, for some reason, they opt for the latter, innovative option (reanalysis), the item will naturally extend its distribution, becoming a possible modifier of any verb type (extension) (Breitbarth, Lucas, and Willis 2013). Alongside this, semantic bleaching occurs, replacing lexical, referential meaning with a semi-grammatical meaning conveying emphatic negation, as with *one bit* in (33).

(33) I don't like this one bit.

The items that have formed the basis of a successful Jespersen cycle have all undergone developments of this type. Incipient Jespersen's cycle is, however, common, and does not guarantee continuation into Jespersen's cycle proper. Various European languages today allow indefinite pronouns in certain contexts where they are no longer direct objects, showing the potential for extension to general use as adverbials:

(34)

¿Quieres	decir	que	no	lloró	**nada**?
want.PRES.2SG	say.INF	that	NEG	cry.PAST.3SG	nothing

'You mean he didn't cry at all?' (*Collins Spanish dictionary*, s.v. *all*)

(35) A:

Eu	fui	ao	Japão	quando	era	novo.
I	go.PAST.1SG	to	Japan	when	be.IMPF.1SG	young

'I went to Japan when I was young.'

B:

Não	foste	**nada**.
NEG	go.PAST.2SG	nothing

'You didn't!' (Portuguese)

(36) Von Freitag auf Samstag hab ich aber fast **nichts** geschlafen.
from Friday to Saturday have I however almost nothing slept
'However, between Friday and Saturday I hardly slept.' (German) (Bayer 2009: 12)

(37) Danes nisem spal **nič**.
today NEG.be.PRES.1S sleep.PP nothing
'Today I didn't sleep at all.' (Slovene)

Estonian has a well-established negative adverb *mitte* (connected to a partitive of 'what', probably used as an indefinite emphasizer 'nothing', Mägiste 1983 s.v. *mitte*, Honti 1997: 164, cf. also Finnish *mitään* below), normally the constituent negator, but used in sentential negation too (Ehala 1996: 20–3), illustrated in (39) as compared to (38). It retains the emphatic quality of early stage II constructions (Tauli 1983: 125).

(38) Ta ei lähe.
he NEG go.CNG
'He's not going.' (Estonian) (Ehala 1996: 21)

(39) Ta ei lähe **mitte**.
he NEG go.CNG NEG.EMPH
'He's not going (at all).' (Estonian) (Ehala 1996: 21)

These all show the potential for full-scale Jespersen's cycle, but are not sufficient for it in and of themselves.

Details of such developments are considered for each individual Jespersen cycle in this volume. Often, contextual restrictions live on for some time: in the case of a negative polarity adverb that originated from a direct object, the adverbial use may be blocked in cases where it could be confused with its former nominal use. Thus, while Venetian *gnente* (section 3.3.2), originally 'nothing' as in (40), may now be used for emphatic negation with an intransitive verb in (41), it is still unavailable with a transitive one in (42):

(40) Nol lavora **gnente**.
NEG.SCL work.PRES.3SG nothing
'He doesn't work.' (Venetian) (Garzonio and Poletto 2009: 133)

(41) Nol dorme **gnente**.
NEG.SCL sleep.PRES.3SG nothing
'He doesn't sleep.' (Venetian) (Garzonio and Poletto 2009: 133)

(42) *Nol leze **gnente** i libri.
NEG.SCL read.PRES.3SG nothing the books
'He isn't reading the books.' (Venetian) (Garzonio and Poletto 2009: 133)

Incipient negators are often permitted in a range of negative polarity environments, such as interrogative clauses, conditional clauses, and comparative clauses, and not merely in strictly negative contexts, thus acting, for a while at least, as negative polarity items (NPIs) (for a full definition of this term, see section 1.8.1 below). The details seem to depend on the properties of the item in question before it was recruited as a negator. Thus, indefinites originally restricted to negative contexts seem to grammaticalize immediately as negative reinforcers, as with English *not* and Middle High German *niht*, while other items may go through a negative polarity stage. Such a stage is documented for French *point* (section 2.2.1), illustrated in (43), Welsh *ddim* (section 7.3.2) and Breton *ket* (section 7.4), all of which subsequently become restricted to negative clauses.

(43) ...il leur fist savoir se nostre dit filz le Dauphin
he them.DAT make.PAST.3SG know.INF if our say.PP son the dauphin
yroit **point** en Normandie.
be.IMPF.3SG POINT in Normandy
'...he let them know if our aforementioned son the dauphin was ever in Normandy.' (Jean Chartier, *Chronique de Charles VII* 101) (Catalani 2001: 362)

While such uses have died out in French and Welsh, they remain available in some (southern) dialects of the generally more conservative Catalan, as the following example demonstrates:

(44) Que ho volia fer ella **pas**?
that it want.PAST.3SG do.INF she EMPH
'Do you think she wanted to do it (by any chance)?' (Dialectal Catalan) (Tubau 2008: 250)

Finally, in the early stages of Jespersen's cycle, the new marker of negation always seems to be emphatic, if only in virtue of being a marked option relative to the original negator alone. This makes sense if we understand linguistic emphasis, following Israel (1998, 2001), as being when the proposition expressed entails what one might normally expect to be expressed, given the context. A new negative construction also frequently cancels a presupposition explicit or implicit in the discourse that the negated proposition was in fact true, see Hansen (2009) and Hansen and Visconti (2009). Hence, the Catalan sentence in (45) with emphatic negator *pas* is appropriate only in a discourse context where it has previously been said, implied, or inferred that 'It was all done badly' (see also Espinal 1993); and the marked Finnish construction with the emphatic negator *mitään* (partitive of *mikään* 'nothing') in (46) is appropriate only if the truth of 'you slept' is contextually presupposed (e.g. in response to 'How long did you sleep for?').[5]

[5] Fronting of the negative auxiliary in (46) also contributes to this meaning. The usual emphatic negative would be as in (i), without the new negator *mitään*, and without fronting of the negative auxiliary *en*:

(i) (Mä) en nukkunut yhtään.
I NEG.1SG sleep.CNG at.all
'I didn't sleep at all.'

(45) No dic **pas** que tot s' hagi fet malament.
NEG say.PRES.1SG NEG that everything REFL have.PRES.1SG do.PP badly
'I'm not (in fact) saying it was all done badly.' (Catalan) (Wheeler 1988: 199)

(46) (En) mä **mitään** nukkunut.
NEG.1SG I NEG sleep.CNG
'I didn't sleep (I was doing something else).' (Finnish) (Silva Nurmio, p.c.)

Whether this is a stage found in all Jespersen cycles is a topic of ongoing investigation. It is well attested for Old French (section 2.2.2) *ne ... pas*, Italian *non ... mica* (section 3.2.1), Catalan, Brazilian Portuguese *não ... não* (Schwenter 2005), and the Spanish-based creole Palenquero *nu ... nu* (Schwegler 1991a: 180–1), and was perhaps also found in Middle English *ne ... not* (Wallage 2012).

The motivation behind the central portion of the cycle also needs to be considered. Once an item has become established as marker of emphatic negation, how and why does it go on to replace the original marker? While Jespersen's presentation of the cycle implied a pull-chain scenario, in which change is motivated by the phonological weakness of the preverbal negation and thus its lack of fitness to act as an effective means of communication, the focus today is much more on the inflationary effects of overuse of the new negator, its initial emphatic pragmatic quality becoming devalued over time until it becomes a plain marker of negation (see Schwegler 1983: 320–1, Detges and Waltereit 2002, Detges 2003, Kiparsky and Condoravdi 2006). Having created a new marker of emphatic negation, speakers appear to overuse it 'in order to pretend that the negative assertion of some state of affairs ... is particularly surprising and relevant' (Detges and Waltereit 2002: 183). Under such a push-chain scenario, the preverbal marker is only lost when the new marker takes over the function of expressing sentential negation. Once the new marker is pragmatically neutral and no longer expresses emphasis, the old marker is functionally redundant and pressure is created for it to be eliminated.

The idea that change is driven by the 'weakness' of the preverbal negative marker overlooks the fact that many languages survive with a non-stress-bearing negator: for instance, the contrast between northern Italo-Romance varieties with Jespersen's cycle and south central ones without it does not correlate with any phonological difference in the preverbal negator (Posner 1985: 188, Breitbarth 2009: 85–6). On the other hand, the absence of Jespersen's cycle in Slavonic could be explained by reference to the phonological strength of the Slavonic negator, which can normally be stressed and can be stranded in clause-final position. Slavonic languages do develop emphatic negators (section 9.3), but none has ever been devalued and participated in a full-scale Jespersen's cycle. It is unclear why inflationary processes that operate in Western Europe should be absent in Slavonic, but it makes sense that a stress-bearing negator might be more resilient in the face of Jespersen's cycle.

'Hybrid' approaches argue that the 'weakening' (in a morphosyntactic, not necessarily phonological, sense) of the preverbal marker and the establishment of a new negator, leading to the demise of the preverbal element, go hand in hand (Wallage 2005, 2008, Breitbarth 2009, Willis 2010: 148–9).

1.5 Progression through Jespersen's cycle: how many stages and how fast?

Different linguists have come to different conclusions about the number of stages that need to be distinguished in Jespersen's cycle. At its simplest, and given in idealized form in (47), we have a three-stage cycle: at stage I, the old marker is present; a new marker emerges at stage II, to give bipartite expression of negation; while, by stage III, this marker has entirely replaced the original one.

(47) stage I NEG VERB
stage II NEG VERB NEG
stage III VERB NEG

Under this conception, a 'stage' refers to a particular construction type, and it makes sense to say that a language at a particular time manifests both stage I and stage II (i.e. the original negator is compulsory and the new one is optional).

Stage II though clearly has a great deal of internal complexity, in both semantic/pragmatic and syntactic terms, with stage II constructions potentially reflecting a number of different grammatical systems. As we have seen, it seems that, in all cases, the new marker is first emphatic or carries some other special pragmatic significance, such as cancellation of presupposition. When it is pragmatically meaningful, it is naturally optional, but as it rises in frequency it becomes devalued, becoming the usual exponent of negation under all circumstances. Syntactically, its category may change from noun or pronoun to adverb, in rough schematic terms:

(48) I did not go [DP a step] > I did not go [AP a-step] > I did not go [NegP a-step]
It did not help [DP a bit] > It did not help [AP a-bit] > It did not help [NegP a-bit]

Its syntactic position may also change, moving from clause-late to an earlier position, a change which may reflect greater integration into the negative system proper (for instance, reanalysis of an adverb as occupying a dedicated position within a negative projection). This change is attested in Welsh (section 7.3.2) and is inferrable for Italo-Romance varieties (section 3.7). Willis (2010, 2011a) treats this as reflecting a reanalysis of the new negator from VP-adverb to dedicated specifier of NegP. Hence, when pattern (49) is replaced by pattern (50) in Welsh, with the new negator now preceding prepositional phrase complements of the verb, this is interpreted as reflecting a new structural position for the negator within NegP, illustrated in the shift from (51) to (52) (assuming movement of the verb to the left of NegP).

(49) NEG waited they for.him NEG

(50) NEG waited they NEG for.him

(51)

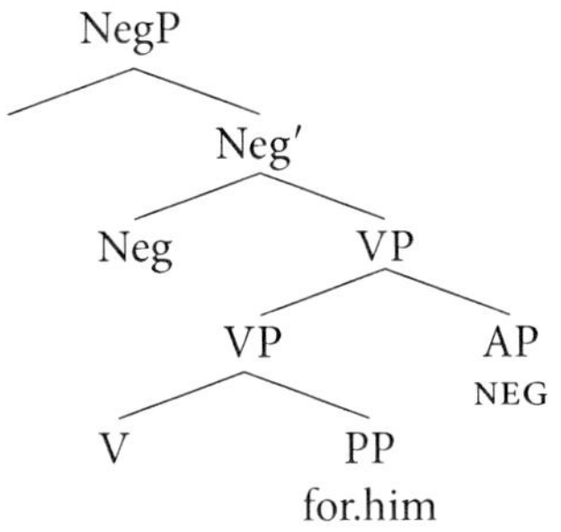

(52)

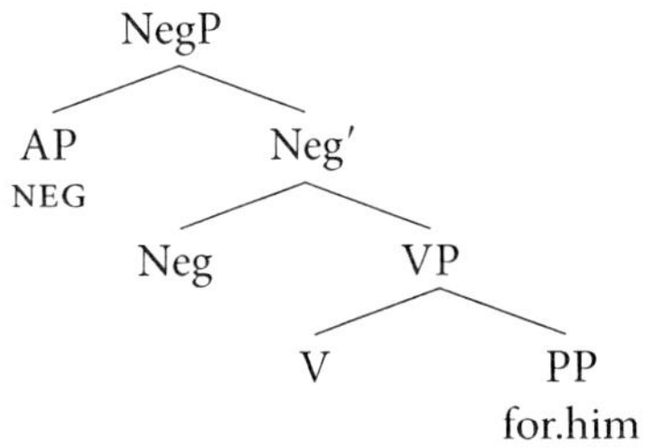

For Zanuttini (1997), this would represent a shift in the locus of negation from a lower to a higher specifier position.

The broadest conception of stage II encompasses all of the following: the initial grammaticalization of a new marker as an optional means of expressing emphatic or presuppositional negation, a shift in the locus of negation from the old to the new marker, and the early signs of the loss of the older marker. Completion of this loss amounts to the transition to stage III. The breadth of stage II under this conception means it is sometimes useful to recognize substages, where the new negator has a different semantic or syntactic status (cf. Schwegler 1988):

(53) stage IIa NEG VERB NEG.EMPH
stage IIb NEG VERB NEG

This, superimposed on (47), would effectively lead to a four-stage cycle.

Other linguists prefer to treat the stages of Jespersen's cycle as characterizing the entire language, so that, for instance, a separate stage is recognized where NEG VERB and NEG VERB NEG coexist. This leads to a five-stage system such as the following, cf. also van der Auwera's (2009: 39) six-stage system:

(54) stage 1 NEG VERB
stage 2 NEG VERB (NEG/NEG.EMPH)
stage 3 NEG VERB NEG
stage 4 (NEG) VERB NEG
stage 5 VERB NEG

Zeijlstra (2004: 56) adopts this articulated system, but with an additional final stage where the new negator attaches to the verb (see section 8.2.3, especially Table 8.1 for details). Another schematization might allow for the possibility of NEG VERB co-existing alongside NEG VERB NEG and VERB NEG, a synchronic combination which cannot be described in (54), and would need an expansion of the system:

(55) stage 3′ NEG VERB ~ NEG VERB NEG ~ VERB NEG

This would be described as a language permitting all stages according to the taxonomy in (47).

To some extent the different configurations are a matter of taste and do not reflect different conceptualizations of the processes involved. For a fuller exposition of the possibilities and linguists that have used them, see van der Auwera (2009: 37–40) and Breitbarth (2009: 81–9). Nevertheless, there are very real differences between different language histories, and any comparative perspective on Jespersen's cycle will have to recognize these differences and seek to account for them. We have not imposed a uniform system on individual chapters, allowing authors to follow whatever system seems appropriate to their language and preferred analysis. Hence, the chapters on German, Dutch and Low German, Brythonic Celtic (Welsh and Breton), and Afro-Asiatic use the classic three-stage system (stages I–III) in (47), while, for French, Italo-Romance, and English, five-stage systems, differing in details that reflect the attested histories of these languages, are adopted (stages 1–5). For Greek, the complexities of the changes involved, combined with gaps in attestation, seem to defy straightforward idealization (section 8.2).

Languages vary considerably in the rate at which they progress through Jespersen's cycle. Indeed, as we have seen, an item which shows signs of incipient grammaticalization as a new negator may remain an optional, pragmatically marked emphasizer indefinitely. For instance, the grammaticalization of Italian *mica* '(not) at all' < 'crumb' as an adverbial reinforcer denying a presupposition dates already to the medieval period, but, in many varieties including the standard language, it has not yet been reanalysed as a neutral expression of negation, and is still subject to pragmatic restrictions (section 3.2.1), see also Cinque (1976) and Visconti (2009).

Not all languages go through a stage where bipartite negation is the norm, hence what is a well-defined stage in the history of one language may be an unstable transitional stage with much variable usage in another. While some languages undergo a complete change in the expression of negation within a relatively short space of time, others stop at varying points in the development and live through periods of relative stability.

For instance, within the Continental West Germanic languages, there are significant differences between Dutch, High German, and Low German in terms of the duration of stage II (stages 2–4 according to (54)). While Jäger (section 5.1.2.2) does not find a clearly delimited stage II in the history of High German, arguing that stage III is essentially reached by 1300, the bipartite construction accounts for the majority of instances of sentential negation in Middle Low German (section 6.2.1) and especially Middle and (Early) Modern Dutch for several hundred years, with southern Dutch varieties being particularly conservative (section 6.2.2). In Welsh, Willis (section 7.3.3) finds a rapid transition from stage I to stage III, with a very poorly attested stage II and coexistence of constructions from all stages for a short period. While stage II is well attested in English, all three types coexisted briefly and the shift from stage I to stage III seems to have proceeded rapidly (section 4.2.1). In French (section 2.2.3), while *pas* 'not' < 'step' had already grammaticalized as the expression of sentential negation by the 14th century at least in some dialects, and pushed other regional reinforcers such as *mie* 'not < crumb' out of use during the 15th century (Catalani 2001), the old preverbal negator *ne* only begins to be dropped in the 19th century (Martinet and Mougeon 2003). Breton (section 7.4), like French, retains stable stage II for several centuries. Concerning Arabic (section 10.2.1.1), the fact that Maltese exhibits exactly the same stage II construction as the North African Arabic dialects to which it is most closely related (most of which have no stage III construction) suggests that this has been a stable feature of these varieties since before contacts between Maltese speakers and speakers of North African Arabic varieties were severed in the 13th century (Lucas 2009).

These diverse facts raise a very real ongoing research question: why do some languages progress quickly through Jespersen's cycle while others have stable stage II, and why do some have quite distinct stages, while others mix all three stages together for a period of time? Prescriptive pressure has been cited as a reason for the slow pace of the French development (cf. Armstrong and Smith 2002: 39–40), but this is unlikely to be the whole story, given the range of Arabic dialects that pattern with French in their leisurely rate of change.

1.6 Croft's cycle

While Jespersen's cycle is the best-known historical pathway for the development of sentential negation, it is not the only cyclic development to be found. Croft (1991) infers the existence of a second negative cycle, dubbed Croft's cycle by Kahrel (1996), on the basis of the typological distribution of certain negative markers. He notes three synchronic relationships between the expression of negation in existentials and that of ordinary sentential negation:

Type A: an existential predicate is negated by the verbal negator
Type B: there is a special form for a negative existential predicate (NEGEX)
Type C: there is a special negative existential predicate, identical to the verbal negator

Some languages also show synchronic variation of types A~B, B~C, and C~A, suggesting a diachronic pathway from type A > B > C > A in a negative-existential

Stage A	Stage B	Stage C	Stage A′
NEG existential	NEGEX	NEGEX	NEGEX existential
NEG V	NEG V	NEGEX V	NEGEX V

FIGURE 1.1 Croft's cycle

cycle. According to this account, a special negative existential form arises (A > B), and comes to be used as an ordinary sentential negator with lexical verbs (B > C), replacing the original negator in that context. Finally, the negative of the existential itself is reformed using the special negative existential, now evidently reanalysed as a simple marker of negation, plus the existential verb itself (C > A′). The ultimate result is that the language returns structurally to its original configuration (symmetric marking of the existential and of other verbs), but has replaced its original negative marker (NEG) with a new one based on the negative existential (NEGEX). This development is schematized in Figure 1.1.

Croft's cycle is not a common development in Europe. Type A is the dominant pattern in Western Europe and generally shows no evidence of having emerged from earlier stages via a Croft cycle. So, in French, existential *il y a* 'there is' and non-existential *je sais* 'I know' form parallel negations *il (n')y a pas* and *je (ne) sais pas* respectively. The transition to stage B has occurred in various Slavonic (see section 9.3) and many Uralic languages. For Uralic, stage B is illustrated in (56) for Hungarian (as discussed by Croft 1991: 8), where affirmative existential *van* is negated, not by the usual sentential negator *nem*, but by replacing it with a suppletive negative existential *nincs*.

(56) **Van** jó vanat?
EX good train
'Is there a good train?' (Hungarian) (Whitney 1944: 12)

(57) Itt **nincs** taxi.
here NEGEX.PRES taxi
'There's no taxi here.' (Hungarian) (Whitney 1944: 32)

Similarly, Udmurt has a special existential negator *övöl* 'there isn't', contrasting with the ordinary verbal negator *u-* (Hamari 2010):

(58) Otyn jegit pi **övöl**.
there young boy NEG
'There is no young boy there.' (Udmurt) (Hamari 2010)

Such forms typically arise via phonological fusion of the usual negator with the adjacent existential verb, a phenomenon found also with other verb types, especially modals, auxiliaries, and copulas. One complication is that special forms of 'be' are often found in the negative, irrespective of whether 'be' expresses existential or copula meaning. This is true, for instance, of Latvian *nav* 'isn't', the negative of *ir* 'is' (Mathiassen 1997: 164) and of Middle Welsh *nyt* (section 7.3.1).

At stage C, this negative existential is reanalysed as a normal negator, but no additional existential verb is required to express a genuine negative existential. This is the case in Tongan (Polynesian), where *'ikai (ke/te)* functions both as a negative existential verb, as in (59), and as a negative marker accompanying a finite verb in (60) (as discussed by Croft 1991: 12).

(59) 'Oku 'ikai ha faiako 'i heni.
PRES NEGEX a teacher at here
'There isn't a teacher here.' (Tongan) (Churchward 1953: 56–7)

(60) Na'e 'ikai [ke] 'alu 'a Siale.
PAST NEGEX go ABS Charlie
'Charlie didn't go.' (Tongan) (Churchward 1953: 56)

Within Europe, some Welsh dialects show a version of the shift to type C, having innovated a new negative auxiliary from a former negative existential (variously *sa*, *so*, *smo* etc. < *nid oes dim o* 'there isn't any…') (Borsley and Jones 2005: 62–4). Udmurt too shows signs of the transition from type B to type C, with *övöl*, used originally as a negative existential and in some other related environments, as seen in (58) above, spreading to a few non-existential contexts, such as the second past tense in (61) (Hamari 2010):

(61) Övöl myn-em-ed.
NEG go-PAST2-2SG
'You didn't go, haven't gone.' (Udmurt) (Hamari 2010)

Hamari notes similar developments in Komi and Mari.

The cycle turns full circle when this new negative marker recombines with the affirmative existential verb, and the symmetrical distribution of a type A language is reasserted, as is happening today in Marathi (Croft 1991: 12), and may have occurred historically with the Erzya Mordvin marker of nonverbal predicates *avol'* (section 11.5.2).

1.7 Other pathways

Another, rather different pathway for change is characteristic of the Uralic languages. These typically express negation via negative auxiliaries, a minor but robustly attested type worldwide, found in 40 (17%) out of 240 languages in Dahl's (1979) sample, in 47 (4%) out of 1159 languages in Dryer's (2011) sample (excluding undetermined cases), and in 16 (5%) out of the 297 languages in Miestamo's (2005) sample. In such languages, while an affirmative clause contains a lexical finite verb, the corresponding negative clause contains a negative auxiliary inflected for person and number plus a non-finite participial (connegative) form. For instance, in Finnish, person–number inflection, such as first person singular *-n* in (62) and second person singular *-t* in (63), appears on the lexical verb in the affirmative but on the auxiliary in the negative.

(62) Luen. En lue.
read.PRES.1SG NEG.1SG read.CNG
'I am reading.' 'I'm not reading.' (Finnish) (Dahl 1979: 84)

(63) Luet. Et lue.
read.PRES.2SG NEG.2SG read.CNG
'You are reading.' 'You aren't reading.' (Finnish) (Dahl 1979: 84)

Across Uralic, the extent to which such verbal categories are expressed on the auxiliary varies from language to language. For instance, in Finnish, while person–number is expressed on the auxiliary, tense is expressed on the connegative; contrast *lue* in the present connegative in (62) and (63) above with *lukenut* in the past connegative in (64).

(64) En lukenut.
NEG.1SG read.CNG.PAST
'I didn't read.' (Finnish) (Dahl 1979: 85)

Originally, Uralic tense, mood, and person–number marking appeared on the negative auxiliary, while the lexical verb appeared in some kind of nominalization in *-k (Honti 1997: 170–1, 249). The negative auxiliary itself probably arose from a negative copula via Croft's cycle (Honti 1997: 173). However, diachronically, the typical pattern is for the number of features expressed on the auxiliary to decline, often but not always with those features coming to be expressed on the lexical participial verb (Tauli 1966, Comrie 1981, Honti 1997, Miestamo 2010). Comrie (1981: 354) posits the hierarchy in (65) regulating which categories appear on the lexical verb and which on the auxiliary. Categories towards the right of the hierarchy are those most likely to be expressed on the lexical verb and diachronically to be lost from the auxiliary.

(65) imperative > tense / person / number > mood > aspect > voice

Standard Finnish, as seen above, retains person–number marking and some tense marking on the negative auxiliary, but has abandoned other categories. Estonian, along with certain Finnish dialects (Miestamo 2011), has taken this process further, and the auxiliary has lost all verbal categories, acquiring an invariant morphological form based on the historical third person singular (Estonian *ei*) (Dahl 1979: 85, Honti 1997: 83). However, in this pattern, illustrated in (66) and (67), the lexical verb has not yet acquired person–number marking, and so presumably should continue to be considered non-finite, with finiteness remaining on the negative auxiliary (Miestamo 2011: 90).

(66) Loen. Ma ei loe.
read.PRES.1SG I NEG read.CNG
'I am reading.' 'I'm not reading.' (Estonian) (Miestamo 2011: 90)

(67) Loed. Sa ei loe.
read.PRES.2SG you NEG read.CNG
'You are reading.' 'You aren't reading.' (Estonian) (Miestamo 2011: 90)

The extreme development is found in Mansi, where the negative auxiliary has become an invariant particle, *at*, and the lexical verb inflects fully in negative clauses (Miestamo 2005: 218–19). The now-extinct Finnish dialect of Värmland underwent the same development, perhaps under contact with Swedish, using fully finite verbs alongside an invariant negative particle *ei*:

(68) Ei minä lyö-n sinua.
NEG I hit-PRES.1SG you.PART
'I will not hit you.' (Värmland Finnish) (Miestamo 2011: 97)

For further details of all these developments, see Tauli (1966), Honti (1997), Miestamo (2000, 2011).

1.8 Indefinites: basic concepts

Cyclic renewal does not only affect the expression of sentential negation. Indefinites, including both indefinite pronouns such as *anything* or *nothing* and indefinite adverbials such as *ever* or *never*, are also frequently subject to directional changes which Ladusaw calls the 'argument cycle':

> The development of negation-expressing argument phrases from regular indefinite arguments has the following stages: first the argument is a regular indefinite argument, then it becomes a co-occurring 'supporter' of the clausal negation, and finally it becomes an independent expressor of negation. We could call these the 'one thing', 'anything', 'nothing' stages of the Jespersen argument cycle. (Ladusaw 1993: 437–8)

That is, indefinites starting out as contextually unrestricted items are often observed to become restricted to 'more negative' contexts, being restricted first to 'weak' negative polarity contexts, such as questions or conditional protases, alongside negative clauses, later being restricted to stronger negative polarity contexts like comparatives, and indirect and direct negation. The development of French *personne* 'no one' and *rien* 'nothing' is a famous case in point, as they developed from originally contextually unrestricted generic nouns, namely Latin *persona* 'person', and *rem* 'thing.ACC' respectively. They are now restricted to negative clauses, but historically were once also attested in weaker negative polarity contexts such as questions (section 2.3). Recognizing that these developments may apply to any indefinite, whether or not it is an argument, we shall refer to this pathway of change as the 'quantifier cycle' (see section 1.9.1 below).

A significant difficulty encountered in discussions of the development of indefinites in the scope of negation is the large amount of associated terminology, much of which is used ambiguously or inconsistently. Before examining typical historical developments in the next section, we will set out the key terms, along with the definitions assumed in the present work, highlighting any variation in usage among the contributors of the various chapters.

1.8.1 Negative quantifiers and negative polarity items

Negative quantifiers are inherently semantically negative indefinites (de Swart and Sag 2002). Classic, uncontroversial examples come from non-negative-concord languages, for example Classical Latin *nemo* 'nobody', standard German *nichts* 'nothing', and standard English *nothing*, *nobody*, and *never*. The presence of one of these items in a clause is always sufficient to render it negative:

(69) John never arrives on time.

The fact that all of these items contain a morphological marker of negation (at least from an etymological point of view) makes their analysis as semantically negative especially uncontroversial, but containing such an overt morpheme is not usually thought of as a necessary feature of negative quantifiers. Indeed, being marked negative morphologically need not be seen as sufficient to guarantee that an indefinite is a negative quantifier either (see section 5.2.4, where Jäger argues against an analysis of the morphologically negative indefinites of High German as negative quantifiers).

Negative polarity items (NPIs), by contrast, are usually assumed not to be negative. The crucial property of an NPI is that it is restricted to appearing in 'non-assertive' contexts such as negation, interrogatives, the protases of conditionals, and comparative clauses. A clear example of an NPI is the English temporal indefinite *ever*. This is evidently not negative and is grammatical in all of the aforementioned contexts; but it is ungrammatical in affirmative declarative sentences:

(70) John doesn't ever arrive on time.

(71) Does John ever arrive on time?

(72) If John ever arrives on time, I'll eat my hat.

(73) John arrived earlier than we could ever have expected.

(74) *John ever arrives on time.

According to the older so-called Fauconnier–Ladusaw Hypothesis (Fauconnier 1975, Ladusaw 1979), the defining property of contexts which permit (license) NPIs is that they are downward entailing (monotone decreasing); that is, they have the property of reversing the entailment relations of expressions in their scope. For instance, while the affirmative *John ran fast* entails that *John ran*, the opposite is true if the sentence contains a downward-entailing operator such as sentential negation: *John did not run* now entails *John did not run fast*. More recently, the notion of downward entailment has been argued to be inadequate to capture the whole range of contexts licensing NPIs and has been challenged by the broader concept of non-veridicality (Zwarts 1995, Giannakidou 1998). A sentence containing a non-veridical operator does not entail the proposition the operator modifies; for instance, an imperative *Tell Paul to wait!* does not entail that the addressee does in fact tell Paul to wait (or that Paul will actually wait). Interrogatives and conditional clauses are also non-veridical. Sentential negation is a subtype of non-veridical operator—an anti-veridical operator, that is, one that entails that the proposition that it modifies is false. This is defined formally in (75).

(75) Let *Op* be a monadic propositional operator. The following statements hold:
a. *Op* is veridical just in case $Op\ p \rightarrow p$ is logically valid. Otherwise *Op* is non-veridical.
b. A non-veridical operator *Op* is anti-veridical just in case $Op\ p \rightarrow \neg p$ is logically valid. (Giannakidou 1998: 106)

Note, however, that many NPIs are licensed in comparative clauses, as with *ever* in (73) above, and these are not obviously non-veridical (cf. Giannakidou 1998: 151–3), though they are downward-entailing (Stechow 1984: 29), suggesting that we need both concepts: while some NPIs require a downward-entailing context, others require a non-veridical one.

Furthermore, some NPIs are restricted to a subset of non-veridical or downward-entailing contexts. NPIs that are licensed only in the context of negation are referred to as 'strong' NPIs (NPIs which are not strong are called 'weak'; Zwarts 1998). An example of a strong NPI is *one bit* in English (contrast the grammaticality of the weak NPI *at all* if substituted into (77) or (78)):

(76) Mary didn't like it one bit.

(77) *Did Mary like it one bit?

(78) *If Mary likes it one bit, then buy it.

(79) *Mary liked it one bit.

Intermediate distributions are also possible, suggesting the need for other means to define the contexts for certain NPIs (van der Wouden 1997).

Historically, items that formerly had no restrictions on their distribution sometimes come to be restricted to negative-polarity contexts and thereby acquire the status of NPIs. During this period, they can be referred to as 'semi-NPIs' (Hoeksema 1994, 2009), that is, items that may occur in veridical, upward-entailing contexts but which are more frequent in downward-entailing contexts, especially in the context of negation (see section 10.3.1 for discussion of a number of such items in Arabic).

Finally, we also need to distinguish in this connection positive polarity items (PPIs). PPIs are items which cannot be interpreted in the scope of a non-veridical/ downward-entailing operator. For instance, (81) is ungrammatical on the most natural interpretation with *some cakes* within the scope of negation; and is only grammatical on a reading with the scope of negation inside that of the indefinite ('There were some cakes that Mary didn't bring.').

(80) Mary brought some cakes.

(81) *Mary didn't bring some cakes.

1.8.2 Negative concord and n-words

Next, we consider the terms 'negative concord' and 'double negation'. Although 'negative concord' has the flavour of a technical term with a precise meaning, in fact it is usually used rather impressionistically, as in the following from Giannakidou (2000: 458): 'Generally, we talk about 'negative concord' in situations where negation is interpreted just once although it seems to be expressed more than once in the clause.' The impression that negation is expressed more than once in such situations is created by the co-occurrence of two or more items which would intuitively be judged to be negative. For instance, Lithuanian is shown to be a negative-concord language in (82), since a negative indefinite *nieko* 'nothing' requires the verb to bear negative marking in the form of the negative prefix *ne-*. Spanish shows the same property in (83).

(82) Jis **nieko** **nesakė.**
he nothing NEG.say.PAST.3SG
'He said nothing.' (Lithuanian) (Mathiassen 1996: 80–1)

(83) **No** dijo **nada.**
NEG say.PAST.3SG nothing
'He said nothing.' (Spanish)

Traditionally (e.g. Jespersen 1917: 62–80), and even today among non-specialists, negative concord tended to be called 'double negation'. In the scholarly literature on negative concord, however, 'double negation' is now used to refer to situations where two (or an even number of) negative expressions in a sentence cancel one another out, such that the sentence is truth-conditionally (but usually not pragmatically) equivalent to an affirmative one. This is as in propositional logic, where two negatives equal a positive: $\neg \neg p \leftrightarrow p$.

Indefinites that appear to be negative, and which participate in negative-concord structures are often called 'n-words' (or 'n-indefinites' or 'n-items'), after Laka (1990: 107–9). Laka coined the term as a label for the Spanish indefinite series containing *nadie* 'n.one' (= 'anyone/no one'), *nada* 'n.thing', *ningún* 'any/no', *nunca* '(n)ever' etc. The label 'n-word' was simply chosen to highlight the fact that most of these items in Spanish, as well as parallel items in Italian, Portuguese, and other Romance varieties, begin with /n/. However, the presence of a morphological expression of negation is not a necessary condition for an item to be an n-word. Laka (1990: 108) makes this clear when she points out that *nadie* and *nada* originate in (*homines*) *nati* 'born men' and (*res*) *nata* 'born thing', respectively, and have thus never been morphologically negative.

Like similar items in other languages, these indefinites manifest a combination of apparently contradictory properties. On the one hand, they appear to be non-negative (and hence negative-polarity items, akin to English *anyone*); on the other they appear to be negative (and hence negative quantifiers, akin to English *no one*).

Consider first their non-negative properties. If n-words were straightforwardly inherently negative items, then the combination of two n-words, or an n-word with a sentential negator, as in the Lithuanian example in (82), would be expected to result in logical double negation, as it does in English (cf. the double-negation interpretation, in standard English, of *He didn't say nothing* as synonymous with *He said something*). However, (82) means 'He said nothing' rather than 'He didn't say nothing.' Furthermore, some n-words can appear in non-negative environments such as comparatives. While this is not the case with n-words in the Slavonic or Baltic languages, it is often the case in Romance and Celtic. Thus Spanish *nunca* '(n)ever' appears in a comparative in (84), a fact that cannot be straightforwardly accounted for if it is negative:

(84) Juan ha llegado más tarde que **nunca**.
Juan have.PRES.3SG arrive.PP more late than n.ever
'Juan has arrived later than ever.' (Spanish) (Herburger 2001: 298)

On the other hand, if n-words were straightforwardly non-negative, other properties would be mysterious. N-words can convey a negative in a fragment answer, and permit modification by 'almost', generally incompatible with existential contexts. Contrast the behaviour of Spanish *nadie* and English *anyone* with respect to fragment answers in (85) and (86); and the behaviour of Spanish *nunca* and English *ever* with respect to the 'almost'-test in (87) and (88).

(85) A: A quién viste?
ACC who see.PAST.2SG
'Who did you see?'

B: A **nadie**.
ACC n.one
'No one.' (Spanish) (Herburger 2001: 300)

(86) Who did you see? #**Anyone**.

(87) Carlos no bebe casi **nunca**.
Carlos NEG drink.PRES.3SG almost n.ever
'Carlos almost never drinks.' (Spanish) (Aranovich 2007: 196)

(88) a. *Mary doesn't almost **ever** drink.
b. Mary almost **never** drinks.

In synchronic work on indefinites, the criteria for considering an item to be an n-word are not always made explicit and applied consistently. A widely adopted definition of n-words, which is consistent with Laka's original usage, is given in (89). This takes the apparent contradiction between negative concord in (82)–(83) and the availability of negative interpretations of fragment answers in (85) to be the central (and sufficient) defining property.

(89) n-word
An expression α is an n-word iff:
a. α can be used in structures containing sentential negation or another α-expression yielding a reading equivalent to one logical negation; and
b. α can provide a negative fragment answer. (Giannakidou 2006: 328)

By these criteria, both Spanish *nadie* and Lithuanian *nieko* are n-words, even though they differ in other crucial properties (e.g. in their ability to appear in various weak negative polarity environments). Consequently, n-words do not necessarily form a natural class. Rather, 'n-word' is a label applied to items which are problematic in that they have one property that is consistent with their being inherently negative—(89b)—and another that apparently is not—(89a). Once an item is identified as an n-word according to the definition in (89), therefore, it is still in need of an analysis as to how it manages to exhibit these apparently contradictory properties. It is for this reason that negative concord, and n-words in particular, are typically seen as a theoretical problem in need of further analysis.

The two simplest analyses of a given series of n-words in a given language are that they are either negative quantifiers (for Spanish and Catalan, Espinal 2000) or NPIs (for Spanish, Laka 1990). Another possible analysis (for Spanish, Herburger 2001) is that they are ambiguous between the two.

Not everyone uses the term 'n-word' in accordance with the definition in (89), however. For example, de Swart (2006) is representative of a number of authors who adopt a wider definition that encompasses uncontroversial negative quantifiers among the class of n-words. Hansen adopts this usage in her discussion of French indefinites (section 3.2.1). Giannakidou (2006) herself, in fact, despite giving the definition in (89), also appears to lapse into this usage, in that she periodically refers to the 'n-words of standard West Germanic languages', despite the fact that the languages in question lack negative concord, that is, none of their indefinites satisfies both clauses of (89).

Another key terminological distinction concerns the two subspecies of negative concord known, following den Besten (1986), as 'negative doubling' and 'negative spread'. Negative doubling, shown schematically in (90), obtains when negation is expressed by both the sentential negator and an indefinite in the scope of negation, that is, when that indefinite is an n-word according to the criteria in (89). The standard cases of negative concord given above in (82) and (83) are instances of negative doubling. In negative spread, if more than one indefinite is present in the scope of negation, then negation is expressed on each and every one of them:

(90) John NEG saw n.thing. (negative doubling)

(91) John saw n.thing n.where. (negative spread)

(92) John NEG saw n.thing n.where. (negative doubling and negative spread)

For instance, Latvian is both a negative-doubling language and a negative-spread language. Example (93) shows that an n-word co-occurs with the ordinary marker of sentential negation (i.e. *nekad* 'never' co-occurs with *ne-* on *nerunā* 'NEG.talk.PRES.3SG'), hence negative doubling is manifested. Furthermore, two n-words may co-occur in the same sentence (*nekad* 'never' and *ne par vienu* 'about no one'), yielding a single negative reading, hence negative spread is manifested. These patterns are obligatory in the language.

(93) Viņs **nekad** ne par vienu **nerunā**.
he never NEG about anyone NEG.talk.PRES.3SG
'He never talks about anyone.' (Latvian) (Mathiassen 1997: 72)

Normally, a language shows either both negative doubling and negative spread or neither, but some Germanic varieties have negative doubling without negative spread or negative spread without negative doubling, as in the following example from West Flemish:

(94) T ee **niemand** **niets** gezeid.
it have.PRES.3SG no.one nothing say.PP
'Nobody said anything.' (West Flemish) (Zeijlstra 2004: 62)

Negative concord (negative doubling) can be further differentiated into 'strict' and 'non-strict' negative concord (Giannakidou 1998, 2000, Zeijlstra 2004). In a strict negative-concord language, ordinary full clauses containing an n-word always contain the sentential negator too, irrespective of the relative position of the two items. On the other hand, in a non-strict negative-concord language, the sentential negator must be omitted when the n-word precedes the finite verb.[6] The same basic

[6] Since this distinction centres on the presence of sentential negation, we are clearly dealing with two varieties of negative doubling, so the term '(non-)strict negative doubling' is more appropriate. Note, however, that the majority of authors, including those in this volume, use the term '(non-)strict negative

observation is made by Haspelmath (1997: 201) in his classification of languages into three types:[7]

(95) (i) NV–NI (negation marker–verb–negative indefinite) (broadly equivalent to strict negative concord)
(ii) (N)V–NI (the negation marker sometimes co-occurs with a negative indefinite) (non-strict negative concord and similar systems)
(iii) V–NI (the sentential negation marker is absent in sentences containing a negative indefinite) (no negative concord).

Strict negative concord is found in today's Slavonic and Baltic languages, as well as in Romanian, Greek, Hungarian, and Maltese. Non-strict negative concord is found today in Spanish, Portuguese, and Italian, while Catalan allows both patterns. So, in Romanian (Falaus 2007: 75), omission of the negative particle *nu* leads to ungrammaticality in both (96) and (97), that is, irrespective of whether the n-word precedes the verb/negative marker, as in (96), or follows it, as in (97).

(96) **Nimeni** **nu** ştie ce se întâmplă.
n.one NEG know.PRES.3SG what REFL happen.PRES.3SG
'No one knows what's happening.' (Romanian)

(97) **Nu** am aflat **nimic** nou.
NEG have.PRES.1SG find.out.PP nothing new
'I didn't find out anything new.' (Romanian)

Contrast this with the Spanish examples in (98) and (99) (from Laka 1990: 107): where the n-word precedes the verb in (98) there is no marker of sentential negation, and its presence would lead to a double-negation interpretation 'No one isn't coming'; where the n-word follows the verb, as in (99), *no* is obligatory.

(98) **Nadie** vino.
n.one come.PAST.3SG
'No one came.' (Spanish)

(99) **No** vino **nadie**.
NEG come.PAST.3SG n.one
'No one came.' (Spanish)

concord' to describe this phenomenon. A closely related distinction is Déprez's (2000) 'symmetrical' versus 'asymmetrical' negative concord. Asymmetrical negative concord (a term employed for Italo-Romance in section 3.8) is identical to non-strict negative doubling. Symmetrical negative concord encompasses both strict negative doubling and negative spread in the absence of negative doubling, as found in informal spoken French.

[7] Note, however, that, for Haspelmath, any indefinite in the scope of negation is a negative indefinite, even if that indefinite is incapable of expressing negation on its own. Hence, applying his definition, English *any*-series items, when in the scope of *not*, instantiate the pattern NV–NI, which we would not consider to be an instance of negative concord.

The non-strict distribution further exacerbates the n-word problem, since, in (98), *nadie* appears to contribute a negative by itself, while, in (99), the negation appears to be contributed by *no* rather than *nadie*. In the current context, there is also the issue of how negative-concord systems develop over time, a question which will be discussed further below (section 1.10).

1.9 Cyclic developments in indefinites

Indefinites normally occur in series, such as the English *any*-series, whose members include *anyone*, *anything*, *anywhere*, and *ever*. These are often morphologically regular and are generally referred to by the element common to all, or most, members of the series. Where the forms are morphologically irregular, a convenient convention is to refer to the series using the member used for persons, for instance, the French *personne*-series includes the items *rien* 'nothing' and *nulle part* 'nowhere', which are morphologically unrelated to *personne* 'no one', the item that gives the series its name. It is often useful to generalize across an entire series, since items tend to behave alike. Diachronically, however, new items sometimes join a series while existing items may leave. Furthermore, during periods of change, not all members of a series change their behaviour at the same time. For instance, French *jamais* 'ever, never' and Catalan *mai* 'ever, never' are rather more conservative than the other items in their series, appearing in a range of non-negative environments from which the other members of the series have disappeared (see section 2.3, especially Table 2.3).

Haspelmath (1997) proposes a semantic map, given in Figure 1.2, regulating the distribution of indefinites in a particular language. Generalizing across the patterns found in a sample of 40 languages, he proposes that all functions of a given item must be located in a contiguous space on the semantic map. Discussion here will be focused on those items involved in negation and negative polarity, that is, the six environments to the right in Figure 1.2, along with items that move into that area during the course of their historical development.

Direct negation refers to clausemate negation, as in (100), while indirect negation refers to a range of rather disparate syntactic environments, including superordinate negation (negation in a higher clause), as in (101a); pseudonegative contexts such as 'without'-clauses, the scope of quantifiers such as 'hardly' or 'few', or the complement

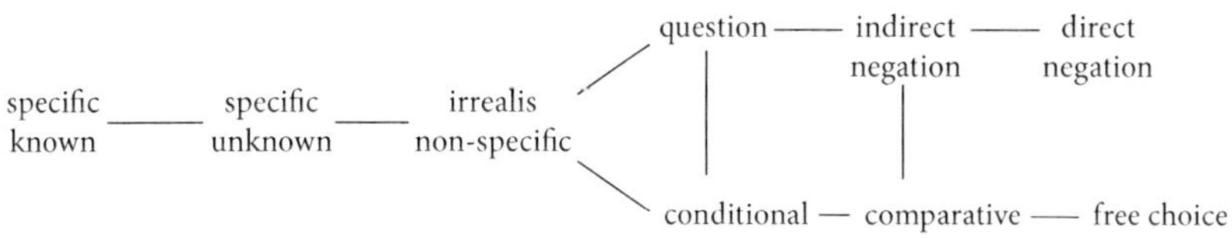

FIGURE 1.2 Haspelmath's (1997) semantic map of indefinites

of adversative predicates such as 'deny' or 'be annoyed', in (101b) (Klima 1964: 314). In English, direct negation allows both the *no*-series and the *any*-series, while the other contexts generally require the *any*-series, illustrated in (102)–(105). Questions and conditionals also allow *some*-series items under certain interpretations. While questions, conditionals and comparatives are syntactic contexts, free choice is a meaning (the proposition holds of an arbitrarily chosen member of a set), and hence may co-occur with some of the other syntactic contexts.

(100) I saw no one. / I didn't see anyone. (direct negation)

(101) a. I didn't admit that I had seen anyone.
b. I denied that I had seen anyone. (indirect negation)

(102) Did you see anyone? (question)

(103) If you see anyone, hide. (conditional)

(104) Mary is taller than anyone. (comparative)

(105) Anyone can ride a bike. (free choice)

Haspelmath's system makes clear diachronic predictions, specifically that 'where markers gradually acquire new functions, they will first be extended to those functions that are adjacent to the original functions on the map, and only later to functions that are further away' (Haspelmath 1997: 63). The two commonest types of development seem to be the quantifier cycle and the free-choice cycle.

1.9.1 The quantifier cycle

In terms of Figure 1.2, the quantifier cycle represents a contraction in the range of environments available for an item, so that it is ultimately available only under direct negation. As it retreats to this environment, some other (possibly innovated) element takes over its former functions. The best-known example of this phenomenon involves the French *personne*-series (section 2.3), a number of whose members were generic nouns in Latin (e.g. *persona* 'person', *rem* 'thing (acc.)', etc.) and early Old French. These items entered the indefinite system when they were reanalysed and recruited as indefinite pronouns in Old French, becoming restricted first to negative polarity environments and then to direct negation (along with some indirect negative contexts). As they became 'more negative', their former functions were taken over by a newly innovated *quelque*-series (e.g. *quelqu'un* 'someone'), based on an original free-relative structure ('whichever one it may be'). Similarly, Middle High German *dehein* 'any' was found in non-negative NPI environments, while its present-day equivalent (Modern German *kein* 'no') is restricted to direct negation, with its former functions having been taken over by the indefinite article *ein* or by other indefinites such as *irgend*(*ein*) 'any' (sections 5.2.2–4).

In general, modern Romance negative indefinites are expressed using composite series of items, constructed from a mixture of etymologically negative and etymologically non-negative sources. The merger into a single series led to the emergence of (broadly) parallel properties, with etymologically negative items initially spreading to weak NPI contexts in the medieval languages (Haspelmath 1997: 232–3). Their subsequent development, however, has generally been towards becoming restricted more and more to negative environments.

For instance, in Italian (section 3.7), an etymologically non-negative item, *alcuno* (< Latin *alicunum* < *aliquem unum* 'any one') became restricted to negative contexts as 'no, no one'. Negative polarity environments have been partially taken over by *qualche* 'any', also originally a free-relative marker. Non-negative uses of the originally negative items *nessuno* 'anyone, no one' and *niente* 'anything, nothing' were permitted more freely in medieval Italo-Romance than today, and they have retreated in particular from conditional clauses.

Spanish n-words (the current *nadie*-series) once showed a wider range of NPI properties than today, appearing, for instance, in conditional clauses, where they are no longer possible:

(106) Si **ningun** **uillano** viniere ala uilla del Rey...
if any commoner come to.the town of.the king
'If any commoner should come to the king's town...' (Old Spanish) (*Fuero General de Navarra*, 13th c.) (Poole 2009: 32)

These items have retreated from conditionals, and to a lesser extent from questions (Herburger 2001: 299), and have acquired a negative interpretation in fragment answers (Poole 2011). Again, this means that items that were originally affirmative, such as *nada* 'anything, nothing' (< *res nata* 'a born thing'), have undergone a straightforward development from generic noun > weak negative polarity item > negative quantifier.[8] As originally negative items, such as *ningún* 'no' (< Latin *nec ūnus* 'nor one'), joined the series, they seem to have taken on the properties of the other members, extending their use to non-negative contexts to match the distribution of the other items in the series (cf. (106) above). Similar kinds of developments, albeit less advanced, are found in Catalan, with items such as *res* 'anything' having gone from being generic nouns to n-words occurring in questions, conditionals, and negative contexts and originally negative *ningú* 'any, no' becoming an n-word (Sandanya 2004).

Countervailing changes in German also seem to be the result of the merger of items with different etymologies. German (section 5.2.2) merges *etwas*, originally an

[8] This of course assumes that n-words in modern Spanish should be analysed as negative quantifiers, as argued, for instance, by Espinal (2000). However, even on other analyses, the broad direction of change is clear. Thus, according to Herburger's (2001) analysis, Spanish n-words are ambiguous between NPIs and negative quantifiers, but historically they began as NPIs and developed as negative quantifiers only later (Herburger 2001: 323–6).

ordinary indefinite, and *jemand*, originally a negative polarity item, creating a new series with the properties of an ordinary indefinite. As a consequence, *jemand* has become 'less negative' in its history, going from negative polarity item to ordinary indefinite.

Some Celtic developments have closely paralleled Romance: the Welsh *neb*-series has innovated new items on the basis of generic nouns such as *dim* 'thing', and the series as a whole has increasingly disappeared from non-negative NPI environments (section 7.6, especially 7.6.5). Broadly similar developments have occurred with the Breton *den*-series, probably accelerated by contact with French (section 7.7). As in French, the new indefinite that replaces the old one in negative polarity contexts (*unan bennak* 'someone, anyone') is derived from a former free relative.

1.9.2 The free-choice cycle

A second common development is the tendency for original free-choice items (cf. (105) above) and free-relative markers to spread, becoming ordinary indefinites, whether unmarked affirmative indefinites, or, of more interest in the current context, indefinites specialized for negation or negative polarity (Haspelmath 1997: 149–50, Willis 2012: 336–9). We have already seen that French *quelqu'un*, Italian *qualcuno*, and Breton *unan bennak* originally participated in free relatives, but have now become ordinary indefinites available in non-negative environments. The West Slavonic *si/ś*-series (Czech *kdo-si*, Polish *ktoś* 'someone') may also be an example of this pathway, having ultimately become an affirmative indefinite (positive polarity item) like English *someone*.

In a second variant, the items enter the negative system, becoming weak negative polarity items. We find repeated examples of this kind of development in Slavonic languages, where an original free-choice item or free-relative marker spreads to all negative polarity environments except negation, producing what is often referred to as a 'bagel' distribution. Synchronically, this is often seen as problematic, since it seems odd that an item should be associated with negation, apparently being a negative polarity item, but absent from the prototypical negative environment; however, it makes sense as the diachronic development of an item extending its distribution from free-choice or free-relative environments, but which has not (yet) reached negative contexts. The general development is perhaps most clearly seen with Polish *kolwiek*-items, such as *ktokolwiek* 'anyone' (section 9.5.7). These began as free relatives 'whoever' etc., but spread to all non-negative NPI contexts in the history of Polish. A range of other examples of this type is discussed in sections 9.5.5 (Russian) and 9.5.8 (Serbian). The general encroachment of the English *any*-series items onto negative clauses, that is, the increasing use, since Middle English, of patterns like (108) in place of patterns like (107) (section 4.4), might also be seen as an instance of the free-choice cycle.

(107) I have seen no one.

(108) I haven't seen anyone.

The question of emphasis has been raised in this context by Kiparsky and Condoravdi (2006) for Greek (see also section 8.2.3). They argue that Greek *típote* 'anything, nothing', historically deriving from *ti* 'something/anything' and *pote* 'ever', a formation often found with free-choice items and free-relative markers (cf. Polish *cokolwiek* < *co* 'what' + *koli* 'ever' + *wiek* 'ever' or English *whatever*), was originally emphatic, but lost its emphatic quality to become the ordinary form of the indefinite pronoun in a development entirely parallel to the inflationary processes in the Jespersen cycle. Such differences may also have driven the expansion of the English *any*-series into negative contexts, with (108) originally being in some sense emphatic in comparison to (107), and perhaps the Slavonic cases too. Old Hungarian underwent a cyclic renewal of indefinites, reminiscent of Jespersen's cycle too, creating reinforced negative indefinites such as *semmi* 'nothing, anything' from *es* 'even' + *nem* 'not' + *mi* 'what' (Kiss forthcoming) (cf. also pan-Slavonic formations of the type Serbian *niko* < *ne* 'not' + *i* 'even' + *kŭto* 'who'). In all these cases, overuse of the emphatic option leads to its devaluation and reinterpretation as the neutral indefinite in negative clauses.

1.9.3 *Motivating the developments*

As with Jespersen's cycle, the most important historical question with all of the patterns of change found in indefinite systems is the question of motivation: does an indefinite like French *personne* come to be 'felt' to be too negative to appear in non-negative environments, requiring speakers to resort to some other element (a pull-chain mechanism); or does *quelqu'un* come to be overused beyond its original free-relative/free-choice meaning, and begin to compete with *personne* once it becomes an unmarked indefinite (a push-chain mechanism)? Comparing diachronies seems to show that both scenarios are possible: with English, only the push-chain seems likely, since *no one* was already limited to direct negation even when it began to be pushed out by *anyone* (cf. the inflationary scenario outlined in section 1.9.2); with other cases, the ambiguity of the old indefinite in conditionals and interrogatives once Jespersen's cycle is underway may favour a pull-chain account. Clauses of the type 'if you see n.thing' will be ambiguous between a negative interpretation 'if you see nothing' and a positive one 'if you see anything'. This ambiguity can be avoided by using another indefinite for the positive interpretation; however, doing so reduces the frequency of the old n-indefinite in conditionals, leading a subsequent generation to fail to acquire this as a grammatical pattern at all.

1.10 Shifts in strict and non-strict negative concord

Today, within Romance, Romanian and Standard French[9] show strict negative concord; Spanish, Portuguese, and Italian show non-strict negative concord; while Catalan has optionally strict negative concord (that is, the sentential negator is optional with a preverbal n-word, and compulsory with a postverbal one). Within Slavonic and Baltic, all standard languages show strict negative concord.

While negative concord was not present in Classical Latin, it developed in informal Latin and early Romance. There was much variation in medieval Romance varieties, with many varieties seemingly like modern Catalan, treating the negative particle as optional when a negative indefinite preceded (in other words, allowing both strict and non-strict negative concord) (section 3.8). Different linguists have assumed different historical pathways: Haspelmath (1997) assumes extensive variation in the medieval languages, but a broad pathway from Latin V–NI to (N)V–NI in the medieval languages and in some modern languages, with the more innovative modern languages having fully implemented negative concord and reached NV–NI. Martins (2000: 193–6) notes the existence of preverbal indefinites preceding the negative marker in all the medieval western Romance languages and concludes that they were once strict negative-concord languages, some of which have shifted to non-strict negative concord. Posner (1984) is more circumspect, noting much medieval variation. While she considers that 'the regular omission of *non* after INDEF-NEG is an innovation of Italian, Sardinian, Spanish and Portuguese' (Posner 1984: 13), for her, this is a development from an initial stage where both patterns were widely attested. Parry discusses the evidence for this in section 3.8, suggesting that both strict negative concord and the mixed system allowing both options were found in medieval Italy.

Earlier stages of Slavonic (section 9.5.1) attest non-strict negative concord alongside the strict pattern, again seemingly like modern Catalan. Foreign influence for the presence of this pattern cannot be excluded, but, if it reflects vernacular usage, this means that there has been a general shift towards strict negative concord in Slavonic. Hungarian presents a very similar picture: non-strict negative concord is also found in Old Hungarian (Kiss forthcoming), although, again, the question of foreign (Latin)

[9] Strictly speaking, negative concord (negative doubling) needs to be defined for each marker of negation. Standard French shows strict negative concord between *ne* and indefinites, and no negative concord between *pas* and indefinites:

(i)	Personne	n'	est	(*pas)	venu.
	n.one	NEG	be.PRES.3SG	NEG	come.PP

'No one came.' (French)

(ii)	Jean	n'	a	(*pas)	vu	personne.
	Jean	NEG	have.PRES.3SG	NEG	see.PP	n.one

'Jean saw no one.' (French)

textual influence makes it difficult to interpret its significance. If we treat the texts at face value, there has been a shift towards strict negative concord. Finally, Ingham (section 4.4) suggests that some non-West Saxon Old English has non-strict negative concord, while West Saxon had strict negative concord.

Haspelmath (1997: 211–12) suggests that a shift from (N)V–NI (non-strict negative concord) to NV–NI (strict negative concord) is motivated by the Neg-First Principle in conjunction with pattern uniformity; that is, NEG ... indefinite is motivated by the need for negation to occur as early as possible, while indefinite ... NEG is motivated by an economy principle, the need to maintain a uniform expression of negation. This makes extension of negative concord the natural development.

Counterdevelopments are found in Welsh and French, but only under rather special circumstances. In French (section 2.3), negative concord between *ne* and n-words was required, but did not develop between *pas* and n-words when *pas* replaced *ne* as the main expression of sentential negation. Similarly, in Welsh (section 7.6.6), there has been a shift towards a form of non-strict negative concord (albeit not the classic one). Welsh once showed strict negative concord between the sentential negator *ni(d)* and n-words. However, it has not developed strict negative concord between the new negator *ddim* and n-words, adopting instead a rule by which there is negative concord between *ddim* and an n-word only if the two would be non-adjacent. Although these cases go against the idea that negative concord tends to extend its domain, this only happens when the language develops a completely new negator that fails to spread to all clauses containing indefinites. These counterdevelopments thus involve interactions with Jespersen's cycle, a topic to which we turn in the next section.

1.11 Interactions between negative concord and Jespersen's cycle

As Ladusaw's term 'Jespersen argument cycle' (Ladusaw 1993: 438) suggests, a close connection is often perceived between Jespersen's cycle and negative concord: just as the expression of negation is reinforced by new negative material, so indefinites in the scope of negation are also affected by cyclic renewal. Some (e.g. Kiparsky and Condoravdi 2006) see n-words explicitly as negation strengtheners on a par with the adverbial elements that feed into Jespersen's cycle, that is, the renewal of the expression of sentential negation.

One interpretation of this relationship is that Jespersen's cycle is a pathway for a language to eliminate negative concord. Absence of negative concord in languages with apparently negative indefinites seems to be a cross-linguistically rare property. Haspelmath (2011) finds 11 languages that use this pattern exclusively and 13 with mixed behaviour in a sample of 206 worldwide. This raises the question of how this property should arise in the first place. Haspelmath (1997: 203–5) observes, building on a proposal by Bernini and Ramat (1996: 184), that, in Europe, absence of negative concord is restricted to languages with postverbal negators which arose under

Jespersen's cycle, namely most West Germanic languages and French. Languages at stage II of Jespersen's cycle often do not use the new marker of negation with indefinites. For instance, standard French, at stage II of Jespersen's cycle, uses *pas* as its new negator, but *pas* does not co-occur with an indefinite pronoun in subject or direct object position. In general, clauses containing indefinites as objects seem to be an unfavourable context for the new negator. Haspelmath links this to the emphatic function of indefinites, suggesting that reinforcement is not needed, hence the new negator is not necessary with indefinites. Consequently we witness the development of the patterns illustrated in (109) and (110) for French.

(109)

	NEG	V	NEG
Je	ne	dis	pas.
I	NEG	say.PRES.1SG	NEG

'I don't say.' (French)

(110)

	NEG	V	N-INDEFINITE
Je	ne	dis	rien.
I	NEG	say.PRES.1SG	nothing

'I say nothing.' (French) (Haspelmath 1997: 204)

It might be added that, since new markers of negation typically arise from reanalysis of direct objects, particularly reanalysis of indefinite pronouns used as direct objects, there is also a syntactic reason why indefinites would initially not co-occur with the new marker of negation. French and, to an extent, Welsh (see section 7.6.6) currently exemplify this configuration, but comparable situations obtained at the relevant stages in the history of various West Germanic languages (see, for instance, section 5.2 on Middle High German and section 6.3.1 on Middle Low German). Middle English, for instance, shows a dispreference for co-occurrence of *not* with *no*-indefinites ('Jack's law', Jack 1978a, 1978b, Iyeiri 2001, cf. also Wallage 2005: 225–6), see section 4.4.2. This notwithstanding, English seems to be a special case, since the loss of negative concord (negative doubling and negative spread) is additionally due to the spread of the *any*-series spread into negative contexts (Iyeiri 2002a, 2002b, Wallage 2005, Nevalainen 2006, see also section 1.9.2 above).

As the language reaches stage III of Jespersen's cycle, losing the old preverbal negator, negative concord (negative doubling) disappears (NEG V N-INDEFINITE > V N-INDEFINITE). The result is a non-negative-concord language. Jespersen's cycle is thus a possible scenario for how such a typologically rare pattern as non-negative-concord can arise.

Another commonly made connection between Jespersen's cycle and negative concord is to relate the syntactic status and diachronic changes in the sentential negator in a language to the availability, rise, or disappearance of negative concord. Jespersen, citing Old English, Russian, Greek, and Hungarian as examples, originally noted that 'repeated negation [negative concord – DW, CL, & AB] seems to become a habitual phenomenon only in those languages in which the ordinary negative element is comparatively small in regard to phonetic bulk' (Jespersen 1917: 71–2), an observation that has come to be known as 'Jespersen's generalization'. Recent treatments have taken this to mean that there is a positive correlation between a language having a negator that is a Neg-head and negative concord (negative doubling). Two specific formulations are available, either as a bidirectional implication (non-phrasal negation ↔ negative concord), as in (111), or a unidirectional one (head negation → negative concord), as in (112).[10]

(111) 'A language is an NC language iff the regular marker of pure sentential negation is not associated with SpecNegP.' (Rowlett 1998: 87, 100)

(112) 'All languages with a negative marker X^0 are NC languages.' (Zeijlstra 2004: 165)

Such generalizations make clear diachronic predictions: both formulations predict that, if a language reanalyses a phrasal negative marker as a head, the language will innovate negative concord (if it does not already have it, which is a possibility allowed under (112)). Furthermore, under (111), if a language replaces an existing head negator with a new phrasal marker, the language should lose negative concord; however, under (112), such a language is free either to retain negative concord or to give it up. As with Haspelmath's proposals above, these approaches suggest that we can expect later stages of Jespersen's cycle to be associated with a move away from negative concord. A return to the start of the cycle, however, is associated with the (re)introduction of negative concord.

Zeijlstra (2004: 278–9) proposes non-standard English negative concord as an example of this last development, innovation of negative concord due to reanalysis of a phrasal marker as a head. Many varieties of English are in fact negative concord languages, allowing patterns such as (113).

(113) I didn't do nothing.

Zeijlstra proposes that this is an innovation triggered by reanalysis of the negative marker from phrasal *not* to a head *-n't* (head status, for instance, being shown by its obligatory attachment as a suffix to an auxiliary). By (112), the end of Jespersen's cycle, namely reanalysis of a negative phrase as a head (cf. discussion of (14) above), automatically triggers introduction of negative concord.[11] For fuller discussion of the historical facts, see section 4.4.2.

[10] The weaker prediction in (112) is to allow for various West Germanic varieties, such as West Flemish and Bavarian, with negative concord (negative doubling), but clearly phrasal sentential negators; see Rowlett (1998: 126–31) and Zeijlstra (2004: 255–7).

[11] Formally, for Zeijlstra (2004, 2008), negative concord obtains in languages with a covert negative operator in SpecNegP. This operator triggers negative agreement on any uninterpretable negation features

French, as already discussed above with reference to Haspelmath's proposals, instantiates the second development. Loss of the head negator *ne* in colloquial French (the shift to obligatory stage III of Jespersen's cycle), as in (114), allows the language to abandon negative spread. Hence, sentences like (115) may be interpreted by some speakers as negative concord (negative spread) with a single semantic negation, while other speakers have abandoned negative concord entirely and treat them as involving two negations that cancel each other out (see section 2.3).

(114) Jean mange rien.
Jean eat.PRES.3SG n.thing
'Jean isn't eating anything.' (Colloquial French) (Zeijlstra 2004: 278)

(115) Personne dit rien.
n.one say.PRES.3SG n.thing
'No one is saying anything.' or 'No one is saying nothing.' (Colloquial French) (Zeijlstra 2004: 278)

Whether Jespersen's generalization holds more widely is a subject that requires further research. Mordvin (section 10.8), along with other Uralic languages, appears to be a clear counter-example, having a negative auxiliary head, but no inherently negative indefinites and hence no negative concord.

1.12 Negative imperatives and prohibitives

Many languages have special ways of negating imperatives, either using a different negative marker in the imperative (and sometimes in other modal contexts) or else disallowing negative imperatives entirely and adopting some kind of alternative expression involving a subjunctive or an infinitive. Forms that are specialized for expressing negative imperatives are termed prohibitives.

In the *World Atlas of Language Structures*, van der Auwera and Lejeune (2011) identify four types of language with respect to prohibitive marking: those with no special marking (normal imperative + normal negative); those with a special negative marker but the normal imperative; those with a special verbal form replacing the imperative but the normal negative marker; and those with both. Worldwide, special prohibitives of some kind predominate, as can been seen from Table 1.2 (based on data from van der Auwera and Lejeune 2011). However, within Europe, they are a

in its scope, including the head of Neg and any indefinites. A separate NegP projection is posited by language acquirers only if necessary, so is absent from languages where the negator is a phrasal adverb, which is assumed to adjoin to VP. So non-standard English has reanalysed VP-adverb *not* as Neg^0 *-n't*, thereby innovating NegP in the language, forcing the postulation of a null negative operator in NegP and the introduction of negative concord. Conversely, loss of a Neg^0 head from a language removes much of the evidence for postulating NegP, leaving open the possibility that it may fail to be acquired, thereby eliminating negative concord. See section 5.2.4 on the application of this analysis to German.

TABLE 1.2 **Frequency of different types of negative marking with imperatives worldwide**

type	no. of languages	%
no special marking	113	23%
normal imperative + special negative	182	37%
special imperative + normal negative	55	11%
special imperative + special negative	145	29%

minority. Two large groups, Germanic and Slavonic, typically have no special marking for negative imperatives and negative imperatives have therefore not been covered extensively in the chapters covering these languages.

A number of languages have negative markers specialized for use in imperatives and often for other semantically related contexts, notably subjunctives. Such a distinction is found in Goidelic Celtic (Irish *ná*, Scottish Gaelic *na*) and historically in Brythonic Celtic too (section 7.3.1), hence must be reconstructed for Common Celtic. In Greek, negative particles found in the imperative (Ancient Greek *mē* and Standard Modern Greek *min*) also have a wide range of uses in subjunctive and subjunctive-like contexts, but are distinct from negation of indicative clauses (section 8.3). In Modern Greek, the imperative is replaced by a subjunctive in the negative with *min*, as in other southern European languages discussed below. The imperative/subjunctive vs indicative contrast has its roots in distinctive marking of the two types in Proto-Indo-European (section 8.4), cf. also Sanskrit (*mā* in imperatives, instead of *na*, Joseph 2002), Latin (*nē* instead of *non* in subjunctives and subjunctives standing in for imperatives in Classical Latin and with true imperatives in early and colloquial Latin, Woodcock 1959: 84–6, 96–7), and Albanian (*mos* instead of *nuk*, Tomić 1999: 204–5), etc. Outside of Indo-European, such a distinction is found in Hungarian (*ne* instead of *nem*) (Kenesei, Vago, and Fenyvesi 1998: 22, Zeijlstra 2006: 418–19), and in some historical and dialectal varieties of Bizkaian Basque (*ze* instead of *ez*) (Trask 1997: 209). In Maltese, which lacks an indicative/subjunctive distinction, imperatives are negated with a postverbal (stage III) construction that is unavailable in other contexts, or (archaically) with a bipartite construction whose preverbal element *la* occurs only with imperatives and coordinated negative sentences, *ma* being the usual preverbal negator (Borg and Azzopardi-Alexander 1997: 27).

In conformity with their typical patterns of negation generally, most Uralic languages retain a special negative auxiliary going back to Proto-Uralic *elV- for use in the imperative (see section 11.3.2) (in the second person singular Estonian *ära*, Finnish *älä*, Saami *alẹ*, Erzya *il'a*, Khanty *ääɬ*, etc.). This auxiliary is today variously followed by an imperative or a connegative form of the lexical verb (the latter often itself historically an imperative) (Erelt 2009: 17, Karlsson 1999 [1983]: 165–8, Abondolo 1998: 73, 112, 380):[12]

(116) Sööda! Ära sööda!
feed.IMP.2SG NEG.IMP.2SG feed.IMP.2SG
Söotke! Ärge söötke!
feed.IMP.2PL NEG.IMP.2PL feed.IMP.2PL
'Feed!' 'Don't feed!' (Estonian) (Viitso 1998: 141)

(117) Laula! Älä laula!
sing.IMP.2SG NEG.IMP.2SG sing.CNG
Laulakaa! Älkää laulako!
sing.IMP.2PL neg.IMP.2PL sing.IMP.CNG
'Sing!' 'Don't sing!' (Finnish) (Miestamo 2011: 88)

The main aspects of this type appear to have been historically stable.

In much of southern Romance, true negative imperatives are today impossible and are replaced by negative subjunctives (Spanish, Catalan, Sardinian) or infinitives (standard Italian, Romanian). This is illustrated for Spanish below, where the expected negation with *no* plus the imperative in (118) is ungrammatical and instead replaced with *no* plus the present subjunctive in (119).

(118) *¡No lee!
NEG read.IMP.2SG
'Don't read!'

(119) ¡No leas!
NEG read.PRES.SUBJUNC.2SG
'Don't read!' (Spanish) (Zeijlstra 2006: 406)

The modern Romance situation emerges from complex patterns of historical and dialectal variation. The general historical direction has been increasing movement away from use of true negative imperatives. Latin already shows signs of this by using periphrases with *noli*/*nolite* 'do not wish (2SG/2PL)' alongside symmetrical negation for imperatives. In the history of Italo-Romance and in today's dialects, negative imperatives are variously formed from negative marker + infinitive, negative marker + imperative, and negative marker + subjunctive, with the infinitive structure dominating (see section 3.6.1). Zanuttini (1994, 1997) correlates the absence of true negative

[12] In (117), the form *laula* does not change when negated, and hence could be the second person singular imperative even in the negated clause. However, in all other person–number combinations, we find *laulako*, which is clearly connegative. Furthermore, *laula* is also the form of the connegative in the present indicative, where it is distinct from the affirmative inflected forms. It thus seems justified to treat *laula* here as a present connegative, rather than as an imperative, hence the glossing (cf. Miestamo 2011: 88).

imperatives with the position and type of the negative marker in a given variety. Within Romance, varieties with postverbal negative markers (Piedmontese, Valdôtain, Milanese) allow true negative imperatives, while those with preverbal ones, with some exceptions, do not. She suggests that preverbal markers are heads that select for a mood phrase with mood features that must be checked, and that the Romance imperative, being a bare form, does not bear the features needed to satisfy this requirement. Such an approach naturally raises historical and comparative issues: this generalization does not hold outside of Romance, where preverbal markers of negation are often compatible with imperatives (cf. Slavonic, Goidelic Celtic). Conversely, Welsh developed special negative marking in imperatives (a new negative auxiliary *paid* followed by *â* 'with' and an infinitive) just after it innovated postverbal negation (see section 7.12).

Languages may develop prohibitive markers without abandoning their regular, symmetrical strategy for negating imperatives. Thus, some South Slavonic languages have developed new prohibitive markers, such as Serbian *nemoj* (< *ne* 'not' + imperative of *moći* 'be able'), and Hungarian has an emphatic prohibitive *nehogy* (< *ne* 'not' + *hogy* 'that (complementizer)':

(120) Nehogy le-masól-d a kulcsot!
NEG.IMP PFX-copy.IMP-DEF.OBJ the key.ACC
'Don't you copy the key, or...!' (Hungarian) (Kenesei, Vago, and Fenyvesi 1998: 22)

In these languages, though, both this and a regularly formed negative imperative are available. These developments may represent a form of renewal of negation that has not yet led to the replacement of the earlier pattern.

From a diachronic perspective, the important question is how and why new prohibitive markers emerge and how and why old ones disappear. Negative imperatives seem to be more subject to the Neg-First Principle than other verbal forms, presumably for the very functional reason that misinterpretation of a prohibition as a positive command could have serious practical consequences. While this is a promising line of reasoning for the emergence of new preverbal marking of prohibition in Welsh, it does little to address the complex historical patterns of variation found in Romance, or the pressure to create new markers of prohibition more generally.

1.13 Contact-induced change in negation

Finally, we turn to the question of to what extent commonalities of patterning in negation and the existence of common historical developments within Europe are due to contact between languages. This is a question which takes us beyond any individual chapter of the current volume, but is one that will be taken up further in volume 2. Bernini and Ramat (1996: 49–51) argue that, since bipartite negation is rare and arose via similar processes in the histories of different languages across Western Europe and North Africa, Jespersen's cycle must be an areal feature, but one of a rather abstract kind, arising via transfer of a structural model without speakers equating particular items in their two languages. The frequency of bipartite negation is no reliable guide to the frequency of Jespersen's cycle in the histories of the world's languages. However, the observed high frequency of Jespersen's cycle itself in Western Europe and North Africa, but not in Eastern Europe, does mean that we should take contact explanations for Jespersen's cycle in Europe seriously. Nevertheless, as Bernini and Ramat concede, the chronology is far from straightforward, with languages undergoing the transition from one stage to another at radically different times. A language whose Jespersen cycle has run to completion (stage III) offers no particular model for another language: it is therefore unlikely that, for instance, the Old Norse Jespersen cycle provided a model for German, Dutch, or English to imitate. Furthermore, the lexical material used often differs from language to language ruling out direct replica grammaticalization in the sense of Heine and Kuteva (2005): the homonymy of Old French *pas* as 'not' and 'step' would provide no particular spur for English to develop *nawiht* 'nothing' into plain 'not'. Contact as an overarching explanation for all of Jespersen's cycle in Western Europe and North Africa is therefore not convincing. However, there are a number of individual cases where a more considered case for the role of contact can be made.

Dutch, Low German, and High German clearly form a continental West Germanic dialect continuum, with High German being most advanced in terms of Jespersen's cycle, and Dutch most conservative. Contact overlays the general spread of Jespersen's cycle from south to north, apparently accelerating innovation in the northeast: the varieties of Low German that reach stage III of Jespersen's cycle quickest are those spoken in Hanseatic cities (Lübeck and Stralsund) in colonial areas formerly inhabited by Slavonic speakers. Here contact between different Low German dialects and the formation of an urban *koine* (Peters 2000a: 1414) accelerated change, as did contact with Scandinavian and High German due to Hanseatic trade (section 6.2.1).

In northern Italy, neighbouring dialects innovate in similar ways, introducing new markers of negation, but from different lexical sources, encompassing minimizers, indefinite pronouns, and the anaphoric negator *nò* 'no' (section 3.4). The prevalence of these developments suggests diffusion of a structural model (reinforcement of negation) even though the lexical means used to implement that model vary, perhaps favoured by conditions of balanced bidialectalism (Trudgill 1994: 19).

Within Celtic, the Welsh Jespersen cycle is probably internally motivated, being chronologically out of step with the English one. However, the Breton Jespersen cycle shows close parallels with French, both in using a range of reinforcers based on minimizers, and in the relative stability of its stage II. Similarities in the use of negative reinforcers between Cornish and Breton suggest transfer from Breton into Cornish (section 7.5). Both Breton and Cornish show direct loans of negative markers and indefinites from Romance (Latin or French) and English respectively (sections 7.7 and 7.8).

In Uralic, the inherited pattern of a richly inflected negative auxiliary with the lexical verb in a special connegative form is best retained in the eastern Finno-Permic languages. In the west, Saami, Finnish and, especially, Estonian have limited the extent of inflection, moving towards an invariant negative particle like most Indo-European languages, and this development may well be due to contact with Indo-European. Hungarian uses a negative particle followed by an ordinary finite verb, the standard Indo-European pattern, which may be an innovation postdating contact with Indo-European (Sammallahti 2011: 206).

In Afro-Asiatic, the evidence points to the occurrence of Jespersen's cycle in a subset of Arabic dialects, Coptic, Berber, and Modern South Arabian all being linked by contact (section 10.2), with internal changes in Modern South Arabian and Coptic triggering parallel developments in the Arabic dialects of the southern Arabian Peninsula and North Africa respectively, and stage II negation then spreading from North African Arabic varieties to Berber. The relative chronologies of Jespersen's cycle in North Africa and Western Europe suggest that there is no link between the two.

The role of contact is of course not limited specifically to sentential negation. As with sentential negation, in those cases where contact influence can be identified in indefinites, it is normally structural, concerning the distribution of items, rather than involving borrowing of individual forms, which seems to occur only under particularly intense contact (Haspelmath 1997: 184).

Such structural borrowing is evident in Celtic. Cornish originally showed no sensitivity to negative polarity in its indefinites, but latterly created a new series of items, the *veeth*-series, showing just such sensitivity, modelled on the English *any*-series (section 7.8). The Welsh *unrhyw*-series is currently adopting the distribution of the English *any*-series (section 7.6.5). Replica grammaticalization (Heine and Kuteva 2005) is evident in the emergence of the Breton *bennak*-series from a free-relative structure mirroring the grammaticalization pathway of French *quelqu'un* 'someone' etc. (section 7.7).

In the Balkans, Romanian negative indefinites have a distribution remodelled on Slavonic, with a central distinction between negative and non-negative (including NPI contexts), and with strict negative concord, rather than the non-strict system found extensively elsewhere in Romance. Romanian never extended generic nouns into the indefinite system, and instead developed a *ni*-series from a mixture of existing material (*nimeni* < Latin *neninem* 'no one (acc.)') and new items parallel to Slavonic (*nimic* 'nothing' < *nec mica* 'not even a crumb') (Haspelmath 1997: 263–5). Conversely, a realignment in the distribution of indefinites in some Bulgarian dialects seems to have been due to contact with Greek (section 9.5.8).

The details of the scenarios by which these structural transfers have taken place have yet to be fully established.

1.14 Summary

This chapter has set out the kinds of development that are characteristic of changes in negative systems. With markers of sentential negation, we see extensive and repeated renewal in Western Europe and North Africa via Jespersen's cycle, while much of Eastern Europe shows relative stability, and the negative auxiliaries found in the Uralic languages go through their own processes, often leading towards the emergence of uninflected negative particles. Detailed comparison of individual cases reveals much commonality, but also significant differences that need to be accounted for.

In indefinite systems, we also see forms of cyclic renewal: in some languages, items become restricted to more negative contexts, being replaced in other contexts by existing or newly created items. New items that interact with the negative system develop from such sources as generic nouns, free-choice items and free-relative markers. The interaction between developments among indefinites and those involving sentential negation is itself complex and will only be understood properly once more language histories have been fully investigated on a comparative basis.

We have also tried to sketch the general concepts used to analyse negation more generally, and to see how those concepts have been applied to propose explanations for the typical historical patterns of development that we witness. These set the context for detailed investigations by individual contributors to which the remainder of this volume is devoted.

2

Negation in the history of French

MAJ-BRITT MOSEGAARD HANSEN

2.1 Introduction

The basic stages in the evolution of what Payne (1985: 198) refers to as 'standard negation' (namely the most basic form of clause negation, which applies in declarative main clauses and does not involve quantifiers) in French are well known. They are set out as the central stages 1–4 in Table 2.1, where stage 0 represents the Latin source construction, and Stage 5 a plausible (if not inevitable, cf. section 2.2.4 below) future scenario.

Negative sentences involving quantifiers (e.g. *rien* 'nothing', *personne* 'nobody', *jamais* 'never'...) follow a largely similar pattern—although with additional complications—in as much as the quantifiers in question originally had positive meaning and required the presence of preverbal *ne*, but—like postverbal *pas*—gradually acquired independent negative meaning.[1]

The basic outline of the evolution of French clause negation is very well known, and has been adduced as a textbook example of grammaticalization (e.g. by Hopper and Traugott 1993: 58). In particular, it is seen as a salient instantiation of what has become widely known as Jespersen's cycle (Dahl 1979: 88), whereby

> the original negative adverb is first weakened, then found insufficient and therefore strengthened, generally through some additional word, and this in turn may be felt as a negative proper and may then in the course of time be subject to the same development as the original word. (Jespersen 1917: 4)

As noted by van der Auwera (2010a), Jespersen's account focuses on the formal properties of negation, and essentially explains the negative cycle in French as having

[1] Note, however, that their exact semantic status in Modern and Contemporary French remains controversial, as will be discussed in section 2.3 below.

TABLE 2.1 The evolution of French clause negation for a sample sentence: 'I do not say...'

Stage	Example	Description
(0. [Classical Latin]	*non dico*	The negator is preverbal)
1.	*je ne dis*	The preverbal negator is phonetically reduced
2.	*je ne dis (pas)*	The preverbal negator is optionally complemented by a postverbal element
3.	*je ne dis pas*	The postverbal element grammaticalizes as part of a discontinuous negator embracing the verb
4.	*je (ne) dis pas*	The original preverbal negator becomes optional
(5. [Future French?]	*je dis pas*	The negator is postverbal)

been triggered by the phonetic weakening of Latin *non* to *ne*. An alternative account was, however, proposed by Jespersen's contemporary, Antoine Meillet (1921 [1912]: 140), who, observing that Latin *non* itself represents the univerbation of an earlier negative marker *ne* and a reinforcing expression *oenum* 'one', pointed to the importance of pragmatics rather than form:

> Les langues suivent ainsi une sorte de développement en spirale: elles ajoutent des mots accessoires pour obtenir une expression intense; ces mots s'affaiblissent, se dégradent et tombent au niveau de simples outils grammaticaux; on ajoute de nouveaux mots ou des mots différents en vue de l'expression; l'affaiblissement recommence, et ainsi sans fin.
>
> 'Languages thus undergo a sort of spiral development: they add extra words to obtain an intensified expression; those words weaken, wear out and are reduced to the level of simple grammatical tools; new or different words are added for expressive purposes; the weakening process begins anew, and so on without end.' (my translation)

As it stands, Table 2.1 is compatible with both accounts, neither of which necessarily excludes the other. Indeed, as one might expect, the history of negation in French is a good deal more complex than the simplified representation in Table 2.1 suggests.

The structure of this chapter is as follows: in section 2.2, I discuss standard clause negation, focusing in particular on stages 2 and 4 in Table 2.1, that is, those stages where variation in the expression of clause negation is found. A certain amount of attention will be paid to recent proposals concerning the pragmatics of variable clause negation. The subsequent section 2.3 deals with negation involving quantifiers, more particularly the synchronic status that can be attributed to the latter and the parallels between the basic negative cycle and the quantifier cycle in French. Negated infinitival clauses, which have special properties in Modern French, are treated briefly in section 2.4, and section 2.5 is a conclusion. For reasons of space, I will deal only with clause negation, leaving constituent negation (cf. Klima 1964) out of the picture.[2]

The exposition will principally be concerned with 'standard' French, understood as the variety that is written and spoken by educated middle-class speakers native to France, and which manifests no salient regional features. Other dialects/sociolects will only be sporadically adduced for comparative purposes. The following periodization will be observed: Old French (9th–13th centuries), Middle French (14th–16th centuries), Classical French (17th–18th centuries), Modern French (19th–20th centuries), Contemporary French (late 20th–21st century).

2.2 Standard negation in the history of French

With respect to what was defined as standard clause negation[3] in section 2.1 above, Table 2.1 simplifies the situation in Old French, represented by stages 1–2 and in Middle French (stage 2) in two respects.

First of all, *ne* did not, in fact, rule supreme as the preverbal negative marker in declarative clauses. Its etymological source, *non*, was still in use, although for most of the period only in a restricted set of contexts where it could be used predicatively, that is, could on its own stand in for the main predicate of a clause or sentence, in opposition to non-predicative *ne* (Moignet 1965). While in the oldest texts, examples like (1) can be found, *non* was otherwise used as a standard negator mainly in minimal clauses, featuring the auxiliary verbs *estre* 'to be', *avoir* 'to have', or the pro-verb *faire* 'to do', whose content represents an explicit correction or rejection of the content of some preceding clause, cf. (2) (Nyrop 1930: 26ff., Foulet 1965: 235ff.). This latter use of *non* decreased gradually over the course of Middle French, and disappeared completely in Classical French:[4]

(1) Elle colpes **non** auret.
SBJ.3FS guilt.SG NEG have.IPFV.PAST.3SG
'She did not have guilt.' (*La séquence de Sainte Eulalie*, v. 20, *c*.880, from Base de français médiéval)

[2] Constituent negation in the history of French is treated in the following studies: Nyrop (1930), Foulet (1965), Moignet (1965), Marchello-Nizia (1997), Buridant (2000), Catalani (2001). For Modern French, see for instance Gaatone (1971), Muller (1991), Mignon (2008).

[3] In the following, I leave out of consideration the evolution of the emphatic adverbs *nullement* and *aucunement* (both 'in no way'), as well as of the approximator *guère* 'barely, hardly', all three of which are of infrequent use and can be considered to belong to elevated registers, where they function largely as stylistic variants of *pas*.

[4] For a recent analysis of Old French *non*, see Larrivée (2011).

(2) « Toutes voies, fet li rois, i essaierez vos se vos la porriez oster. »
'"In any case, says the King, try if you can pull it out."'

— « Sire,	fet	il,	**non**	ferai	ge. »
Sire	do.PRES.3SG	SBJ.3MS	NEG	do.FUT.1SG	SBJ.1SG

'— "Sire, says he, I will not do [it]."' (*La queste del Saint Graal*, p. 6, *c.* 1220)

Apart from such cases, a phonetically 'intermediate' form, *nen*, is found as in (3) before verbs beginning in a vowel, up until the twelfth century. As this form does not have predicative function in examples like (3), it must be considered a phonologically conditioned variant of *ne* (Moignet 1965: 58).

(3)

En	la	citet	**nen**	ad	remes	paien /
In	DEF.DET.FS	city.SG	NEG	have.PRES.3SG	remain.PP	heathen.SG

Ne	seit	ocis	u	devient	chrestien.
NEG	be.PRES.SUBJUNC.3SG	kill.PP	or	become.PRES.3SG	Christian.MS

'In the city there did not remain a heathen who was not either killed or converted to Christianity.' (*La chanson de Roland*, vv. 101–2, *c.*1090)

Secondly, and more importantly for the negative cycle, there was no unique post-verbal marker, either, *pas* (< Lat. *passu(m)* 'step') being in competition with various other particles, most saliently *mie* (< Lat. *mica(m)* 'crumb'), but also *point* (< Lat. *punctu(m)* 'point'), and to a lesser extent, *goutte* (< Lat. *gutta(m)* 'drop'), and yet others besides (for the latter, less grammaticalized, items, see Möhren 1980).

2.2.1 *The emergence of negative reinforcement*

As seen above, Jespersen (1917) proposes that the change from stage 1 to stage 2 can be explained largely by the phonetic weakening of the Latin negator *non* to *ne* /nə/, creating an increased need for reinforcing expressions. There are problems with this hypothesis, however. First of all, it does not explain why *ne* nevertheless remained perfectly capable of negating clauses without the help of a reinforcing element for centuries following the sound change in question, cf. (4).

(4)

Selon	la	commune	opinion,
according.to	DEF.DET.FSG	common	opinion

les	connins	**ne**	touchent	aux	raisins…
DEF.DET.PL	rabbit.PL	NEG	touch.PRES.3PL	to+DEF.DET.PL	grape.PL

'According to general opinion, rabbits don't touch grapes…' (Olivier de Serres, *Le théâtre d'agriculture et mesnage des champs*, t. 1, p. 449, 1603, from Frantext)

Secondly, according to Martineau and Mougeon (2003: 123–4), *ne* actually only became unstressed at the end of the Middle French period, and the schwa-deletion that is now common in spoken French (as in *Je n'sais pas* 'I don't know') thus did not set in prior to that time (see also Moignet 1965: 58). Thirdly, other Romance languages in which Latin *non* was either not phonetically reduced at all (Italian) or was not reduced to the same extent as in French (Catalan) have nevertheless developed cognate forms of reinforced negation as alternatives to the simple preverbal *no(n)*, as shown by the contrasting Catalan examples in (5)–(6) and the similar Italian examples in (7)–(8) (see further section 2.2.2 below). Indeed, reinforced negation is found in Latin, from the pre-Classical period onward, in texts having an oral and/or colloquial tenor, for instance, the comedies of Plautus (cf. (9)). Thus, the origins of the Old French particles can in all probability be traced back to Latin usage in contexts like (9)–(10).

(5) L. Wittgenstein **no** va ser un lingüista.
L. Wittgenstein NEG go.PRES.3SG be.INF INDEF.DET.MSG linguist.SG

(6) L. Wittgenstein **no** va ser **pas** un lingüista.
L. Wittgenstein NEG go.PRES.3SG be.INF NEG INDEF.DET.MSG linguist.SG
'L. Wittgenstein was not [*pas*] a linguist.' (Espinal 1993: 354)

(7) **Non** fa freddo fuori.
NEG do.PRES.3SG cold.MSG outside

(8) **Non** fa **mica** freddo fuori.
NEG do.PRES.3SG NEG cold.MSG outside
'It's not [*mica*] cold outside.' (Manzotti and Rigamonti 1991: 284)

(9) AM. Haec sacerdos Veneris hinc me petere aquam iussit a vobis.
SC. At ego basilicus sum:
quem nisi oras, **guttam** **non** feres.
REL.MSG.ACC if.not entreat.PRES.2SG drop.SG.ACC NEG carry.FUT.2SG
'AM. That priestess of Venus ordered me here to get some water from you.
SC. But I'm an important person: if you don't entreat me, you'll not take a drop away.' (Plautus, *Rudens* act 2, scene 4, 3rd–2nd century BC, from Perseus)

(10) quinque dies aquam in os suum non
five day.PL.ACC water.SG.ACC in mouth.SG.ACC POSS.MSG.ACC NEG
coniecit, **non** **micam** panis
put.PFV.3SG NEG crumb.SG.ACC bread.SG.GEN
'for five days he didn't put any water in his mouth, not a crumb of bread' (Petronius, *Satyricon*, section 42, 1st century AD, from Perseus)

Given examples like (9)–(10), it is plausible to assume a three-stage development from Latin to medieval French. At the initial stage, we have a variety of ordinary lexical items with no inherent preference for either positive or negative contexts, but which—because they happened to denote minimal quantities of something—would have lent themselves quite naturally to underscoring the negative content of a clause with a semantically compatible verb, following a rhetorical strategy that Detges and Waltereit (2002: 177) formulate as follows:

If you want to express in a strong way that some state of affairs did not take place at all, say that the state of affairs in question did not even take place to the smallest degree imaginable.

The assumption is then that prior to the situation observed in Old French, the reinforcing elements must have been confined to contexts equivalent to 'I don't walk a step', 'I don't eat a crumb', etc.

By the Old French period, use of *pas*, *mie*, and *point* had, however, been generalized, and they were thus no longer constrained to occur with semantically compatible verbs, cf. (11)–(13) (Moignet 1976: 277).[5] They had also largely lost their nominal properties in favour of a particle-like status, occurring in the vast majority of cases without determiners or any other form of modification.

(11) Soleill n' i luist ne blet **n'** i poet **pas**
sun.SG NEG LOC shine.PRES.3SG nor corn.SG NEG LOC can.PRES.3SG NEG
creistre.
grow.INF
'The sun does not shine, and corn cannot [*pas*] grow.' (*La chanson de Roland*, v. 980, *c.*1090)

(12) Dist Clarïen: 'Dame **ne** parlez **mie** itant!'
say.PFV.PAST.3SG Clarien lady.SG NEG speak.IMP.2PL NEG so.much
'Clarien said, "My Lady, don't [*mie*] talk so much!"' (*La chanson de Roland*, v. 2724, *c.*1090)

(13) La vostre gent **ne** puet il
DEF.DET.FSG POSS.2PL.SG people.SG NEG can.PRES.3SG SBJ.3MSG
point amer.
NEG love.INF
'Your people he cannot [*point*] love.' (*Le couronnement de Louis*, v. 829, *c.*1130)

Some traces of their original nominal status remained, however: *point*, in particular, and to a lesser extent, *mie*, could be used with a partitive PP, as in (14),[6] and all three particles enjoyed a relative freedom of position, occasionally preceding *ne*, as in (15).

[5] *Goutte* was largely confined to the verbs *voir* 'see', *entendre* 'hear', and *comprendre* 'understand', and will not be dealt with further in this chapter, but see Price (1990) for a more detailed analysis of its use in medieval French.

[6] Although this construction appears to have been almost non-existent with the particle *pas* in Old French, it is the origin of the Modern French use of the reduced partitive article following negation, as in *Je n'ai pas acheté de pommes* 'I didn't buy apples' (Foulet 1965: 269).

(14) ... je suiz cil qui l' em porterai:
SBJ.1SG be.PRES.1SG DEM.SG SBJ.REL OCL.3MSG away carry.FUT.1SG
car aussi **n'** ai je **point** **d'** escu.
for indeed NEG have.PRES.1SG SBJ.1SG NPI of shield.SG
'... I am the one who will take it with me: (lit.) for I don't have a bit [*point*] of a shield.' (*La queste del Saint Graal*, p. 27, *c*.1220)

(15) Iluec trova le riche rei Glafre;
Pas **nel** salue.
NEG NEG+OCL.3MSG greet.PRES.3SG
'Here he found the mighty King Galafre; He doesn't [*pas*] salute him.' (*Le couronnement de Louis*, vv. 448–9, *c*.1220)

Furthermore, in medieval texts, the particles sometimes occur without a preverbal *ne* and with non-negative meaning:

(16) S' il daigne **pas** parler ancor.
if SBJ.3MSG deign.PRS.3SG NPI speak.INF still
'If he still deigns to speak at all [*pas*].' (*Le roman de Renart*, vv. 12636–7, 12th–13th cc.) (from Price 1993: 193)

(17) Tut seie fel, se jo **mie** l'
all be.PRES.SUBJUNC.1SG disloyal.MSG if SBJ.1SG NPI OCL.MSG
otrei!
grant.PFV.PAST.1SG
'I would be a complete traitor, if I granted that in the least [*mie*]!' (*La chanson de Roland*, v. 3897, *c*.1090)

(18) Resanble je **point** a celui
look.like SBJ.1SG NPI to DEM.MSG
Qui sol... vos secorut a cel besoin?
SBJ.REL alone.MSG OCL.2PL help.PFV.PAST.3SG in DEM.MSG need.SG
'Do I look at all [*point*] like him who alone... helped you in this need?' (*La folie de Tristan*, B390, late 12th c.) (from Eckardt 2006: 136)[7]

Marchello-Nizia (1997: 306) and Eckardt (2003: ch. 4) point out that the contexts in which this occurs are those that are usually classified as negative polarity contexts, which leads Eckardt to argue that, at stage 2 of Jespersen's cycle, the French negative reinforcers are, in fact, better analysed as negative polarity items than as inherently

[7] For a discussion of Old and Middle French uses of *pas/point* without *ne* in interrogatives, see Price (1993).

negative particles. A similar conclusion is reached by Winters (1987), who, however, focuses not so much on the use of *pas/mie/point* without *ne* as on the fact that they still retained some nominal properties.

I find it entirely plausible, and convincingly supported by Eckardt's analysis of the *ne*-less uses of the particles, that there should have been an intermediate stage of grammaticalization where *pas, mie,* and *point* would have been polarity-sensitive items. However, their non-negative uses are comparatively rare in Old French. It is true that the more 'noun-like' uses exemplified in (14)–(15) are common with *point*, but this particle is far less frequently used in the medieval period than were *pas* and *mie*, with which partitive modification and preposing are sporadic at best. Indeed, these particles can occasionally be found alongside NPI-like elements denoting minimal quantities, as in (19)–(20), which strongly suggests that they must have been perceived by speakers as part of the negation itself.

(19)

Tuit	vos	Franceis	**ne**	valent	**pas**	meaille.
all	POSS.2PL.PL	Frenchman.PL	NEG	be.worth.PRES.3PL	NEG	small.coin.SG

'All your Frenchmen aren't [*pas*] worth a dime.' (*Le couronnement de Louis*, v. 2433, *c*.1130)

(20)

Trestuz	les	alters	**ne**	pris	jo	**mie**	un	guant.
all	DEF.DET.PL	other.PL	NEG	value.PRES.1SG	SBJ.1SG	NEG	INDEF.DET.MSG	glove.SG

'All the others I don't [*mie*] consider worth a glove.' (*La chanson de Roland*, v. 3189, *c*.1090)

In my view, the seemingly contradictory data can perhaps best be explained as evidence of divergence, that is, the fact that, in grammaticalization, older uses of an item or construction are not necessarily discarded when that item/construction takes on new functions, but may continue to be instantiated alongside the new one (Hopper 1991: 24). In the case of *pas* and *point*, it is quite clear that the full nouns continue to exist with their original meanings ('step' and 'point') alongside the negative particles in Contemporary French, and there is no reason why the same could not have been the case with the intermediate-stage NPI-uses until the latter finally disappeared from the language in Classical French.

2.2.2 *The pragmatics of negative reinforcement at stage 2*

The literature has traditionally considered that, already by the Old French period, the reinforcing particles were grammaticalized to a sufficient degree for there to have been little or no difference in meaning between simple and bipartite negation (e.g. explicitly to this effect, Perle 1878: 5, Sten 1938: 30, Togeby 1974: §258, and implicitly,

Yvon 1948, Price 1962, Foulet 1965, Harris 1978, Winters 1987).[8] As argued in Hansen (2009) and Hansen and Visconti (2009), however, a fine-grained qualitative analysis of several Old and Middle French texts supports the hypothesis that there was, in fact, a functional differentiation between the two constructions for several centuries preceding the eventual generalization of *ne...pas* as the standard form.[9]

As already mentioned, a number of contemporary Romance vernaculars exhibit competition between a canonical, plain preverbal form of standard clause negation and a bipartite construction featuring, in addition, a postverbal particle (cf. (5)–(8) above). Now, several studies have suggested that the use of Italian *non...mica* (a cognate of Old French *ne...mie*), Catalan *no...pas* (a cognate of French *ne...pas*), and Brazilian *naõ...naõ* is subject to discourse-functional constraints linked to the presupposed or otherwise 'given' nature of the negated proposition or its underlying positive counterpart (e.g. Schwegler 1988, Espinal 1993, Bernini and Ramat 1996, Zanuttini 1997, Schwenter 2006). According to Zanuttini (1997: 61), for instance, in Italian, only (22) below would constitute a felicitous exchange with the addition of *mica*; (21) would not.

(21) A. Chi viene a prenderti?
'A. Who's coming to pick you up?'

B. Non so.	Ma Gianni **non** a	(***mica**) la
NEG know.PRES.1SG	but Gianni NEG have.PRES.3SG NEG	DEF.DET.FSG
macchina.		
car.SG		

'B. I don't know. But Gianni doesn't [**mica*] have the car.'

(22) A. Chi viene a prenderti—Gianni?
B. Non so. Ma Gianni **non** a **mica** la macchina.
'A. Who's coming to pick you up—Gianni?
B. I don't know. But Gianni doesn't [*mica*] have the car.'

Schwegler (1988: 45–6) suggests that occurrence of the bipartite construction in Old/Middle French was similarly restricted to presupposed propositions.[10] The data presented in Hansen (2009) and Hansen and Visconti (2009) show, however, that this

[8] It should be noted that some of the above-mentioned authors do suggest that the choice between the forms may have been influenced by syntactic factors or—in the case of poetry—metrical considerations, e.g. Perle (1878: 5), Foulet (1965: 262), Buridant (2000: 605).

[9] The remainder of this chapter will not consider reasons for the eventual obsolescence of *ne...mie*, but some suggestions can be found in Hansen and Visconti (2009: 148–9). *Ne...point* became fashionable in Renaissance French, and is still in (infrequent) use in relatively elevated registers. For an analysis of the differences between *pas*, *mie*, and *point* in Old French, see Hansen (2011).

[10] In a not unrelated vein, Guiraud (1964), Martin (1972), and Marchello-Nizia (1999: 114) analyse the difference between plain *ne* and *ne...mie/pas* in semantic terms, as involving the virtuality vs actuality of the proposition. These notions, which are crucial to their accounts, remain, however, undefined, hence rather too vague to be really useful. Furthermore, Guiraud's and Martin's accounts are mutually contradictory.

restriction is too strong. Instead, the latter studies propose that clauses negated by *ne... mie/pas* were constrained to be discourse-old, as defined by Birner (2006).[11] Thus, while a proposition expressed by an Old/Middle French clause negated by a bipartite construction need not be presupposed, let alone believed, it should be such that the speaker could assume that it was either already activated in the short-term memory of the hearer or accessible to activation based on other propositions thus activated.

Broadly, four types of use of reinforced negation can be discerned from the examples in Hansen's (2009) and Hansen and Visconti's (2009) database:[12]

a. Examples where the *ne... mie/pas*-marked clause represents a denial or rejection of part of the preceding text, as in (23);
b. Examples where the *ne... mie/pas*-marked clause represents a repetition or paraphrase of part of the preceding text, as in (24);
c. Examples where the *ne... mie/pas*-marked clause represents either the expression or the denial/rejection of a (pragmatic) presupposition, as in (25);
d. Examples where the *ne... mie/pas*-marked clause represents either the expression or the denial/rejection of another type of inference warranted by the previous text, as in (26).

(23) «Dame, soffrez que nostre noviaus chevaliers viegne avec nos a la cort mon seignor le roi. [...]
'"My Lady, allow our new knight to come with us to the court of our Lord the King. [...]"'
– « Sire, fet ele, il **n'** ira **pas** ore... »
Sire do.PRES.3SG SCL.3FSG SCL.3MSG NEG go.FUT.3SG NEG now
'– "Sir, says she, he'll not [*pas*] go now..."' (*La queste del Saint Graal*, p. 3, *c.*1220)

(24) ...de dous mille et huit cens chevaliers que li roys mena en Egypte, ne l'en demoura que sept cens... [...]
'...out of two thousand and eight hundred knights that the King brought with him to Egypt, only seven hundred remained... [...]'
...pour ce que il **ne** li
for DEM.SG COMP IMPERS.SBJ.3MSG NEG DAT.OCL.3SG
estoit **pas** demouré la tierce partie de
be.IPFV.PAST.3SG NEG remain.PP DEF.DET.FSG third part.SG of
ses gens...
POSS.3SG.PL people.PL
'...because not [*pas*] even a third of his men had remained with him...' (Joinville, §§147 and 149, 1298–1309)

[11] Schwenter (2006) proposes a similar explanation for the use of *naõ... naõ* in present-day Brazilian Portuguese.

[12] *Ne... point* is not represented in the two studies cited. However, Hansen (2011) appears to confirm that the uses of this negator in medieval French fall into the same four pragmatic categories.

(25) « Sire, coment avez vos non ... »
'"Sir, what is your name ..."'

— « De	mon	nom,	fet	il,	**ne**	puez
of	POSS.1SG.SG	name.SG	do.PRS.3SG	SBJ.MSG	NEG	can.PRES.2SG

tu	**mie**	savoir ... »
SBJ.2SG	NEG	know.INF

'— "Of my name, says he, you cannot know [*mie*] ..."' (*La queste del Saint Graal*, p. 29, *c.*1220)

(26) Turpins de Reins, quant se sent abatut, De .IIII. espiez par mi le cors ferut, Isnelement li ber resailit sus; Rollant reguardet, puis si li est curut, E dist un mot:
'Turpin of Reims, when he finds himself knocked down, his body pierced by four spears, immediately, the brave man stands up again; he looks around for Roland, then runs to him, and says one word:'

«**Ne**	sui	**mie**	vencut ... »
NEG	be.PRES.1SG	NEG	defeat.PP

'"I'm not [*mie*] defeated ..."' (*La chanson de Roland*, vv. 2083–7, *c.*1090)

As the unmarked form, plain preverbal *ne* could, of course, also be used to negate discourse-old information, but the data adduced in Hansen (2009) show that, in the course of Old French, *ne* becomes increasingly confined to clauses expressing discourse-new information, as in (27), and that when it does occur with discourse-old propositions, the discourse salience of the information tends to be reduced due to the non-declarative, non-referential, and/or irrealis nature of the context. This is illustrated by (28), where the clause negated by plain preverbal *ne* directly denies a preceding clause, but where both these clauses function as antecedents of conditional constructions.

(27) Or avint ainsi que Oliviers de Termes, qui bien et viguerousement s'estoit maintenus outre mer, lessa le roy et demoura en Cypre,
'Now, it happened thus that Olivier de Termes, who had held his own well and vigorously overseas, left the King and stayed in Cyprus,'

lequel	nous	**ne**	veismes	puis	d' an	et	demi	après.
OBJ.REL	SBJ.1PL	NEG	see.PP.1PL	subsequently	of year.SG	and	half	after

'and we didn't see him again for a year and a half after that.' (Joinville, §16, 1298–1309)

(28) A œuvre devons-nous mettre ceste menace que Diex nous a faite, en tel maniere que, se nous sentons que nous aiens en nos cuers et en nos cors chose qui desplaise à Dieu, oster le devons hastivement; et quanque nous cuiderons qui li plaise, nous nous devons esforcier hastivement dou penre. Et se nous le faisons ainsinc, Nostre-Sires nous donra plus de biens en cest siecle et en l'autre que nous ne sauriens devisier.

'We should apply the warning that God has given us in such a way that if we feel that we have in our hearts and in our bodies anything which displeases God we must quickly get rid of it; and whenever we think there is anything that pleases Him we must hasten to set our hands to it. And if we do this, Our Lord will give us more blessings both in this world and in the next than we can imagine.'

Et se nous **ne** le faisons ainsi, il fera
and if SBJ.1PL NEG OCL.MSG do.PRES.1PL thus SBJ.3MSG do.FUT.3SG
aussi comme li bons sires
thus as DEF.DET.MSG.NOM good.MSG.NOM master.SG.NOM
doit faire à son mauvais sergant;
shall.PRES.3SG do.INF to POSS.3MSG bad.SG.OBL servant.SG.OBL
'And if we don't do this, he will do what a good master should do to his bad servant;'

car après la menace, quant li mauvais serjans ne se veut amender, li sires le fiert ou de mort ou de autres greignours meschéances, qui piz valent que mort.
'for if after a warning, the bad servant will not mend his ways, the master punishes him either with death or with other greater afflictions that are worse than death.' (Joinville, §41, 1298–1309)

Evidently, if the above analysis is correct, a markedness reversal must have taken place at some point, in order for *ne...pas* to eventually have become grammaticalized as the new canonical form of clause negation in French, replacing the plain *ne*. Two factors may plausibly have been instrumental in bringing about such a reversal: first, as noted by Hansen (2009), the numerically most important of the four categories of uses of reinforced negation distinguished above is category (d), where the greatest amount of inference is required to establish the negated proposition as discourse-old. It is conceivable that not all hearers may have gone to the trouble of performing the necessary inferences on every occasion, and that this may have contributed to the reinterpretation of the bipartite forms as pragmatically unrestricted. Secondly, as Hansen (2009) and Hansen and Visconti (2009) show, the negative reinforcers were frequently used in medieval French in contexts that were, so to speak, Janus-faced, for while the negated clause was backwards-oriented, that is, oriented towards prior discourse, in the sense of being discourse-old, it could at the same time be forwards-oriented by expressing a contrast with the immediately following clause,[13] as exemplified by (29).

[13] Offord (1976: 333) notes the use of negative reinforcers in contexts marked by the contrastive/adversative conjunctions *mes* and *ainz* 'but (rather)', but not the Janus-faced nature of the contexts in these examples. Moreover, such Janus-faced contexts need not be marked by contrastive/adversative connectives.

(29) «Biau Sire, fet Gauvains, donc me poez vos bien dire, s'il vos plest, en quoi sui tiex come vos me metez sus.»
'"Good Sir, says Gawain, then you can surely tell me, if you please, in what way I am that which you accuse me of."'

— « Je	**nel**	vos	dirai	**mie,**
SCL.1SG	NEG.OCL.MSG	DAT.OCL.2PL	say.FUT.1SG	NEG
fet	cie,			
do.PRES.3SG	DEM.MSG			

'— "I'll not [*mie*] tell you, he says,

mes vos troveroiz par tenz qui le vos dira. »
but soon you'll find one who'll tell you."' (*La queste del Saint Graal*, p. 52, *c*.1220)

Here, the first (negated) clause of the second speaker's reply constitutes a denial of Gawain's immediately preceding request, but simultaneously stands in contrast to the contents of the second clause. In such cases, depending on which of the two relations appeared more salient to them, hearers would have been free to interpret the reinforced negator as marking a discourse-old proposition, that is, as fulfilling what Hansen (2009) and Hansen and Visconti (2009) argue was its original pragmatic function, or alternatively, to reinterpret it as an unmarked form of negation, not subject to any particular pragmatic constraints.

2.2.3 Ne*-deletion*

By the seventeenth century, bipartite negation with *ne...pas* (and, to a lesser extent, *ne...point, ne...mie* having more or less dropped out of use) had established itself as the normal way to express standard clause negation in French. Plain preverbal *ne* remained (and remains to this day) an option in only a handful of clearly delimited syntactic context types, and principally in formal registers. In other words, the language had by this time reached stage 3 of Jespersen's cycle. There is some evidence that *pas*, rather than *ne*, was already felt by speakers to be the principal exponent of negation, in as much as there is sporadic evidence of non-occurrence of *ne* in utterances produced by children and lower-class speakers. However, up to the nineteenth century, that is, the Modern French period, *ne*-deletion remained highly infrequent (Ayres-Bennett 1994, Martineau and Mougeon 2003).[14]

[14] In Late Middle French and Early Classical French, direct interrogatives with *point* and (to a lesser degree) *pas*, but without *ne* were relatively common, as in:

(i) Pensez	vous	**point**	de	les	me	rendre?
think.PRES.2PL	SBJ.2PL	NPI	of	OCL.3PL	DAT.OCL.1SG	give.back.INF

'Do you plan to give them back to me at all?' (*La farce de Maistre Pathelin*, l. 1460) (Price 1993: 192)

However, Price (1993) argues convincingly that the postverbal particle in this construction originally had negative polarity meaning (≈ 'at all'), and only took on actual negative meaning in the 16th century, before dropping out of fashion by the 18th century. In other words, there is reason to believe that the evolution of this construction is separate from that of *ne*-deletion in declarative clauses.

By the nineteenth century, French was clearly at stage 4 of the negative cycle, and despite the presence of dissenting voices (Blanche-Benveniste and Jeanjean 1987; Coveney 1996: 90), there can probably be little doubt that *ne*-deletion has been consistently on the rise for the past two hundred years. This is supported by a number of studies showing that, in both real (Ashby 2001, Armstrong and Smith 2002, Hansen and Malderez 2004) and apparent (Pohl 1968, Ashby 1981, Coveney 1996: ch. 3) time, the use of preverbal *ne* continues to decline in France, and that in at least some varieties outside of France, it is close to non-existent in conversational registers. Thus, with respect to Québec French, only 1.5% of the negative clauses in a sizeable corpus of conversational Montreal speech contained *ne* (Sankoff and Vincent 1977: 252). Likewise, a recently constituted corpus of informal Swiss French exhibited a mere 2.5% rate of *ne*-retention (Fonseca-Greber 2007: 256).

Although levels of *ne*-retention in conversation do remain higher in France, Ashby's (2001) analysis of two highly comparable sociolinguistic corpora gathered in the city of Tours (traditionally considered to be a region where the language spoken approaches most closely to the standard) in 1976 and 1995, respectively, showed a drop from 37% to 18% in overall use of *ne* among his informants. In roughly the same time period, Hansen and Malderez (2004: 16) similarly observed an overall drop in *ne*-retention among Parisian speakers from 15.8% to 8.2%. Even in the relatively formal genre of radio speech, Armstrong and Smith (2002: 30) found a highly significant decline in *ne*-retention, from 92.6% to 72.5% between 1960 and 1997. Although, in Ashby's (2001) study, *ne*-deletion was most advanced among younger lower-middle-class women, no social group, gender or age group was immune to the change; indeed, the decline in *ne*-use between 1976 and 1995 was sharper among older, (upper-)middle-class, and male informants. Hansen and Malderez (2004: 17ff.) report a similar finding for older as compared to younger speakers in their data, whereas gender and social class appear to play little role.

The studies cited draw attention to a variety of linguistic factors which seem to either favour or disfavour *ne*-deletion. For details, the reader is referred to the individual studies. Chief among the factors that show a significant correlation with *ne*-deletion in all the studies, however, is the presence of a clitic subject in the negated clause. Accordingly, several studies (Harris 1978: 26, Ashby 1981: 681, Armstrong and Smith 2002: 34, Martineau and Mougeon 2003: 140) suggest that increasing fusion of the French subject clitics with the finite verb may have been what triggered the rise of *ne*-deletion from the nineteenth century onwards, as this would have caused the intervening preverbal *ne* to be 'squeezed out', as it were. Martineau and Mougeon (2003: 140ff.), in particular, make a plausible diachronic case for this explanation, adducing a number of syntactic clues that subject clitics began to move towards affixal status at precisely the time when the phenomenon of *ne*-deletion becomes noticeable. A potentially confounding factor, however, is Coveney's (1996: 76–7) observation that in his data the occurrence of non-subject preverbal clitics appears to slightly favour the retention of *ne*, a finding which is, indeed, replicated in Ashby's (1981: 679) corpus. Moreover, *ne* is of course itself a proclitic element, and in principle nothing should prevent sequences of clitics from occurring with one and the same verb.

Typological explanations for the increasing disappearance of *ne* have also been proposed: based on the theory of Lehmann (e.g. 1973), both Vennemann (1974: 366ff.) and Harris (1978: 23ff.) suggest that more generalized word-order changes were responsible for the shift from preverbal to embracing—and increasingly just post-verbal—negation. However, these authors agree neither on the categorial status of the negators nor on the basic typological status of the different stages of French. Thus, according to Vennemann, the negating particles are adverbs, and

> French was developing from the SXV type, which in its consistent realization has adverbs before the verb, to the SVX type, which in its consistent realization has adverbs after the verb. (Vennemann 1974: 367)

For Harris, on the other hand, as indeed for Lehmann (1973), negation is a sentential operator. As such, it should therefore, according to Lehmann's theory, occur *after* the verb in an OV language (i.e. what Vennemann refers to as a SXV language in the above quotation), while preceding it in a VO language. In fact, Harris explains the negative cycle in French as prompted by a typological shift in a completely different direction from that hypothesized by Vennemann, interpreting Old French as an SVO language, and contemporary spoken French as a VSX language (Harris 1978: 118), in which the subject and object clitics have become prefixes on the verb, forcing *ne* out (thus making the link with the hypothesis presented above). This, however, raises the question of why *pas* has not then become preverbal in contemporary spoken French.

Finally, Armstrong and Smith (2002: 40) propose that, although it must clearly have been initially triggered by other factors, the acceleration in *ne*-deletion over the past fifty years may be attributable to broader societal changes, towards more egalitarian and youth-oriented, hence less formal, modes of interaction, which go hand in hand with less prescriptive attitudes to language.

Whatever the case may be, the ultimate cause of *ne*-deletion is likely to be the fact that, since bipartite *ne...pas* acceded to the status of standard clause negator in Classical French (following the bleaching of its hypothesized original pragmatic function), the function of being the principal negative element has gradually been transferred from *ne* to *pas*, possibly at least in part due to the latter's greater phonological prominence. As Rowlett (1998: ch. 1) shows, *ne* is insufficient to mark negation on its own in Modern and Contemporary French, except in a very restricted set of contexts in elevated registers. At the same time, *ne* has expletive uses in certain contexts, where it does not have negative force, as in (30) below. *Pas*, on the other hand, has inherent negative meaning in all registers, thus effectively rendering the presence of *ne* redundant.

(30) Je crains que Pierre **ne** vienne.
SCL.1SG fear.PRES.1SG COMP Pierre EXPL.NEG.CL come.PRES.SUBJUNC.3SG
'I fear that Pierre may come.'

2.2.4 *The future of standard clause negation in French*

Now, given the above observations, the obvious question is, of course, whether the evolution of standard clausal negation in French will eventually accede to stage 5, where only the postverbal negative particle remains. The rates of *ne*-deletion in contemporary conversational French cited above would seem to suggest that it will. Two factors may, however, retard or even prevent the eventual disappearance of *ne*.

First, several scholars have pointed out that, even among those speakers who use it the least, retention of preverbal *ne* may still serve a stylistic purpose, given that its occurrence seems to correlate both with that of various other markers of more formal register and to be linked to certain, more formal, conversational topics such as education or religion (Pohl 1968: 1358, Sankoff and Vincent 1977: 252ff., Coveney 1996: 89, Hansen and Malderez 2004: 27).

Secondly, it is not impossible that *ne* may be in the process of developing a new pragmatic function in contemporary spoken French: thus, a recent paper by Fonseca-Greber (2007) suggests that, in conversational Swiss French at least, bipartite negation may express heightened foregrounding of, hence emphasis on, the negated clause, as compared with the more neutral plain postverbal *pas*. While Fonseca-Greber's proposal needs to be corroborated by additional data from other dialects, it is of potentially great interest in two ways: first of all, such a development would constitute a case of pragmaticalization, within the same conceptual domain, of a grammatically increasingly obsolescent item, supporting the idea that grammaticalization and pragmaticalization are different types of change (Erman and Kotsinas 1993, Dostie 2004, Hansen 2008: ch. 3.2). Secondly—although the acquisition of a new

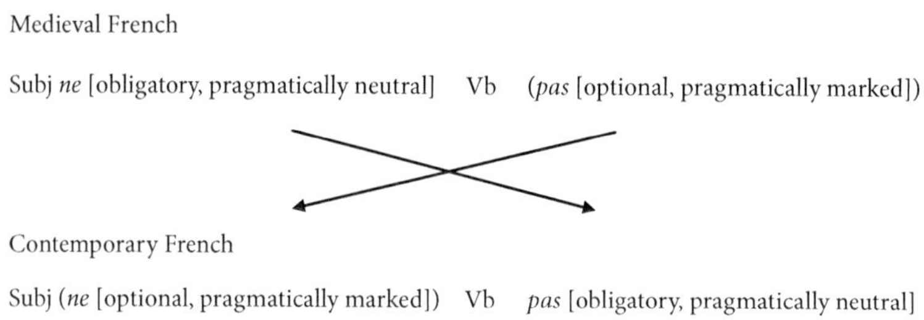

FIGURE 2.1 A pragmatic cycle in French?

function for preverbal *ne* would prevent the completion of the negative cycle as traditionally understood, that is, as laid out in Table 2.1 above—if seen in conjunction with the pragmatic analysis of bipartite negation in medieval French sketched in section 2.2.2, the suggested development would instantiate a diachronic cycle of a different kind, namely one where an obligatory, hence purely grammatical, item and an optional, pragmatically laden item have gradually come to swap functions while remaining in their original syntactic slots, returning the bipartite construction as a whole to something closely resembling (if not identical to) its original meaning, as illustrated in Figure 2.1.

2.3 The quantifier cycle in French

As observed in section 2.1 above, the evolution, in French, of clause negation involving quantifiers is largely parallel to that of standard negation, in as much as the diachronic sources for the quantifiers that are in use in Modern French were, for the most part, originally positive in meaning[15] and were used as reinforcing elements in connection with preverbal *ne*. Over time, the items in question increasingly took on negative polarity uses, and subsequently, came to carry negative meaning on their own, as so-called n-words. Table 2.2 gives an overview of the general evolution.

The above account is, however, highly simplified in two ways: first, the individual quantifiers are very different in nature, some having nominal, some pronominal, and some even adverbial origins. Hence, they have different diachronic trajectories—reinforced by the fact that they developed neither at the same time nor at the same pace—and it is not clear that they are fully identical in terms of their synchronic

TABLE 2.2 **The evolution of French clause negation with quantifiers**

Stage	Example	Description
Stage 1	*Je ne dis* (*rien*) 'I do not say (a thing)'	A positive NP optionally accompanies preverbal *ne* to make the scope of the negation explicit
Stage 2	*Je ne dis rien* 'I don't say anything'	*ne* + negative polarity item
Stage 3	*Je (ne) dis rien* 'I don't say anything/I say nothing'	N-word optionally accompanied by preverbal *ne*
(Stage 4 [Future French?]	*Je dis rien* 'I say nothing'	Negative quantifier)

[15] For some of the exceptions, i.e. *nul* 'no[ne]' and—it seems—*jamais* 'never', see Ingham (2011) and Hansen (2012).

status in contemporary French. Secondly, even to the extent that they are comparable, their precise synchronic status is in fact a matter of some controversy.

Contemporary reference grammars of standard French operate, on the one hand, with a set of three negative adverbs, two temporal adverbs, *plus* ('no more'/'no longer' < Lat. *plus* 'more') and *jamais* ('never' < Lat. *iam magis* 'now more'), and one place adverb, *nulle part* ('nowhere' < Lat. *nulla parte* 'no place'), and on the other hand, with a set of four negative indefinite pronouns, namely *personne* ('nobody' < Lat. *persona* 'character'), *rien* ('nothing' < Lat. *rem* 'thing'), *aucun* ('no(ne)' < Lat. *aliquis* + *unus* 'someone'), and *nul* ('no(ne)' < Lat. *nullum* 'no(ne)'), the latter two of which also function as negative determiners.[16]

In standard French, all of these must, like the standard postverbal negator *pas*, co-occur with a preverbal *ne* (irrespective of their syntactic function and position in the clause), except in elliptical utterances such as (31)–(32), but like *pas*, they often occur on their own in informal discourse (particularly in the spoken channel, but also, for instance, in internet chat), cf. (33)–(36).

(31)	Qui	t'	a	vu?	—**Personne**!
	INTERROG.SBJ	OCL.2SG	have.AUX.PRES.3SG	see.PP	nobody

'Who saw you? —Nobody!'

(32)	Qu'est-ce que	tu	as	vu?	—**Rien**!
	INTERROG.OBJ	SCL.2SG	have.AUX.PRES.2SG	see.PP	nothing

'What did you see? —Nothing!'

(33)	Je	**n'**	ai	**rien**	vu.	(Standard French)
	SCL.1SG	NEG.CL	have.AUX.PRS.1SG	nothing	see.PP	

'I didn't see anything. / I saw nothing.'

(34)	**Personne**	**ne**	m'	a	vu.	(Standard French)
	nobody	NEG.CL	OCL.1SG	have.AUX.PRES.3SG	see.PP	

'Nobody saw me.'

(35)	J'	ai	**rien**	vu.	(Informal French)
	SCL.1SG	have.AUX.PRES.1SG	nothing	see.PP	

'I didn't see anything. / I saw nothing.'

(36)	**Personne**	m'	a	vu.	(Informal French)
	nobody	OCL.1SG	have.AUX.PRES.3SG	see.PP	

'Nobody saw me.'

[16] Medieval French had additional items at its disposal, e.g. *onques* 'never', *niant* 'nothing', and *ame* 'nobody', which were more or less synonymous—hence in competition—with some of the members of the currently existing inventory, and which gradually became obsolete as a result. These items will not be discussed further in this chapter (but for *onques*, see Hansen 2012).

Empirical studies show, however, that *ne* is less often deleted in negative sentences involving quantifiers than in standard negative clauses with *pas*, and that, furthermore, the different quantifiers do not all appear to favour *ne*-deletion to the same extent (Ashby 1981: 678, Coveney 1996: 76, Hansen and Malderez 2004: 23).

Generally speaking, French features so-called negative concord, that is, it allows several quantifiers to co-occur within one and the same clause without giving rise to double-negation readings, cf. (37). Double-negation readings are, however, possible with some—but not all—of the quantifiers, depending on context and the use of focalization devices such as prosody, cf. (38). In standard French (but not in certain regional and/or substandard varieties, which can be considered to have preserved more archaic features), combinations of *pas* with one or more quantifiers always result in a double-negation reading, cf. (39).

(37) **Personne** n' a **rien** dit.
nobody NEG.CL have.AUX.PRES.3SG nothing say.PP
'Nobody said anything.'

(38) **Personne** n'a **RIEN** dit.
(Preferred reading) 'Nobody said nothing.' = 'Everybody said something.'

(39) **Personne** n' est **pas** venu.
nobody NEG.CL be.AUX.PRES.3SG NEG come.PP
'Nobody didn't come.' = 'Everybody came.'

2.3.1 *NPIs or n-words?*

The key question with respect to these quantifiers in Modern and Contemporary French is how exactly to interpret the behaviour illustrated by the above examples: do the quantifiers in fact still essentially have the status of negative polarity items (Muller 1991: 263, Ladusaw 1993: 444, Rowlett 1998: 167, 205), or have they by now reached the stage of being inherently negative indefinites (Déprez 2000: 264–5, de Swart and Sag 2002: 373)? If the former, the default interpretation of (37) would come about as a result of the first quantifier in the string (in this case *personne*) being governed, and thus negativized, by the preverbal negation *ne* (i.e. 'not somebody' > 'nobody'), while any further indefinite quantifiers following it (in this case *rien*) would retain positive meaning (Muller 1991: 250). This would, however, leave the possibility of the interpretation in (38) unaccounted for, and would further raise the question of how *personne* comes to be governed by *ne* even when it precedes the latter in the string, or alternatively, if it is assumed to be inherently negative when not governed by *ne*, why it would ever need to be so governed in order to receive a negative interpretation.

If, on the other hand, the quantifiers are n-words, the difference between (37) and (38) can be explained as a difference between two different types of quantification, respectively, 'resumptive' and 'iterative' quantification (Déprez 2000: 269, de Swart and Sag 2002: 383–4, both following May 1989). Informally put, if two or more n-words function resumptively, quantification is polyadic, that is, the quantifiers have a shared scope, which allows a single feature of negativity to 'spread across' the pair (or series) of quantifiers, the result being negative concord. If quantification functions iteratively, on the other hand, the individual, monadic, quantifiers will enter into a scopal relationship with one another, and a double (or multiple) negation reading will ensue.

The fact that combinations of a quantifier with *pas* always lead to double negation in standard French, and that some combinations of quantifiers are more likely to do so than others, is explained by assuming that resumptive quantification will only obtain if there are sufficient syntactic and semantic similarities between the n-words involved. Thus, a double-negation reading is favoured in (40), for instance, because, as a pronoun, *personne* always has argument functions, while *aucun* here functions as a determiner, that is, part of a DP. The two are therefore likely to be perceived as too dissimilar to yield the resumptive reading.

(40) **Personne** n' a mangé **aucun** gateau.
nobody NEG.CL have.AUX.PRES.3SG eat.PP no.MSG cake.SG
Preferred reading: 'Nobody ate no cake.' = 'Everybody ate at least one cake.'
(Déprez 2000: 310)

Déprez (2000) and Déprez and Martineau (2004) further relate the change in the quantifiers from NPI to n-word status to a broader syntactic change that took place towards the end of the Middle French period and the beginning of the Classical period, from which time on the language became increasingly intolerant of bare NPs. The hypothesis put forward is that, at stage 2 of the quantifier cycle in Table 2.2 above, the quantifiers function like bare NPs, which at that time were common in French. As such, they had the status of variables, capable of being bound by a negative marker higher in the clause. Over the course of time, however, and presumably triggered by the aforementioned change in the acceptability of bare NPs, they have acquired semantic autonomy by turning into a type of numeral determiner akin to *zero*, which in combination with other like items can function either resumptively or iteratively.

Although probably superior to the NPI-account, the n-word account too leaves a residue of problems. For one thing, it does not account for the fact that n-words in standard French do retain positive readings in a number of contexts that are not negative as such, but which are commonly classified as being of negative polarity, for instance, (41)–(42).

(41) Max connaît la Chine mieux que **personne.**
Max know.PRES.3SG DEF.DET.FSG China well.COMPAR than anybody
'Max knows China better than anyone.'

(42) J' aimerais bien visiter l' Inde,
SCL.1SG like.COND.1SG well visit.INF DEF.DET.FSG India
mais aurai- je **jamais** l' occasion
but have.FUT.1SG SCL.1SG ever DEF.DET.FSG occasion.SG
de le faire?
INF OCL.MSG do.INF
'I'd like to visit India, but will I ever have the opportunity to do so?'

Secondly, the nature of the syntactic–semantic similarities necessary for two or more items to enter into resumptive quantification is unclear: to take a few simple examples, as already observed, combinations of other n-words with *pas* always result in a double-negation reading in standard French, but not necessarily in non-standard varieties, as shown in (43) below. On the other hand, combinations of other n-words with *plus*—which distributionally resembles *pas* rather than *personne* or *rien*, for instance, can never yield double negation, cf. (44).

(43) ...je ne loupe **pas** **aucun** fait divers...
SCL.1SG NEG.CL miss.PRES.1SG NEG no.MSG news.item.SG
'...I don't miss any item of news...' (Corpus Elicop, Orléans, t014.txt)

(44) **Personne** n' a **plus** d' argent.
nobody NEG.CL have.PRES.3SG no.more of money.SG
'Nobody has any more money.'
*'Nobody has no more money.' = 'Everybody still has some money.'

Thirdly, elegant though it is, it is unclear how the hypothesis of a change in the internal structure of the quantifiers put forward by Déprez (2000) and Déprez and Martineau (2004) would apply to those quantifiers that are clearly adverbial in nature, namely *plus* and *jamais*.

There is no doubt that in medieval French many of the quantifiers were not yet n-words.[17] Examples of the use of *rien*, *personne*, and *aucun* in both positive contexts and in negative polarity contexts (as opposed to negative contexts *per se*) are numerous. Thus, for instance, *rien* is used as a regular positive noun in (45) and as a negative polarity item in (46).

[17] Hansen (2012) argues that *jamais* constitutes an exception, providing evidence of having been lexicalized, via univerbation of its etymological sources *ja* (< Lat. *iam* 'as of now') + *mais* (< Lat. *magis* 'more'), as a negative indefinite and having developed its range of NPI uses as subsequent extensions. Ingham (2011) shows how *nul* 'no[ne]', a direct descendant of the clearly negative Latin indefinite *nullus*, similarly acquired NPI uses in medieval French.

(45) Et si vous dirai une **rien**.
and so DAT.OBJ.2PL say.FUT.1SG INDEF.DET.FSG thing.SG
'And so I'll tell you a thing.' (Béroul, *Tristan*, v. 179, *c.*1170) (from Buridant 2000: 610)

(46) As tu **riens** fet?
have.AUX.PRES.2SG SBJ.2SG anything do.PP
'Have you done anything?' (Rutebeuf, *Le miracle de Théophile*, mid 13th c.) (from Buridant 2000: 610)

There are, however, clear signs that the Modern French n-words have been undergoing grammaticalization, particularly in the case of those that have nominal origins. Thus, *rien* and *personne* have lost their substantival properties, and can no longer co-occur with determiners or be directly modified by an adjective (cf. (47)). Furthermore, they have changed their originally feminine gender to masculine, as shown by the choice of anaphor in (48) and by past-participle agreement (masculine singular *fait*) in (49):

(47) Je n' ai vu rien
SCL.1SG NEG.CL have.AUX.PRES.1SG see.PP nothing
*(**d'**) intéressant.
of interesting.MSG
'I didn't see anything interesting. / I saw nothing interesting.'

(48) Personne$_i$ n' a dit qu' **il**$_i$ ne
nobody NEG.CL have.AUX.PRES.3SG say.PP COMP SCL.3MSG NEG
voulait pas venir.
will.IPFV.PAST.3SG NEG come.INF
'Nobody$_i$ said that they$_i$ didn't want to come.'

(49) Pour l' instant, rien n' a été
for DEF.DET.MSG moment nothing NEG.CL have.AUX.PRES.3SG be.PP
fait.
do.PP.MSG
'For the time being, nothing has been done.'

Similarly, Prévost and Schnedecker (2004) show that from Middle French onwards, *aucun* has gradually reduced its morphological, distributional, and referential range, with the result that in contemporary French, it occurs almost exclusively as a singular determiner with indeterminate reference, and the same appears to be true of *nul*.

In general, the negative polarity contexts in which most French n-words can appear with positive meaning seem to be increasingly restricted, and not all n-words show the same potential, as shown in Table 2.3 (adapted from Muller 1991: 265).

TABLE 2.3 **Negative polarity uses of French n-words**

n-word	*jamais*	*rien, aucun, personne*	*nulle part*	*nul*	*plus*
negative polarity context					
after *sans* 'without'	✓	✓	✓	✓	✓
after *plus que* 'more than'	✓	✓	✓	✓	×
after *trop pour* 'too . . . for/to'	✓	✓	✓	×	(✓)
complement or infinitival clauses following a negated matrix verb	✓	✓	✓	×	×
complement or infinitival clause following a semantically negative matrix verb	✓	✓	×	×	(✓)
after *avant* (*que/de*) 'before'	✓	✓	×	(✓)	×
after *peu* 'little, few'	✓	×	×	×	×
direct (rhetorical) question	✓	(✓)	×	×	×
conditional	✓	×	×	×	×
indirect interrogative	✓	×	×	×	×

Moreover, positive interpretations are in some cases limited to frozen expressions, cf. the contrasts between (50) and (51), or between (52) and (53).

(50) Si **jamais** tu changes d' avis, tu
if ever SCL.2SG change.PRES.2SG of opinion.SG SCL.2SG
peux venir me voir.
can.PRES.2SG come.INF OCL.1SG see.INF
'If you ever change your mind, you can come to see me.'

(51) *Si tu avais **jamais** changé d' avis,
if SCL.2SG have.AUX.IPFV.PAST.2SG ever change.PP of opinion.SG
tu aurais pu venir me voir.
SCL.2SG have.COND.2SG can.PP come.INF OCL.1SG see.INF
'If you'd ever changed your mind, you could have come to see me.'

(52) J' aime le champagne **mieux** **que**
SCL.1SG love.PRES.1SG DEF.DET.MSG champagne.SG well.COMPAR than
rien au monde.
anything in+DEF.DET.MSG world.SG
'I like champagne better than anything in the world.'

(53) Une bière, c' est **mieux** **que**
INDEF.DET.FSG beer.SG IMPERS.SCL.3SG be.PRES.3SG well.COMPAR than
rien.
nothing
'A beer is better than nothing.'

As suggested by Ladusaw (1993: 445) and Zeijlstra (2008), the diachronic evolution of the quantifiers can be assumed to be closely linked to that of standard clause negation, in as much as the former are recruited to support the preverbal negator *ne* at a stage where *ne* is itself sufficient to express negation, just like what took place with postverbal *pas* and *point*. Given this parallelism, it is hardly surprising that the quantifiers should also eventually have been reanalysed as having negative semantic content of their own. Those positive uses that still remain can plausibly be explained in the same way as the similarly puzzling uses of *pas*, *mie* and *point* in Old French, namely as a case of divergence in grammaticalization.

2.4 Negation in infinitival clauses

Unlike participial clauses, in which the relative positions of *ne*, the verb, and the 'reinforcing' negative element are the same as in finite clauses (cf. (54)), infinitival clauses in Modern French generally have *ne* and the 'reinforcing' element occurring together preceding the infinitive and any preverbal clitics, as in (55).

(54) **Ne** l' ayant **pas** vu depuis
NEG.CL OCL.3MSG have.AUX.PRESP NEG see.PP since
des années, je ne sais pas du tout
INDEF.DET.PL year.PL SCL.1SG NEG.CL know.PRS.1SG NEG at.all
où Pierre habite actuellement.
LOC.INTERROG Pierre live.PRES.3SG currently
Lit. 'Not having seen him for several years, I don't know at all where Pierre lives these days.'

(55) Ce serait dommage de **ne** **pas** le voir.
DEM.CL.SG be.COND.3SG shame.SG INF NEG.CL NEG OCL.3MSG see.INF
'It would be a shame not to see him.'

This was not always so, however. While negated infinitives appear to have been avoided in Old French, and to have been negated by *non* rather than *ne* when they did occur (Moignet 1965: 54), they are common from Middle French onwards, by which they are standardly negated by *ne... (pas)*. At the time, however, their preferred word order was similar to that of finite clauses, the reinforcing element (in so far as one was present) occurring after the infinitive, as in (56). Although the modern word order was already a possibility in Middle French, it is only in Classical

French that the reinforcing element starts to move leftward with greater frequency, initially to the position seen in (57), between non-subject clitics and the infinitive, and subsequently to the currently preferred position illustrated in (55) (cf. Martineau 1994: 56).

(56)

et	de	**ne**	s'	en	empescher	**point**
and	INF	NEG	OCL.3SG.REFL	IND.OCL.3SG	prevent.INF	NEG

'and not to stop himself' (Commynes, *Mémoires*, late 15th c.) (from Martineau 1994: 56)

(57)

de	**ne**	le	**pas**	aimer
INF	NEG	OCL.3MSG	NEG	love.INF

'not to love him' (LaFayette, *La Princesse de Clèves*, 1678) (from Martineau 1994: 56)

When the negation of an infinitival clause involves quantifiers, the individual quantifier will determine the type of construction used: thus, one group of quantifiers (*plus, jamais, rien*) occurs in pre-infinitival adverbial position, and the other (*personne, aucun, nul, nulle part*) in post-infinitival argument position, cf. (58)–(59):[18]

(58)

Je	m'	ennuie	à	**ne**	**rien**	faire.
SCL.1SG	OCL.1SG	bore.PRES.1SG	to	NEG.CL	nothing	do.INF

'It bores me not to be doing anything.'

(59)

Je	m'	ennuie	à	**ne**	voir	**personne**.
SCL.1SG	OCL.1SG	bore.PRES.1SG	to	NEG.CL	see.INF	nobody

'It bores me not to see anyone.'

As in finite clauses, *ne* can be dropped before infinitives in colloquial registers of Modern and Contemporary French, cf. (60). Thus, if we consider only the most basic form of negative infinitives, namely those that do not involve quantifiers, it is in fact here, rather than in finite clauses that Jespersen's cycle finds its clearest expression, for here the erstwhile postverbal reinforcing element has not only become established as the principal exponent of negation, but it has actually moved into the preverbal slot, where it is capable of negating the infinitival clause in the absence of the original negator:

(60)

Ce	serait	dommage	de	**pas**	voir	Pierre.
DEM.CL.SG	be.COND.3SG	shame.SG	INF	NEG	see.INF	Pierre

'It'd be a shame not to see Pierre.'

This infinitival cycle can thus be represented as in Table 2.4.

[18] Although *rien* has argument functions, it is probably felt—perhaps due to its being monosyllabic and semantically non-transparent to contemporary language users—to be more strongly grammaticalized, and hence more adverbial in nature, than *personne*, which is still in use as a normal (feminine) noun alongside its n-word use.

TABLE 2.4 The infinitival cycle in French

Stage	Example	Description
Stage 1	*ne dire*	The negator is preverbal
Stage 2	*ne dire (pas)*	The preverbal negator is optionally complemented by a postverbal element
Stage 3	*ne pas dire*	The postverbal element becomes obligatory and moves to a preverbal position
Stage 4	*(ne) pas dire*	The original preverbal negator becomes optional
(Stage 5 [Future French?]	*pas dire*	The original preverbal element drops out leaving the erstwhile postverbal marker—now in preverbal position—as the sole exponent of negation)

2.5 Conclusion

In this chapter, I have discussed the syntax, semantics, and pragmatics of clause negation from medieval to Contemporary French. I have shown that three separate negative cycles can be identified, depending on whether the negated clause is finite or non-finite, and on whether or not it involves quantifiers. Although separate, the three cycles are clearly linked, in as much as the quantifier cycle and the infinitival cycle both depend on the standard-negation cycle to trigger them. In all three cases, the tabular representations have been shown to be simplified ones which ignore a great deal of internal variation at each individual synchronic stage. Nevertheless, the overall trends appear, in all cases, to be clear enough that the cycles can reasonably be said to be valid in a long-term diachronic perspective.

Primary sources

Base de français médiéval. Electronic database, <http://bfm.ens-lyon.fr/>.
La chanson de Roland, ed. Gérard Moignet. 1985 [*c*.1090]. Paris: Bordas.
Corpus Elicop. Electronic database, <http://bach.arts.kuleuven.be/elicop/>.
Le couronnement de Louis, ed. Ernest Langlois. 1961 [*c*.1130]. Paris: Honoré Champion.
Frantext. Electronic database, <http://www.frantext.fr/>.
Joinville, Sire Jean de. 1874 [1298–1309]. *Histoire de Saint Louis*. Credo et lettre à Louis X. Ed. by Natalis de Wally. Paris: Firmin Didot Frères.
La queste del Saint Graal, ed. Albert Pauphilet. 1967 [*c*.1220]. Paris: Honoré Champion.
Perseus Digital Library. Electronic database, <http://www.perseus.tufts.edu/hopper/>.

3

Negation in the history of Italo-Romance

MAIR PARRY

3.1 Introduction

The cyclical evolution of the expression of negation, responsible for the merger of the Latin sentential negator *ne* with the reinforcer *oenum* 'one' to give *non* (Jespersen 1917: 14)[1] and widely known as 'Jespersen's cycle', has affected Latin's descendants in Italy to different degrees depending on their geographical location and structure. Indeed, such is the variation and diversity of Italo-Romance negative structures that together they present a textbook illustration of all the various stages recognized for Jespersen-type cycles.[2] They also afford an incomparably rich database for theoretical research into the typology, structure, and development of negative constructions which has inspired major synchronic studies such as Bernini and Ramat's (1996) typological survey of European languages and Zanuttini's (1997) in-depth generative analysis of Romance negation, with particular reference to Italo-Romance. From a historical perspective Molinelli's (1988) overview of the development of negation in Italian and other Italo-Romance varieties takes a functional approach, highlighting the pragmatic significance of negation and its close interaction with focus, as well as drawing attention to sociolinguistic factors. An outline of the structural variation resulting from Jespersen's cycle in Italo-Romance is found in Parry (1996), whilst extensive modern dialectal data are now available in Manzini and Savoia (2005), together with a generative interpretation that differs significantly from that of Zanuttini. Recent research has focused on the synchronic analysis of modern

[1] See van der Auwera (2010a) for the three main historical sources of negators, this being an example of the third: univerbation of an original negator with a frequently collocated form.

[2] So as to account for synchronic variation, six stages are identified by van der Auwera (2009) but, as he illustrates, models vary greatly in the number of stages recognized. I shall continue to use the traditional Neg1, Neg2, Neg3 labels to refer not to stages but to the three basic structural types of preverbal, discontinuous (pre- and postverbal), and postverbal negation (Bernini and Ramat 1996: 13).

varieties, but advances in syntactic, semantic, and pragmatic theory are now being applied to diachronic investigation.

The chapter is structured as follows: section 3.2 offers an overview of Jespersen's cycle in Italo-Romance; sections 3.3 and 3.4 present formal and functional analyses respectively of the various negative strategies in finite declarative clauses; section 3.5 considers interrogatives and exclamatives, as well as 'expletive' negation; section 3.6 examines negative imperatives or prohibitives, as well as negated infinitives; section 3.7 discusses negative pronouns and adverbs, section 3.8 the phenomenon of negative concord, section 3.9 constituent negation, while section 3.10 draws a conclusion.

3.2 Jespersen's cycle in Italo-Romance

The modern areal distribution of predicate negation is broadly as follows: preverbal negation only (Neg1), continuing Latin morphology and syntax, is found in the standard language, Italian, in the Florentine dialect on which it is based, as well as in the other central and southern dialects of Italy, including Sardinia.[3] This majority strategy is also found in the northeast and northwest: Venetan, Friulian, and some Ladin varieties, as well as Ligurian. In the central Po area, Piedmontese and Lombard have mainly Neg3, in which a postverbal element, originally used to 'reinforce' the preverbal negative, has taken over its functions; this can be a noun with general reference merged with a negative, Pied. *nen*(*t*) 'nothing' < Lat. *ne gente*(*m*) 'no people' (Rohlfs 1968: 218); or non-negative minimizers such as Lomb. *mi*(*n*)*ga* (< *mica*(*m*) 'crumb'), which acquired negative meaning through frequent collocation with the preverbal negative in an intermediate discontinuous strategy, *n* ... *nen*/*minga* (Neg2), or less frequently, a resumptive holophrastic negator added clause-finally, Lomb. *nò*. The transitional Neg2 type is still found in Emilian as well as in some alpine Lombard (Ticinese) and border Piedmontese-Ligurian varieties, while Venetian seems to be currently developing a form of Neg2 (see section 3.3.2):

(1) a. Italian **non** dormirò
NEG sleep.FUT.1SG

b. Ligurian **nu** durmiò
NEG sleep.FUT.1SG

c. Emilian (Modena) a **n** dorum **briza**
SCL NEG sleep.FUT.1SG NEG

d. Piedmontese dürmirai **nen**
sleep.FUT.1SG NEG

[3] The so-called 'dialects' of Italy evolved alongside Italian, from which they are structurally distinct; they can be as different from Italian as Spanish or Romanian. Gallo-Romance settlements in southern Italy, e.g. Guardia Piemontese, show the Neg3 strategy of the original dialect area (see Jaberg and Jud 1928–40: map 653, point 760).

e. Lombard dòrmaro **nò**
sleep.FUT.1SG NEG
'I shan't sleep.' (Jaberg and Jud 1928–40: map 653)

These three structural types do not correlate with homogeneous geographical areas, since two or even three types may coexist in the same dialect, as is to be expected in the case of gradual syntactic evolution. In transitional areas in particular, microvariation involves two or even all three strategies, as in dialects of the Val Bormida (Parry 1997), where Neg2 prevails, but Neg1 is found in irrealis clauses and in structures such as *n... âtr* 'only',[4] while Neg3 favours two main contexts: (a) with the verbs 'to be' and 'to have' (frequent as perfective auxiliaries) and (b) in the presence of preverbal complement clitics, particularly nasal ones. Following van der Auwera (2010a), this contemporaneity of variants may be captured by the five-stage evolutionary model in Table 3.1 representing Piedmontese as an example of the most innovative dialect type (schematic models, of course, fail to convey the relative frequency and acceptability according to text type, register, and linguistic context).

TABLE 3.1 **The development of Piedmontese negation**

Stage 1	Stage 2	Stage 3	Stage 4	Stage 5
no(n) V	*ne* V	*n(e)* V		
	ne V *nen*	*n(e)* V *nen*	*n* V *nen*	
		V *nen*	V *nen*	V *nen*

3.2.1 *Stage 2 Medieval Italo-Romance*

All early Italo-Romance varieties continue the Latin construction of predicate negation deriving from Lat. *non* (stage 1) and have strategies that reinforce the preverbal marker in particular pragmatic contexts, for instance, by means of an adverbial phrase:

(2) Campanian (Naples)
lo regno de Thesalia **non** doventerrà **de nullo** nostro nemico
the kingdom of Thessally NEG become.FUT.3SG of nothing our enemy
'The kingdom of Thessally will in no way become our enemy.' (*Libro de la destructione de Troya*, p. 51, 14th c.)

The unmarked position for negative reinforcers is postverbal and there is a significant degree of regional variation in the forms used. There are two main types: negative quantifiers (indicating a zero quantity and resulting from univerbation with a negative

[4] See Zeli (1968) for Ticinese varieties.

particle), for instance, *niente,*[5] *nulla* 'nothing', and non-negative nominal expressions referring to minimal units (a phenomenon already prevalent in Latin), for instance, *mica* originally 'crumb' in Latin, which acquired a quantificational interpretation by being used idiomatically to express the lowest point on a pragmatic scale, that is, 'not even a crumb' (Haspelmath 1997: 115). *Mica* and *niente* are attested all over Italy, but are particularly frequent in the north with the latter characteristic of the northwest; *negota, neota* 'nothing' and other reflexes of *nec gutta(m)* 'and no drop' are concentrated in Lombardy and the Veneto, with *punto* 'point' in Tuscany.[6] The range of possible minimizers used as polarity items is vast, for instance, *un figo seco* 'a dry fig' (13th-c. Verona), *un festugo de palia* 'a straw' (13th-c. Lombard), *fiore* 'flower' (14th-c. Tuscan), *stiça* 'drop' (14th-c. Liguria),[7] but only a few grammaticalized as negative adverbs.

The quantified reading allowed minimizers to extend beyond their original semantic context (bleaching, as in (3a)) and occur, like the negative quantifiers, with a prepositional complement (3b), (3c).

(3) a. Piedmontese

e	cercà	amont	e	aval	e	non	trovè	**mia**
and	searched.3SG	high	and	low	and	NEG	found	MICA

'and he searched high and low and found nothing at all' (*Sermoni subalpini,* p. 251, ll. 34–5, 12th–13th cc.)

b. Lombard

Lá	no se	sente	**miga de male.**
there	NEG REFL	feels	MICA of ill

'There not a bit of ill is felt.' (Pietro da Bescapè, *Sermone*, p. 72, l. 2435, 13th c.)[8]

[5] *Niente* itself is presumably a late Latin fusion of a negative particle with a generic noun. (See Iliescu (2011) for convincing arguments within a pan-Romance comparative perspective in support of a derivation from Lat. *ne(c) gente(m).*) The latter etymology was proposed by Rohlfs (1968: 214) and reflects that proposed for the cognate Gallo-Romance forms, whereas most Italian etymological dictionaries favour the less plausible *nec ente(m)* 'no being'. The loss of the feature [+animate] in the development of a negative quantifier is found in other languages, both Romance and non-Romance: German *nicht* 'nothing' > 'not' < *ni* + *wiht* 'no creature', Spanish *nada* 'nothing' < Lat. *nata(m)* 'born (being)'.

[6] *Mancu* (< Lat. *mancu(m)* 'maimed') is found in early Sicilian, mainly with the meaning 'less', but occasionally showing semantic bleaching as a negative reinforcer 'not less' > 'not even (anything)':

(i) Sicilian (Messina)

Ma però	ca foru	fimini	eu	**nu**	ndi	maravilyu	**mancu**
but because	that were.3PL	women	I	NEG	of.it	marvel	NEG-R

'But because they were women I'm not at all surprised.' (*Libru di Valeriu Maximu*, p. 226, l.18, 14th c.)

[7] See also Molinelli (1988: chap. 4), who also cites numerous Latin examples; Rohlfs (1969: 302–5), Marcato and Ursini (1998: 191) for Venetan; Ledgeway (2009) for Neapolitan.

[8] All the examples of *miga* in the OVI database of medieval texts occur in negative polarity contexts, but Vai (1995: 161) cites a non-negative use from the end of the 15th century:

(i) Milanese

On sté	de scisceri	e	**miga**	de vin	d'	intrà
one bushel	of chickpeas	and	MIGA	of wine	of	income

'One bushel of chickpeas and a little of wine as income . . .' (*Lancino Curti* 6.14)

c. Sicilian

oy que lu inimicu non appi **nienti** **di pagura** di quisti
or that the enemy NEG has nothing of fear of these
'or that the enemy has not a bit of fear of these' (*Libru di Valeriu Maximu*, p. 82, l. 33, 14th c.)

Grammaticalization eventually led to the loss of the preposition (e.g. *miga paura* 'no fear'), causing structural ambiguity in transitive structures:

(4) Venetian

Fioli mie', **non** abié **miga** paura.
sons mine NEG have.IMP MICA.DET/ADV fear
'My sons, do not fear at all.' (*Navigatio Sancti Brendani*, p. 74, ll. 22–3, 13th c.)

Optionally transitive verbs would have provided another syntactically ambiguous context that favoured adverbial reanalysis of the quantifier object since, for example, expressions corresponding to It. *non mangio mica/niente* 'I'm not eating a crumb/ anything' could be interpreted as stronger versions of *non mangio* 'I'm not eating' (see Willis 2011 for Welsh examples). The process of grammaticalization predates the first texts, which show unambiguously adverbial uses of minimizers with no lexical restrictions (5a) and in thirteenth-century texts they occur with direct objects (5b).

(5) a. Venetian

tal hom cre' aver fiiolo, q' el **non** è **miga** so
such man believes have.INF son that he NEG is MICA his
'a man may believe that he has a son, but he isn't his at all' (*Proverbia que dicuntur*, p. 548, l. 591, 12th c.)

b. Lombard

Cotal menestra 'l patre **no** aspegiava **miga**
such soup.DO the father NEG expected MICA
'His father did not expect such a dish at all! (quite the contrary)' (Bonvesin, *Vulgare de elymosinis*, p. 269, l. 868, 13th c.)

The following example illustrates all three of the grammaticalization stages, nominal quantifier, determiner, and adverbial, of different 'reinforcers':

(6) Marchigiano

Se nne fo pentuto, **non** era **mica** paczo! /.../
REFL of.it was repented NEG was MICA.ADV stupid
Che **non** magniaro **niente** pane né companagio, /
that NEG ate.3PL nothing.DET bread nor accompaniments
Et loro bestie **non** abero **punto** **de** veveragio
and their animals NEG had.3PL point (noun) of beverage

'He repented of it, he wasn't stupid! . . . that they ate no bread nor accompaniments and their animals had not a drop to drink.' (Buccio di Ranallo, *Cronaca*, p. 158, l. 881, 14th c.)

The medieval vernaculars may further reinforce the negative value of *mica* by means of *né* < *nec* '(and) not', for instance, Venetian *né miga* '(and) not even the smallest quantity' (cf. Mod. It. *nemmeno* 'not even' < *né meno* 'not less', Lombard *no miga*), although univerbation did not ensue, unlike earlier formations *niente* and *negota*, etc. which had lexicalized with the meaning 'nothing':

(7) a. Venetian
e no 'n reman **né-miga**
and NEG of.it remains and.not-MICA
'and there doesn't remain a bit of it' (*Proverbia que dicuntur*, p. 536, l. 320, 12th c.)

b. Tuscan
ed alcuno altro **non** à **né** **mica** donna
and some other NEG has and.not MICA woman
'and others do not have even a wife.' (*Egidio Romano*, pp. 229–30, ll. 3–6, 13th c.)

Medieval vernacular texts show that the reinforcement of the preverbal negative was an option available to all Italo-Romance vernaculars, occurring when the usual predicate negator was considered pragmatically insufficient. This was achieved by quantifying the scale of the rejection, (8a) and (8b), or the refusal of a proposition (8c), while negative indefinites based on minimal units were typically used to contradict presuppositions, (5b), (8d), and (8e), as *mica* in modern Italian (Cinque 1976). Following Schwegler (1988) and Schwenter (2006), Hansen and Visconti (2009) insist that the rejected proposition must have been activated (explicitly or not) by the preceding discourse or speech-act context:

(8) a. Piedmontese
Lo premer, qui est ric, **no** li vol **nient** aier.
the first who is rich NEG him want nothing help.INF
'The first, who is rich, does not want to help him in any way.' (*Sermoni subalpini* 7, p. 239, ll. 12–13, 13th c.)

b. Hybrid vernacular
k' eu fithança **non** avea **niente** / de vinire ad unu cun la çente
that I faith NEG had nothing / of come.INF to one with the gentle(woman)
'that I didn't have any confidence at all to get together with the lady' (end of 12th-c. poem found in Ravenna, p. 615, ll. 8–9) (Stussi 1999: 615)[9]

[9] This is the earliest 'Italian' love poem in the vernacular, only recently discovered by Alfredo Stussi (see Stussi 1999), and in his opinion its language contains a mix of northern and central-southern features.

c. Sicilian
dichendu ki **in nullu modu** mictissiru lu cavallu dintra
saying that in no way put.3PL the horse inside
'saying that in no way should they put the horse inside' (*Istoria di Eneas*, p. 32, l. 8, 14th c.)

d. Piedmontese
Il **no** 'l diseron **mia** per zo que il lo savesen
they NEG it said MICA for that that they it knew
certanament, **si no** per suspeita.
certainly if not out.of suspicion
'They did not say it because they knew it for certain, but because they suspected it.' (*Sermoni subalpini*, p. 231, ll. 16–17, 13th c.)

e. Ligurian
E zo **non** dixea ello **miga** per compassiom, ch' ello
and that NEG said he MICA for compassion that he
avesse delli poveri ma perzò ch' ello ne vorea invorar la
had for the poor but for.that that he of.it wished steal.INF the
dexena parte,
tenth part
'He didn't say that out of pity for the poor, but because he wanted to steal a tenth of it,' (*Passione*, p. 28, ll. 19–20, 14th c.)

f. Sicilian
Nìn incuminzirò **mica** da lu plù meskinu, anti incuminzirò
NEG begin.FUT.1SG MICA with the most abject, but begin.FUT.1SG
da quillu, lu quali intra poki era tinutu la plù beatu
with that one the which among few was held the most fortunate
'I won't begin with the most abject but on the contrary with the man who was considered the most fortunate of the few' (*Libru di Valeriu Maximu*, p. 232, ll. 13–14, 14th c.)

Intrinsically negative indefinites, such as *niente, neota* 'nothing' may occur without a preverbal negative if the verb is outside the scope of the negation:

(9) a. Florentine
Ma il loro podere fu **niente** appo la forza de' Romani…
but the their power was nothing next.to the force of.the Romans
'but their might was nothing compared to the force of the Romans' (Villani, *Cronica*, p. 144, l. 27, 14th c.)

b. Sicilian

et	chilli	**che**	**pareno**	**nienti**	a	lo	mundo	so	multo	accepti
and	those	that	seem	nothing	to	the	world	are	very	accepted

davanti	a	Dio
before	to	God

'and those who seem nothing to the world are very welcome before God'
(*Dialagu de sanctu Gregoriu* p. 16, ll. 6–7, 14th c.)

3.2.2 *Stages 2–5: from the Middle Ages onwards*

There is a widespread view that the main driver of the negative cycle is the continual erosion of the semantic force of negation that prompts speakers to find more expressive ways of conveying a message relating to such a pragmatically marked domain (Givón 1978).[10] Schwegler (1988: 41) notes that in particular the contradiction of discourse-active presuppositions or expectations (explicit or otherwise) encourages inflationary language use. This may be exploited for rhetorical purposes, as in the *Sermoni subalpini*:

(10) Piedmontese

Si	ben	vols	ofrir,	dreitament	deis	partir:	**né**	retenir	**mia**
if	well	wish.2SG	offer	justly	must.2SG	divide	and.not	keep	MICA

la	meillor	partia...
the	best	part

'If you wish to make a good offering, you must divide things fairly: by no means keeping the better part...' (*Sermoni subalpini*, p. 221, ll. 15–18, 13th c.)

Frequent use may lead to these reinforcing expressions becoming grammaticalized and replacing standard forms of negation, but this is not a necessary development, as shown by the persistence of reflexes of simple preverbal *non* in many parts of Italy. Southern Italian dialects in particular show a far lower degree of extension of minimizers beyond the original semantic context (Ledgeway 2009: 687–89), while those varieties that have progressed beyond the optional reinforcement of stage 2 are the Gallo-Italian varieties. These are characterized not only by syllabic and vocalic weakening (*non* > *no, ne, n*), which is a characteristic of several southern varieties also, for instance, Abruzzo and Campanian (see negative forms in Manzini and Savoia 2005: 132), but also by the development of subject clitics. Consequently, other factors believed to have contributed to the cyclical development of negative structures are phonetic erosion, which encourages speakers to adopt with increasing frequency more substantial, alternative expressions, and structural constraints on the number of preverbal clitic elements permitted (to be discussed below in section 3.4). A brief review of the history of negation in Piedmontese, as documented in (i) the late

[10] See van der Auwera (2009) for an overview of the various accounts of the negative cycle.

12th-/13th-century *Sermoni subalpini*,[11] (ii) a 16th-century farce in the dialect of Asti, (iii) 17th-century Turinese popular songs, and (iv) the Turinese late-17th-century play *Il Conte Pioletto* (but not printed till 1784), serves as illustration of the different stages schematized in Table 3.1. It concludes with examples of a different type of negative reinforcer from those discussed above, a type that within the Italo-Romance area is mostly associated with Lombard dialects.

3.2.2.1 Jespersen's cycle in Piedmont The preverbal negative in the early *Sermoni subalpini* shows more or less regular phonetic conditioning (pre-vocalic *non* ~ pre-consonantal *no*, which can appear reduced to *ne*, especially before complement clitics, as in over 65% of cases). In the early 16th-century dialect of Asti *ne* was the typical negator, with reinforced negatives (NEG-R) also common for emphatic objections (*nent, za*)[12] and rejection of presuppositions (*mia*):

(11) Astigiano

a. Basta, che te **n'** an sarai **nent**
enough that you NEG of.us be.FUT.2SG NEG.R
'Enough, you definitely won't be one of us.' (G. G. Alione, *Comedia de l'homo*, p. 430) (Bottasso 1953)

b. vôi **n'** an sarei **mia** exempt / **ni** venrei **za**
you NEG of.it be.FUT.2SG NEG.R exempt / nor come.FUT.2SG ever
a côl vostr intent / d' esser d'i çinq
to your intent / of be.INF of.the five
'You wouldn't escape that fate (contrary to what you think!) nor would you ever achieve your objective of being one of the five.' (G. G. Alione, *Comedia de l'homo*, p. 634) (Bottasso 1953)

Four 17th-century Turinese popular songs reveal a discontinuous structure, *n…pa* (12a), as the main negative strategy, the intriguing feature being its non-indigenous form (cf. indigenous *pas* 'step' in (12b)); simple *n* (Neg1) is reserved for irrealis clauses, some modal verbs, and fixed phrases as in (12b), while the two instances of Neg3 also anticipate the distribution found in modern Cairese (with preverbal complement clitics (12c) and with imperatives (12d)):[13]

[11] The rich collection of sermons known as the *Sermoni subalpini* are the earliest vernacular text from the north-west area, probably composed around the turn of the 12th/13th century in a busy religious community located in the now Provençal-speaking Susa valley (Gasca Queirazza 1996). Although much debate has focused on their hybrid language, which contains elements of Gallo-Romance (both northern French and Provençal) as well as Piedmontese, it seems reasonable to assume that as far as negative structures are concerned, it provides as accurate a representation of local usage as do early texts from other regions.

[12] *Za* 'ever' < Lat. *iam* 'now', It. *già* 'already', cf. *Sermoni Subalpini*, 4, l. 10 *no morrà ia* 'will never die'.

[13] A lone example of simple postverbal *nent* occurs, however, in a 15th-century Turinese sermon (Clivio 1976: 41).

(12) a. Turinese

a-**n**	sà	**pa**	scasi	ont	viresse
SCL-NEG	knows	NEG.R	almost	where	turn.INF.REFL

'one almost doesn't know where to turn' (Clivio 1976: 50)

b.

Ch-é-lo	ch'	**n**-é-lo,	slongand'	un	**pas**...
that-is-it	that	NEG-is-it	lengthening	a	step

'Suddenly, lengthening her step...' (Clivio 1976: 54)

c.

ch'	s'	fà	**pa**	a	i-atr	bai
that	REFL	does	NEG	at	the-other	balls

'that's not done at the other balls' (Clivio 1976: 47)

d.

mostre	**nent**	ch'	o-i	pòrte	amor
Show	NEG	that	you-to.them	carry	love

'do not show your love to them' (Clivio 1976: 38)

In many Piedmontese dialects *pa* nowadays serves a similar function to *mica* in Italian, that is, denial of an activated presupposition, whilst the unmarked negator is *nent* (Parry 1996: 248, Zanuttini 1997: 67), whereas more western Gallo-Romance dialects, for instance, the Francoprovençal of the Val d'Aosta have generalized *pa*. Although Bernini and Ramat (1996: 223, fn. 1) consider the borrowing of negative morphs between languages as rare, they cite Breton loans from French: *nompas, poent, pas*. That speakers in contact situations may resort to borrowing to reinforce their denials in emotional contexts is confirmed by current colloquial instances of the English idiom *no way* replacing the corresponding Welsh negative phrase (*does dim ffordd* 'there's no way') clause-initially:

(13) Modern Welsh

a	**no way**	wneith	nhw	aros	yn	yr	un	lle	drwy	dydd
and	no way	do.FUT.3SG	they	wait.INF	in	the	same	place	through	day

'and no way will they stay in the same place all day' (<http://maes-e.com>, 2006)

By the end of the 17th century, the city dialect of Turin, as reflected in the play *Il Conte Pioletto*, presages the modern situation, with *n*...*nen* as the unmarked negative strategy (14a) and *n*...*pa* used in heightened emotional situations, including in the clefted example in (14b).

(14) Turinese

a.

Pare	**'n**	veul	**nen**,	ch	'i	scota...
father	NEG	wants	NEG-R	that	I	listen

'Father does not want me to listen...' (C. G. Tana, *Il Conte Pioletto*, p. 30)

b. 'N è **pa**, ch' i abbia po, ma...[14]
NEG is NEG-R that 1SG.SCL have fear but
'It's not that I'm afraid, but...' (C. G. Tana, *Il Conte Pioletto*, p. 44)

In this text Neg1 is rare, used only in so-called 'expletive' contexts (see below, section 3.5) or in fixed expressions:

(15) për ampacè / Ch' voi **'n** abbie caich lite
to prevent that you NEG have some lawsuit
'to prevent your having a lawsuit' (C. G. Tana, *Il Conte Pioletto*, p. 56)

Simple postverbal *nen*, on the other hand, is fairly frequent, particularly with the verb 'to be' and in contexts involving proclitic pronouns (here only a third of the imperatives show Neg3):[15]

(16) a. a son **nen** mie
3PL.SCL are NEG mine
'They are not mine.' (C. G. Tana, *Il Conte Pioletto*, p. 7)

b. E già ch' voi 'm volì **nen** sposè
and since that you me want NEG marry.INF
'And since you don't want to marry me...' (C. G. Tana, *Il Conte Pioletto*, p. 78)

The five interrogatives (with subject–clitic inversion) in the play are divided between *nen* (2) and *pa* (3); none has a preverbal negative, thus revealing a similar trend to that noted in the history of French by Martineau and Vinet (2005: 198–200), who show that polar interrogatives used as exclamations or conveying surprise frequently lack *ne*, unlike declaratives and imperatives. This may be due to the fact that such structures lack a negative interpretation, although one could argue that the postverbal markers have a negative value which generates the sense of the actual situation being contrary to some expectation, as proposed by Zanuttini and Portner (2000) for so-called 'expletive' negation (see section 3.5.1):

(17) a. L'aje **pa** dit, Ch' cost era un mariagi, ch' era scrit?
it have.I NEG said that this was a marriage that was written
'Didn't I say this was a marriage that was predestined?' (C. G. Tana, *Il Conte Pioletto*, p. 83)

[14] Bernini (1992: 210) notes that It. *mica* is not found with clefted *non è che* in his data.

[15] The statistics for *Il Conte Pioletto*'s 83 declarative sentences are: *n... -nent*, 55%; *nent...*, 28%; *n*, 5%; *n...pa*, 9%; the latter is further strengthened in: *n...pa-nen*, 1%; *pa-nen*, 1%; *n...pa* + minimizer, 1%. There is also one example of *pa* on its own in a mixed (Italian and Turinese) sentence.

b. V' halo ancor **nen** parlà?
to.you has.he yet NEG spoken
'Hasn't he spoken to you yet?' (C. G. Tana, *Il Conte Pioletto*, p. 29)

In modern Piedmontese, as in French, the nominal origin of the sentential negator persists in the obligatory partitive construction for mass and indefinite plural NPs after a negative:[16]

(18) Modern Turinese
I l' hai **nen** mangia **'d pom** ancheuj
SCL SCL have NEG eaten of apples today
'I haven't eaten apples today.'

Remnants of a typologically different reinforcer may also be found in Piedmont: in the Monferrato and on the southern border, in the Val Bormida and especially in the Terra Brigasca, on the border with France, where a now-unstressed *nu* cliticizes to the end of the verb phrase (Parry 1996: 232):

(19) Brigasco
ti në dèvu-**nu** fà cuscì
you NEG must-NEG do thus
'You musn't do that.' (Massajoli and Moriani 1991: 296)

This postverbal type is characteristic of Milanese, where until the 17th century the evolution of the negative cycle followed a similar pattern to the one described above for Piedmontese (Vai 1995), with preverbal *no* being reinforced by *minga* (< *mica(m)*, other Lombard variants being *miga/mia;* also *nagott* < *ne gutta(m)*). As in Piedmont, the 18th century saw a significant increase in the loss of the preverbal negative, though variation still characterized the poetry of Carlo Porta at the end of the century (Vai 1995: 164). In the early 17th century, however, there had appeared the holophrastic reinforcer *nò* 'no!', presumably originally adjoined to the end of the clause, and representing a reiteration of the rejection (Schwegler 1988: 38):

(20) Milanese
no farò da corù nò
NEG do.FUT.1SG like him NEG
'I won't do like him.' (*Prissian da Milan*, p. 60, 1606) (Vai 1995: 162)

The postverbal *nò* type also occurs in central Ladin varieties, where it retains its pragmatically marked nature (Siller-Runggaldier 1985: 74–5).

[16] As in French; see Roberts and Rousseau (2003) for a generative analysis based on Kayne's (1984: 48ff.) interpretation of the DP as containing a null negative determiner.

3.3 Structural analyses

3.3.1 The preverbal negative

A particulary interesting feature of the development of negation in northern Italian dialects, which has stimulated much theoretical discussion, is the way in which the preverbal negative marker came gradually to interact with other preverbal clitic elements. Up to and including the 16th century the former always follows subject clitics and precedes complement clitics in all Italo-Romance texts, as in modern French. However, 17th-century Paduan, for example, shows innovations resulting in word-order variation: the preverbal negative now *precedes* all subject clitics, except for the vocalic clitic *a* that is not marked for person or gender agreement (Benincà 1983: 30, n. 4), while early 17th-century Milanese texts show the preverbal negative still following the second person singular subject clitic, but *preceding* third person singular subject clitics occurring with auxiliary verbs. Here too, by the end of the century negation precedes the second person singular subject clitic (Vai 1995: 161–3). As late possibly as the 20th century, a similar development affected Ligurian subject clitics:

(21) Ligurian (Genoese)

se	**no**	**ti**	**l'**	**accapisci**
if	NEG	2SG.SCL	her	understand

'if you do not understand her.' (Pasolini 1982: 15)

According to Toso (1997: 229) the relative position of second person singular subject clitic and preverbal negative is still variable in Genoese.

The above interaction has led within generative studies to the identification of separate structural positions for different types of subject clitics (Poletto 1993, 2000), with the change in clitic order relative to the negative marker attributed to their incorporation into the verbal inflexion (Rizzi 1986).[17] Examples of even more unusual preverbal ordering of clitics and negation are discussed in Parry (1997) in relation to dialects in the Piedmontese–Ligurian transitional area (Val Bormida), where the preverbal negative follows not only all subject clitics, but also all first- and second person complement clitics, singular and plural, as well as all reflexive clitics.[18] Since these are dialects characterized by discontinuous negation, Zanuttini (1997) argues that there are two types of preverbal negative markers: a 'strong' one, which negates the clause on its own, represents an independent syntactic head located

[17] Recently, van Gelderen (2008: 200–1) interprets this type of change as another example of a general cognitive principle, namely the Head Preference Principle (HPP), whereby 'pronouns change from emphatic full phrases to clitic pronouns to agreement markers', just as negative adverb phrases become negation markers.

[18] A pattern found also in the dialect of Vermes in the Jura (Butz 1981).

higher than the Agreement projection and prone to interaction with subject clitics, as seen above; the other, a 'weak' one, which needs an adverbial reinforcer, raises with the verb from a structurally very low NegP (within VP) to the Tense projection, and may therefore occur in between complement clitics. Data such as the following, which show two preverbal negative markers, confirm the availability of two preverbal structural positions:

(22) Carcarese (Val Bormida)

e-**n**	**ten**	capisc
1SG.SCL-NEG	you.NEG	understand

'I do not understand you'

The two generalizations made regarding the interaction on the one hand of the 'strong' negator with subject clitics and on the other the 'weak' negator with object clitics are largely upheld by the data, but need further refinement since there exist dialects with Neg2 that show re-ordering involving subject clitics (Neg–Scl) (23a) and dialects with Neg1 where the negative follows some complement clitics (23b) (Parry 1997 and, for more examples, Manzini and Savoia 2005: 286):

(23) a. Varzi (Emilian)

en	**te**	m'	ö	**mía**	dit	k'	**en**	**te**	sarís	stat	a	ka
NEG	2SG.SCL	me	have	NEG	said	that	NEG	2SG.SCL	would.be	been	at	home

'You didn't tell me that you wouldn't be at home.' (Zörner 1994: 89)

b. Bedizzano (Northwest Tuscany)

i	**tte**	**ne**	l'	da /	i	**n**	**te**	**ne**	l'	da
3SG.SCL	you	NEG	it	gives	3SG.SCL	NEG	you	NEG	it	gives

'He does not give it to you.' (Manzini and Savoia 2005: 297, IPA in the original)

Apparently, negation may occasionally be lexicalized in all the preverbal positions in which a negative marker may potentially occur, up to three times (i.e. before the subject clitic, between this and the complement clitic, and after the latter):

(24) Carcarese (Val Bormida)

en	t **en**	t **en**	lovi	(**nent**)
NEG	2SG.SCL.NEG	yourself.NEG	wash	NEG

'you don't wash yourself' (Manzini and Savoia 2005: 299, IPA in the original)

Whether the above repetition of preverbal negation was ever linked to emphatic or discourse-sensitive negation is unclear. In central Ladin dialects, however, a different type of preverbal reduplication of negation is associated with a pragmatically marked, 'emphatic' context (Siller-Ringgaldier 1985: 75):

(25) a. Gardenese: Ie **no** **ne** ciante

b. Ampezzano:

Io	**no**	**no**	cianto
I	NEG	NEG	sing

'I'm definitely not singing.'

Patruno and Sgarioto (2004) argue that structures such as these are different from the preceding examples in that they represent the raising of postverbal *no* (originally holophrastic and thus a maximal projection) to a specifier projection in CP (as proposed by Poletto and Zanuttini 2003: 193–4).

3.3.2 *The postverbal negative markers*

We have seen that the syntax of the *nò* type is different from that of other postverbal sentence negators such as *pa*, *mica*, *nent*, but these negators also exhibit syntactic differences. As mentioned above, a key factor in their grammaticalization seems likely to have been the fact that the sense of a sentence containing an optionally transitive verb and a negative indefinite object meaning 'nothing' is equivalent to the negation of the event ('I eat nothing/not even a crumb' = 'I don't eat'). Willis (2006) considers this type of reanalysis to have been the main factor in the evolution of Welsh *dim*, and the distinction between lexical and functional item may still be ambiguous in some contexts (Berruto 1990: 14, n. 26, regarding Pied. *nen*). Garzonio and Poletto (2009) examine the current incipient adverbial uses of *gnente* 'nothing' in Venetian, noting its ambiguous status with the verb 'eat', but also identifying an interesting lexico-aspectual restriction on its spread (in addition to the expected incompatibility with a definite nominal direct object at this stage of development). Adverbial *gnente* occurs with intransitive activity verbs only, including those with bare indefinite objects, as in (26c):

(26) Venetian (Garzonio and Poletto 2009)

a.

Nol	magna	gnente
NEG.3SG.SCL	eats	nothing

'He doesn't eat anything / at all.'

b.

Nol	dorme	gnente
NEG.3SG.SCL	sleeps	nothing

'He doesn't sleep at all.'

c.

Nol	me	leze	gnente	libri,	sto	fio
NEG.3SG.SCL	me	read	nothing	books	this	boy

'He doesn't read any books at all, this boy.'

d.

*Nol	me	leze	gnente	i	libri,	sto	fio
NEG.3SG.SCL	me	read	nothing	the	books	this	boy

'He doesn't read the books at all, this boy.'

It is well known that negative quantifiers functioning as direct objects tend in some varieties to float next to the main verb in compound modal and aspectual verb structures (for French, e.g. *il n'a rien vu* 'he has seen nothing', see Kayne (1975), and Garzonio and Poletto (2009) for early Florentine/Italian):

(27) a. Piedmontese

e	dis	que no	poea	**mia**	trover
and	said	that NEG	could	MICA	find

'and he said that he couldn't find anything at all' (*Sermoni subalpini*, p. 251, l. 33, 12th/13th cc.)

b. Tuscan (Siena)

E'	pare	che	noi	non	abbiamo	**niente**	fatto
it	seems	that	we	NEG	have	nothing	done

'It seems that we have done nothing.' (*Fatti di Cesare*, p. 159, l. 9, 13th c.)
(vs Mod. It. non abbiamo fatto **niente**)

Since this is the position in which these indefinite quantifiers came to be reanalysed as adverbs, there can be a clear difference in position in modern Piedmontese dialects between the adverb *nen* and the current direct object negative quantifier *nen* (or more frequently the Italianism, *gnente*), for instance, in compound tenses, direct objects follow the past participle, but the negators precede it, like certain adverbs (e.g. *già* 'already', *subit* 'immediately'):

(28) Modern Piedmontese

i	l'	ai	**nen**	mangià	vs	i	l'ai	mangià	**nen**
1SG.SCL	SCL	have	NEG	eaten		1SG.SCL	have	eaten	nothing

'I haven't eaten.' vs 'I have eaten nothing.'

The syntactic behaviour of the Lombard *nò* type betrays its later and different origin: Milanese *minga* precedes a predicative adjective, whereas *nò* follows it:

(29) Milanese (Vai 1995: 167)

l'	è	**minga**	bel	vs	l'	è	bel	**nò**
3SG.SCL	is	NEG	beautiful		3SG.SCL	is	beautiful	NEG

'It's not nice.'

From a diachronic perspective, however, *nò* has moved forward in the sentence from its original clause-final position, since in modern Milanese it follows the past participle but precedes the direct object:

(30) Milanese (Vai 1995: 168)

u	**minga**	vist	la	tuza	vs	u	vist	**nò**	la	tuza
have.1SG	NEG	seen	the	girl		have.1SG	seen	NEG	the	girl

'I haven't seen the girl.'

To the south of Milan, in the transitional dialect area (Lombard/Piedmontese/Emilian) and in Pavia, dialect speakers optionally place *nò* before the past participle, possibly due to contact influence from the *minga* type:

(31) Borgoratto (Alessandria) (Zörner 1994: 88)

ta	m'	è	**no**	dit	ke	t	sarè	**no**	a	ka
2SG.SCL	me	have	NEG	said	that	2SG.SCL	will.be	NEG	at	home

'you haven't told me that you won't be at home'

(32) Pavese (Zanuttini 1997: 91)[19]

La	Maria	l'	ha	**no**	mangià	la	carne.
the	Mary	3SG.SCL	has	NEG	eaten	the	meat

'Mary hasn't eaten meat.'

In studies such as Zanuttini (1997) and Manzini and Savoia (2005), the respective positions of postverbal negators in modern varieties have been plotted against the universal hierarchy of adverbial positions proposed by Cinque (1999), for instance, Pied. *pa* occupies a higher position in the sentence than *nen*, the higher position (which also hosts It. *mica* and Mil. *minga*) being associated by Zanuttini with the contradiction of discourse-activated presuppositions. Manzini and Savoia (2005: 209, 216–18; 2002 (in 2008 reprint): 79–97), however, hold that all negative adverbs (both pre- and postverbal) are still essentially nominal, since they interact syntactically with nominal elements and therefore must occupy nominal positions; also some northern dialects, especially Piedmontese, require indefinite direct objects in negative sentences to appear as partitive structures (see above, example (18)).

An interesting hypothesis that correlates the syntactic distribution of the different types of negation with their etymological origin is proposed by Poletto (2008). Classifying the negation markers found in the four NegP positions identified by Zanuttini (1997) as 'scalar negation' (< Lat. *non*), 'minimizer negation' (*mica*, *brisa*, *pa*, etc.), 'quantifier negation' (*nent*, *nia* < indefinite pronouns meaning 'nothing'), and 'focus negation' (Mil. *nò*), the analysis concentrates on their contemporary syntactic differences and possible co-occurrences, and argues for viewing NegP as a set of multiple projections, each realized by a different lexical and etymological type, which form a complex 'circuit' whose activation provides the negative meaning of the sentence.[20] Pursuing this line of diachronic investigation and concentrating on the two intermediary types, minimizers (m-negation) and negative quantifiers (q-negation), Garzonio and Poletto (2009) argue that, although both types originally

[19] See also Manzini and Savoia (2005: 190).

[20] The labels are not to be taken literally, since it is not just the preverbal *non* types that are scalar in function and, although there is often a correspondence between these four etymological types and different structural positions, Zanuttini (1997: 88) proposed that Valdotain sentential negator *pa* (as well as possibly French *pas*) can also occur in the same position as Pied. *nen* (with a non-presuppositional value).

functioned as direct objects, their evolutionary paths were not identical and they have come to occupy distinct structural positions because of the difference in their original meaning and category (quantity-noun vs quantifier).

3.4 Functional analyses of the development

The evolution and distribution of negative structures in Italo-Romance may be linked to a combination of factors operating on several linguistic levels:

(i) pragmatic: inflationary tendency to reinforce negative expressions in particular discourse contexts (Bernini and Ramat 1996: 3) (all Italian dialects);
(ii) phonetic: syllabic weakening, possibly due to Celtic substratum (for a critical view, see Tagliavini 1969: 139), which distinguishes Piedmontese, Lombard, and Emilian from the other northern Italian dialects, although Vai (1995: 164–5) is sceptical whether phonetic erosion contributed to the loss of the preverbal negative in northern Italy;
(iii) morphosyntactic: restrictions on the occurrence of the preverbal negative following the grammaticalization of independent subject pronouns as agreement clitics on the verb (northern Italian dialects);
(iv) syntactic: the influence of typological consistency in relation to the drift away from SOV word order (Venneman 1974): the grammaticalization of erstwhile direct objects as negative reinforcers provides the opportunity for postverbal negation, which matches the SVO Operand + Operator sequence (Molinelli 1988);
(v) syntactic: 'natural syntax' which favours introducing the negative marker as early as possible: Jespersen's Neg-First Principle (Horn 1989: 293);
(vi) areal: diffusion of a structural model due to language contact (Bernini and Ramat 1996: 50–1).

Whereas Jespersen (1917) saw the origin of the negative cycle in the second factor, the primacy of the first was recognized by Gardiner (1904) and Meillet (1912), as noted by van der Auwera (2009). Regarding the fourth factor, there is little consensus on the categorization of structural elements (see Hansen, this volume), while Ramat (2006a) points out the need to recognize the continuous interplay of opposing forces on languages, so that 'improvement' on one plane may upset another:

> The preverbal NEG is typologically inconsistent in VO languages, but if we adopt 'natural syntax' as a parameter, this has the advantage of having the most relevant part of the predication, namely the verb, in its scope and thus of offering an easier processing of the sentence. [...] Conversely, postverbal NEG is typologically consistent in VO languages and has a stronger pragmatic emphasis. (Ramat 2006a: 367)

'Natural syntax' thus encourages the cyclical return to preverbal after a period of postverbal negation, as can be seen in the new position of *mica* that is gaining ground in informal Italian, especially in the north (Bernini 1992, Ramat 2006a):

(33) Italian
mica potevo entrare
MICA could.1SG enter
'I couldn't enter.'

The incidence of this type, unaccompanied by *non*, is as yet low (Bernini 1992, Visconti 2009). In southern dialects similar pragmatically marked constructions are found with Neap. *manco*, Sic. *mancu* (via *non manco* 'not less, not even'; cf. n. 8):

(34) Neapolitan (De Filippo 1973: 280)
E io **manco** tengo 'o riesto dint' a sacca
and I NEG hold the change in the pocket
'And I don't have any change in my pocket (what do you expect?!).'

The concentration of the highly marked Neg2 structure in a fairly restricted part of Europe, with some extension towards Arab areas, led Bernini and Ramat (1996) to propose the areal diffusion of a structural model from a Gallo-Romance epicentre, but van Gelderen (2008) demonstrates that the negative cycle is far more pervasive among different language types than previously thought.

Recent functional diachronic studies have sought to obtain a better understanding of the grammaticalization of discontinuous negative structures by building on the insight that contemporary Italian (*non*) *mica* is used to contradict a proposition activated explicitly or implicitly by the preceding discourse or speech-act context (Cinque 1976). Following Schwegler (1988) and, in particular, Schwenter's (2006) investigation of the extent to which the various negative strategies are governed by information-structure constraints, Hansen and Visconti (2009) aim to refine the traditional view whereby reinforced or metaphorical usage simply loses its original innovatory value through the law of diminishing returns (Haspelmath 1999). They document the grammaticalization of discontinuous negative structures in French and Italo-Romance, that is, the gradual loss of dependency on discourse-old information (see Hansen, this volume). The practice of following the denial with adversative statements (introduced by *ma* 'but', *anzi* 'on the contrary') that counter previous explicit or implicit propositions (Schwegler 1988: 46) is seen as a key factor that establishes links with discourse-new information, thus reducing in their view the contextual restrictions on the reinforced strategy. On the basis of a statistical analysis of written and spoken Italian textual corpora, Visconti (2009: 937) argues that the use of *mica* shows 'a cline, from a cluster of monologual contexts in which p is discourse-old by virtue of explicit textual evocation, to a cluster of dialogual cases in which an increasing amount of inferencing is required'. Such a development is deemed to represent 'an increase in "intersubjectivity", in the sense of an increasing coding of the speaker's awareness of the interlocutor's attitudes and beliefs'. Whether this is a real change in the value of the discontinuous negatives over time, or an effect of the different nature of the early and modern texts, which vary in style and genre, needs further investigation. The reinforced structures, introduced for pragmatic and semantic reasons arising from the controversial nature of denying an active expectation to the hearer's face (although the proposition can belong to a third party), are bound to lose their impact with increased use (Dahl 2001) and eventually lead to their being used for less contentious acts of denial (i.e. less discourse-active), especially if other (e.g. structural) factors make the traditional negator a less efficient negating strategy.

Once a given strategy loses its pragmatically marked value, alternative expressions normally take its place: most modern dialects, whether they are predominantly Neg1, 2, or 3, express discourse-active presuppositional negation with a dedicated negative structure, as *non... mica* in Italian, for instance, Pied. *pa* vs normal negator *nen* (Zanuttini 1997: 67), *menga* vs *brisa* in Modena (Manzini and Savoia 2005: 150), but some northern dialects use the same form for both, especially if derived from *mica* (Manzini and Savoia 2005: 153), but also Valdotain *pa* (Zanuttini 1997: 100). Unlike the normal adverbial negator (in Neg 2 and 3 areas), the pragmatically marked type (indicated by the diacritic @ before the English translation) can freely co-occur with negative indefinites (35), as well as in some varieties with the normal sentential negator (36).

(35) Piedmontese (Zanuttini 1997: 77)
A veddu **pa** **gnun**
SCL see NEG no.one
'@I don't see anyone.'

(36) Piedmontese (Lanzo) (Zanuttini 1997: 75)
Fa **pa** **nen** sulì!
do NEG NEG that
'@Don't do that!'

Bernini (1992) provides a very interesting contrastive pragmatic analysis of the *non... mica* strategy with another non-standard variant found in modern informal Italian: the originally metalinguistic negative strategy, *non è che...* 'it is not that...', which has also come to be used for rejecting discourse-active presuppositions, but in a more indirect (and therefore more courteous) fashion than *non... mica*.

While pragmatic studies throw light on the emergence and extension of reinforced negative structures, structural analyses may explain why they come to be grammaticalized in certain language varieties but not others, such as central and southern

Italo-Romance. The availability of alternative strategies permits speakers of varieties whose preverbal negative is weakened or threatened in various ways to revitalize the expression of negation. Phonetic weakening through syllabic reduction was followed in the Gallo-Italian vernaculars not only by the frequent reduction of two common preverbal clitics to forms similar to or even homophonous with the preverbal negative (the 1PL complement clitic < *nos* and the partitive < *inde* can both be reduced to [n]), but also by the emergence of compulsory subject clitics (dialects vary as to which persons were affected). The idea that the development of subject clitics contributed to the negative cycle in French and other Romance varieties (Harris 1978, Ashby 1981, Posner 1985) seems to be confirmed by evidence of co-occurrence restrictions on preverbal elements in northern Italian dialects (Parry 1997). In certain Friulian dialects subject clitics do not appear with complement clitics or the preverbal negative (Benincà 1986: 468). The historical Piedmontese data seen above, together with contemporary microvariation in the Val Bormida, show that if grammaticalized discontinuous structures exist, the preverbal negative is most likely to be dropped when other proclitics occur, thus avoiding the more costly operation of substitution:

(37) Cairese (Parry 1997)

u	l'	è cin	~	u	**n'è**	**nen** cin	>>	u	l'è	**nen** cin
SCL	SCL	is full		SCL	NEG-is	NEG full		SCL	SCL.is	NEG full
'It is full.'				'It is not full.'				'It is not full.'		

3.5 Interrogatives and exclamatives

In section 3.2.2.1 we noted the tendency for interrogatives to lack the preverbal negator *n*.[21] The extent to which this phenomenon may relate to the general loss of *n* in present-day northern Neg3 dialects (and French) is of particular interest, so that contemporary microvariation affords scope for closer examination. In some north-eastern Italian dialects inversion is incompatible with the preverbal negative alone,[22] and the following examples illustrate the nuances of meaning of the various possible structures (P. Benincà, p.c.):

(38) Venetan (Paduan)

a. Nol	vien?	b. *No	vien-lo?
NEG.he	comes	NEG	comes-he

'Is he not coming? (What a pity!)'

but:

[21] See Piedmontese data above and historical French data (Martineau and Vinet 2005).

[22] See Zanuttini (1997: 7, 42ff., and 55) for a structural explanation according to which the negative marker itself can check the interrogative feature in C^0, making it unnecessary for the verb to move there.

c. No l vien miga?
NEG he comes MICA
'He's not coming?!'
(surprise and I think he should)

d. No vien-lo miga?
NEG comes-he MICA
'Isn't he coming?'
(surprise and stronger disapproval)

e. Vien-lo miga?
comes-he MICA
'Is he coming, by any chance?'

From a functional viewpoint, it seems that, at this stage of development, the preverbal negative can be omitted in this context (38e) to reduce the negative force of the proposition, with *miga* contributing the scalar interpretation that allows consideration of various options. On the basis of similar data, Zanuttini (1997: 42, 55) distinguishes between 'strong' (independent) preverbal negative heads and 'weak' clitic negators. As it allows inversion in interrogative sentences, the first element of the pragmatically marked discontinuous negative structure (*no... miga*) is interpreted as a clitic that moves to C^0 along with the verb (like French *ne*). In another Venetan dialect (Sant'Anna), Neg2 *ne-mina* has been grammaticalized as the usual negator, but only *mina* occurs in interrogatives. Benincà and Poletto's (2005) interpretation is that S. Anna's *ne* remains an independent head and so cannot combine with inversion.

Paduan, however, does allow inversion in exclamative clauses with simple preverbal *no* (Benincà 1996: 30):

(39) Paduan (Portner and Zanuttini 2000)
No ga-lo magnà tuto!
NEG has-he eaten everything
'He's eaten everything!'

This leads Portner and Zanuttini (2000) to argue that the same language variety can have two, formally identical, preverbal sentential negative markers, one an independent head, as in (38a) above, that prevents inversion, and the other a clitic on the verbal head (39) that allows it. Paduan exclamatives with an apparent *non-negative* value (thus often referred to as 'expletive' negation, see below, section 3.5.1) do not involve negation of the proposition itself, but merely of a contrary expectation (Benincà 1996: 30), for instance, regarding a child in (39) who normally eats very little. Historical attestations from other dialects with Neg1 show inversion with preverbal negation in similar exclamatory contexts:

(40) 17th-century Ligurian (Taggia) (Forner 1998: 327)
no follo bravo? (fo-llo)
NEG was.he good
'Wasn't he good!'

The conclusion is thus that it is not preverbal negation as such that blocks inversion, but the type of negative marker and its structural position (these being determined by the semantico-pragmatic value of the negation).

Benincà and Poletto (2005) also draw interesting parallels between the diachronic development of negation and that of two other key areas of Romance syntax, namely interrogative structures and personal pronouns. On the basis of microvariation in northern Italian dialects, they link the emergence of clitic forms of the preverbal negative to the development of doubling structures in the other two domains.

3.5.1 'Expletive' negation

In their synchronic studies of Paduan microvariation, Zanuttini and Portner (1996) and Portner and Zanuttini (2000) argue that the negative found in exclamations like (39) is not actually 'expletive', but has a real semantically negative value that generates a characteristic scalar implicature similar to that produced by *even*. For example, in (41) the implication is that he told him all sorts of things and there was nothing of note that he didn't tell him:

(41) Paduan (Portner and Zanuttini 2000: 193)
Cossa **no** ghe dise-lo!
what NEG to.him tells-3SG.SCL
'The things he's telling him!'

That 'expletive' or negative polarity uses of sentential negators still have an intrinsically negative value also emerges from Nocentini's (2003) diachronic study of five different types of so-called expletive negation (found in questions expressing doubt, exclamations, comparisons, complements of verbs of fearing, denying, preventing, etc., and after *finché* 'until'). Drawing on earlier studies, especially Ageno (1955), his functional analysis maintains that the negation in questions and exclamations has the pragmatic value of rejecting one of the options, while the negation in clauses introducing terms of comparison and following verbs of fearing etc. derives from the original paratactic constructions where it was semantically motivated, since they involved the rejection of a presupposition (e.g. 'I fear; may it not be true'; 'the street is more dangerous; you do not believe it'). In the case of comparisons, Nocentini argues that the early semantic negation has been reanalysed for the pragmatic purpose of attenuation (conveyed also by subjunctive modality), that is, to reduce the implicit confrontation:

(42) Modern Italian (Nocentini 2003: 78)
La strada è più pericolosa di quel che tu non creda.
the street is more dangerous of that which you NEG believe.SUBJUNC
'The street is more dangerous than you believe.'

As for verbs of preventing and fearing, which in Latin took a complement clause introduced by the negative complementizer *ne* (replaced in the vernacular by *che non*), tighter syntactic subordination to the main clause made the negative marker increasingly redundant so that it is omitted in modern usage (it was also potentially ambiguous):

(43) a. Genoese
de defender che li mercanti toeschi **no** zeyssen a Venexia
to prevent that the merchants German NEG went to Venice
'to prevent the German merchants from going to Venice' (*Proposizioni fatte dal Comune di Genova*, 24: 24–5, 14th c.)

Compare:

b. Modern Italian
per impedire che i mercanti tedeschi andassero a Venezia
to prevent that the merchants German went.3PL to Venice
'to prevent the German merchants from going to Venice'

3.6 Prohibition and negated infinitives

3.6.1 Negative imperatives

As with interrogatives, the focusing of the verb (raising to C^0 in generative terms) in imperatives may interact with preverbal negation and, consequently, the cyclic evolution of negation seen above has had consequences for the expression of negative imperatives or prohibition (Zanuttini 1997). For the second person singular especially (for 2PL structures, see Parry 2010), Italo-Romance prohibition shows an interesting range of options that diverge from the usual structures found in classical Latin texts—these may be illustrated with the verb *facere* 'to do'. The imperative preceded by a dedicated negative marker (44a) was replaced in Classical Latin by the following formations: the same preverbal negative marker followed by either the present subjunctive (44b) or, more commonly, the perfect subjunctive (44c), or by a negative auxiliary verb plus an infinitive (44d) (Pinkster 1990). Not until late Latin does the most common Italo-Romance type appear, formed by the preverbal declarative negative and the infinitive (44e) (Bourciez 1930):[23]

[23] For example: *non tangere* (4th c. AD *Mulomedicina* 129) 'Don't touch!' For attested examples of the other constructions, see Pinkster (1990: 198–9).

(44) Latin

a. ne fac (limited to poetry after Plautus and Terence)
NEG do.IMPER.2SG

b. ne facias
NEG do.2SG.PRES.SUBJUNC

c. ne feceris
NEG do.2SG.PERF.SUBJUNC

d. noli facere (considered the most polite strategy)
want.AUX.NEG do.INF

e. non facere
NEG do.INF
'Don't do . . . !'

The Italo-Romance developments are explored in Parry (2010) on the basis of data from a wide range of early vernaculars and dialects, there being three main options for second person singular imperatives:

(i) Neg1–infinitive, (45), (46), as in modern Italian (the majority option)
(ii) Neg1–imperative, (47), (48), and (49), as in early Latin, but with the usual declarative negative
(iii) Neg1–subjunctive, as was common in Latin (as in Sardinian and Spanish, see below (51c)).

(45) Veneto
Fa-te en dre', **no** me **vegnir** sora!
make-you in back NEG me come.INF above
'Get back, don't come near me!' (*Atti Lio Mazor*, p. 68, l. 30, 14th c.)

(46) Campania
De octubro . . . porri **no** **mangiare**
of October leeks NEG eat.INF
'In October . . . do not eat leeks.' (*Regimen Sanitatis* 580, l. 577, 13th c.)

Negated 'true' imperatives[24] are found in some medieval Ligurian, Lombard (47), and Tuscan texts, and in the 16th-century Piedmontese of Asti, (48).[25]

[24] 'True' imperatives are forms unique to the imperative in that person (Rivero 1994), while suppletive imperatives use indicative or subjunctive, or non-finite forms (infinitive or gerund). The label 'true' is not meant literally in so for as Italian 2PL *cantate* 'Sing!', which derives from the Latin true imperative, is not considered a 'true' imperative as it also stands for the 2PL present tense, and this permits it to occur with *non*. In contrast, the 2SG imperative is only a dedicated form in the case of the first conjugation and a few irregular verbs, yet it is unacceptable with preverbal negation regardless of the conjugation (as noted by Zanuttini 1997: 174, n. 3).

[25] As in Old French:

(i) «Rois, **ne** **va** plus avant, car tu nel doiz pas fere!»
king NEG go.IMP more forward for you NEG-it must PAS do

(47) Lombard

Taia	lo	pan	per	ordene,	**no**	**va'**	taiand	per	tuto,
cut.IMP	the	bread	by	order	NEG	go.IMP	cut.GER	for	all

No	**va'**	taiand	dal	parte
NEG	go.IMP	cut.GER	by.the	sides

'Cut the loaf neatly, don't cut it all up anyhow, don't cut bits off the sides.' (Bonvesin, *De quinquaginta curialitatibus*, p. 318, ll. 91–2, 13th c.)

(48) Piedmontese (Asti)

e	tôrna	an	leg,	**n'**	i	**pensa**	pu
and	return.IMP.2SG	in	bed	NEG	to.it	think.IMP.2SG	more

'and get back into bed, don't think about it any more' (G. G. Alione, *Farsa del braco*, 16th c.) (Bottasso 1953: 216)

Negated true imperatives survive to this day in a few Ladin and Friulian dialects:

(49) Ladin (Cortina d'Ampezzo)

No	tóma!
NEG	fall.IMP

'Don't fall!' (Jaberg and Jud 1928–40: map 1621, point 316)

although they are more frequently found in periphrases formed on auxiliary verbs, in particular the verb *stare* 'stay, be', which is also widely used to form the progressive (+ infinitive or gerund) in Italo-Romance. In Friulian, for example, the use of *stâ* 'stay, be' has been grammaticalized as a negative imperative auxiliary:[26]

(50) Friulian

No	sta	(a)	cródi!
NEG	stay.IMP.2SG	(to)	believe.INF

'Don't believe (it)! (lit. Don't be believing it!)' (Marchetti 1977: 293)

Examples such as (49) contravene a correlation proposed by Zanuttini:

In Romance, preverbal negative markers that can negate a clause by themselves do not co-occur with true imperative forms of main verbs, but do co-occur with true imperative

'"King, do not go any further forward, for you must not (contrary to what you think)!"' (*La Queste del Saint Graal*, p. 85, 1225)

[26] E.g. 2PL:

(i)

no	stàit	a	fâ	sunsûr
NEG	stay.IMP.2PL	to	make.INF	noise

'Don't make a row!' (Marchetti 1977: 293)

Ligurian dialects also often resort to periphrases with *stâ*, but there the modern form of the 2SG. imperative is homophonous with the infinitive (see Parry 2010: 156, fn. 6 for details).

forms of auxiliaries. Post-verbal negative markers do not exhibit any such restriction.
(Zanuttini 1997: 121)

Thus, while Piedmontese and French use identical forms for second person singular positive and negative imperatives (51a), Italian and Romanian use infinitives (51b), while Sardinian and Spanish use present subjunctive forms (51c):

(51) a. Fr. Chante! ~ (Ne) chante pas! Pied. Canta! ~ Canta nen!
b. It. Canta! ~ Non cantare! Rom. Cântă! ~ Nu cânta!
c. Sard. Canta! ~ Non cantes! Sp. Canta! ~ No cantes!
'Sing' ~ 'Don't sing!'

The fact that true imperatives freely occur with post verbal negation leads Zanuttini to conclude that the above restrictions are not semantic in origin but syntactic. Within the generative framework adopted, she argues that in imperative structures the illocutionary force of the sentence must be checked in C^0 by the verb *or* by the preverbal negative (head). The latter case requires that the Mood projection be checked by a complement marked overtly or covertly for Mood (this excludes true imperatives that are composed of the verbal root + thematic vowel only, with occasional number marking). To accommodate preverbal negatives with true imperatives of auxiliary verbs, as in Friulian, it is argued that mood marking may be supplied either morphologically (e.g. by a finite subjunctive form) or syntactically by an auxiliary (Zanuttini 1997: 119–21). The admittedly rare cases of negated non-auxiliary imperative forms thus remain unaccounted for.[27] This problem is addressed in Zeijlstra (2006), who posits an interesting correlation in a range of languages between the grammaticality of true negative imperatives (TNIs) and 'strict' negative concord (NC), that is, the compulsory marking of negation on the verb (even when this is preceded by another negative nominal or adverbial element). Early Italo-Romance vernaculars that have TNIs indeed show preverbal negative concord (see below), but both phenomena are optional and a given text need not exemplify the correlation. The explanation offered by Zeijlstra for the fact that strict-NC languages allow their preverbal negative to raise with the imperative verb to C^0 is that such a negative has an uninterpretable feature, which prevents its having scope over the illocutionary force of the imperative (Zeijlstra 2006: 417).[28] Instead, negative scope must be provided by a null operator in SpecNegP. Yet, although one might therefore

[27] Early Friulian also has distinct 2PL present indicative (*-s*) and 'true' imperative (*-t*) forms of full verbs (Benincà and Vanelli 2005); here too the prohibitive is unambiguously composed of preverbal negation + imperative (see Parry 2010).

[28] Zeijlstra adopts Han's (2001) theory that the reason why negative imperatives are impossible in some languages is because the operator encoding the illocutionary (imperative) force would be c-commanded by the preverbal negation, and so 'interpreted as being negated'.

expect a strict-NC language like Venetian to allow TNIs, it has infinitival negative imperatives, as do most Italo-Romance varieties.[29]

In an attempt to account for the range of diachronic data, Parry (2010) argues that diverse pragmatic factors linked to the particular emotive situations in which prohibitives are used could have favoured the adoption of alternative periphrases instead of the original negated imperatives, considered to be either too abrupt or to lack impact. Instead, speakers resorted to face-saving, deferential expressions on the one hand or, on the other hand, more dynamic and forceful turns of phrase. Here we have another type of negative cycle, which appears to be already completed in Latin (Pinkster 1990: 192, 198–9). Thus the colloquial connotations of TNIs (as attested for modern Romance, e.g. Romanian) could explain their absence from classical Latin texts, but more research is needed to see whether modern Romance TNIs are indeed a continuation of non-standard Latin usage. If they are, it may be that their retention in Lombard, Piedmontese, and French, together with the rejection of the periphrastic alternatives found in medieval texts, has less to do with the position or syntactic status of the sentential negator, as argued by Zanuttini, than with the avoidance of person ambiguity. In Piedmontese (as in French first conjugation verbs), the second person plural imperative became homophonous with the infinitive (e.g. Pied. *canté/scrive/finì*), thus cancelling the formal difference between singular prohibitives based on the infinitive and plural prohibitives. Ambiguity, though not a necessary trigger for change, could in this case of rival alternatives have encouraged the avoidance of infinitival prohibitives in favour of the more colloquial second person singular 'true' negative imperative forms. In Lombard, no person ambiguity arises, but the third conjugation infinitive is homophonous with the second person singular imperative. In this case, analogy could have promoted the spread to other conjugations of a symmetrical pattern of positive and negative imperatives:

(52) Lombard

Met!	~ No	met	mia! ⇩
put. IMP	NEG	put.INF/IMP	MICA
'Put!'	~ 'Don't put!'		

Párla!	~ No	parlà	mia!	>>>	Párla!	~ No	párla	mia!
speak.IMP	NEG	speak.INF	MICA		speak.IMP	NEG	speak.IMP	MICA
'Speak!'	~ 'Don't speak!'				'Speak!'	~ 'Don't speak!'		

Interestingly, in a few modern Emilian varieties the 'postverbal' minimizer negator has been fronted and occurs with the suppletive infinitival negative imperative:

[29] Romanian is another strict-NC language that normally has infinitival negative imperatives. However, TNIs are found in non-standard Romanian with a few basic verbs (Parry 2010: 158).

(53) Emilian (Bologna)
briza movrat
NEG move.INF.yourself.2SG
'Don't move!' (Jaberg and Jud 1928–40: map 1647, point 456)

It is probable that the infinitival construction represents a reduction of an original prohibitive periphrasis involving an initial auxiliary (see Friulian ex. (50) above). Indeed, such is also the case for the unusual southern (especially Puglia, Calabria, and Basilicata) second person singular prohibitive type formed from Neg + *gerund* (Rohlfs 1969: 110–11). The selection of infinitive or gerund thus depends on the type of dialect:

(54) Puglia (Trani)
non bartènnë
NEG leave.GER
'Don't leave!'

Some dialects still have non-reduced prohibitives:

(55) Tarantino
No **scé** scennë
NEG go.INF go.GER
'Don't go!' (Zanuttini 1997: 124)

3.6.2 Negated infinitives

The diachronic aspects of negated infinitives in Italo-Romance have received less attention: *non*-type markers remain preverbal, whatever the form of the verb, whereas the postverbal type shows evidence of a gradual cyclic promotion to preverbal position before the infinitive in certain dialects and loss of the first element of discontinuous structures (for some exceptions, see Manzini and Savoia 2005: 240):

(56) Turinese
për **nen** criè
for not shout.INF
'not to shout' (*Il Conte Pioletto*, p. 80, 17th c.)

Some Neg2 dialects retain the discontinuous structure here too, for instance, Romagnolo (Parry 1996: 249, Manzini and Savoia 2005: 245–6), but the trend conforms to Jespersen's Negative First Principle. The behaviour of the more recent *nò* type lags behind that of the minimizer type, as illustrated by the contrast between the two negators *minga* and *no* in Milanese in (57). To the southwest of Milan, transitional dialects such as that of Breme (58), however, are again at the forefront of developments, with *nò* occurring before the infinitive (see the data in Manzini and Savoia 2005: 229–38).

(57) Milanese

a. de **minga** credeg	but	b. de credeg	**no**	(*de **no** credeg)
of NEG believe.INF.it		of believe.INF.it	NEG	
'not to believe it' (Vai 1995: 168)		'not to believe it'		

(58) Breme (Pavia)

per **no** vògti
for NEG see.INF.you
'not to see you' (Manzini and Savoia 2005: 231)

In Zanuttini's (1997) generative analysis, the developments are explained in terms of the infinitive raising to different, but generally low, positions within the core sentence structure, according to the dialect. However, the enclisis of complement clitics on the infinitive and inconsistencies related to adverbial positions lead Manzini and Savoia (2005: 254ff.) to consider the infinitive as lexicalized in the high pre-sentential modal area C, an analysis deemed to be supported by the existence in some dialects, regardless of the type of negation found with finite verbs, of a dedicated infinitival negator, often a non-reduced form *nun* (for this form, see also Parry 2005: 263).

3.7 Negative pronouns and adverbs

Medieval texts show a range of pronominal formations that continue earlier univerbation processes attested by Classical Latin *nemo* (< *ne-homo) 'no one' and *nihil* (< *ne-hilu 'no thread') 'nothing'. The former survives in early Tuscan *nimo* (59), and modern Sardinian *nemos* (60), but more commonly 'no one' is signified by reflexes based on the numeral 'one': Lat. *nec unu(m)* 'and not one' > *neun, niun, niunu, neù, negun, nigun, negono*; Lat. *nec-ullu(m)*, dim. of *unu(m)* > *nullu* > *nul, nullo, nullu, nu omo* 'no man'; Lat. *nec ipsu(m) unu(m)* 'and not itself one' > *nesun, nisun, nixun, nexunu*. These combinations of negative scalar focus particle (*nec* 'nor'), the minimal number *unu*, and the intensifier *ipse* all contribute a scalar value that emphasizes the negative sense 'not even one':

(59) Tuscan (Volterra)

non c' enterrà **nimo**
NEG there will.enter no.one
'No one will enter.' (*Confessione di prete Berto*, p. 24, 1326)

(60) Sardinian

nemos iskit kando morit
no.one knows when will.die.3SG
'No one knows when they will die.' (Jaberg and Jud 1928–40: map 1672)

(61) Lombard

negun no te pò avé
no.one NEG you can have
'No one can have you.' (*Disputatio roxe et viole,* p. 108, l. 204, 13th c.)

(62) Neapolitan

ca **nullo** pote vivere un' ora solamente
for no.one can live one hour only
'for no one can live just one hour . . . ' (*Regimen sanitatis,* p. 564, l. 39, 13th c.)

As adjectives the above forms vary for person and gender, for instance:

(63) Lazio

per **nulla** raczone **nullo** homo lo potte tenere de annare ad la vattalia
for no reason no man him could keep from go.INF to the battle
'There was no way anyone could persuade him from going into battle.' (*Storie de Troia e de Roma,* p. 36, ll. 14–16, 13th c.)

Inanimate reference ('nothing') is represented by reflexes of *ne*(*c*) *gente*(*m*) 'nor people' *nient, niente, nigente, nente*; *nec gutta*(*m*) 'nor a drop' > *negota, nigota, neota, nuta, nigò*; *nulla* (*res*) 'no thing' > *nulla* or combined with the noun 'thing' *nulla cosa*:

(64) Piedmontese

e' **no** soi aizià de doner te **nient** aora a questa fiaa
I NEG am able to give you nothing now at this time
'I cannot give you anything at this moment.' (*Sermoni Subalpini,* p. 238, ll. 16–17, 13th c.)

(65) Lombard

Lló **negota** 's perde, **negota** g' invedrisce
there nothing REFL loses nothing there grows.old
'There nothing is lost, nothing grows old.' (Bonvesin, *De scriptura aurea,* p.155, l. 125, 13th c.)

(66) Campanian (Naples)

e no nce lassaro **cosa** **nulla** de valore
and NEG there left thing none of value
'They left behind nothing of value.' (*Libro de la destructione de Troya,* p. 256, ll. 30–1)

In the early vernaculars non-specific indefinite pronouns corresponding to 'someone', 'anyone', and 'anything' also functioned in conjunction with the sentential negator: *alcuno* (< *alicunum*) 'someone, anyone', *veruno* (< *vere unu*(*m*)) 'truly one' > 'someone', 'anyone' (cf. Romanian *vreun*(*ul*)), *chivelli* (< *q*(*u*)*i* / *c*(*u*)*i velles*) 'who(ever) you wish', *covelle* (< *quod velles*) 'what(ever) you wish' (Rohlfs 1968: 221–2):

(67) Lombard (Cremona)

No se dé **alcun** laudar de soa propïa boca.
NEG REFL should anyone praise of his own mouth
'No one should praise himself.' (Patecchio, *Splanamento*, p. 563, l. 65, 13th c.)

(68) Tuscan (Pisa)

In sua p(re)sentia **alcuno** **no(n)** dè essere lodato.
in his presence anyone NEG should be praised
'No one should be praised to his face.' (*Trattati di Albertano*, p. 5003, ll. 144–5, 13th c.)

Alcuno may also feature in negative polarity (69a) and positive contexts (69b):

(69) a. Tuscan (Florentine)

Et se **alcuno** domandasse per qual cagione
and if someone asked for what reason
'And if someone asked why…' (Brunetto Latini, *Rettorica*, p. 153, l. 10, *c.*1260–1)

b. Tuscan

Alcuno à furato d' una chiesa uno cavallo.
someone has stolen from a church a horse
'Someone has stolen a horse from a church.' (Brunetto Latini, *Rettorica*, p. 84, ll. 16–17)

(70) Tuscan (Assisi)

se a coloro, coi quagli vive e more, **no** lassasse **covelle**
if to those with whom lives and dies NEG left whatever
'if to those with whom he lives and dies, he were not to leave anything' (*Statuti della Confraternita dei Disciplinati*, pp. 172–3, 1329)

From a diachronic perspective there has been a reduction in the number of indefinite pronouns (e.g. the *chivelli/covelle* type has disappeared) as well as in polar versatility (Martins 2000: 192), so that modern Italian singular *alcuno* only occurs postverbally and then only with a negative sense, while the positive use found in the early vernaculars (as in (69b)) has been replaced by *qualche*, *qualcuno* (< *quale che (sia)—the 'it may be' type of Haspelmath 1997: 135). It may occasionally be found in the literary language in interrogative and conditional sentences, but only when followed by a noun (Zanuttini 2010). In Venetian, which requires the preverbal negative marker even when a negative indefinite precedes the verb, *algun*(*o*) has disappeared (Marcato and Ursini 1998: 187).

In contrast, early Tuscan *veruno*, which like *alcuno* frequently occurred in modal/irrealis contexts (71a), acquired intrinsic negative meaning through frequent collocation with negative markers (as in 71b). This can be seen as a parallel semantic development to that of the bare nouns seen above (*mica*, *brisa*, etc.) and its negative value is proved by the coordinated negative in (71c) (Ramat 1998: 406).

(71) Tuscan

a. Prato

se	**veruno**	della	detta	Compagnia	infermasse
if	anyone	of.the	said	company	became.ill

'if anyone of the said Company were to become ill…' (*Capitoli della Compagnia della Santa Croce*, p. 445, l. 28, 1295)

b. Prato

che	**veruno**	**non**	lavori	nè	apra	bottega	il	venerdì	sancto
that	anyone	NEG	work	nor	open	shop	the	Friday	holy

'that no one work (n)or open shop on Good Friday' (*Breve dell'arte de' calzolai*, p. 22, ll. 13–14, 1347)

c. Pisa

sicché	**veruno**	si	può	lamentare,	**né**	dire:	Io…
so.that	no.one	REFL	can	complain	nor	say	I

'so that no one can complain, nor say: I…' (D. Cavalca, *Epistola di S. Girolamo*, p. 424, l. 231, 1342)

Despite having absorbed a negative element, *nessuno* 'no one' and cognates also conveyed a non-negative existential value ('any') in modal/irrealis contexts such as interrogatives and conditional clauses, as did *niente, nulla* 'nothing':

(72) Venetian

Doma(n)dà	s'	el	li	vit	**arma**	**nesuna**
asked	if	he	to.him	saw	arm	none

'He asked whether he saw he had any weapon on him.' (*Atti Lio Mazor*, p. 70, l. 26, 1312–14)

The following example nicely contrasts non-negative and negative uses:

(73) Lombard (Milan) (*Elucidario*, p. 160, ll. 11–12, 14th c.)

a.

D. Noxe	**niente**	a li	fantin	si li	in nadi	de no-licito	matrimonio?
Harms	nothing	to the	children	if they	are born	of non-legal	matrimony

'D. Does it harm children at all, if they are born of an illegal marriage?'

b.

M. Se	illi	in batezadi	zo	**no**	noxe	**niente**	a	loro
if	they	are baptized	that	NEG	harms	nothing	to	them

'M. If they are baptized it does not harm them at all.'

(74) Umbrian

Avereste	tu	parlato	**niente**	de	alcuno	donçello?
would.have	you	spoken	nothing	of	any	young.man

'Would you have spoken at all about any young man?' (*Perugia e Corciano*, p. 120, ll. 21–4, 14th c.)

(75) Campanian (Naples)
Et eo plu tosto lo vorria patere a la mia persone, davante che
and I more soon it would.wish suffer.INF at the my person before that
nullo ambassatore patesse **nulla** vergogna oy offensa inde la corte mia
no ambassador suffered no shame or offence in the court my
'And I would prefer to suffer it personally, rather than for any ambassador to suffer any shame or insult in my court.' (*Libro de la destructione de Troya*, p. 130, ll. 4–6, 14th c.)

Modern usage is more restricted, with conditional contexts becoming less acceptable than interrogative ones (for Italian, see Longobardi 1988: 667):

(76) Modern Italian (Longobardi 1988: 667)
a. Hai visto **niente**?
have.2SG seen nothing
'Have you seen anything?'

b. ??Se dovesse succedere **niente** di rilevante, telefonami
if should happen nothing of relevant phone.me
'If anything relevant should happen, phone me.'

According to Martins (2000: 192) this restriction in usage is due to 'the fact that less salient polar contexts are not able to constitute robust enough evidence for acquisition'. Modern Venetian is deemed to have evolved further in this direction, sharing with Romanian the feature of 'not allowing negative indefinites in non-negative modal contexts' (Martins 2000: 197). Yet my informants, like Marcato and Ursini (1998: 190), accept negative indefinites as alternatives with different pragmatic implications from positive indefinites:[30]

(77) Venetian (G. Lepschy, p.c.)
a. Ti vedi **gnente**? ~ Ti vedi **qualcossa**?
you see nothing you see something?
'Can you see anything / something?'

b. Se ti vedi **nissun**, dimelo. ~ Se ti vedi **qualchidun**, dimelo.
If you see no.one tell.me.it if you see someone tell.me.it
'If you see anyone / someone, tell me.'

Interesting developments regarding morphological agreement may be found in some Italian dialects, for instance, *nessuno* 'no one' may be marked for number as well as gender:

[30] It seems that the negative indefinites are more likely to expect a negative answer.

(78) Modern Venetan vs Italian

a. no go trovà **nissuni** vs b. non ho trovato **nessuno**
NEG have.1SG found no.one.PL NEG have.1SG found no.one.SG
'I've found no one.'

while, even more unusually, the original (non-negative) nominal minimizer *punto* 'point' has acquired adjectival number and gender marking in the regional Italian and dialects of Tuscany (for a diachronic analysis, see Garzonio 2008):

(79) Tuscan

a. in **punte** strade
in PUNTO.FPL streets
'in no streets' (Rohlfs 1968: 217)

b. Quante matite hai? Non ne ho **punte**
how.many pencils have.2SG NEG of.them have.1SG PUNTO.FPL
'How many pencils have you? I haven't any.' (Longobardi 1988: 657)

Niente 'nothing' may also be used as a determiner, but it is invariable:

(80) Modern Italian
niente lunghe code
no long queues
'no long queues' (*Il sole24ore*, 3 April 2010)

As elsewhere in modern Romance, the aspectual adverb *mai* 'ever' (< *magis* 'more') has also acquired a negative temporal value through frequent collocation with the sentential or constituent negator, so that it now functions on its own in preverbal position:

(81) Modern Italian
Mai avrebbe pensato di riuscirci...
never would.have.3SG thought of succeeding.there
'Never would he have thought he would succeed...'

In 15th-century Tuscan texts, both *mai non*, (82), and *mai* alone are found preverbally, the latter encouraged by the asymmetric patterning of the negative indefinite pronouns: *non* V *nessuno* vs *nessuno* V, etc. In the medieval vernaculars *mai* is often reinforced by reflexes of *iam* 'now, already' or residual *unque, unca* (< *unquam* 'ever'), as in (83). It is always accompanied by an overt negative, whether pre- or postverbal, even in areas that lack strict negative concord, e.g. Neapolitan (84):

(82) Lombard (Milanese)
La summitá del napo col polex **mai no toca**
the top of.the goblet with.the thumb ever NEG touch.2SG.IMP
'Never touch the top of the goblet with your thumb.' (Bonvesin, *De quinquaginta curialitatibus*, p. 321, ll. 173–4, 13th c.)

(83) Ligurian

L' animo to ni lo voler / **unca mai** fim **no** pò aver[31]
the soul your nor the will / ever more end NEG can have
'Neither your soul nor your will can ever have an end.' (*Anonimo Genovese*, p. 176, l. 644, 14th c.)

(84) Campanian (Naples)

che **iamay** li ochy suoy **no** le levava de sopre
for evermore the eyes her NEG him.DAT lifted from upon
'...for never did she take her eyes off him.' (*Libro de la destructione de Troya*, p. 101, l. 10, 14th c.)

Derivatives of *plus*, on the other hand, have not become semantically negative and must occur in the scope of another negative, such as *mai*, for instance, Sardinian:

(85) Sardinian

mai prus faco cussu
never more do.1SG that
'Never again will I do that.' (Jones 1993: 23)

or have merged with the predicate negator, as in Piedmontese *papi, pinen* 'no more'.

3.8 Negative concord

Negative concord, 'the co-occurrence of more than one negative element in the same clause with the interpretation of a single instance of negation' (Zanuttini 1997: 9), is a well-known feature of Italian and Italo-Romance varieties. It includes the 'negative doubling' and 'tripling' phenomena seen above, and involves the co-occurrence of negative marker and negative constituent as well as that of several negative constituents. It was not acceptable in Classical Latin, but its existence in informal early and late Latin texts of the entire Romance area reflects the tendency to reinforce denial or rejection by negating at both constituent and sentence level. It is particularly evident in late Latin statutes and ordinances embodying prohibitions of various sorts (Molinelli 1988: 38). Whereas modern Italian and central and southern Italo-Romance varieties show asymmetric negative concord whereby a postverbal negative indefinite (NI) must always occur in the scope of a sentential negative or other negative element, but a preverbal one *cannot* occur with the sentential negator, in the medieval period many regions seem to have allowed the latter combination (Molinelli 1988: 55), so-called 'strict' negative concord. In Piedmont, as in Gallo-Romance, 'strict' negative concord was the rule (see (86)), but elsewhere the sentential negator seems optional when the NI is preverbal (as in (87)–(92)). Campanian attestations appear to be restricted to subordinate clauses involving prohibitions reinforced by strong negative prepositional phrases (94):

[31] In the TLIO database, *unca* occurs only in the north and in Sicily (Fr. *onques*).

(86) Piedmontese

Nula ren que viva **no** 'l po perpenser né dire
no thing that lives NEG it can understand nor say
'No living thing can understand or describe it.' (*Sermoni Subalpini*, p. 259, ll. 35–6, 13th c.)

(87) Ligurian

a. ché **nixun no** pò ben finir
for no.one NEG can well end
'...for no one can come to a good end.' (*Anonimo Genovese*, p. 543, l. 11, 14th c.)
vs

b. e **nixum** gi pò fuzì
and no.one him can escape
'...and no one can escape him.' (*Anonimo Genovese*, p. 204, l. 10, 14th c.)

(88) Venetian

Niente non m' è romagnudo
nothing NEG to.me is remained
'Nothing is left to me.' (*Fr. Grioni, Santo Stady*, p. 77, l. 1078, 14th c.)

(89) Tuscan (Siena)

Perciò che **neuno non** à paura sed elli non vede alcuno pericolo aparére
for.thatthat no.one NEG has fear if he NEG sees any danger appear
'Because nobody is afraid, if he does not see any danger (appear).' (*Egidio Romano*, p. 43, l. 26, 1288)

(90) Romanesco

...e **nullo non** volea gire in Yspangia
and nobody NEG wanted go.INF in Spain
'...and nobody wanted to go to Spain.' (*Storia de Troia e de Roma*, p. 194, ll. 20–1, 13th c.)

(91) Abruzzese

pur planien[d]o **nunquam no** finaro
though crying never NEG they.stopped
'though they never stopped crying' (*Transito della Madonna*, p. 21, ll. 39–42, 14th c.)

(92) Sicilian (Messina)

Statua indaurata... **nullu homu non** la vitti
statue golden no man NEG it saw
'A golden statue..., no man had seen...' (*Libru di Valeriu Maximu*, pp. 65–6, 14th c.)

(93) Sardinian

a. ...e pro dinari **nexunu** **non** campit
and with money no.one NEG escape
'...and no one shall be exempt through paying (a fine).' (*Carta de Logu*, p. 44, 14th c.)

is semantically equivalent to

b. ...e **non** campit pro dinari **alcunu** (*Carta de Logu*, pp. 36, 38, 14th c.)

(94) Campanian (Naples)
mantenente ordenao... che per **nullo** muodo **non** ricipessero Dyomede
immediately ordered that by no way NEG receive.3PL Diomedes
'Immediately he ordered...that they should in no way receive Diomedes.'
(*Libro de la destructione de Troya*, p. 277, ll. 10–11, 14th c.)

Strict symmetric negative concord is a regular feature of modern Venetian and appears to be optional in modern Friulian (Faggin 1997: 117). According to Haspelmath (1997: 202), marking negation on the verb (NV–NI, i.e. Negative Verb and Negative Indefinite,[32] *irrespective of the order of the elements*) is the default option cross-linguistically. The much less common V–NI type (necessarily with a semantically negative pronoun) is largely restricted to Europe, and often a consequence of Jespersen's cycle. In his opinion, V–NI involving a *postverbal* NI is doubly aberrant from a functional viewpoint, since it violates both the isomorphic principle of marking predicate negation on the verb (the head of the verb phrase) and that of Neg First. He posits a unidirectional development from the asymmetric type (N)V–NI (i.e. NV–NI but NI–V) to the symmetric type NV–NI, so that the verb was negated whatever the position of the negative indefinite and Neg First satisfied. So, in the case of Romance, the decline of the verb-final constraint in Latin syntax would have led to instances of aberrant V–NI, but speakers would tend to remedy the situation by negating the verb when the pronoun was postverbal (thus (N)V–NI). Then, this pattern of verbal concord would have been analogically extended to preverbal NIs, as in the Slavic languages, thus satisfying the isomorphic principle. If, however, as intimated in Posner (1984) and Martins (2000), NV–NI (including strict negative concord) characterized all proto-Romance, a problem arises with modern central and southern Italo-Romance varieties, as well as Spanish, since they would represent the inverse development (i.e. from NV–NI > (N)V–NI). Haspelmath (1997: 212–13) therefore doubts whether proto-Romance was uniformly NV–NI, and suggests that there was instead a development from Latin V–NI to

[32] 'Negative indefinite pronoun' is used by Haspelmath in the sense 'indefinite pronoun that has direct negation as an important function' (Haspelmath 1997: 199), thus including forms corresponding to Eng. *anybody* and *anything*, as well as *nobody* and *nothing*.

(N)V–NI, with for example central and southern Italian varieties stopping at this stage, while other varieties continued to NV–NI (a similar progression is attested in new English dialects, such as African-American dialects, although sentential negation is not compulsory when the indefinite is in preverbal position, Haspelmath 1997: 213). The fact remains that, as seen above, strict negative concord NV–NI *is* attested optionally in most regions of medieval Italy, including in those where it is no longer allowed, for instance, in Tuscany, Lazio, Sicily, and Sardinia. This may indicate a fairly general extension of the late Latin trend, which, however, remained optional and was eventually aborted in most varieties for reasons of economy (and possibly prescriptivism) in structures where negation was already expressed preverbally.[33] Nowadays, the preverbal expression of negative indefinites may be felt to be somewhat unnatural, since they are often placed postverbally by informants when responding to questionnaires (Benincà, p.c.). In early Romance the existence of a sentence-initial informational-focus position made them more likely to occur preverbally (raised from lower down the sentence structure in the wake of the verb raising to the complementizer position in a V2 structure).

There has been much debate from a synchronic perspective about the status of the above forms in Romance, whether they are negative quantifiers (as in Zanuttini's 1997 influential treatment) or indefinites (e.g. Acquaviva 1992, Manzini and Savoia 2005). Martins (2000) offers a comparative diachronic analysis of Romance negative pronouns (including Italian and Venetian) that distinguishes between weak and strong polarity items in respect of a theory of semantic features involving the notion of underspecification. She argues that weak polarity items become strong through the language-acquisition process: variable underspecified negative polarity items, such as early Italo-Romance *nessuno*, *niente*, were increasingly identified as negative in unambiguous negative contexts, while their ambiguity in modal polarity contexts reduced their survival chances there (as mentioned above for Italian). According to this view, negative polarity items in most modern Italo-Romance varieties have become sufficently 'strong' to convey negation on their own, dispensing with the sentential negator when they occur preverbally. In central and southern Italo-Romance varieties, they cannot do this when postverbal and must stand in a relation of negative concord to the normal sentence negator. More research is necessary on

[33] Similarly, the negative concord associated with negative coordinating particles in the medieval vernaculars (a more frequent phenomenon than that involving preverbal indefinites) has been lost:

(i) Florentine
Tu non la dovei punire **né** **non** convenia ad te punirla di ciò
you NEG her should.PAST punish nor NEG was.appropriate to you punish.her of that
'YOU should not have punished her nor was it appropriate for you to punish her for that.' (Brunetto Latini, *Rettorica*, p. 155, ll. 9–10, 13th c.)

the early use of these negative indefinites, many of which were etymologically negative, as shown above: thus it may be that the intrinsic negative value of formations incorporating a negative particle was exploited in irrealis contexts, as in the so-called expletive uses of the sentential negator, to widen the scope of reference by generating a scalar implicature.

Déprez (2000) takes a different view of the evolution of Romance negative indefinites, adopting a syntactic, rather than lexical approach, but rejecting Zanuttini's model according to which the distribution of negative constituents hinges on the position in the sentence structure and nature of the sentential negator (head vs specifier) in favour of one based on the internal syntax of the n-words themselves. From a diachronic perspective she argues that via the grammaticalization process an indefinite pronoun or bare noun (as French *personne*) can become a pronominal quantifier, acquiring intrinsic quantificational force through internal head-movement from NP to the head of DP, which means that it is no longer a variable requiring a marker of sentential negation to bind it. Piedmontese and Milanese negative indefinites would presumably have undergone a similar development to French, while asymmetric-concord languages, like Italian and other central and southern Italo-Romance varieties, represent a situation in which this development is context-dependent: postverbal n-words are not affected. Although syntactically licensed by the verb, these still require semantic binding by an overt negative operator, since they lack intrinsic quantificational force. In preverbal position, however, their null D^0 cannot be licensed by the verb which does not c-command them, so internal head-movement must apply as a last resort; this has the semantic consequence of endowing the preverbal negative indefinites with intrinsic quantificational force that prevents them from functioning as variables bound by a negative operator (Déprez 2000: 291). Like Martins (2000), Déprez presupposes an evolution in a contrary direction from that proposed by Haspelmath (i.e. NV–NI > (N)V–NI).

Although northern varieties have lost preverbal negation and their n-words can convey negation independently, the cycle of negative strengthening to convey more robust denial continues with the emergence of *postverbal* negative concord, confirming the already noted tendency to mark negation at both sentence and constituent level (i.e. head- and dependent-marking). In this case all elements appear postverbally, but in many dialects there are structural constraints on co-occurrence that produce the following typical hierarchy: modal/aspectual periphrases > compound tenses > prepositional phrases, with the contexts on the left of the hierarchy most favourable to negative concord between the postverbal negator and the negative indefinite (Parry 1996: 247):[34]

[34] See Manzini and Savoia (2002) for an analysis of combinatory possibilities with negative indefinites that distinguishes between 'nothing'-type and bare-noun sentential negation.

(95) Modern Turinese
a veul **nen** mangé **gnente**
SCL wants NEG eat.INF nothing
'(S)he doesn't want to eat anything!'

(96) Modern Milanese
a. l' ha **minga** mangià **nisün**
SCL has NEG eaten nobody
'Nobody has eaten!'
b. ghe 'l do **no** a **nisün**
to-him it give.1SG NEG to nobody
'I'm not giving it to anyone!'

3.9 Constituent negation

Whereas Classical Latin had a dedicated marker for constituent negation (*haud*), early Italo-Romance used the same negator as for sentence negation, namely variants of *non*. In some modern Venetan dialects, *non* has evolved to distinguish between sentential and constituent negation, e.g. Paduan [no] and [ɔ] respectively ([no] is also used before quantifiers) (Benincà, p.c.). The medieval *non* forms show similar pragmatic reinforcement with minimizers as in their sentential use:

(97) Piedmontese
e vivre si cum raisonavoil, **no mia** cumma bestie
and live as like rational not MICA like animals
'and live as rational beings, not like animals' (*Sermoni subalpini*, p. 265, l. 30, 13th c.)

(98) Lombard (Milanese)
ma **no minga** da tuti
but not MICA by all
'but not by all' (*Elucidario*, p. 177, ll. 18–19, 14th c.)

(99) Sicilian (Messina)
E acustandu l' unu a l' autru, **nonià** per longy passi, ma multu adaiu.
and approaching the one to the other, not now with long steps but very slowly
'And, approaching each other, not with long steps, but very slowly.' (*Istoria di Eneas*, p. 204, ll. 23–5, 14th c.)

These structures are similarly affected by Jespersen's cycle, with northern Neg2 and Neg3 varieties eventually losing the first element; for instance, by the end of the 17th century nasal-initial *nen* and *no pa* are attested in Piedmont (*Il Conte Pioletto*, pp. 37 and 55), while in Milanese simple *minga* is attested at the end of the 18th century (Vai 1995: 163).

3.10 Conclusion

The history of negation in the Italo-Romance varieties presents a rich tapestry of variation that graphically illustrates the cyclical reinforcement of both form and meaning, the marked nature of negation encouraging its constant renewal in particular pragmatic contexts as a counterbalance to formal erosion and semantic bleaching in the wake of grammaticalization. Of note is the interaction between dialect type and negative structures, with the northern Gallo-Italian dialects showing the most advanced structural variation. Not only are all the various stages of the classic Jespersen cycle of predicate negation reflected in the geographical distribution of modern Italo-Romance negative expressions (with some dialects presenting several stages contemporaneously), but negative concord as well as negative imperatives (or prohibitives) also show cyclical patterns of development and renewal, which in turn produce synchronic dialectal variation. Other sentence types appear to show non-negative use of negative structures, so-called 'expletive' negation, but recent studies show that the original negative value still plays a significant semantic role in the creation of scalar implicatures.

Acknowledgements

This chapter draws on many years of research and fieldwork in Italy during which I have benefited from discussions with scholars too numerous to thank individually. I am, however, particularly grateful to my colleagues at the AHRC-funded (2000–4) SAVI (Sintassi degli antichi volgari d'Italia) project on the early medieval vernaculars of Italy, to all my dialect informants, and to Paola Benincà and Johan van der Auwera for stimulating comments on earlier versions of the chapter. I also thank the editors for the invitation to contribute to this volume and for their constructive advice and suggestions.

Primary sources

All bibliographical references for the examples from the early vernaculars relate to editions found in the invaluable online database, *Opera del Vocabolario italiano* (OVI), Chicago and Florence, Accademia della Crusca, <http://www.lib.uchicago.edu/efts/ARTFL/projects/OVI/>, apart from the following:

Bottasso, Enzo (ed.). 1953. *Giovan Giorgio Alione, L'opera piacevole.* Bologna: Libreria Antiquaria Palmaverde.

Casula, Francesco C. (ed.). 1995. *La "Carta de logu" del Regno di Arborèa.* Sassari: Delfino.

De Filippo, Eduardo. 1973. *I capolavori di Eduardo*, vol. 1. Turin: Einaudi.

Pasolini, A. 1982. *A Cattainin de nisseue.* Genoa: Nuova Editrice Genovese.

Pauphilet, Albert (ed.). 1949. *La Queste del Saint Graal.* Paris: Champion.

Tana, C. G. 1784. *Il Conte Pioletto, commedia piemontese.* Turin: Briolo (anastatic reprint, Turin: Viglongo, 1965).

4

Negation in the history of English

RICHARD INGHAM

4.1 Preliminaries

4.1.1 Introduction

Over the 1200 years or so of its recorded history, English has been a language in which the syntactic expression of negation appears remarkably unstable. Jespersen's (1917) cycle of negation, which modelled changes in the syntax of negation in various European languages, provides five stages for English, whereas the other languages that have undergone cyclical change in negation typically needed no more than three. Sentential negation, where proclitic *ne* gave way to postverbal *not*, and then to enclitic negation in the form *-n't*, was entirely refashioned. Changes have also occurred with indefinites under the scope of negation, leaving the language with two systems in this context, the *any/ever* series and the *no/never* series. The overall scale of these developments well exceeds those that took place in Slavic, Romance, and other Germanic languages, even though some of them were also affected by Jespersen's cycle.

This chapter surveys the main changes in the forms and distribution of items that have contributed to the expression of negation as attested by the English textual record. It is organized as follows: first the principal features of the grammar of negation in Modern English are outlined, so as to fix the current point of arrival of the changes to be described. Thereafter we consider the forms and patterns of English negation across the centuries of its recorded history, and demonstrate the principal changes that have occurred. Finally, we consider what factors may account for the changes that took place. The structural analysis is framed in terms of a theoretical framework initiated by Klima (1964), and developed by Pollock (1989), who proposed that a sentence negator occupies a NegP constituent, rather than appearing as a simple adverb. The role of NegP is central to a structural analysis, as is also the question of what sort of elements may introduce negation to a clause, that is, 'expressing' negation in the sense of Ladusaw (1992), making it essential to consider the semantic properties of the elements involved, not just their syntactic arrangement.

Until the late 1980s, as pointed out by Tottie (1991), the grammar of negation in the history of English had not formed the object of any published study, and this remained the case in the following decade (see Iyeiri 2005, introduction). However, over the years, and increasingly in recent decades, a large number of journal articles and book chapters have appeared, to which we refer as appropriate. Mazzon (2004) offers a philologically oriented overview of negation in the history of English which can be consulted on relevant issues, though its approach is not always easy to relate to the concerns of linguistic investigation as represented in the present volume.

4.1.2 *Textual and historical background*

A brief outline is first presented here of the conventionally understood periods of the history of English and of the type of sources used in describing it. The Old English period is taken to have extended until the mid 12th century, the Middle English period from then until about 1500, after which the next 250 years form Early Modern English. The syntax of negation in English did not evolve in step with these conventionally recognized periods: insofar as it has any linguistic basis, the conventional periodization of the history of English uses the evolution of inflectional morphology (Lass 2000), which did not affect negation as such. There was thus little difference between the syntax of negation in Old English prose and in early Middle English prose, but between the Early and Late Middle English periods sharp differences appear. In addition, the Old English period presents traces of developments that took place during the Anglo-Saxon era (van Kemenade 2000, Ingham 2007). Successive stages in the evolution of negation will therefore be treated in their own terms, rather than generalizations in terms of, for instance, 'negation in Middle English'.

Data sources become more plentiful and diverse in genres as time proceeded, by and large, though there are noticeable gaps in the textual record between the 12th and 14th centuries, when Latin and then French were extensively used as languages of record. From the 15th century onwards, however, textual deficiencies place no serious limits on our understanding of how negation evolved, at least in the written medium. Old English sources are less diverse than those for the Early Modern period, but they are without doubt the most plentiful of any early medieval European vernacular language. They fall into two broad categories, poetry and non-fictional prose, and negation is realized rather differently in these two types of data source. Few Old English manuscripts predate the 10th century, and the majority are from the years around AD 1000. They include most of the non-fictional prose, which is thought to have been composed in the later 10th and early 11th centuries, as well as most of the verse sources, even though the dates of composition of the latter are in many cases believed to have been much earlier than most of the prose. A fair number of works from the later period, prose as well as verse, were originally created in other dialects and are considered to have been rewritten by West Saxon scribes. Thus it is not very easy to get a sequential picture of changes from early to late Old English.[1] The Middle English period has the opposite problem: there are such great differences between early Middle English of the 13th century and late Middle English of the 15th century that we are really looking at two different states of language. In addition, as already mentioned, the Early Middle English period suffers from source limitations: prose works of the 13th century are very restricted in terms of genre (almost entirely religious texts) and dialect (Southeast and West Midlands). The object of study usually assumed in diachronic linguistics, change in the vernacular, is in any case only imperfectly reflected in written texts. When these offer only limited resources to investigators, the task is made even more challenging, and in what follows the limitations of the textual record will need to be borne in mind. In short, there are challenges which confront any piece of research that attempts to deal with diachronic changes in how English syntax developed, but the amount and timespan of available materials is not unsatisfactory when compared with the situation obtaining in the history of other languages.

4.1.3 *The elements of negation in contemporary English*

It is useful to have a brief summary of the semantics of negation as a reference point for the syntactic analysis: we will provide this in terms of the syntax of contemporary English, the (no doubt provisional) point of arrival of the changes to be discussed in this chapter. To say that a clause is negative is to say that that the proposition that it denotes is assigned a negative truth value, symbolized in formal logic as $\neg p$. Most of the time in this chapter we will be concerned with forms and patterns that in the history of English expressed such a meaning, as in the Modern English sentence:

(1) John did not see Mary this morning.

This negates the proposition that 'John saw Mary this morning'. In context, the user of (1) may wish to communicate that there was an event of John's seeing Mary, but it occurred on another occasion than this morning, or that there was an event of John's seeing someone this morning, but it involved another person. Such uses, which would typically deploy special prosody when spoken, could be considered to partially negate (1). However, as far as syntax is concerned it is the whole proposition that is negated, whereas in (2) we have negation of a constituent, not of a clause:

(2) John saw Mary not this morning but last night.

[1] A small number of texts, such as the Anglo-Saxon Chronicle, do allow this perspective to be taken (cf. Bean 1983).

The negator *not* is the same in both cases, except that in the clausal case it can be contracted to *-n't*. In earlier stages of the language, as we shall see, the two types of negation were distinct.

Negative meaning is also introduced into the clause through the use of what are often taken to be negative quantifying expressions, such as *no one*, and the correlative coordinators *neither... nor*, for instance:

(3) a. No one saw Mary this morning.
  b. Neither John nor Bill saw Mary this morning.

In (3a), the set of individuals said to have 'seen Mary this morning' is empty, while, in (3b), the propositions that 'John saw Mary this morning' and that 'Bill saw Mary this morning' are each denied.

In earlier stages of English, as we shall see, words equivalent to *no one* or *never* co-occurred under a single negation interpretation. To label them as negative quantifiers is problematic (Haegeman 1995), since, if each had been inherently negative, co-occurrence of such items in a clause should have produced double rather than single negation, whereas, in fact, the literal equivalent of *No one never saw Mary* asserted a single negative proposition. To avoid prejudging the issue of their exact status they will be referred to as 'n-items' in the following discussion.

We also adopt the approach taken by Haegeman (1995), following Klima (1964), that a sentence that is syntactically negative can take a positive tag question, for instance:

(4) a. John didn't see Mary this morning, did he?
  b. No one saw Mary this morning, did they?
  c. Speaker A: John didn't see Mary this morning.
   Speaker B: Nor/Neither did Bill, did he?

In contrast, morphological negation, as in *unimportant*, *inessential*, *dislike*, etc. although semantically negative, does not involve negation in the structure of the clause, as the corresponding tag question has to be negative, for instance:

(5) John dislikes Mary, doesn't he?

Anaphoric negation, in which a proposition is denied by the use of the particle *no*, is also independent of clause syntax, as it often stands alone and can be followed by a negative question tag:

(6) Speaker A: Did John see Mary this morning?
  Speaker B: No(, he didn't).

In this chapter, the focus will be on items that have been involved in changes in clause structure in the expression of negation over the historical duration of English. These will be clause negators, indefinite items, and coordinators, featuring the forms that through the various stages of development of negation in English have corresponded to those in (1) and (3).

4.2 The development of clause negation

4.2.1 Jespersen's cycle: from ne *to* not, *to* -n't

We first consider the various ways in which clause negation has been expressed over the course of the history of English. This topic formed the basis of Jespersen's (1917) account of cyclical negation in English and other languages, and has been extensively treated by Beukema and Tomić (1996), Iyeiri (2001), etc. The basic facts are straightforward: a sentence could be, and normally was, negated by the presence of the element *ne* until about the 14th century. From that time onwards negation became increasingly expressed by *not* (in a variety of spellings which for simplicity will be represented below using the modern form), and later by its reduced form *-n't*. In the medieval period from the 13th to the 14th centuries, preverbal *ne* and postverbal *not* very commonly co-occurred.

Van Kemenade (2000) argued that there is evidence for a very early stage of negation in Old English in the epic poem *Beowulf*, which contains a good number of clauses negated by an initial adverb *no*, with SV order:[2]

(7) No ic me on herewæsmun hnagran talige… þonne Grendel.
NEG I me in war-strength inferior count than Grendel
'I do not count myself less in war-strength… than Grendel.' (*Beowulf*, 675)

No substantial OE prose texts predate the late ninth century, by which time clausal negation is usually in the form of *ne*, the Old English counterpart of clause negators in other Germanic languages, for instance OHG *ni*. Van Kemenade saw *no* negation in *Beowulf* as the forerunner of *ne* negation in classical Old English. However, as shown by Ohkado (2005), her analysis suggests a stage when the earlier and later structures were combined in the form of *no* directly preceding the finite verb, followed by inversion. Without such a stage, according to Ohkado, the required cliticization onto the verb could not have taken place. But this stage, with inversion after initial *no*, is not attested in *Beowulf*, or indeed anywhere else in OE. It seems more likely that *ne* is a reduced version of the negator *ni* found in other early Germanic languages, and that initial *no* negation was a blind alley in the development of negation in English. *Ne* always immediately preceded the main verb, regardless of the latter's syntactic position, and should doubtless be seen as proclitic on it.

[2] Old and Early Middle English examples have been glossed word for word and translated; Late Middle English examples, where glossing is normally superfluous, have been translated.

The origin of *not* is taken to lie in the use of the Old English *nawiht* (lit. 'not a wight'), as a reinforcing expression, though evidence for this in Old English is very slim; its normal use was in argument position, as an expression meaning 'nothing' (Mazzon 2004). As shown by van Kemenade (1999), Old English had a fairly common reinforcing negator in the shape of the adverb *na*, which shows signs of incipient grammaticalization in so far as it generally occupied a regular syntactic linear position immediately following the finite verb, and could assume sentential scope rather than negating a phrasal constituent. Other reinforcers were *nateshwon*, and *nanra þinga* (both meaning 'in no way'), which did not survive productively into the Middle English period. In the major 13th-century prose works, *not* is well established as a sentence negator, while in the later 14th century it becomes near-obligatory. Ingham (2005a) showed that even in later-14th-century texts purposely selected for a high level of *ne* retention, *not* was supplied 95% of the time by this point.

Not occurred rarely in Old English as a clausal negator, and when it did, its syntactic position was not fixed. It could occur separated from the finite verb:

(8) a. Næs he æðelboren, ne him naht to þam cynecynne ne gebyrode
NEG.was he noble.born nor him NEG to the royal.race NEG belonged
'He was not of noble birth nor did he belong to the royal race.' (*Ælfr. Hom.* I 80, 33)

b. Heo þonne þæs bearnes noht lata ne wæs
she then the child.GEN NEG late NEG was
'She then was not late in child-bearing.' (NB: Previous context suggests she was late.) (*BHom.* 163, 8)

In such cases, its link to a constituent is often quite strong: it precedes a constituent which it can be seen as negating. Alternatively, it may be interpreted as taking scope over the whole clause (e.g. in (8b), 'it was not the case that she was late') and it may then be that this constructional ambiguity is why it began to assume a role as sentence negator. A transitional context of this type, according to grammaticalization theorists such as Heine and Kuteva, is a key step in the process. In any case, the move towards *not* becoming a sentential negator was no more than embryonic in Old English texts and it took centuries to establish itself as such, at least in the written record. In Old English it was of negligible importance compared with *na*. The other reinforcers also shared the variable positioning of Old English *not*.

By the 13th century, the position had changed noticeably. Frequency counts of *ne* and *not* in Early Middle English (henceforth EME) prose works reported below show very clearly that the process leading to the grammaticalization of *not* is well under way by this time. Instances were counted of clause negation, excluding all cases where

TABLE 4.1 **Frequencies of *ne* and *ne . . . not* in early-13th-century English prose texts**

	ne		*ne . . . not*		Total
	n	%	n	%	
Southeast	129	54	108	46	237
West Midlands	132	49	138	51	270
Total	261	51	246	49	507

there were already other negative expressions, notably indefinites, reinforcers *nawiht*, *na*, or clausal constituents conjoined by the conjunctions *ne . . . ne*. In Table 4.1, the frequencies of sentential negation patterns thus obtained in nine early-13th-century prose works[3] are shown, divided into the two main dialect areas represented in prose texts of this period.

Of the 507 cases in Table 4.1, *not* was present in almost half. There was no substantial difference between the two dialect areas represented. However, these figures do not show a significant fact, namely that in main clauses *not* occurred two thirds of the time, whereas in subordinate clauses it occurred scarcely more than one third of the time, as shown in Table 4.2.[4]

TABLE 4.2 **Frequencies of *ne* and *ne . . . not* in main and subordinate clauses in Early Modern English prose**

	ne	*ne . . . not*	Total
Main clauses			
West Midlands	38	81	119
Southeast	42	65	107
Total	80 (35.4%)	146 (64.6%)	226
Subordinate clauses			
West Midlands	88	57	145
Southeast	86	43	129
Total	174 (63.5%)	100 (36.5%)	274

[3] These were as follows: Southeast: *Vices and virtues*, *Trinity homilies*; West Midlands: *Ancrene Riwle*, MS T. (Morton, pp. 1–300 only), *Saint Margaret*, *Saint Juliana*, and *Saint Katherine*, *Sawles Warde*, *Hali Meiðhad*, *Lambeth homilies* 7–8, 14–17.

[4] The totals do not quite reach the same overall figure as Table 4.1 since, in Table 4.2, a few clauses introduced by the items *þonne*, *forðon*, etc. which were ambiguous between main and subordinate clauses (Mitchell 1985) were dropped.

On grounds of obligatoriness in negative clauses, it was clearly *ne*, rather than *not*, that was the grammaticalized negative form in the varieties of Early Middle English represented by the texts from this period, where it was never dropped in the presence of *not*. *Not* was essentially optional, and indeed was present only a minority of the time in subordinate clauses. In addition, it could be preposed:

(9) a. Nawt nis hit swa.
not NEG.is it so
'It is not so.' (AR(T) M218 71, 20)

b. Ne nawt nart thu wummon other wummon ilich.
and not NEG.are you woman or woman like
'And you are not a woman or like a woman.' (*Seinte Marherete* 30, 23)

Positional inflexibility is a hallmark of a grammaticalized element in a strict-word-order language such as English, so the grammaticalization of *not* appears not yet to have been complete, on this criterion as well. However, its medial position in the clause with respect to other elements was strictly determined. In clauses containing an auxiliary, as shown by Haeberli and Ingham (2007), it always preceded an object nominal, as in (10a–b), whereas ordinary adverbs, such as *þus*, could follow an object nominal, in (10b):

(10) a. ...þe ne wilen noht here sinnes forleten
who NEG want not their sins renounce
'who do not want to renounce their sins' (PPCME-2, CMTRINIT, 83.1110)

b. ...þt naueð naut þe heorte þus afaitet
who has not the heart thus disciplined
'who has not disciplined his heart in such a way' (PPCME-2, CMANCRIW, II 208.2992)

Not never occurred following direct objects, adverbs, and prepositional phrases in the prose works studied.

By comparison, *nowiht*, which continued to appear in early Middle English as a reinforcing expression, had a more flexible distribution[5] than *not*: it could be separated from the *ne*-plus-finite-verb complex by a prepositional phrase:

(11) Ne ic ne cume to heom nawiht
and I NEG come to them not
'I do not come to them at all.' (*Lamb. Hom.* 31, 18)

In short, it seems that English underwent a process, similar to that in French, in which early in the Middle English period a number of reinforcing negators competed, and that by the 13th century in prose texts the grammaticized form *not* had emerged as the clear winner. While the unreduced form *nawiht* temporarily retained a

[5] Bybee (2010: 110) seeks to derive the clausal negator use of *not* from *na wiht* used in object position.

TABLE 4.3 **Percentage realization of *ne* and of *not* in negative clauses without indefinites in Middle English verse**

	13th century		14th century	
	first half	second half	first half	second half
ne	100	96	78	51
not	48	50	69	75

marginal existence as a reinforcing adverb, *not* went on to establish itself eventually as the sole clause negator, at the expense of *ne*.

Some idea is given of the gradual unfolding of this later process by verse data in Ingham (2005b). Table 4.3 shows percentages of negative-clause contexts (excluding indefinites) in which *ne* and *not* each appeared, across four time periods representing manuscripts compiled in the 13th century, first half and second half; and the 14th century, first half and second half. Cases of *ne*... *not* contributed to both percentages. Verse data, as noted earlier, offer only fairly weak evidence of the timing of syntactic change, but they nevertheless suggest a phase when both *ne* and *not* were present two-thirds of the time or more in clausal-negation contexts. Since *not* was becoming realized the majority of the time even in the presence of *ne*, it was adopting the role of the main negator, leaving *ne* with no more than a pleonastic function. The parallel drawn by Jespersen and others with modern French in this respect thus seems well supported. Rissanen (2000) proposed that the sentential negator *ne* started to decline only when *not* had become almost compulsory, but these verse data covering the period in question show the decline of *ne* starting while *not* was well short of obligatoriness. The role of *not* prior to becoming the grammaticized negative marker seems to have been as an illocutionary force marker, and is discussed below in section 4.2.3.

As *not* established itself as a negative marker, the role of *ne* became threatened. No evidence appears in the early 13th century for the use of unsupported *not* as a clause negator; thereafter, the dearth of prose texts from the mid 13th century until late in the following century complicates matters. Ingham (2005b) found the earliest examples of unsupported *not* in later-13th-century verse manuscripts. Though rare at that point, in the first half of the 14th century such examples become much commoner, and account for nearly half of cases in the second half of the 14th century. This was the point at which prose data again become fairly plentiful, and when Ingham (2005a) found that negation by *not* alone accounted for well over half of prose data, for instance, both examples of clause negation in:

(12) Ghif thei heeren not Moyses and prophetis that spaken by God, thei shall not trowe to dede men.

'If they do not listen to Moses and prophets that spoke from God, they will not believe dead men.' (*Wyclif sermons* I, 3)

4.2.2 *Later developments in clause negation*

The main later developments were the erosion of the postverbal negator *not* to *-n't* and the rise of auxiliary *do* in negative clauses, which took place during the 16th and 17th centuries. Rissanen (2000) proposed that *not* became an enclitic negator in Early Modern English, mentioning mid-16th-century data, but examples can in fact be found in early-16th-century drama:

(13) a. Syr, wyll not ye for my trouth undertake? (*Fulgens & Lucres*, 636) (*c*.1497, printed 1513–19)

b. Am not I a goodly person? (*Hick Scorner*, 157) (*c*.1510)

Here the auxiliary + *not* is positioned before the clause subjects *ye* and *I*: assuming syntactic movement of auxiliaries, the negator must have cliticized to the auxiliary in order to move above the clause subject. It was therefore a weak form, for which the spelling *-n't* would later come into use. Religious dramatic verse before 1500 (miracles, mystery plays, etc.) completely lacks evidence of this type for enclitic *-n't*.[6] It should be noted that the frequency analysis offered by Ellegård (1953) reveals that a very sharp rise took place in the use of *do* in negative interrogatives at this point, in the first quarter of the 16th century: with the loss of verb movement, auxiliary *do* increasingly hosted enclitic *not* negation.

Evidence for the loss of verb movement can be seen in the preceding century (Kroch 1989). The Paston letters show the first cases of *never* + finite verb from about 1450 onwards, as in (14b); cf. the earlier finite verb + *never* order in (14a):

(14) a. He meved never this matter.
'He never brought forward this matter.' (*Paston* D 423, 98 (1426))

b. Ye never received peny.
'You never received a penny.' (*Paston* D 313, 57 (1479))

They also contain an instance of a construction that has been held by Ukaji (1992) to bridge medieval and modern negative syntax, in which the negator *not* precedes the finite verb:

(15) I seyde I cowde not tellyn that I **not** herd.
'I said I could not tell what I didn't hear.' (*Paston* 705, 51 2)

[6] As pointed out by van Gelderen (2009), *not* in late Middle English can be found written without a space after a preceding auxiliary.

As pointed out by Iyeiri (2005), however, this construction is extremely rare outside early modern verse, suggesting that it was stylistically restricted. The encliticization of *not* in progress at this time doubtless made it a rather marked construction since it lacked a host for *not*.

Thus between about the late 13th and the beginning of the 16th century, English appears to have gone through a relatively short-lived stage when its main negator was a free postverbal form. Thereafter a significant change took place as the negator weakened to become an enclitic: in clauses where no auxiliary would otherwise have been present, *do* was increasingly used to support clitic *-n't*, and negation by *not* alone gradually died out. The process was not particularly rapid, and the latter pattern was still not uncommon in the 18th century, at least in formal registers, for instance:

(16) I know not if she will ever more come out of her chamber. (Boswell, *Life of Johnson*, p. 550)

Indeed *not* remains a full form with contracted operator verbs such as *be* (e.g. *I'm not sure*). Before and after the 1300–1500 period, nevertheless, the predominant patterns in clausal negation seem to have involved a clitic form of the negator.

The legacy of developments in the medieval period and after has been to leave English with two negation patterns with contractible auxiliaries:

(17) a. She isn't going./She's not going.
b. He hadn't been available./He'd not been available.
c. We haven't got time./We've not got time.

The pattern exemplified by the second of each pair, according to Britain (2007), is more typical of northern England, though whether this is relatable to the earlier loss of *ne* in northern Middle English than in southern dialects is unclear.

4.2.3 *The grammaticalization of* not

The replacement of *ne* by *not* has generally been seen as a classic instance of grammaticalization, where what had been a reinforcing adverb was semantically bleached and became the grammatical element expressing negative polarity. There is little question that *not* arose as a reinforcing element, but the reasons why *ne* needed reinforcing have been debated. One view (Kiparsky and Condoravdi 2006) is that there is a tendency for 'emphatic' negation to replace original 'plain' negation, as speakers seek ways of rendering their meaning more expressively. Jespersen's suggestion was that *ne* was eroded because of its phonetic weakness. *Ne* is known to have contracted to *n-* before vowels (see e.g. Levin 1958), for instance, *net* (*ne et*) (*ate*), *nam* (*ne am*), *nis* (*ne is*), and before non-obstruent consonants, especially /w/ and /h/, for instance, *nist* (*ne wist*), *nabbe* (*ne habbe*), so the vowel was unquestionably vulnerable to erosion.

TABLE 4.4 **Percentage realization of *ne* in Middle English verse 1200 to 1400**

	13th century		14th century	
	first half	second half	first half	second half
with *not*	100	96	76	25
with an n-item	99	90	64	28

This view of how *ne* negation declined receives further support from a reanalysis we have carried out of the verse data in Ingham (2005b). If *ne* had weakened only thanks to competition for expressivity with the reinforcing negator *not*, *ne* would not be expected to have declined simultaneously in contexts where it accompanied n-item indefinites, in the absence of *not*,[7] yet this is in fact what verse data show in the period 1200–1400. As can be seen in Table 4.4, the percentage of omission of *ne* was very similar across the two types of contexts in all four sub-periods. These figures lend support to Jespersen's contention that *ne* declined because it was phonetically weak, rather than that it was simply pushed out by competition from *not*. The decline of *ne* in contexts with an indefinite clearly cannot be attributed to competition from *not*.

The other main factor involved in the rise of *not* is the role it is presumed to have played in reinforcing the expression of negation pragmatically. Reinforcers commit the speaker, or writer of a written text, to the falsity of a proposition more strongly than ordinary negation does. This relates to the notion of illocutionary force (Searle 1975): in the case of an assertion, the speaker is committed to the truth of the proposition, and in the case of a negative assertion to its falsehood. In a subordinate clause, however, the speaker is very often not committed to its truth status, cf. the anomaly of uttering (18a) by comparison with (18b).

(18) a. #Cabbages don't grow in damp soil, but that's absurd.
b. John said that cabbages don't grow in damp soil, but that's absurd.

If *not* had been fully grammaticalized as a negator in early Middle English, we would expect it to appear in subordinate clauses as frequently as in main clauses. However, the breakdown into main and subordinate clause contexts in the chief prose works of the early 13th century conducted above (Table 4.2) revealed that main clauses, which tend to convey the speaker/writer's illocutionary act, were far more likely than subordinate clauses to contain the secondary negator *not*. In LME, this disparity had disappeared (Ingham 2005a) and *not* was almost obligatory in subordinate clauses too.

[7] In such contexts, according to Jack (1978a), *not* was normally absent, at least until late Middle English (Ingham 2000), and indeed no uses of *not* accompanying an n-item were found in the verse data surveyed.

The link between illocutionary force and the occurrence of *not* was explored further by consideration of the pragmatics of negative clauses. An examination of early-13th-century prose was carried out with a view to investigating whether clear differences in the supply of *not* would appear in clause types differing in their illocutionary force. The first type consisted of negative imperatives, which have a strong negative illocutionary force as prohibitions, for instance:

(19) Ne forlet tu me nawt.
NEG leave you me NEG
'Do not abandon me.' (*St. Marg.* 56, 28)

Next were investigated clauses whose negative polarity was contrasted with an affirmative polarity clause, for instance:

(20) a. ȝef þu wult... ȝef þu nult no
if you wish if you NEG.wish NEG
'If you will ...; if you will not...' (*St. Jul.*, p. 10, ll. 1–3)

b. Þeh he beo all ichefte lauerd... notheles he nis
though he be PRES.SUBJUNC all created.GEN lord nevertheless he NEG.is
nawiht alle monne lauerd
NEG all men GEN lord
'Though he is the lord of all creation... still he is not the lord of all men.' (*Lamb. Hom.* 77, 5)

c. Ðanne ðu ðus dest... Ghif ðu ðis ne diest
when you thus do if you this NEG do
'When you do thus... If you do not do thus...' (*Vic. Virt.*, p. 41, l. 1)

To be counted in this category, the clause featuring the reinforcing negator had to have the same verb lexeme as the affirmative clause with which it contrasted.

Finally, negative rhetorical questions were examined. These are pragmatically assertive, for instance:

(21) Nabbe ȝe teð ba & tunge to sturien ?
NEG.have you teeth both and tongue to move
'Don't either of you have teeth and a tongue in your head?' (*St. Kath.* 1266–7)

TABLE 4.5 **Presence of a reinforcing negator in three contexts, Early Middle English prose**

	+ *not/nawiht*	%	− *not/nawiht*	%	Total
imperative	40	95	2	5	42
contrastive	25	81	6	19	31
rhetorical question	0	0	65	100	65

A total of 138 clauses belonging to these three types was examined in the same early-13th-century prose texts surveyed for the purpose of the analysis in Table 4.2. The use of reinforcing negation (counted not only in terms of the presence of *not* but also of *nawiht*) was distributed as in Table 4.5.

The distribution of reinforcing negation was thus far from random, but depended on the discourse-pragmatic status of the context. Where the illocutionary force was prohibitive or where the discourse contrasted a negative as against an affirmative proposition, the presence of a reinforcing negator was strongly favoured. Where the illocutionary force was assertive, as in the case of a rhetorical question, *not* was always absent.

The above results make it quite clear that in the first half of the 13th century *not* served to express pragmatic purposes, rather than the grammatical function of expressing the polarity of the clause, a role still retained by *ne*. By the later 14th century, however, a rhetorical question typically contains *not*:

(22) What! Nis þis nouȝth Josepes son, þe carpenter, & Maries?
'What! Isn't this the son of Joseph the carpenter and Mary?' (*Pepysian Gospel Harmony* 26, 5)

By this stage, a speaker no longer, it seems, had the option of omitting *not* from a negative clause if the illocutionary force of the clause was not negative.

The development of imperative clauses suggests that the relation of *not* to illocutionary force developed early in the transition to Middle English, since in Old English they were commonly negated by *ne* alone, though they could be reinforced by *na*, for instance:

(23) a. Ne fare ge on hædenra manna wege
NEG travel you on heathen men.GEN way
'Do not travel the way of heathen men.' (*Ælfr. Hom.* 68.47)

b. Ne gang þu na on godes hus, þu hafast besmitene handa
NEG go you NEG in God.GEN house you have dirty hands
'Do not go in to God's house, you have dirty hands.' (Helsinki Corpus, comartyr, R 282)

Negative imperatives appear most commonly without reinforcement in Early Middle English texts in the Helsinki corpus.

4.2.4 Constituent negation

Phrasal constituents could never be negated by *ne*. In Old English, *na* and *nalles* were used as constituent negators. By the 13th century, *not* had already assumed this function:

(24) Ant meiden stont þurh heh lif i þe tur of ierusalem; Nawt of lah on eorðe, ah of þe hehe in heouene.
and maiden stood through high life in the tower of Jerusalem NEG of

low on earth but of the high in heaven
'A maiden stands, through her exalted life, in the tower of Jerusalem, not from a low place on earth but from the high tower of heaven.' (*Hali Meiðhad* 5)

Since *not* appears modifying a constituent in an affirmative sentence, it is clearly able to introduce a negative meaning independently of the structure of the clause. It appears, then, that *not* was already at this stage an independent negative element, not an n-item.

4.3 Changes in the syntax of negative clauses

4.3.1 A structural analysis

Since *not* originated as an adverbial element, it is natural to suppose that its structural analysis underwent a shift from being an adverb adjoined to the verb phrase, to being a functional element in a higher position. This kind of syntactic evolution is typical of the grammaticalization process (Roberts and Roussou 2003), and van Gelderen (2008) has indeed suggested an analysis along such lines. Following Pollock (1989) and much subsequent work, she argued that, once it had become the sentence negator, *not* occupied SpecNegP, in a structure such as the following, where TP stands for Tense Phrase, vP for verb Phrase and NegP for the phrase accommodating *not*:

(25) [$_{TP}$ … *ne*-V [$_{NegP}$ *not* ~~ne-V~~ [$_{vP}$ ~~ne-V~~ …]]]

(NB: It is assumed that the finite verb, with proclitic *ne*, originates in the vP and moves successively to Neg^0 and T^0.) Prior to this development, *not*, *nawiht* and other reinforcers are supposed to have been adverbs left-adjoined to vP. Haeberli and Ingham (2007) showed that in 13th-century data *not* indeed differs from ordinary adverbs in its distribution with respect to object nominals, which it always preceded:

(26) Þu … Keiser, nauest nawt þis strif rihtwisliche idealet
you emperor NEG.have NEG this contest justly assigned
'You … Emperor, have not fairly apportioned this contest.' (*St. Kath.* 36, 750)

However, this was not the case with ordinary adverbs, which could either precede or follow objects in OV order. This fact seemed to these authors to support the claim that *not* was not an adverb but occupied a position in syntactic structure dedicated to negation.

The question then arises of whether a NegP constituent should be posited for Old English. The structure of negation in languages such as the Slavic languages has been analysed by Zeijlstra (2004) using the notion of a null negative operator which occupies SpecNegP; such languages have a preverbal clitic negator as the main negative element. This analysis would work well in principle for Old English, where negation was expressed by *ne*, and a secondary negator had not yet grammaticalized. The relevant structure would then be:

(27) [$_{\text{TP}}$ *ne*-V [$_{\text{NegP}}$ NegOP ~~*ne*-V~~ [$_{\text{vP}}$ ~~*ne*-V~~ ...]]]

The process of grammaticalization therefore can be seen as involving the optional availability of *not* in the place of a negative operator in SpecNegP. As time goes on, it becomes the dominant and then the only choice, leading to the loss of the null negative operator from the grammar. This is a significant step, because it is the presence of a negative operator that licenses multiple negation in the approach taken by Zeijlstra. As we shall see in section 4.4, negative concord weakens in LME, as would be expected if the loss of a negative operator in SpecNegP occurred at this time.

4.3.2 *Neg V1 order*

Old English negative root clauses very commonly exhibit VS syntax, as in example (28).

(28) Ne gesomniu ic gesomnunge heara
NEG assemble I meetings of.them
'I shall not assemble their meetings.' (Vesp. Psalter 15, 4)

Note that a subject pronoun did not normally undergo inversion in Old English V2 contexts (Haeberli 2002), so we are dealing with a property of negative clauses distinct from ordinary V2. Accordingly, this construction will be referred to here as NegV1.

NegV1 has been relatively little studied in the history of English, though see Andrew (1940) and van Kemenade (1997). It was extremely common in Old English and continued as a frequently taken option well into the 13th century. Table 4.6, using the same early Middle English sources as Table 4.1, shows that, out of 130 possible contexts for NegV1, it appeared in 72, or 55% of the time.

Not was somewhat more common in uninverted (78%) than in inverted clauses (61%), but, even in the latter, *not* is supplied nearly two-thirds of the time, so the fact of inversion seems independent of the presence or absence of *not*.

TABLE 4.6 **Inversion with and without *not* in non-conjoined negative main clauses with overt subjects, early-13th-century prose texts**

	inversion			non-inversion			overall
	ne	*ne... not*	total	*ne*	*ne... not*	total	total
Southeast	7	10	17	4	9	13	30
West Midlands	21	34	55	9	36	45	100
Total	28	44	72	13	45	58	130

In Late Middle English texts from about 1370 on, NegV1 was found to be extinct by Ingham (2005a), who also noted that the decline of VS in negated main clauses took place roughly in the same time period as the rise of secondary negator *not*. However, this could be coincidental: the decisive factor might not have been the rise of *not*, but the decline of *ne*. Arguably, the presence of *ne* could have been required for inversion to take place. Accordingly, Ingham (2005b) looked at the role of both negative elements in Middle English verse in order to see whether NegV1 declined in parallel with the demise of *ne*. A sharp drop was found in the early 14th century, when *ne* was still well preserved but *not* was strongly in the ascendant, suggesting that the loss of NegV1 was not a matter of the loss of *ne* as such. Rather, it followed from the switch to *not* as the element responsible for introducing negation to a negative clause, similar to the development of French negation towards making *pas* the dominant negator (Rowlett 1998).[8] Haeberli and Ingham (2007) therefore saw NegV1 as an archaic feature of Early Middle English, retained from Old English, in which *ne* introduced negation to a negative clause. Its extinction in 14th-century English has something of the character of 'catastrophic' grammar change argued for by Lightfoot (1999): there was a decisive shift towards adopting *not* as the obligatory clause negator, and the elimination of the now archaic grammar in which it was a reinforcing element. Since the data in Table 4.6 show that there was no link between the presence of *not* and NegV1, it seems it was not the mere presence of *not* that mattered, but rather its grammatical status: once this changed to being the grammaticized exponent of clause negation in the 13th–14th centuries, NegV1 quickly declined and disappeared.

4.3.3 *The syntax of negated subjects and objects in Old and Middle English*

Old English was characterized by very considerable word order flexibility, which in Middle English was considerably reduced, and, by the modern period, English had become a fairly strict SVO language in terms of permitted surface orders. Declarative main clause XVS order was very common in the earlier periods, as was OV order. By the 15th century, these patterns were dying out, with the exception of cases where a subject or object was negated. Examples are still frequently seen of a negated subject embraciated between a finite auxiliary and the lexical verb:[9]

(29) Withowten hem þer may no man resceyve is saviour.
'Without them no man may receive his saviour.' (*MES* 282, 7)

[8] A count was made of VS order and the use of *not* in contexts for inversion in the 13th-century Middle English verse data analysed by Ingham (2005b). Here it was found that neither inversion nor SV order favoured the use of *not*.

[9] Other than in verb-second clauses, post-finite negated subjects were restricted to contexts with impersonal *there* (cf. Ingham 2003).

This type of construction might be thought to represent a kind of transitive expletive construction. But the equivalent of such word orders with ordinary (non-negative and non-quantified) subjects was not found by Ingham (2000, 2003). Likewise, negated objects can also commonly be found in OV order in auxiliated clauses, despite the general loss of OV order by this time:

(30) I may no leysour haue.
'I may have no leisure.' (*Paston* D 182, 48 (1465))

It was proposed by Ingham (2000) that these patterns share a common structural feature, placement of the negated constituent in SpecNegP. They both decline and disappear in the early 16th century (Moerenhout and van der Wurff 2005), long after generalized VS in declarative XVS structures and OV order had been eliminated from the language. Examples of postfinite negated subjects with unambiguously impersonal *there* appear in late-13th- and early-14th-century verse in embedded contexts where *there* occupied SpecTP, for instance:

(31) . . . þat þer may no drop passi fro Lazar and come þe to
'. . . that no drop may pass from Lazarus and come to you.' (*SEMP* 130, 1318)

Here the structural position of the associated subject *no drop* can therefore be analysed as SpecNegP. The alternative would be to suppose that it stood in SpecvP, and that we are dealing in (29) and (31) with a type of transitive expletive structure. But this predicts that ordinary subjects could also be found in the string *þere*–v–Subj–V, but Ingham (2002) found no evidence of ordinary subjects in this construction in Middle English verse, or in late-14th-century prose or in 15th-century private correspondence. Ordinary subjects always preceded the finite verb in embedded clauses. Post-finite negated subjects, however, were productive in all three sets of data in both matrix and embedded clauses. There is plentiful empirical support, therefore, for seeing negated constituents standing between the finite verb and a non-finite main verb, as in examples (29)–(31), as instances of movement to NegP, rather than of a generalized transitive expletive construction.

A question examined in Ingham (2007) is whether the same structural analysis can be applied to Old and Early Middle English, that is, whether negated subjects and objects moved to SpecNegP, whereas ordinary constituents in inversion and OV orders occupied some other position. If negated object constituents were found to stand in every linear position that ordinary subjects and objects could, the case for attributing a special structural analysis to them, in terms of NegP, would clearly be hard to make. But this is almost exactly what is found. In simple V-final clauses in OE, negated objects could directly precede V:

(32) . . . þeh hie ðær nan licgende feoh ne metten
though they there any lying wealth NEG found
'. . . though they found no wealth lying there.' (*Orosius* 116, 31–2)

They could be separated from V by an intervening sentence constituent:

(33) ...þæt hi nanes ætes on ðære fare ne onbirigdon
that they any food on the way NEG taste.PAST
'...that they tasted no food on the way.' (*Ælfr. Hom.* II 168)

In auxiliated clauses with vV order, the same is true, the negated object either following the tensed verb directly, or being separated from it by an intervening constituent:

(34) a. ...þæt hi ne mihton nænige monn þær ghefon
that they NEG may.PAST any man there give
'...that they could not give to any man there.' (*Wærferth* 141, 4)

b. ...þæt hie ne mehton from him nænne flan asceotan
that they NEG may.PAST from him any arrows shoot
'...that they could not shoot out any arrows from it.' (*Orosius* 294, 24)

Negative objects as well as ordinary objects could scramble over a vP adverb:

(35) a. Ac we ne magon nænne sædere godes lare rihtlicor undergytan
but we NEG may any sower God's lore rightly.COMPAR understand
'But we cannot more rightly understand any sower of God's precepts.' (*Ælfr. Hom.* 90, 8–9)

b. ...mid þam we magon ure mod gastlice gegladian
with which we may our mind spiritually gladden
'...with which we may spiritually gladden our minds.' (*Ælfr. Hom.* II 62, 35)

Ordinary objects may likewise remain immediately adjacent to the main verb (as also in (34b)):

(36) Swa mon ðonne sceal fullfremedlice **Godes fiend** hatigean
so one then ought perfectly God's enemy hate
'We ought then to hate God's enemies so perfectly.' (*Cura Past.* 352, 7)

In auxiliated clauses, Pintzuk (2005) noted a constraint in Old English against *V v NegObj in an apparently head-final TP structure. That is, negated objects did not move rightwards, whereas ordinary objects could. However, this observation does not show where a negated object is located, and indeed Pintzuk did not claim that it indicated a special position for negated objects.

In fact, another Old English construction type for negated objects in auxiliated clauses can be found which is not mentioned by Pintzuk (2005), that is, NegObj v V, as in:

(37) ...þæt mon nane burg ne mehte ieð mid feo geceapian
that one any city NEG may.PAST easily with wealth buy
'...that one could not buy any city with wealth easily' (*Orosius* 228, 20)

Once again, however, ordinary objects can also be found preceding v V:

(38) forðan ðe se metoda drihten ure gecynd hæfde
because the doom lord our nature had
on him sylfum genumen
on him self taken
'because the Lord Creator had taken his nature upon himself' (*Ælfr. Hom.* 380, 6–7)

The situation in Old English was very different, then, from that in LME: as shown by Pintzuk, negated objects enjoyed no special distributional privileges, so no surface distributional case can be made for placing them in SpecNegP. Given the clear parallelism between the clause-medial positions in which ordinary objects and negated objects are found, positing movement of objects containing n-item indefinites to NegP would be redundant in Old English, since movement operations needed for ordinary objects quite adequately take care of the observed surface orders with negated objects as well. Unlike Late Middle English, Old English would have presented no cues that the latter had a different structural position in learnability terms (cf. van der Wurff 1999). Movement of negated XPs to NegP in Old English is also ruled out within Zeijlstra's (2004) framework. Old English was a negative-concord language, so a negative operator obligatorily occupied SpecNegP, which cannot be doubly filled.

In Early Middle English, as in Old English, ordinary objects could still be scrambled to a medial position in clauses containing an auxiliary:

(39) He hefde his ranceun fulleliche ipaighet
he had his ransom fully paid
'He had fully paid his ransom.' (AR II 101.1228)

The same was true of negated objects:

(40) Ne schulde ha nane wunne lihtliche wilnin.
NEG should she no joy easily desire
'She should not easily desire any joy.' (*St. Jul.*, MS Royal 14.1)

Negated objects were found in no position in Early Middle English in which ordinary objects were not. Thus, it is clear that a change took place in the syntax of Late Middle English such that negated objects, as in (30) above, unlike ordinary objects, could undergo clause-internal movement.

4.3.4 Coordinate negation

At issue in the syntax of coordinate clauses is the nature of the coordinator, and whether a clause negator and/or n-items co-occur with it. In Old English, at least by the time of the classical 9th–11th-century texts, the coordinator *ne* could not by itself negate a coordinated clause, which additionally required the normal sentential negative particle *ne*, for instance:

(41) Ealle þas þinc synd fram eagan gewitene, **ne** hit nu nan
all these things are from eyes witnessed, nor it now no
þinc þære sawle **ne** helpeð.
thing the.DAT soul NEG helps.
'All these things are eyewitnessed, and it will not now help the soul at all.'
(Helsinki Corpus, COEPIHOM, p. 165)

The second conjunct could also, as we see from (41), contain n-items such as *nan* in addition to *ne*. As this example shows, it was possible to conjoin a negative clause with a preceding affirmative one (Ingham 2009). This pattern continued into Early Middle English. In Late Middle English, the sentential negator *ne* was being lost, replaced by *not*, and coordinator *ne* also declined, in favour of *ner*, later *nor*. Otherwise matters remained as before in that the clause was negated internally, not by the coordinator, for instance:

(42) The best of alle and the fayrest ys cleymyd, ner yt ys not in hys jnventory.
'The best of all and the most beautiful is claimed, and it isn't in this inventory.'
(*Paston* vol. I, 373)

Eventually, in the 16th century, *nor* started to introduce negation by itself, without any negative element in the second conjunct clause:

(43) Sir P C could not bring the same matter to good effect, nor was there any Man so mete to bring it good effect.
'Sir P C could not bring the matter to a good conclusion, nor was anyone able to do so.' (Helsinki Corpus, CETRI1, I,69.C2)

In this context, subject and verb always inverted, whereas clauses with another negative element did not show inversion (Jacobsson 1951). This development appears to have gone hand in hand with the decline of multiple negation in Early Modern English, which we shall consider in section 4.4.

The beginnings of coordinate negation in English are hard to determine, but a line from a probably pre-9th-century poem known as the *Leiden Riddle* may show *ni* as a negative clausal coordinator with no negative element in the body of the clause:

(44) Uundnae me ni biað ueflæ, ni ic uarp hafæ

wound me NEG are woofs nor I warp have
'Woofs are not wound for me, nor do I have a warp.' (Helsinki Corpus, CONORTHU, *Leiden Riddle*)

Whereas the first instance of *ni* appears to be the usual preverbal negative particle, the second stands apart from the verb *hafæ* 'have' and seems to function as a coordinator introducing negation by itself, which was not normally the case in Old English.

In addition, passages in *Beowulf* might suggest that in early Old English a correlative form of coordination was available in the form of *no…no*. Each introduced negation into the clause it preceded:

(45) No he wiht fram me flodyþum feor fleotan meahte,
NEG he at.all from me sea.waves.DAT far float may PAST
hraþor on holme; no ic fram him wolde.
swifter on water NEG I from him wish.PAST
'Not at all far from me was he on the sea-waves, he could float, he was swifter on water, nor did I wish to part from him.' (*Beowulf* 541)

To the extent that such cases are informative,[10] it can be said that coordinate negation in English has also undergone a cyclic process in which a single-negation construction developed towards multiple negation, and then in the modern period single negation was re-established.

4.4 The development of indefinites in the scope of negation

A central issue in the evolution of English negation was discussed by Tottie (1991), who investigated the choice between *not* negation and *no* negation, as she called it, as, for instance, in the Modern English:

(46) a. I did not see any ships in the harbour.
b. I saw no ships in the harbour.

The alternative here is between negating a clause with the sentence negator or with an n-item indefinite, such as *no*. Tottie's research documented the disparity in the modern language between spoken register data, which makes relatively little use of

[10] An additional problem with assessing the informative value of the *Leiden Riddle* for Old English negation is that it is thought to be a translation of a Latin original, and classical Latin avoided multiple negation.

no negation, and written register data, in which it is common. In discussing the diachronic development of these alternatives, she made the point that it was the *no* series of indefinites that greatly predominated in the medieval period. Negation by a clausal negator has of course been present throughout the history of English: it is negation by a *no*-series element standing alone which requires comment. Until the modern period the *no* series most often tended to be accompanied by a clause negator, as in the following West Saxon example:

(47) a. Ne mæg nan man twam hlafordum þeowian.
NEG may any man two masters serve
'No man may serve two masters.' (*WS Gosp.*, Matt. 6: 24)

b. ... þaet hie nænig mon siþþan findan ne meahte
that they any man after find NEG may.PAST
'... that they then could not find anybody' (Helsinki Corpus, COCHROA2, 418)

Pre-modern stages of English are thus generally reckoned to have had negative concord (NC) grammars.

However, Ingham (2006a) brought to light dialectal differences in the expression of negation with indefinites by examining texts known to have non-West Saxon associations, notably in their lexis. In these sources, *ne* is always represented if an n-item indefinite follows the finite verb, for instance:

(48) Nes nefre in his muðe nympðe Crist
NEG.was ever in his mouth except Christ
'He always spoke of Christ.' (*St. Chad* 239)

But the cases where the finite verb stands after an n-item indefinite often lacked *ne*:

(49) ... þæt he nane þinga him andwerdan wolde
that he no things to.him answer would
'... that he would in no wise answer.' (*Wærferth* 122, 16)

West Saxon texts, on the other hand, display symmetrical use of *ne* in clauses containing a negative indefinite, regardless of whether the latter is pre- or postverbal. In the texts with non-West Saxon associations, *ne* omission with preverbal indefinites was over 60%, whereas with postverbal indefinites *ne* was always present, just as in the West Saxon texts. Users of Old English not employing West Saxon would seem, on the basis of these results, to have had a negative-concord variety in which negation could optionally be expressed by a preverbal n-item, without a negative verbal particle, but required n-items standing after the verb to be supported by a negative particle, for instance:

(50) a. ... þæt þær ne mihte nænig hrof on beon on þære cirican

that there NEG may.PAST any roof on be on that church
'...that no roof could be on that church.' (*Martyrol.* 74, 21)

b. ...þæt ic ne meahte nænize þinga aræfnan
that I NEG may.PAST any things tolerate
'...that I could not tolerate in any wise.' (*Wærferth* 89, 26)

This variety manifested the same property as Romance negative-concord varieties, such as modern Italian, in which the sentence negator is present with n-items standing after the finite verb but absent with those standing before the finite verb. This property is referred to as 'non-strict' negative concord by Zeijlstra (2004). The West Saxon variety, on the other hand, belonged to the strict type of NC language, having symmetrical use of *ne* with both pre- and postverbal n-items. Consequently, it appears that NC between *ne* and a negated indefinite was not generally a matter for free variation in Old English: the variation in question should be recognized as having been between West Saxon and other varieties, concerning preverbal n-items.

A third type of dialectal variation in Old English is exemplified in texts from Northumbria, especially the Lindisfarne glosses, which display an early form of *not* negation, that is, the negator *ne* is accompanied by the ancestor of *any*, *ænig*:

(51) Gefea iuer ne nimeð ænig fro iuh
joy your NEG takes any from you
Latin: Gaudium vestrum nemo tollit a vobis.
'No one takes your happiness away from you.' (*Lindisfarne*, John 16: 22)

Ænig is clearly a polarity indefinite, found in all stages and in all dialects of Old English quite generally in affective contexts, not just in negative clauses:

(52) Hwæt segst þu, sceaphyrde? Hæfst þu ænig gedeorf?
what say you shepherd have you any toil
'What do you say, shepherd? Do you work hard?' (*Aelfric's Colloquy* 12)

Ænig appears with *ne* even in preverbal position, cf. (43) above:

(53) Ænig mon ne mæg tuæm hlaferdum hera.
any man NEG may two lords.DAT serve
Latin: Nemo potest duobus dominis servire.
'No one can serve two masters.' (*Lindisfarne*, Matt. 6: 24)

This construction is the only one found with indefinites in negative clauses in verse known to be of an early date, such as the Genesis A poem. It appears, therefore, that all three possibilities with respect to negative concord—strict NC, non-strict NC, and lack of NC—were exemplified in different areas of Anglo-Saxon England.

4.4.1 *Indefinites under negation in Middle English*

Middle English from the late 12th century to the end of the 14th century systematically demonstrated negative concord in prose data contained in the Penn–Helsinki Parsed Corpus of Middle English (Kroch and Taylor 2000). This was established by searching for instances of the non-assertive indefinites *eni, eauere, eahwer,* etc. in clauses containing a NEG parse label, and comparing them with frequencies of their n-item counterparts *nan, neauere, nahwer,* etc. The outcome can be very simply stated: out of dozens of examples, the negated series was used in every case but one. The only example where a non-assertive indefinite was found within a negated clause was:

(54) þing þt ich ne mahte nawt bringe to eni ende.
thing that I NEG may.PAST not bring to any end
'a thing that I might not bring to any end' (Helsinki Corpus, CMSAWLES 173.94)

However, even here *eni* is not found in the other manuscript of this work, which has *bringe to ende* only. Otherwise, early Middle English displayed strict negative concord within a single clause, as exemplified by:

(55) a. þus ne did neauer nan dedlich mon.
thus NEG do.PAST never any mortal man
'Thus never did any mortal man.' (*St Kath.* 1047)

b. þt nan ne seide na wiht
that any NEG said any thing
'so that none of them said anything' (*St Kath.* 1252)

The sentential negator *ne* was as good as never dropped, even when a negated expression preceded the finite verb, for instance:

(56) a ...þat neauer mi sawle ne beo mit sunne isulet
that never my soul NEG be PRES.SUBJUNC with sin soiled
'...that my soul may never be stained with sin.' (*St Marg.* 25)

b Na mon ne mei iuggi wel blod
no man NEG may judge well blood
'No one may judge blood well.' (AR f32b, 9)

Negative concord normally failed to arise between main clause and subordinate clause, for instance:

(57) Nule ich naut that ani seo ow bute leaue
NEG.wish I NEG that any see.PRES.SUBJUNC you except leave
habbe of ouwer special meister

have PRES.SUBJUNC of your special master
'I do not wish anyone to see you unless he has leave of your particular master.' (PPCME-2, CMANCRIW, II.47.438)

Non-assertive polarity items were almost always used in such instances. The PPCME-2 corpus provided eleven such cases as (57). Two more cases are found in a decree of Henry III from the later 13th century (Hall 1920). Of these 13 cases, only one showed an n-word in a subordinate clause.[11] The two exceptional cases we have seen are both of questionable status in that they involve discrepant readings between manuscripts. They thus appear so isolated that symmetric negative concord must be seen as quasi-absolute in Early Middle English.

Negator–indefinite dependencies into a non-finite clause in EME were rare, as there were few infinitival complement clauses before the Late Middle English period. One example is:

(58) Ne we ne beoð iboren for to habbene nane prudu.
nor we NEG are born for to have any pride
'And we are not born to have any pride.' (PPCME-2, CMLAMBX1,7.72)

Likewise in the Old English texts sampled, only the *any* series items appeared possible:

(59) Ne ic naefre git nyste thaet aenig other byrig
nor I never yet NEG.know.PAST that any other town
us waere gehende.
us be PAST.SUBJUNC near
'I never knew before that any other town was near to us.' (*Ælfric Lives* 23, 542, De septem dormientibus)

The situation in early Middle English thus appears to have been a continuation of the West Saxon strict NC grammar in that n-items had to be accompanied by the negative particle *ne*, and NC relations were clause-bound.

4.4.2 Indefinites under negation in Late Middle English

In Late Middle English, *ne* has largely been lost. In theory, *not* could have taken its place and systematically accompanied n-item indefinites, but this did not happen: n-item indefinites became able to introduce negation by themselves, as in Modern

[11] But in this instance there is again an issue as to which manuscript version is used. One version has *eni*, whereas the other simply cancels *eni*.

English. In addition, in most sources until about 1500, negative concord remains very common, though not absolute, between the n-expressions themselves:

(60) a. He should never come into non other woman's bedde.
'He should never have entered another woman's bed.' (*Brut* 319, 8)

b. þese swyn mowe not be i-kept by no manere craft.
'These pigs cannot be kept by any kind of skill.' (Trevisa 361, 3)

Any-series items also started to appear in negative clauses, accompanied by *not*. Ingham (2006b) found that this pattern occurred first in texts with northern associations, in the late 14th century, for instance:

(61) ...þat he be not grevyd ony time be our euyl dedis.
'...that he should not be grieved at any time by our evil deeds.' (*Prose rule of St Benet* 1, 12)

Very soon, however, and particularly in 15th-century legal documents (Rissanen 2000), the use of *any* in negative clauses spread to texts written in the south. It is undoubtedly significant that *ne* was lost earlier in northern texts than in southern ones, suggesting that a reanalysis away from the earlier NC grammar took place once *ne* and the negative operator it identified (Zeijlstra 2004) had disappeared.

For a while, n-items were able to introduce negation themselves, and also to appear in clauses with multiple negation. In the latter context, the n-item was syntactically licensed by a c-commanding negative expression. Given the erosion of *ne*, this analysis would be virtually forced on a language learner hearing data such as (60a) and (60b) lacking *ne*. However, English already had indefinites for use in contexts accompanying a negative element: the *any* series. The effect of the optional reanalysis of n-items as negative quantifiers was to create a set of lexical doublets running alongside the *any*-series indefinites; see Martins (2000) for this proposal, to account for a similar situation in medieval Romance. The situation in LME was thus as represented in Table 4.7.

The co-occurrence of these two systems is well attested by the mid 15th century, when the *any* series was infiltrating negated clauses in private correspondence:

Table 4.7 **Multiple lexical entries for n-items in Late Middle English**

negative quantifiers	n-item indefinites	*any*-series indefinites
none	*none*	*any*
nothing	*nothing*	*anything*
nowhere	*nowhere*	*anywhere*
etc.	etc.	etc.

(62) I herde neuyre syn that tyme **any worde** owt off Norffolk.
'Since then I never heard anything from Norfolk.' (*Paston II* 263, 4)

In 15th-century correspondence, furthermore, a negative-concord relation begins to appear across a main–subordinate-clause boundary, for instance:

(63) He kan not thenk that sche hath **non ryth** to have jt.
'He can't think she has any right to have it.' (*Paston* 128, 76 (1448))

In Middle English up to the 14th century, the *any*-series indefinites, it will be recalled, had been rigorously excluded from clauses containing the negator *ne*. In the 15th century, with the loss of *ne*, the contexts supporting the second and third series became identical. An indefinite c-commanded by a negative element such as *not* could appear in the form of items from either series, contrast (60b) and (61). This applied in intraclausal and in cross-clausal contexts, cf. (62)–(63). Effectively, then, there was no functional reason to differentiate the two series in negative contexts.

However, *not* negation, with the use of the *any*-series indefinites, did not quickly displace *no* negation. In 16th-century correspondence, the *no* series indefinite series was strongly preferred. The loss of negative concord in educated English thus could have aligned English with continental Germanic languages having *no* negation as the favoured option. Why this did not happen is an interesting question. Its dominance in spoken data in contemporary English, as shown by Tottie (1991), may be linked to the rise of the negated auxiliaries, especially high-frequency items such as *don't* and *can't*. In 16th-century English, as negative concord appears to have receded, the increasing use of negated auxiliaries would have favoured increased use of the *any* series.

English negation in the 15th and early 16th centuries was clearly undergoing a stage of flux, which can be compared to the situation in Old English in that negative concord and non-NC grammars coexisted. However, in the Old English period, it seems to have been largely a matter of dialectal variation, whereas in Late Middle English the evidence of correspondence data (Kallel 2005) is that individual writers alternated between the two grammars. Subsequently, negative concord became confined to non-standard English. However, since the earliest traceable stigmatizing pronouncements against multiple negation came in the later 18th century (Nevalainen 2006),[12] the demise of NC in educated writing is hard to attribute to prescriptivism. It is fair to say that, for the West Saxon scribes who regularly 'corrected' originally non-West Saxon texts by inserting *ne* into clauses where it was 'missing' (see the start of section 4.4), the later grammarians' condemnation of multiple negation would have been a very unexpected development indeed.

[12] As shown by Kallel (2005), so-called 'Chancery Standard' scribes, often argued to have been a standardizing influence in the 15th century, alternated freely between NC and non-NC.

4.5 Summary and conclusions

The main changes in the development of clausal negation in the history of English can therefore be summarized as follows. Firstly, as regards clause negation, the clitic negator *ne* was lost, replaced by *not*. *Not* cliticized to *-n't* in the early Tudor period. Secondly, indefinites under negation saw largescale change. At an early stage in Old English, the indefinite-polarity-item system cognate with *not... any*, inherited from early Indo-European, was challenged by a negative-concord system in which indefinites under negation had to be items belonging to the n-item series. Then, from around 1400 onwards, as the proclitic negator *ne* was lost, n-items became intrinsically negative, as did coordinator *nor*, and *any*-series polarity items were generally admitted to negative contexts: NC as a constraint on indefinites in negative clauses was lost. In terms of word order in negative clauses: NegV1 was lost with the grammaticalization of *not*, and negative inversion arose in the early modern period. Neg movement to SpecNegP arose in the Late Middle English period, then was lost. The factors triggering these various changes can be grouped under three headings:

(i) pragmatic reinforcement of, as well as phonetic erosion of, the proclitic negator;
(ii) a widespread tendency among early Germanic languages for a polarity item to incorporate the clausal negative particle, creating n-items when used together with the clausal negative particle;
(iii) an n-item is reinterpreted as inherently negative, e.g. *ne... none* is replaced by *none*.

A recurrent issue throughout these developments is whether negation is expressed singly or in the form of one or more elements constituting multiple negation; this issue has appeared in the development of sentence negation, of indefinites, and of coordinate sentences. A way to bring together these disparate phenomena using the structural analysis proposed by Zeijlstra (2004) was adopted by Ingham (2007), taking a NegP constituent to be the locus of clausal negation in states of language with negative concord. As was discussed in section 4.3.1, NegP in Old and Early Middle English contained a null negative operator identified by the particle *ne*. After the elimination of *ne* made its identification impossible, the null negative operator was lost, but *not* was for a while able to take its place in SpecNegP, so that, in keeping with the theoretical position of Zeijlstra (2004), Late Middle English remained a negative-concord language. However, the weakening of *not* to *-n't* at the end of the medieval period suggests that by then it was losing its status as a specifier (van Gelderen 2008). How the subsequent history of negation in Modern English comports with Zeijlstra's account is less clear, and requires further research. The character of modern Standard English as a non-negative concord language would have developed if

NegP was no longer projected as a syntactic constituent hosting a null or overt negative operator.[13] In this way, the cyclical nature of change in sentence negation and changes in indefinites and coordination could be captured. However, the fact of language standardization and the persistence of negative concord in some registers in Modern English leads to complexities which remain to be fully unravelled.

A final observation needs to be made. The structural analysis of a negative clause in terms of a NegP, taken as a given from Pollock (1989) onwards, could be contested not only, as Zeijlstra (2004) has done, by reserving NegP for NC languages. It could also be disputed whether such a constituent needs to be posited for any language (Biberauer and Roberts 2011). Arguably, the morphosyntactic exponents of negation could be handled either as adverbs or as part of a verbal-clitic or inflectional complex, as indeed they typically were prior to Pollock (1989). NegP is often motivated in terms of the semantic contribution it makes to the interpretation of the clause (Logical Form in Minimalist terms), but in this chapter we have seen some evidence, especially in section 4.3.3, that a NegP constituent has had clear syntactic justification, at least in later Middle English, in terms of the distribution of negated subjects and objects. Another argument for postulating NegP in terms of its syntactic relevance emerges when we compare diachronic changes in negation with those of elements in other closed-class categories, especially conjunctions and prepositions. English has seen very substantial renewal in its stock of such elements over the centuries, but in neither case have lexical changes created a change in the syntactic structure of the clause. With negative elements, as we have seen in this chapter, it has commonly happened that renewal of negative lexical items has gone hand in hand with changes in clause syntax. If English is anything to go by, there thus appears to be a tight linkage between negators and the structure of the clause in which they stand.

Primary sources

Ælfr. Hom. — *The homilies of the Anglo-Saxon church: The first part, containing the Sermones catholici, or Homilies of Ælfric*, ed. Benjamin Thorpe. London: Ælfric Society, 1846.

Ælfric Lives — *Ælfric's Lives of Saints Vol. I*, ed. William Skeat. Early English Text Society, Original Series no. 82. London: Trübner, 1881.

Ælfric's Colloquy — *Ælfric's Colloquy*, ed. G. Garmonsway. London: Methuen, 1947.

[13] This conclusion is somewhat speculative, but one way to implement the proposal would be along the lines of Penka (2010), who provides for a semantic negative operator (not positionally linked to a NegP structure) corresponding to any morphosyntactic element that introduces negation, in the sense adopted in the present article. Thus a negated auxiliary such as *don't* would identify a negative operator, which Penka adjoins to vP. In our approach, following Zeijlstra, the null NegOP stood in NegP in NC states of the language, so we might say that it was adjoined to vP at other periods.

AR	*The English text of the Ancrene Riwle: Ancrene Wisse*, ed. J. R. R. Tolkien. Early English Text Society, Original Series no. 249. London: Oxford University Press, 1962.
AR(T)	*The English text of the Ancrene Riwle*, ed. Frances Mack. Early English Text Society, Original Series no. 252. London: Oxford University Press, 1963.
Beowulf	*Beowulf: Reproduced in facsimile from the unique manuscript, British Museum MS. Cotton Vitellius A xv*, rev. and ed. Norman Davis. Early English Text Society, Original Series no. 245. London: Oxford University Press, 1959.
BHom	*Blickling Homilies*, ed. R. Morris. Early English Text Society, Original Series no. 63. London: Trübner, 1880.
Boswell, Life of Johnson	*Boswell's Life of Johnson*, ed. A. Birrell. London: Constable, 1896.
Brut	*The Brut or the Chronicles of England*, by F. Brie. London: Kegan Paul Trench, 1906.
Cura Past.	*King Alfred's West-Saxon version of Gregory's Pastoral care*, ed. Henry Sweet. Early English Text Society, Original Series no. 45. London: Trübner, 1871.
Fulgens & Lucres	*In Tudor Plays: An Anthology of Early English Drama*, ed. Edmund Creeth. New York: Anchor, 1966.
Hali Meiðhad	*Hali Meiðhad*, ed. B. Millett. Early English Text Society, Original Series no. 284. London: Trübner, 1982.
Helsinki Corpus	M Kytö, *Manual to the Diachronic Part of the Helsinki Corpus of English Texts: Coding Conventions and Lists of Source Texts.* Helsinki: Department of English, University of Helsinki, 1996.
Hick Scorner	In *Two Tudor Interludes*, ed. Ian Lancashire. Manchester: Manchester University Press, 1980.
Lamb Hom.	*Old English Homilies Vol 1*, ed. R. Morris. Early English Text Society, Original Series no. 29. London: Trübner, 1868.
Lindisfarne	*Lindisfarne Gospels*, ed. Walter W. Skeat. Cambridge: Cambridge University Press, 1871–87.
Martyrol	*Das altenglische Martyrologium*, ed. Günter Kotzor. Munich: Verlag der Bayerischen Akademie der Wissenschaften, 1981.
MES	*Four Middle English sermons*, ed. Veronica O'Mara. Middle English Texts no. 33. Heidelberg: Winter, 2002, pp. 1–47.
Orosius	*King Alfred's Orosius, Part I*, ed. Henry Sweet. Early English Text Society, Original Series, no. 79. London: Trübner, 1883.
Paston	*Paston letters and papers of the fifteenth century*, ed. Norman Davis. Oxford: Clarendon Press, 1971.

Pepysian Gospel Harmony	*Harmony of the life of Christ*, ed. M. Goates. Early English Text Society, Original Series no. 157. London: Trübner, 1919.
PPCME-2	Kroch, Anthony, and Taylor, Ann (eds). 2000. The Penn–Helsinki Parsed Corpus of Middle English. University of Pennsylvania, Philadelphia. <http://www.ling.upenn.edu/hist-corpora/>
Prose rule of St Benet	*Three Middle English versions of the Rule of St Benet*, ed. Ernest Kock. Early English Text Society, Original Series no. 120. London: Trübner, 1902.
St. Chad	*The Life of St. Chad: An Old English homily*, ed. Rudolf Vleeskruyer. Amsterdam: North Holland, 1953.
St. Jul.	*Þe liflade of St. Juliana*, ed. Oswald Cockayne. Early English Text Society, Original Series no. 51. London: Trübner, 1872.
St. Kath.	*The life of St. Katherine*, ed. Eugen Einenkel. Early English Text Society, Original Series no. 80, London: Trübner, 1884.
St. Marg.	*Saint Margaret, text and translation*. In *Medieval English prose for women*, ed. Bella Millett and Jocelyn Wogan-Browne. Oxford: Oxford University Press, 1990.
Seinte Marherete	*Seinte Marherete*, ed. Oswald Cockayne. Early English Text Society, Original Series no. 13. London: Trübner, 1866.
SEMP	*The South English Ministry and Passion*, ed. O. S. Pickering. Middle English Texts 16. Heidelberg: Winter, 1984.
Trevisa	*John Trevisa's translation of the Polychronicon of Ranulph Higden, Book VI*, ed. Ronald Waldron. Middle English Texts 35. Heidelberg: Winter, 2004.
Vesp Psalter	In *The oldest English texts*, ed. Henry Sweet. Early English Text Society, Original Series no. 83. London: Trübner, 1885, pp. 461–652.
Vic. Virt.	*Vices and virtues. Part 1.* ed. F. Holthausen, Early English Text Society Original Series no. 89. London: Trübner.
Wærferth	*Bischof Wærferths von Worcester Übersetzung der Dialoge Gregors des Grossen*, ed. Hans Hecht. Bibliothek der Angelsächsischen Prosa 5. Leipzig: G. H. Wigand, 1900.
WS Gosp.	*The Gospel according to Saint Matthew in Anglo-Saxon, Northumbrian, and Old Mercian versions*, ed. Walter W. Skeat. Cambridge: Cambridge University Press, 1887.
Wyclif sermons	*The English works of Wyclif hitherto unprinted*, ed. F. D. Matthew. Early English Text Society, Original Series no. 74. London: Trübner, 1880.

5

Negation in the history of (High) German

AGNES JÄGER

During the thousand years and more of the attested history of German, the marking of negation in the language has changed in many respects. Most noticeable are the changes of the negative particle along Jespersen's cycle and the change from a language exhibiting negative concord (NC) to a non-NC language. Four major historical stages of German are generally distinguished. The oldest attested stage, referred to as Old High German (OHG) dates from around AD 750 to 1050. The ensuing period from around AD 1050 to 1350 is called Middle High German (MHG). Early New High German (ENHG) dates from approximately 1350 to 1650 and is followed by Modern German. The main developments with respect to negation can be observed from OHG to MHG. In MHG, the most frequent syntactic patterns of negation already correspond to those found in Modern German both regarding the negation particle as well as negative indefinites. Accordingly, the present chapter focuses on OHG and MHG, with comparison of ENHG and Modern German. Section 5.1 discusses the development of the negative particle from preverbal clitic to bipartite negative particle to single verb-independent negative particle. The history of clauses containing indefinites in the scope of negation, including the phenomenon of negative concord, is presented in section 5.2. Section 5.3 investigates further ways of marking negation, such as narrow focus of negation and constituent negation, as well as negative subjunctions and disjunctions. The concluding section 5.4 summarizes the main findings.

5.1 Development of the negative particle

5.1.1 Old High German

5.1.1.1 The preverbal clitic negative particle In OHG sentential negation is basically always marked by the preverbal clitic negative particle *ni* (cf. also Behaghel 1918: 229,

Paul 1920: 330, Dal 1966: 163, Lockwood 1968: 207, Keller 1986: 207, Admoni 1990: 46, Schrodt 2004: 135), as illustrated in (1). This particle was inherited from Germanic *ni.

(1) sí **ni** mohta inbéran sin
she NEG could do.without him
'She could not do without him.' (Otfrid I. 8, 3)

In late OHG, the vowel was reduced and the negative particle occurred as *ne*, for instance in Notker's works. In a corpus of several longer OHG texts, namely the OHG Isidor translation (around AD 800), the Tatian translation (around AD 850), Otfrid's gospel book (AD 863–71), and Notker's Psalm translation and commentary (before AD 1020),[1] over 90% on average of all negated clauses contain the preverbal negative particle (cf. Jäger 2008), compare Table 5.1. Thus preverbal *ni* is really the major means of neg-marking in OHG. In the vast majority of cases, namely around 80% of all negated clauses in Isidor, Tatian, and Otfrid, the preverbal clitic negative particle is even the only marker of negation in the entire clause, as was seen in (1). This percentage decreases slightly towards late OHG. Furthermore, *ni* co-occurs with other neg-markers, notably n-indefinites (see section 5.2.1. below).

The fact that the verbal clitic negative particle was the major means of marking negation in OHG is also particularly evident from the translated OHG texts. Where these deviate from the Latin originals with respect to negation, it is mostly in that they add the preverbal negative particle where the Latin original did not contain *non*. Consider the following example from Tatian:

(2) (quia nemo est / In cognatione tua qui uoc&tur / hoc nomine)
nioman **nist** / In thinemo cunne thie thar genemnit sî / thesemo namen
nobody NEG.is in your family who PTCL called be this name
'There is nobody in your family who is called by this name.' (Tatian 30, 26–8)

This example also illustrates another noteworthy characteristic of *ni*: it is often written as one word together with the finite verb, suggesting that the scribes

TABLE 5.1 **Preverbal neg-particle *ni* (or *ne*) in Old High German negated clauses**

	Isidor	Tatian	Otfrid	Notker
negated clauses including negation particle *ni*	94% (47)	91% (873)	93% (93)	88% (88)
ni as the only neg-marker in the clause	78% (39)	81% (770)	85% (85)	65% (65)

[1] The corpus includes all negated clauses from Tatian and Isidor as well as the first 100 negated clauses each from Otfrid and Notker and all negated clauses from the minor OHG texts of the *Hildebrandslied*, the Wessobrunn Sermon (*Wessobrunner Predigt*), and the Paris Conversations (*Pariser Gespräche*). For details of the corpus, see Jäger (2008: 9).

perceived of this combination as one unit and thus corroborating its status as a clitic.[2] In contrast to the generalization given in Wheelock LaBrum (1982: 204), spelling as one word is not only attested with verbs beginning in /i/ but found with various initial vowels and consonants.[3]

(3) (quia / nemo potest uenire ad me. / nisi...)
uuanta / neoman **ni**mag quemen zi mir / niba...
because nobody NEG.may come to me unless
'because nobody can come to me unless...' (Tatian 125, 11–13)

Fusion and vowel-contraction or vowel loss also occur, but, in contrast to Old English, these are indeed only attested with verbs beginning in /i/ as in the case of *nist* in example (2) above vs for instance *niougent* (Tatian 246, 5). Even with initial /i/ contraction is mostly, but not consistently, found, consider *niist* (Tatian 221, 3) or *ni intratent* (Otfrid I. 1, 98).

The negative particle *ni* basically always appears as a proclitic to the finite verb. This generalization holds for Isidor, Tatian, Otfrid, and Notker and is supported by data from various minor OHG texts: all negated clauses in the *Hildebrandslied*, the Wessobrunn Sermon (*Wessobrunner Predigt*), and the Paris Conversations (*Pariser Gespräche*) have *ni*/*ne* as a proclitic on the finite verb. In OHG infinitival constructions, too, it is the finite, not the non-finite verb that *ni* attaches to, compare example (4) as well as (1) and (3) above:[4]

[2] This graphematic fact is, however, often obscured in the classic 19th-century editions of the OHG texts. For example, the more reliable 1994 edition of Tatian by Masser has *ni* and the verb as one word in a vast number of cases where Sievers, in his 1892 edition, which is still widely used for linguistic investigations, consistently splits this unit into two separate words.

[3] Wheelock LaBrum's (1982: 204) incorrect generalization concerning the cliticization of *ni* is the result of the analysis of a sample of negated clauses from Otfrid according to the 1856 edition by Kelle. This underlines the methodological need to use editions that are true to the original manuscripts (such as Masser 1994 for Tatian, or Kleiber 2004 for Otfrid), if not the original manuscripts themselves, for investigations of historical spelling.

[4] There are very few exceptions to this rule. Thus *ni* occasionally occurs as a proclitic on a participle in participial constructions in Tatian, e.g.:

(i) (amice quomodo / huc Intrasti non habens / uestem nuptialem.)
friunt uuvo / giengi thu hera In nihabenti / giuuati brutlouftlih.
friend how went you here in NEG.having robe wedding
'Friend, how did you get in here not having a wedding robe?' (Tatian 206, 24–6)

Such constructions, however, constitute cases of loan syntax from Latin. Yet even here, the translators felt the need to use the preverbal negative particle. Another rare case is the following example from Notker:

(ii) ih ne irsterben muge.
I NEG die may
'I may not die.' (Notker, Psalter 3, 7 (= 13, 27–14, 1))

This pattern is possibly due to verb-complex formation. However, generally, cliticization to the finite verb is also the rule in verbal complexes in Notker.

(4) in ánder gizúngi firnéman iz **ni**.kúnni
in other language understand it NEG.can
'(he) could not understand it in another language' (Otfrid I. 1, 120)

In the corpus from Isidor, Tatian, Otfrid, and Notker, as well as the minor OHG texts mentioned above, *ni*/*ne* never attaches as an enclitic to an element other than the verb (in contrast to MHG, see section 5.1.2.1 below). Cliticization to the finite verb is not restricted by any morphosyntactic specifications of the verb. Even negated imperatives are attested in OHG:

(5) **Ni** fórhti thir bíscof!
NEG fear thee bishop
'Do not be afraid, bishop!' (Otfrid I. 4, 27)

(6) (neminem concutiatis / neque calumniam faciatis,)
niomannen **ni** bliuu& / noh harm **ni** tuot
nobody NEG beat nor harm NEG do
'Do not beat anyone or harm them.' (Tatian 46, 31–2)

As with Modern German, OHG clauses consist of several positional or topological fields according to the so-called topological model. The verb may be positioned either in the right sentence bracket or frame at the end of the clause, or in the left sentence bracket or frame in the left periphery of the clause. The left sentence frame, which corresponds to C^0 in a generative syntactic analysis, may alternatively be taken up by a complementizer, forcing the verb to remain in its clause-final base position, the right sentence frame. The topological field between both parts of the sentence frame or bracket is referred to as the middle field (*Mittelfeld*) and may contain any number of syntactic constituents. The topological field in front of the left sentence frame is referred to as the prefield (*Vorfeld*) and may generally contain only one constituent, but not necessarily the subject. Ordinary declaratives thus take the form of verb-second (V2) clauses. This verb-second characteristic of German is basically already present in OHG (apart from occasional cases of V1 or V3 declaratives), cf. Axel (2007). Phonologically heavier constituents, in particular, as well as subordinate clauses may be extraposed to the postfield (*Nachfeld*), the topological field after the right sentence frame. However, in OHG, MHG, and partly still in ENHG, extraposition was not restricted to heavy constituents only, but also occurred with short adverbs or pronouns.

Crucially for the question of the phrase-structural status of the negative particle (see discussion section 5.1.4. below), OHG *ni* attaches to the finite verb irrespective of verbal placement: we find *ni* as a proclitic on the finite verb in the clause-final base position (Ve (=verb-end)), i.e. the right sentence frame, as well as on the finite verb in the left sentence frame, that is, in second position (V2), or clause-initial (V1) position:[5]

(7) thaz thu irrímen **ni** máht   Ve
that you tell NEG may
'that you cannot tell' (Otfrid I. 11, 52)

(8) (et portę non claudentur.)   V2
endi dor **ni** uuerdant bilohhan.
and doors NEG become closed
'And the doors will not be closed.' (Isidor III, 2)

(9) (si dauid mentiar: semen eius in eternum manebit.)   V1
ni liugu ih dauid, sin samo ardot in æuuin.
NEG lie I David his semen remains in eternity
'I do not lie to David: His semen will remain in eternity.' (Isidor IX, 1)

Examples such as (7) and (9) contradict the generalization formulated by Abraham (2003: 351) for OHG and MHG that the preverbal clitic neg-particle is not sufficient for sentential negation in clauses with an overt middle field: both examples contain an overt middle field (*thu* in (7) and *ih dauid* in (9)), yet the clitic neg-particle *ni* obviously suffices as the sole marker of negation. More generally, (1), (4), (5), (7), and (9) contradict Abraham's (2003: 343–4) 'Asymmetrical Neg-Criterion', according to which *ni* could supposedly only appear if a phrasal neg-marker (n-indefinite or second neg-particle) was present.[6] In this respect OHG as well as MHG and partly ENHG (see sections 5.1.2.1 and 5.1.3 below) differ from languages such as West Flemish, as described in Haegeman (1995), where the preverbal clitic neg-particle only occurs if there is some negative phrase in the clause. In my OHG corpus, *ni* occurs on average about 15 times as often on its own as together with an n-indefinite.

With its omnipresent verbal clitic negative particle *ni*, OHG is clearly a stage-I language in terms of Jespersen's cycle. However, in the course of the OHG period, a second negative particle eventually begins to be grammaticalized from

[5] In the context of so-called residual V2 word order (Rizzi 1996), it has been noted that other older Germanic languages such as Old English (van Kemenade 1999 and others) or Old Norse (Eythórsson 1995, 2002) as a general rule observe Neg–Vfin–Subj word order in negated (main) clauses (cf. also Behaghel 1932: §1428), assuming that the finite verb is attracted across the subject position to C by a negative operator in SpecCP or by a negative feature in C. There is no evidence that any such rule played a role in OHG. In OHG, the V2 grammar is already so firmly consolidated that generalized verb movement to C is a basic rule in main clauses. Subjects occur both before and after the finite verb in negated clauses. There is no evidence for a greater frequency of V1 word order (*ni*–Vfin–Subj) as occasionally suggested in the literature (most recently in Axel 2007: 62, but without concrete quantitative data). Most declaratives in OHG are V2 clauses, whether negated or not.

[6] In phrase-structural terms, the 'Asymmetrical Neg-Criterion' states that SpecNegP must be filled in order to realize Neg^0 but not vice versa (based on the assumption, that is not shared here, that besides the second neg-particle, n-indefinites also occupy SpecNegP).

previously optional emphatic adverbial elements often also referred to as 'negation strengtheners'.

5.1.1.2 The 'negation strengtheners' nieht, wiht, *and* drof *and the beginnings of the grammaticalization of a verb-independent second neg-particle* In OHG, several elements may 'strengthen' negation, i.e. emphasize it in a way similar to English (*not*) *at all* or (*not*) *a bit*. With some of these items used for emphatic negation, one can observe the beginning of a process of grammaticalization that turns them into a second negative particle, so that eventually stage II of Jespersen's cycle is reached.

Among the cross-linguistically relevant types of element that form the diachronic source of a second, verb-independent negative particle, there are (i) n-indefinites, (ii) non-negative indefinites (esp. 'thing', 'being'), and (iii) so-called minimizers (e.g. 'jot', 'drop', 'crumb') (cf. section 1.4). All three types are evidenced in OHG.

The first type, an n-indefinite being used as a 'neg-strengthener' and subsequently as a neg-particle, is found in late OHG. Notker very occasionally uses the n-indefinite *nieht* (< *niouuiht*), which otherwise appears as an argument meaning 'nothing', in adverbial, non-argument use meaning 'in nothing, not at all', or arguably already grammaticalized as a second neg-particle.[7] Consider the following example:

(10) Ih **ne**hábo / <u>nîeht</u> in geméitun sô uîlo geuuêinot.
I NEG.have not.at.all/NEG in vain so much cried
'I did not cry that much in vain.' (Notker, Psalter 6, 11(= 20, 23–4))

Indeed, this element eventually became the standard second neg-particle that we still find in the form of *nicht* as the neg-particle in Modern German today (see section 5.1.3 below). This grammaticalization of original 'nothing' into a new negative particle within Jespersen's cycle is also found in other Germanic and non-Germanic languages (cf. Jespersen 1917: 16), compare English *nought/nawiht* > *not*, Dutch *niet* > *niet*, Old Norse *eittki* > *ekki*, Danish *ikke*, Swedish *icke*, Danish/Swedish *inte*, Greek *ouden* > *den* (see section 1.3 for Scandinavian, section 3.3 for northern Italian dialects, section 4.2.1.

[7] This is the case in four of the 100 negated clauses in my corpus from Notker. Behaghel (1918: 230) states that in Williram's version of the Song of Songs, the use of *niht* as a neg-particle is already a basic rule. The first attestation of adverbial *niowiht* has been argued to be the following example from Otfrid (cf. Behaghel 1918: 230, Lockwood 1968: 207, Wheelock LaBrum 1982: 210):

(i) ni zaweta imo es niawiht
NEG succeeded him it nothing/not.at.all
'He did not succeed (at any of it/at it at all).' (Otfrid II 5, 12)

However, Donhauser (1996: 207) points out that the valency of the verb *zawen* 'succeed', which is attested only in Otfrid, is not entirely clear so that it cannot be excluded that *niawiht* is an argument here, as in all other cases in Otfrid. Furthermore, my data show that Otfrid used *uuiht* 'anything' or *drof* 'a drop' rather than *niowiht* 'nothing' as an adverbial 'neg-strengthener'.

for English, section 6.2 for Dutch, section 7.3.2. for Welsh, section 8.2 for Greek, and section 10.2.1. for Arabic and neighbouring languages).

The n-indefinite *nio* (*in altere*) 'never' is also occasionally used as a 'neg-strengthener' or second neg-particle without its temporal meaning. Consider the following example from Tatian:

(11) (nequaquam minima es / In principibus Iuda.)
nio In altere bist thu minnista / In then heriston iudeno.
never in age/NEG are you smallest in the princes Jewish
'thou art not the least among the princes of Juda' (Tatian 39, 27–8)

Semantic bleaching of the temporal n-indefinite and tendencies towards grammaticalization into a neg-particle can also be found in historical (cf. Jespersen 1917: 16) and present-day English as well as in later stages of German.

Evidence for the second type of element mentioned above, namely a non-negative indefinite that is used as a 'neg-strengthener' and subsequently as a second neg-particle comes from Otfrid's use of the non-negative pronoun *wiht* '(any)thing', adverbially 'at all':[8]

(12) thaz ér mir hiar ni dérre, ouh uuíht mih **ni** gimérre.
that he me here NEG let.wither also at.all/NEG me NEG obstruct
'that he will not let me wither here and not obstruct me at all.' (Otfrid I. 2, 30)

While the use of its neg-marked counterpart *niht* became more widespread and was already the general standard by MHG times (see section 5.1.2.3 below), the use of a non-neg-marked neg-particle also continued in MHG and even until today in some regional varieties (see sections 5.1.2.3 and 5.1.3 below).

The third type of element, that is, a minimizer as a 'neg-strengthener' or neg-particle,[9] is also attested in Otfrid, who uses *drof* 'drop' in the same way as *wiht* to emphasize negation.[10] Note that *drof* is used here as a non-argument and thus shows evidence of grammaticalization (cf. section 1.4):

[8] According to Kelle (1881: 681–3), there are 18 examples of adverbial *wiht* in the entire text of Otfrid. However, according to my results, book one contains more occurrences than are listed by Kelle, so that the total number is presumably higher. Among the first 100 negated clauses from Otfrid, I found seven examples including adverbial *wiht*.

[9] Still in MHG, a wide variety of neg-strengthening minimizers occurs, such as *ein har* 'a hair', *ein bon* 'a bean', *ein stro* 'a straw', *ein ber* 'a berry', *ein stoub* 'a (grain of) dust', *ein wint* 'a wind', etc., cf. Grimm (1890: 706–9), Paul (2007: 388–9).

[10] Among the first 100 negated clauses from Otfrid, there are two examples including *drof* (Otfrid I. 5, 28; Otfrid I. 4, 27), cf. Jäger (2008). Kelle (1881: 78) notes 18 occurences of 'neg-strengthening' *drof* in the entire text, the same number that he gives for *wiht* as a neg-strengthener, so Otfrid uses both emphasizing strategies to the same extent.

(13) a. drof ni forahtet ir iu
drop/NEG NEG fear you you
'Do not be afraid at all.' (Otfrid III 13, 9)

b. uuiht ni forahtet ir iu
at.all/NEG NEG fear you you
'Do not be afraid at all.' (Otfrid III 8, 29)

Minimizers used as 'neg-strengtheners' and subsequent neg-particles are of course well known from the history of French (e.g. *pas* 'step', *mie* 'crumb', *point* 'point', *goutte* 'drop'). Otfrid's home Weißenburg in Lorraine (now Wissembourg) was not far from the linguistic frontier so that *ni...drof* could have been inspired by French *ne...goutte* ('not...a drop'), cf. also Lockwood (1968: 208), Wheelock LaBrum (1982: 221). Similarly, the heavily Romance-influenced OHG text of the so-called Paris Conversations contains two occurrences of *netrophen* (NEG-drop) as translations of Latin *nihil* 'nothing' and *quid* 'anything' in the scope of negation (Paris Conv. sentences 48 and 74). However, Grimm (1890: 704, 724) implies that the influence took place in the opposite direction and that the French 'neg-strengtheners' *mie* and *goutte* were inspired by German equivalents. This hypothesis is supported by distributional facts: *goutte* remains an argument and is used more or less exclusively with 'see' and 'hear' (cf. also French, section 2.2.1., note 5), while *drof* is not restricted in this way, as (13a) illustrates. Thus, *drof* appears to have progressed further along the grammaticalization path towards a neg-particle than its French counterpart.

5.1.2 *Middle High German*

By the time of Middle High German, the grammaticalization of the second negative particle is largely complete: besides occasional attestations of the verbal clitic negative particle in its phonetically weakened form of *en/ne*, we mostly find the verb-independent adverb-like negative particle *niht* (or, more rarely, *iht*). The verb-independent and the preverbal clitic negative particle may occur on their own or in combination.

5.1.2.1 The preverbal clitic negative particle In contrast to Old High German, where the clitic neg-particle occurred in virtually every negated clause, Middle High German *en/ne* is considerably rarer. In a corpus of the first 100 negated clauses each from the *Prose-Lancelot* (*c.*1250), the *Nibelungenlied* (1190–1200; ms. A: before 1280, including comparison with the B and C mss.: second quarter of the 13th century; for datings, see Schulze 1997), and the sermons of Berthold von Regensburg (*c.*1275), less than half of all negated clauses still contain *en/ne*, in Berthold even less than a tenth (cf. Jäger 2008), compare Table 5.2.

TABLE 5.2 **Preverbal neg-particle *en/ne* in Middle High German negated clauses**

	Prosalancelot	*Nibelungenlied* (A)	Berthold
negated clauses including negation particle *en/ne*	44% (44)	26% (26)[a]	7% (7)
en/ne as only neg-marker in the clause	2% (2)	7% (7)	3% (3)

Notes [a] Manuscript B 40%, manuscript C 48%

As in OHG, the clitic neg-particle generally occurs as a proclitic to the finite verb in V1, V2, as well as Ve clauses. This is also the case in infinitival constructions where *en/ne* cliticizes to the finite verb rather than to the infinitive:[11]

(14) er enkvndez niht verenden
he NEG.could.it NEG accomplish
'He could not accomplish it.' (*Nibelungenlied* (A) III, 96, 4)[12]

(15) das ir uns dheyne hant gethan noch fúrbas enmúßent thun
that you us none have done nor further NEG.must do
'that you have not done any to us nor may do yet' (*Prosalancelot* 28, 109)

In V2 clauses, *en/ne* may secondarily attach as an enclitic to the element before the finite verb, notably to (personal) pronouns, adverbs, or particles such as *so, do, ia, nu.* As a result, the vowel of *en/ne* may be dropped. Among the texts considered here, only the *Nibelungenlied* contains this pattern, and there is even variation among the

[11] However, according to Behaghel (1918: 227), *en* occasionally occurs on the infinitive in infinitival constructions, especially when the infinitive stands immediately to the right of the second negation particle *niht*:

(i) daz ich ein kunicriche fur ir minne niht ennemen wolde
that I a kingdom for her love NEG NEG.take want
'that I would not want to take a kingdom instead of her love' (Ms. H. I, 55a; from Behaghel 1918: 227)

In the corpus from *Nibelungenlied*, the Prose-Lancelot, and Berthold's sermons, there is only a single example matching Behaghel's description of *niht en*-Vinf:

(i) Enwollent irs aber nicht enthun, so mus ich...
NEG.want you.it but NEG NEG.do so must i
'If you do not want to do it, I shall have to...' (*Prosalancelot* 50, 252)

In this case, *en* bizarrely appears on both the finite verb and the infinitive. In all likelihood, the doubling of *en* in this case is a mistake. In all other infinitival constructions with Neg^0 (15 each in the *Nibelungenlied* and Prose-Lancelot, 1 in Berthold), *en/ne* appears on the finite verb but not on the infinitive, just as would be expected.

[12] *Nibelungenlied* mss. B/C: *ern chvndez/kundes niht verenden.*

different manuscripts as to whether the clitic neg-particle procliticizes to the finite verb or encliticizes to the element before it, as with example (16):[13]

(16) a. [manuscript A:]

er	**en**kvndez	niht	verenden
he	NEG.could.it	NEG	accomplish

b. [manuscripts B/C:]

er**n**	chvndez/kundes	niht	verenden
he.NEG	could.it	NEG	accomplish

'He could not accomplish it.' (*Nibelungenlied* III, 96, 4)

Syntactically, the clitic is arguably still linked to the finite verb. Only once the finite verb has moved *en/ne* along with it to the left sentence periphery may this clitic neg-particle encliticize to the element before it. Regarding the position of *en/ne*, the crucial generalization is that it always occurs immediately left-adjacent to the finite verb.

In MHG, *en/ne* is very occasionally still used as the only marker of negation in a clause:[14]

(17)

sô	hât	er	dirz	gestoln,	dû	**en**weist	hiute	wie
so	has	he	you.it	stolen	you	NEG.know	today	how

'He has stolen it from you and you don't know today how.' (Berthold I, 126 (22))

In other words, there is still evidence for stage I in terms of Jespersen's cycle. This pattern is rather rare: it is found in seven, two, and three per cent, respectively, of all negated clauses in my corpus from the *Nibelungenlied*, the Prose-Lancelot, and Berthold. It is widely assumed that the type of verb plays a role (cf. Behaghel 1918: 230, Paul 2007: 389–90, de Boor and Wisniewski 1998: 187, Dal 1966: 164): *en/ne* may be used as the only marker of negation when it occurs on modals, on *lazen* 'to let/leave', *tun* 'to do', *ruochen* 'to care', and *wizzen* 'to know',[15] cf. example (17), notably with a complement clause. My data suggest that verbal placement is another crucial factor: in verb-final clauses, *en/ne* on its own does not suffice to mark sentential negation. By contrast, V2 clauses of a special type, namely complementizerless conditional subordinate clauses with V2 word order conveying the meaning of 'unless . . . ' are typically negated by simple *en/ne*, and not by *en/ne . . . niht*, as in example (18); see also Lockwood (1968: 207–8), Paul (2007: 402–3), de Boor and Wisniewski (1998: 190).

[13] Of the 26 examples containing the clitic neg-particle among the first 100 negated clauses of the *Nibelungenlied*, 14 have the negation clitic on the verb and 12 on the word before it.

[14] This fact once more contradicts Abraham's (2003: 343–4) Asymmetrical Neg-Criterion.

[15] Jespersen (1917: 14) also observes that clitic negation is kept longer with the verb 'to know' as well as 'to want' in a number of languages. He mentions Latin *nolo* and *ne scio*, French *je ne sais*, English *nill*, and also MHG *en will*.

(18) miern zerinne miner frivnde, in wirt arebeit bechant.
me.NEG vanish my friends them becomes trouble known
'Unless my friends should vanish, they shall get to know some trouble.'
(*Nibelungenlied* (A) IV, 170, 4)

Another context for simple *en/ne* is expletive negation in a subordinate clause dependent on an adversative or negated matrix predicate (paratactic negation in the terminology of Jespersen 1917). This phenomenon is, however, extremely infrequent in MHG.[16]

(19) die hoh gemvoten degne wolden des niht lan, / [**sin** drvngen
the high spirited warriors wanted that NEG omit they.NEG penetrated
da si sahen die minnechlichen meit].
there they saw the lovely maiden
'The bold warriors did not want to refrain from this / they forced their way through to where they saw the lovely maiden.' (*Nibelungenlied* (A) V, 290, 3)[17]

5.1.2.2 *The bipartite negative particle* Apart from the rare cases in which *en/ne* occurs on its own, it co-occurs with other neg-markers such as the disjunction *noh* (see section 5.3.2. below), n-indefinites (see section 5.2.2. below), or with the second, verb-independent neg-particle *niht* forming a bipartite neg-particle. There is thus also evidence for stage II of Jespersen's cycle in MHG:[18]

(20) daz ich drizic pfunt **niht** **en**naeme Ve
that I thirty pound NEG NEG.take
'that I would not take thirty pound' (Berthold I, 176 (30))

(21) "Ich **en**wil es **niht** erwinden", sprach aber der chune man. V2
I NEG.want it NEG omit said but the brave man
'"I do not want to omit it", said the brave man.' (*Nibelungenlied* (C) III, 117, 1)[19]

Niht is fully grammaticalized as a second neg-particle in MHG.[20] In all texts considered here, the frequency of the neg-particle *niht* is higher than that of *en/ne*. Roughly half or more of all negated clauses contain the neg-particle *niht* (see Table 5.4 below).

[16] In my MHG corpus, there are only two examples, both from the *Nibelungenlied*.

[17] B/C: ... *sine drvngen*....

[18] According to my data (cf. Jäger 2008: 146), and contra Behaghel (1918: 245), it is mostly V2 clauses that contain the bipartite neg-particle instead of simple *niht*.

[19] A/B: "Ich ne wils niht erwinden", sprach der kuone man.

[20] At the same time, *niht* continues to be used as an n-indefinite meaning 'nothing' (see section 5.2.2. below).

TABLE 5.3 **Bipartite neg-particle *en/ne... niht* in Middle High German negated clauses**

	Prosalancelot	*Nibelungenlied* (A)	Berthold
en/ne... niht	27% (27)	13% (13)	4% (4)

As there is hardly any evidence for Jespersen's stage II in OHG or ENHG, let alone in Modern German, the bipartite neg-particle is taken to be a typical characteristic of MHG by most historical grammars and textbooks. They usually describe it as the standard in this period, for instance, Wolf (2000: 1356): 'Im Mittelhochdeutschen ist die doppelte Negation *ne* + *niht* geradezu die Norm' 'In MHG, double negation with *ne* + *niht* is really the norm', cf. also Paul (2007: 389), Dal (1966: 164), Schmidt (1993: 276). However, my corpus data, given in Table 5.3, reveal that only a minority of negated clauses actually contain both particles. None of the texts predominantly employs this pattern to express sentential negation.[21]

One surprising conclusion from these data is thus that there is accordingly no evidence for a stable stage-II period in the history of German, a conclusion reminiscent of the results in Frisch (1997) for Middle English. Even the transitionary period between OHG and MHG cannot be characterized as a stage-II system, as an additional investigation of the early MHG text of the Wiener Genesis (AD *c*.1060–80) indicates. This early MHG text still shows stage I, as in OHG, for the most part: the verbal clitic neg-particle appears on its own five times as often as either the bipartite neg-particle *ne... niht* or *niht* on its own. In classical MHG, on the other hand, when the second neg-particle *niht* had been fully grammaticalized and really increased in use, it already occurred mostly on its own (cf. section 5.1.2.3), i.e. *en/ne* was already on its way out.[22] The loss of the verbal clitic neg-particle took place earlier in Upper German dialects, where *en/ne* is already rare by the end of the 13th century (Lehmann 1978: 103), whereas it is kept throughout the MHG period in Central and Low Franconian (Behaghel 1918: 246). According to Dal (1966: 164) and Lockwood

[21] The rate of *en/ne... niht* is slightly higher in the older manuscripts of the *Nibelungenlied* than in the A manuscript given in Table 5.3 (B: 20 %, C: 25 % of negated clauses containing the bipartite neg-particle), but still lower than in the Prose-Lancelot.

[22] Frisch (1997) argues that the bipartite negative particle does not constitute an independent system but rather the temporal overlap of two competing systems. This could also be argued for German: early MHG was still stage I, classical MHG mostly stage III, the bipartite neg-particle, i.e. stage II, was never a majority pattern. Compared to Frisch's data for Middle English, MHG had even progressed much further in the development towards using the verb-independent neg-particle only. For the period from 1220 to 1290 that is roughly comparable to my MHG corpus, Frisch (1997: 32) reports that *ne* as the only neg-marker is almost twice as common as *ne* + *not*, which in turn is more than ten times as common as *not* only. At the same time, simple *en/ne* is generally the minority pattern in MHG and *niht* as the only neg-marker is used in the majority of cases, a stage that is only reached in English in the second half of the 14th century according to Frisch (cf. also chapter 4).

TABLE 5.4 **Verb-independent neg-particle *niht* in Middle High German negated clauses**

	Prosalancelot	*Nibelungenlied* (A)	Berthold
negated clauses incl. neg-particle *niht*	56% (56)	48% (48)	50% (50)
niht as the only neg-marker in the clause	28% (28)	35% (35)	45% (45)

(1968: 207–8), *en* had virtually become extinct by 1300. Ebert et al. (1993: 426) state that *en* disappeared in the 16th century, except in a few dialects. According to the corpus investigation of various ENHG and early Modern German texts by Pensel (1976), there are remains of *en ... nicht* in Low German and partly also in Central German dialects around 1500, whereas around 1700, *nicht* alone was used in the entire German-speaking area. For more details of the development in Low German, see section 6.2.1.

5.1.2.3 The verb-independent negative particle MHG had already progressed to being mostly a stage-III language. *Niht* could occur on its own from the 12th century onwards, and was increasingly used that way from the middle of the 13th century according to Donhauser (1996: 211). In my corpus (cf. Jäger 2008: 144), most occurrences of *niht* appear in constructions corresponding to Jespersen's stage III. Between almost a third and almost half of the negated clauses in Prose-Lancelot, *Nibelungenlied*, and Berthold contain *niht* as the only neg-marker, i.e. much more than *en*/*ne* on its own but surprisingly also more than the bipartite neg-particle, cf. Table 5.4.

Whereas the neg-particle *en*/*ne* procliticizes to the finite verb and moves with it to the left periphery in V2 clauses, *niht* is placed in a basically fixed, verb-independent position in the topological middle field and is unaffected by verb movement:

(22) a. und da er wúst das er **nit** dot was — Ve
and since he knew that he NEG dead was
'and since he knew that he was not dead' (*Prosalancelot* 16, 31)

b. und als er wusste, dass er **nicht** tot war

(23) a. "Des ist mir **niht** ze mvote", sprach aber Sifrit — V2
that is me NEG to mind said but Siegfried
'"That is not on my mind", said Siegfried.' (*Nibelungenlied* III, 61, 1)[23]

b. "Dessen ist mir **nicht** zu Mute", sprach aber Siegfried.

The Modern German translations in (b) demonstrate that these syntactic patterns of negation are exactly the same as they still are in Modern German.

[23] Manuscripts B/C: "Des enist mir niht ce/ze mvote", sprach abr/do Sivrit/Sifrit.

While *niht* is the most prominent neg-particle in MHG, its morphologically non-negative counterpart *iht* is very occasionally also used as a verb-independent neg-particle. *Iht* thus continues grammaticalization tendencies observed above (section 5.1.1.2) for its OHG predecessor (*io*)*wiht* 'anything', adverbially 'at all'. Consider the following example, in which *iht* constitutes the only marker of negation and is clearly used as a non-argument. In cases like these, *iht* instead of *niht* was arguably used as a neg-particle:[24]

(24)	Wir	sulen	den	iungen	herren	enphahen	dester	baz, / daz
	we	shall	the	young	lord	receive	all.the	better that
	wir **iht**	verdienen	des	snellen	rechen	haz.		
	we NEG	deserve	the	bold	warrior	hatred		

'We shall receive the young lord all the better, so that we do not deserve the hatred of the bold warrior.' (*Nibelungenlied* (A) III 105, 2)

In sum, MHG shows remains of Jespersen stage I and also has a certain quantity of stage-II constructions, but the majority of cases already correspond to stage III in using a verb-independent neg-particle only. Another major means of marking negation in MHG is of course neg-marking through n-indefinites, cf. section 5.2.2. below.

5.1.3 *Early New High German and Modern German*

In ENHG, verb-independent *nicht* is the predominant negative particle. According to Pensel's (1976) data from 1470 to 1530, 87% of the negated clauses including a neg-particle contain *nicht* as the only marker of negation. 11.4% of the clauses still include the preverbal clitic *en*. *En* is very rarely used as the only marker of negation (1.5%). 4.2% of the clauses contain preverbal *en* together with an n-indefinite. Occasionally, *en* co-occurs with *nicht* forming a bipartite neg-particle (5.7%). While ENHG even more than MHG mostly observes stage III of Jespersen's cycle, there is thus still rare evidence for both stage I and stage II. As already mentioned above, *en* is kept longer in Low German (see also section 6.2) and West Central German, but is already largely lost in Upper German and even more so in East Central German.

In the early Modern German texts from 1670 to 1730 that Pensel investigated, *nicht* is the only neg-particle that is attested in all regional varieties of German. *En* has finally become extinct. *Nicht* occurs as the only marker of negation in 90% of the

[24] MHG *iht* < OHG *iouuiht* (cf. Schmeller 1872, I: 30, Grimm and Grimm 1877: 2033–5, Grimm 1890: 47, Paul 2007: 229), but also < OHG *uuiht* (cf. Schmeller 1872, I: 30, Grimm 1890: 714, Wheelock LaBrum 1982: 221).

clauses containing a neg-particle, otherwise it co-occurs with an n-indefinite (see section 5.2.3. below). Right up to the present day, *nicht* is the only neg-particle in German. *Nicht* and regional varieties thereof (e.g. Bavarian *ned*, Saxon *ni*) are also used in all High German dialects. There is no evidence for any difference in syntactic behaviour as compared to MHG and ENHG *ni(c)ht*: *nicht* is still a verb-independent neg-particle. Jespersen's cycle has not yet come full circle in German. Modern German consistently displays stage III of Jespersen's cycle.

In some Upper German varieties, one also finds a morphologically non-negative neg-particle *it* or *et*, for instance, in certain Bavarian (Schmeller 1872: 30, 176), southeastern Swabian (Grimm 1890: 714), and northeastern Swiss dialects:

(25) Des ka **it** sei.
that can NEG be
'That's not possible.' (Swabian)

This verb-independent neg-particle developed from the NPI equivalent of *nicht*, OHG *(io)wiht* > MHG *iht* 'anything'. Recall that this item had already been occasionally used adverbially in OHG and MHG (see sections 5.1.1.2. and 5.1.2.3). This rare pattern has thus also been grammaticalized and survived in regional varieties down to the Modern German period.

5.1.4 Syntactic analysis

According to Pollock (1989) and much subsequent work, sentential negation resides in a functional projection NegP. There is cross-linguistic variation as to whether the negative particle occupies the head or the specifier position, or whether there is a particle in both positions (cf. Ouhalla 1990, Haegeman 1995). Assuming NegP, one has to determine the phrase-structural status and position of the individual historical neg-particles within that projection.[25]

The fact that the neg-particle OHG *ni*, MHG *en/ne* attaches to the finite verb and moves with it, i.e. occurs in final position in Ve and left-peripheral position in V1 and V2 clauses (see sections 5.1.1.1 and 5.1.2.1) shows that it is the head Neg^0 (cf. Weiß 1998, Abraham 2003, Jäger 2005, 2008). There are no instances in which the finite verb moves across this neg-particle and appears in front of it. The lack of such cases is to be expected since the verb has to move through all c-commanding functional projections on its way to the left-peripheral C position in V1 or V2 clauses according

[25] Alternatively, one could assume that the clitic neg-particle is attached to the verb in the lexicon, and that the verb-independent neg-particle simply takes up an adverbial position and adjoins to some verbal projection (cf. Jacobs 1982 and Haider 1997 for Modern German). However, the differing syntactic behaviour of the neg-particles and in particular the diachronic changes of Jespersen's cycle are very neatly captured under the assumption of a functional projection NegP.

to the Head Movement Constraint. In Neg^0, it attaches to *ni/ne/en*, and the entire complex moves on higher in the clause. Neg-particle and finite verb accordingly take up one head position together—in terms of the topological model, the left or the right sentence frame. Crucially, *ni/ne/en* cannot be regarded as a full phrase (*Satzglied*), as is occasionally implied in the literature (e.g. Lockwood 1968: 207, Admoni 1990: 78, Schrodt 2004: 199). Clauses with initial *ni/ne/en*+Vfin syntactically pattern with affirmative V1 clauses; clauses beginning in XP-*ni/ne/en*+Vfin behave like affirmative V2 clauses.[26]

The verb-independent neg-particle *niht* that was newly grammaticalized in Jespersen's cycle, in contrast, is not affected by verb-movement: the verb can move across it and accordingly appears after *niht* in Ve clauses, but in front of it in V1 or V2 clauses (see section 5.1.2.3). One may therefore conclude that it is not a c-commanding functional head.[27]

Weiß (1998), in (26), and Abraham (2003), in (27), suggest analysing *niht* as a second Neg^0/NegP. They differ, however, as to which position this second Neg^0/NegP takes up. Weiß analyses OHG *ni*/MHG *ne* as an orphan Neg^0 which is head-adjoined to V^0, while MHG *niht* projects a separate NegP which is left-adjoined to VP:

(26)

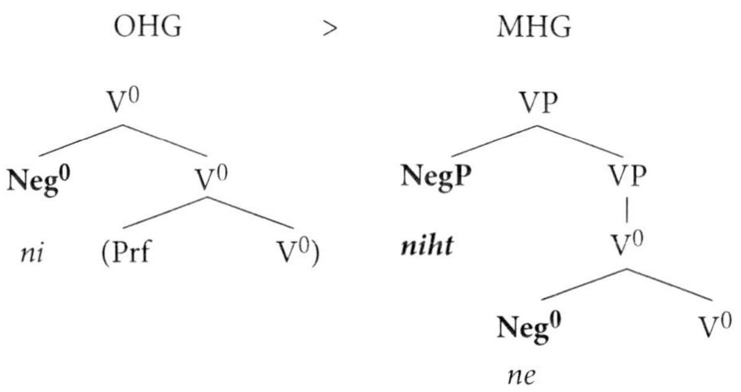

Abraham, on the other hand, takes OHG *ni*/MHG *en/ne* to be the head of a head-initial NegP1 dominating VP and TP, while MHG *niht* appears as the head of a NegP2 which is right-adjoined to VP:

[26] For instance with respect to word order in clauses introduced by certain complementizers, cf. Jäger (2005, 2008: 84–6) for *wanta* 'because' clauses.

[27] Just as in Modern German where topicalization of *nicht* to the prefield, i.e. SpecCP position in V2 clauses, is ungrammatical or at most marginally possible, fronting of *niht* is not attested in my MHG corpus data.

(27)
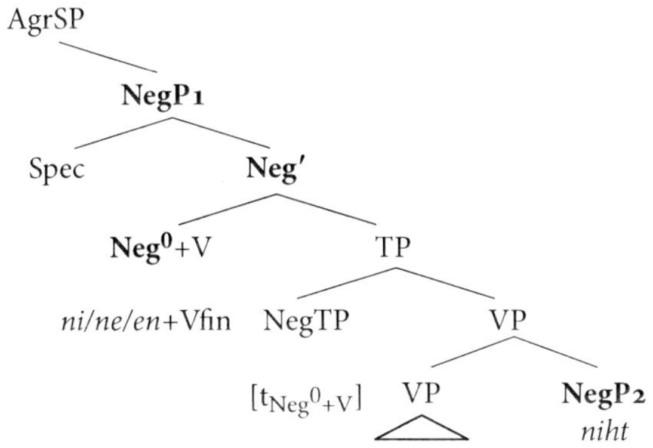

However, due to the assumption of a position of Neg1^{0} left of VP and NegP2 at the right edge of VP, this syntactic structure fails to account for the observed word order in verb-final negated clauses in OHG and MHG: the structure in (27) wrongly predicts that, in all negated clauses, the complex of clitic neg-particle and finite verb occurs left of VP, and thus fails to derive simple verb-final negated clauses, notably ones with VP-internal material preceding *ni/ne/en*+Vfin. The orders [... VP-internal material–*ni*+Vfin], cf. (4), and [... *niht*–VP-internal material–*en/ne*+Vfin], as well as more generally [... *niht–en/ne*+Vfin], cf. (20), and even [... *niht* Vfin], cf. (22), cannot be derived. Furthermore, the wrong order of *niht* and VP-internal material in V1 and V2 clauses, cf. (23), is predicted, indicating that, instead, a position left of VP needs to be assumed for *niht*.

More generally, the assumption of one Neg^{0}/NegP for a single neg-particle and two Neg^{0}s/NegPs for a bipartite neg-particle as in Weiß (1998) and Abraham (2003) seems rather surface-oriented. One may use the concept of NegP to its full potential and analyse diachronic variation along the lines of cross-linguistic variation by assuming instead that *niht* constitutes the specifier of the same NegP that *ni/ne/en* heads (cf. Jäger 2005, 2008). This matches various analyses for other languages with bipartite neg-particles (Pollock 1989 for French, Haegeman 1995 for West Flemish, van Gelderen 2004 for Old and Middle English). According to this analysis, the basic syntactic structure remained unchanged with respect to negation throughout the history of German. Only the lexical filling of the head and specifier position of NegP changed between Old High German in (28) and Middle High German in (29):

(28)

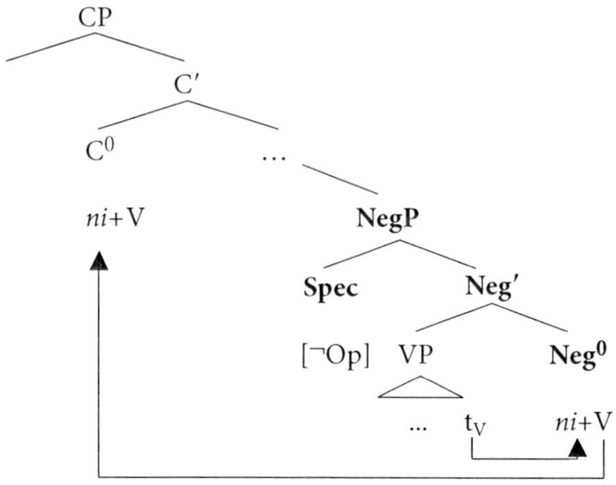

(V-movement in V1/V2-clauses)

(29)

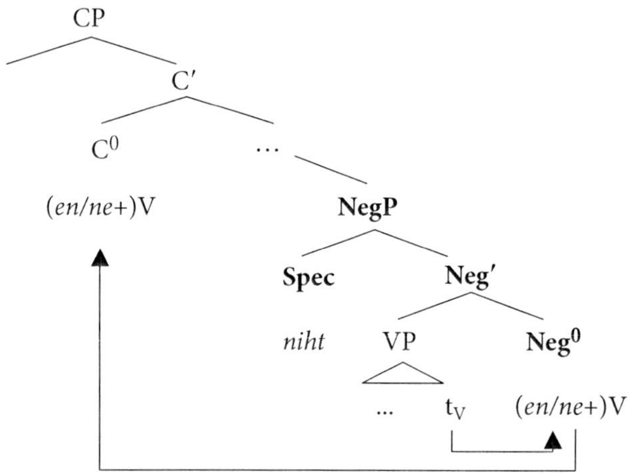

(V-movement in V1/V2-clauses)

Note that NegP in German is assumed to be head-final (for discussion see Jäger 2008: 56, 86–92). On the one hand, this reflects standard assumptions on the headedness of German INFL projections. More significantly, in contrast to (27), it correctly accounts for verb-final negated clauses. The finite verb moves to the right and attaches to *ni*/*ne*/*en*.[28] In V1 and V2 clauses, the resulting complex of Neg^0 and Vfin moves as a whole to the left sentence frame, i.e. C^0.

[28] Alternatively, one may assume that the complex of *ni*/*ne*/*en* and the finite verb is formed in the lexicon and then checked against Neg^0.

The specifier position of NegP that has been occupied by *niht* since late OHG/early MHG crucially precedes VP, thus avoiding the wrong prediction of a consistently clause-final placement of *niht* made by a structure such as (27). Already in the OHG examples from Notker including *nieht* as a second neg-particle, there is evidence that it was placed to the left of VP, compare example (10) above, with two VP-adverbials and a non-finite verb that forms the right sentence frame following *nieht*. Throughout the history of German, the verb-independent neg-particle *ni(c)ht* consistently precedes VP-internal material and the right sentence frame, i.e. the base position of the finite verb and position of any non-finite verbs. Note that VP constituents such as definite DPs etc. may undergo scrambling and therefore secondarily appear in front of SpecNegP *ni(c)ht* (cf. for Modern German, Webelhuth 1990 and Büring 1994).

In classical OHG, SpecNegP was not overtly filled. In (25), an abstract negative operator [¬Op] is assumed to reside in this position (cf. Jäger and Penka 2012). The reason for this analysis lies in the observation that the placement of *ni* in clause-final or in C position depending on the placement of the finite verb to which it attaches does not affect the reading, i.e. the semantic scope of negation. Accordingly, *ni* can be argued not to be the bearer of semantic negation but of a purely formal, uninterpretive negative feature [uNeg] (cf. also for n-indefinites, section 5.2.4. below). The semantically relevant or interpretable negative feature [iNeg] is contributed by a covert negation operator [¬Op] in SpecNegP, as is also assumed in cross-linguistic research for other languages (cf. Ouhalla 1990, Haegeman 1995) and for historical stages of related languages such as Old English (cf. van Kemenade 2000, van Gelderen 2004) or Old Dutch and Old Saxon (see chapter 6). On the other hand, *ni(c)ht* constitutes an overt negative operator as the semantic scope of negation indeed depends on the placement of *ni(c)ht* (consider *nicht alle* 'not all' vs *alle nicht* 'all not' etc.).

In terms of this NegP-analysis, Jespersen's cycle can be understood as a change from a NegP with only the head *ni* being overt in classical OHG as in (28), to a stage that is partly evidenced in MHG with both an overt Neg^0 head *en/ne* and an overt specifier of NegP *ni(c)ht*, to a stage where Neg^0 becomes optional and finally disappears so that just the overt specifier *ni(c)ht* remains, as in the majority of cases in MHG, as in (29).[29] Modern Standard German is still at this last stage with just the overt specifier of NegP. The next step in Jespersen's cycle, the grammaticalization from SpecNegP to Neg^0, has not taken place so far—*nicht* has not (yet) turned into a head as it still does not interact with verb movement from clause-final position to C^0 in V1 and V2 clauses.

As far as the grammaticalization of the verb-independent second neg-particle is concerned, one may assume that this process started from the adverbial use of

[29] Compare also the analysis in Rowlett (1998) for historical French and van Gelderen (2004) for historical English.

niowiht/nieht in which this n-indefinite would occur as a VP-adverb. As such, it was adjacent to the phonetically empty position SpecNegP and could be reanalysed as occupying this position instead, as illustrated in (30).[30]

(30)

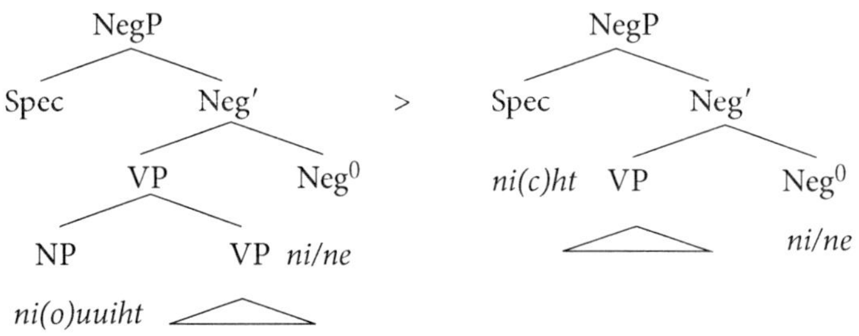

5.2 Indefinites in the scope of negation and negative concord

5.2.1 *Old High German*

In OHG, there is evidence for a largely intact three-set system of indefinite determiners, pronouns, and adverbs with respect to polarity comparable to the tripartition among English *something*, *anything*, *nothing*, etc.: there are normal or PPI (positive polarity item) indefinites notably of the *etes*-series, NPI (negative polarity item) indefinites notably of the *io*-series, and morphologically negative indefinites (n-indefinites), of the *ni(o)*-series.

TABLE 5.5 The Old High German system of indefinite pronouns and adverbs

	determiner	'entity'	'person'	'time'	'place'
'normal'/ PPI-indefinite, 'some-'	*sum(ilih); eteswelih, eteslih/ etilih; ein*	*ete(s)waz, waz*	*ete(s)wer, wer*	*ete(s)wenne*	*etewar*
NPI-indefinite, 'any-'	*dihein/dohein/ dehein(ig), einig/einic(h)*	*(io)wiht/ ieht*	*ioman*	*io (mer/ wanne/ in altere)*	*(io)wergin/ iergen, ioner io(gi)war*
n-indefinite 'no-'	*nihein(ig)/ nohein(ig)*	*ni(o)wiht*	*ni(o)man*	*nio (mer/ in altere), ni(e)wanne*	*(niowergin/ niergent), nioner*

[30] The alternative assumption of a grammaticalization starting from previous movement of the n-indefinite to SpecNegP could not explain why exactly the same process is found for the non-n-indefinite but NPI indefinite *(io)wiht* > *iht* > *it/et*, which is still evidenced as a neg-particle in some varieties of Modern German (see section 5.1.3.). The same applies to other neg-particles grammaticalized from NPIs and minimizers such as French *pas*. Furthermore, there is independent evidence against the underlying assumption of obligatory movement of n-indefinites to SpecNegP (see section 5.2.4. below).

When an indefinite determiner, pronoun, or adverb occurred in the scope of negation in OHG, three basic syntactic patterns were possible: besides marking negation both on the indefinite as well as through the clitic neg-particle on the verb, negation could be marked by the neg-particle only, using a type of indefinite other than an n-indefinite (notably an NPI-indefinite), or negation could be marked on the indefinite only and the neg-particle was lacking:

Pattern A: negative concord (Neg^0 on V + n-indefinite)
Pattern B: negation marked by neg-particle only (Neg^0 on V + (NPI) indefinite)
Pattern C: negation marked by n-indefinite only (n-indefinite, no Neg^0 on V)

These patterns are illustrated in examples (31) to (33):

(31) Pattern A (negative concord)
(& precepit / illis. ne cui dicerent.)
gibot her / in tho thaz sie niheinagamo **ni**sagatin
told he them then that they nobody NEG.told
'Then he told them not to tell anybody' (Tatian 130, 15–16)

(32) Pattern B (neg-particle only, non-negative indefinite)
(In qua sententia nemo dubitet secundam esse personam.)
In dhesemu quhide **ni** bluchisoe eoman, ni dhiz sii chiuuisso…
in this saying NEG doubt anybody NEG this be certainly
'Nobody shall doubt that in this saying, it is certainly…' (Isidor III, 6)

(33) Pattern C (n-indefinite only, no neg-particle)
(In quo nondum quisquam / positus fuerat.)
Inthemo noh nu nioman / Ingisezzit uuas.
in.which still now nobody put was
'in which nobody had been put yet' (Tatian 322, 5–6)

The frequencies of these patterns among all clauses with indefinites in the scope of negation in the OHG corpus from Isidor, Tatian, Otfrid, and Notker are given in Table 5.6.

TABLE 5.6 **Distribution of negation patterns A, B, and C in Old High German (cf. Jäger 2008)**

		A (NC)	B (neg-particle only)	C (n-indefinite only)
OHG	Isidor	18% (2)	82% (9)	0% (0)
	Tatian	87% (60)	9% (6)	4% (3)
	Otfrid	35% (6)	65% (11)	0% (0)
	Notker	85% (11)	8% (1)	8% (1)
	average	56%	41%	3%

Pattern A, that is, co-occurrence of the preverbal clitic negative particle and an n-indefinite in a negative-concord construction (more specifically, the NC subtype called Neg-Doubling), was the majority pattern in OHG. Contra Lehmann (1978), it occurred with preverbal as well as with postverbal indefinites. Where the translated OHG texts deviated from their Latin originals in negated clauses containing indefinites, it was mostly in that they added the negative particle and created Neg-Doubling, which must accordingly have been a genuine part of OHG grammar. Note, however, that there is great variation among the individual texts. The main pattern competing with NC was to mark negation only through the neg-particle and not to use an n-indefinite, but a non-negative (normal or NPI) indefinite instead. In some texts, this pattern is even more common than NC. This is particularly the case in the earlier OHG translation of Isidor. Here, pattern B is even frequently introduced against the Latin original. Note that both pattern A and B are ungrammatical in Modern Standard German, which normally displays pattern C in this type of context, i.e. it marks negation by means of an indefinite only. In classical OHG, pattern C is hardly found at all. In late OHG, on the other hand, it is becoming relatively more frequent—a development that continues in MHG.[31]

While negative concord in the form of Neg-Doubling is a very common phenomenon in OHG, NC in the form of so-called Neg-Spread, namely, co-occurrence of several n-indefinites, is barely attested at all (cf. Behaghel 1918: 240, Donhauser 1998: 297, Jäger 2005, 2008: 213–4). If several indefinites appear in the scope of negation, only one is generally an n-indefinite. As in Modern German, it is mostly the first one that is neg-marked:

(34) (Deum nemo uidit umquam.)
got nioman **ni**gisah io in altere
God nobody NEG.saw ever in ages
'Nobody has ever seen God.' (Tatian 45, 21)

However, very rare attestations of Neg-Spread suggest that this kind of construction was perhaps optionally possible too, yet it was clearly not the rule. In my corpus of over 1200 negated OHG clauses, there is only a single case of Neg-Spread. However, the occurrence of several indefinites within one negated clause is of course generally very infrequent anyway.

(35) (cui nemo unquam / hominum sedit)
in theme neoman neo in aldere / manno saz
in which nobody never in ages of.man sat
'in which no man ever sat' (Tatian 189, 6–7)

[31] The choice of pattern B (or C) over A does not correlate with the syntactic function of the indefinite, *in situ* vs moved position of the indefinite, or adjacency to the finite verb (contra Donhauser 1998). There is a tendency, but no strict rule, to choose pattern A in clauses with preverbal indefinites (cf. also Behaghel 1918, Donhauser 1998, Neg-First Principle in Jespersen 1917, Horn 1989).

5.2.2 *Middle High German*

The same basic syntactic patterns observed in OHG are also found in MHG:

(36) Pattern A (NC):
Da **enwart** nymand konig, (er enwúrd darzu erkorne.)
there NEG.became nobody king he NEG.was to.it chosen
'Nobody became king there, unless he was chosen.' (*Prosalancelot* 10, 9)

(37) Pattern B (neg-particle only, non-negative indefinite):
Was die koniginne sprach, die jungfrauw sprach ein wort **nicht**;
what the queen said the maiden said a word NEG
'Whatever the queen said, the maiden did not say a word.' (*Prosalancelot* 46, 226)

(38) Pattern C (n-indefinite only, no neg-particle):
Und sie hatten nymant miteinander gewunnen dann ein
and they had nobody with.each.other won than a
junges knebelin kleyn
young boy small
'And they had no child with each other apart from a small boy'
(*Prosalancelot* 10, 3)

The frequencies of these syntactic patterns in the MHG corpus are, however, quite different from the situation in OHG, as Table 5.7 illustrates. This fact is partly related to changes in the indefinite system (cf. Jäger 2010): in the course of the MHG period, the former three-set system of indefinites ('normal'/PPIs, NPIs, n-indefinites) is largely reduced to a two-set system with just the opposition between 'normal' indefinites and n-indefinites. The original NPI indefinites either die out as eventually in the case of *iht* 'anything',[32] change towards 'more positive' and

TABLE 5.7 **Distribution of negation patterns A, B, and C in Middle High German (cf. Jäger 2008)**

		A (NC)	B (neg-particle only)	C (n-indefinite only)
MHG	*Nibelungenlied*	17% (8)	4% (2)	79% (37)
	Lancelot	37% (16)	2% (1)	61% (26)
	Berthold	9% (4)	0% (0)	91% (42)
	average	21%	2%	77%

[32] *Iht* partly also showed tendencies to turn into an n-indefinite in MHG. Adverbially, it had been very occasionally used as a neg-particle since OHG and survives as such in some Upper German dialects (see section 5.1.3).

become 'normal' indefinites as in the case of *ioman* 'anybody' > 'somebody', or change towards 'more negative' and become n-indefinites as in the case of *dehein/ kein* 'any' > 'no'. The latter item could still appear as an NPI in MHG, as in (39),[33] but also already as the only marker of negation in a clause, as in (40), which constitutes the decisive criterion for n-indefinite status (cf. Jäger 2007, 2010):

(39) *kein* as an NPI (conditional clause):

Nu	dunckest	du	mich	als	wise,	sol	<u>kein</u>	<u>man</u>	radt
now	seem	you	me	as	wise,	should	any	man	advice
darzu	geben,	das	thust	auch	du.				
to.that	give	that	do	also	you				

'Now you seem to me just as wise. If any man should give advice on this, so should you.' (*Prosalancelot* 88, 521)

(40) *kein* as an indefinite:

das	sol	das	buch	vil	wol	hernach	gesagen,	wann	wir	haben
that	shall	the	book	very	well	later	tell	because	we	have
es	yczo	<u>keyn</u>	<u>stadt</u>							
it	now	no/any	place							

'The book shall tell that later, because we do not have any room for it now' (*Prosalancelot* 10, 3)

Table 5.8 gives an overview of the diachronic changes in the indefinite system. With the virtual loss of NPI indefinites, the former optionality between an NPI-indefinite or an n-indefinite in the scope of negation that accounted for the high frequency of pattern B in OHG was lost. Pattern B consequently drastically decreased and is scarcely attested any more in MHG (on average 2% of all clauses with an indefinite in the scope of negation, cf. Table 5.7). Since the MHG period, an indefinite in the scope of negation as a rule has had to take the form of an n-indefinite in German. This is still the case today. The decrease of pattern B thus led to an increase in pattern C with neg-marking by an n-indefinite, but interestingly not an increase in pattern A of NC between an n-indefinite and a negative particle. Far from it, NC in fact drastically decreased in MHG. While pattern C, i.e. neg-marking just by the n-indefinite (the only grammatical pattern in Modern Standard German) is already the majority pattern amongst clauses with an indefinite determiner, pronoun, or adverb in the scope of negation in MHG (on average 75%), NC has decreased to an average of 23%. This is a reduction by more than half compared to the average found for the OHG data.

[33] In so-called weak NPI contexts such as conditionals, standards of comparison, occasionally clauses dependent on matrix negation/adversative predicates, and in the context of 'rarely', 'hardly', and 'before'.

TABLE 5.8 The development of the indefinite system from Old High German to Modern German

	determiner		'entity'		'person'	
'normal' / PPI	*sum(ilih)* *ein* → *eteslih*	*ein*	*ete(s)waz* →	*etwas*	*ete(s)wer*	*jemand*
NPI	*einig* *dehein(ig)*		*(io)wiht*		*ioman* →	
n-indef.	*nihein(ig)*	*kein*	*ni(o)wiht* →	*nichts*	*ni(o)man* →	*niemand*

	time		place	
'normal'/ PPI	*ete(s)-wenne*	*irgendwann*	*etewar*	*irgendwo*
NPI	*io* →	*je*	*(io)wergin* *ioner*	
n-indef.	*nio* →	*nie*	*niowergin* → *nioner*	*nirgends* *nirgendwo*

In contrast to the literature, which usually takes NC to be a particular trait of MHG,[34] NC is thus in fact more typical of OHG and already diminishing in MHG so

[34] Cf. statements such as Admoni's (1990: 103) that in MHG there is 'in most cases doubling of *ne* on indefinites [Neg-Spread—AJ] or on indefinite and verb [Neg-Doubling—AJ]' ('in den meisten Fällen Dopplung von *ne* an Indef. oder an Indef. und V').

that its eventual extinction can be seen as a natural development that started well before the influence of prescriptive grammars. The quantitative analysis of OHG and MHG data accordingly demonstrates that, as is quite logical, MHG takes up an intermediate position between OHG and Modern German, and—just as with respect to the negative particle—is already a lot closer to Modern German than has widely been assumed. These results also contradict the hypothesis that German developed NC under Greek influence and lost it under Latin influence during humanism, i.e. after the Middle Ages, as proposed by Pensel (1976). NC was a genuine part of OHG grammar, as in other Old Germanic languages. The drastic decrease of NC took place much earlier and was directly related to the loss of the preverbal clitic neg-particle through Jespersen's cycle rather than caused by Latin influence. In fact, at a time when the influence of Latin on the German language was particularly strong, namely during the OHG period, NC was most common—even in translated texts in direct opposition to the Latin original.

The loss of NC is basically caused by the loss of the preverbal clitic neg-particle:[35] as in OHG, the NC in MHG is mostly of one type, Neg-Doubling between the preverbal clitic neg-particle and an n-indefinite, as in (36) above. With the loss of the clitic neg-particle, NC disappears. According to my corpus data, this process is already basically complete in Berthold.[36] One might have expected an increase in NC instead, because more types of NC should be possible in MHG, notably Neg-Doubling involving the newly grammaticalized neg-particle *niht*. Yet this type of NC is found rarely or not at all. This state of affairs is reminiscent of Standard French, where the Neg^0 particle *ne*, but not the SpecNegP particle *pas*, occurs in Neg-Doubling constructions with an n-indefinite (cf. section 2.3 on French, and section 6.4.2. on a similar pattern in Middle Low German).

There are no examples of Neg-Doubling between the neg-particle *niht* and an n-indefinite among the first 100 negated clauses in the *Nibelungenlied* and only one each from the Prose-Lancelot and Berthold. Both of these, however, involve the indefinite *dehein/kein*, which behaves differently, as it is still in transition from an NPI indefinite towards an n-indefinite. Therefore these cannot be counted as clear evidence. My data concur with Behaghel's (1918: 241–2) findings: he states that Neg-Doubling of the neg-particle *niht* with an n-indefinite is lacking in the *Rolandslied*,

[35] A similar correlation has been suggested for English (Frisch 1997: 33, contra Ingham 2006a). These findings are also interesting with respect to the link that has been established typologically between Neg^0 neg-particles and NC (Zeijlstra 2004): languages with a Neg^0 neg-particle will display NC; NC may or may not occur in languages without a Neg^0 neg-particle.

[36] In Middle English, this process apparently took slightly longer. According to Ingham (2006a), this type of Neg-Doubling was lost in the 14th century, whereas co-occurrence of several negative XPs (Neg-Doubling with *not*, Neg-Spread) is attested for at least a century after that.

Berthold, Füetrer's Lancelot and Eilhart-Prose.[37] The fact that Neg-Doubling with the phrasal neg-particle is so rare could be explained on the basis of economy: *niht* stepped in to additionally identify negation mainly in those cases where negation was not already sufficiently identified by an n-indefinite. This resulted in a virtually complementary distribution of *niht* and n-indefinites.[38]

As in OHG, NC of the Neg-Spread type is also virtually lacking in MHG. In my corpus there are no cases involving original n-indefinites. Paul (2007: 391) and Behaghel (1918: 241–2) mention this kind of construction and also give a few examples with original n-indefinites.[39] While Neg-Spread might thus have been marginally possible in MHG, it was certainly not obligatory in MHG either, as the following examples with several indefinites in the scope of negation with only one appearing as an n-indefinite indicate:

(41) wir heten **ninder** einen zagen.
we had nowhere a hesitation
'We did not hesitate at any point' (*Nibelungenlied* (A) IV, 231, 4)

(42) ich wene **nie** ingesinde groezer milte ie gepflac.
I think never attendants greater mildness ever cultivated
'I think that no attendants ever acted with such great mildness' (*Nibelungenlied* (A) II, 43, 4)

MHG differs from related languages such as early Middle English for which Ingham (2006a) states that all indefinites in the scope of negation had to be strictly neg-marked (cf. also section 4.4). In sum, NC between several negative XPs (Neg-Doubling incl. *niht*, Neg-Spread) is virtually unattested in MHG, cf. Jäger (2008: 294–7). As mentioned above, the only noteworthy exception to this rule is found in clauses containing *dehein/kein* 'no' which is still in transition from an NPI to an n-indefinite. As an NPI, it may of course co-occur with the verb-independent XP neg-particle *niht*, even with the bipartite neg-particle, or with an n-indefinite:[40]

[37] In contrast to my data, Behaghel (1918: 241) and Paul (2007: 391), however, also mention a few MHG examples of original n-indefinites co-occurring with the neg-particle *niht*, namely *niemen* 'no one' or *nie/nimmer* 'never' + *niht*.

[38] Under the Neg-Criterion approach, the lack of this type of Neg-Doubling is due to the blocking of the supposed neg-movement of the n-indefinite into SpecNegP by the neg-particle (cf. Zanuttini 1997). Zanuttini (1997) predicts that there will be no NC in languages in which the negation particle occupies SpecNegP. However, that leaves NC in languages such as Yiddish or Bavarian unaccounted for.

[39] However, most examples given in Paul (2007) for the co-occurrence of several n-indefinites are in fact not Neg-Spread but Neg-Doubling constructions with the neg-particle *niht* and the former NPI *kein* 'no, any'.

[40] Very rarely, the pattern *dehein/kein* + n-indefinite is even combined with Neg-Doubling with *en/ne*:

(i) Da macht sie so großen jamer das nye keyn man merern jamer endorfft gesehen
then made she so big mourning that never no/any man more mourning NEG.may see
'She was in such great mourning that nobody may ever see greater mourning.' (*Prosalancelot* 46, 229)

Yet this is by no means necessary. In this respect, MHG differs from languages such as Polish and Russian where Neg-Spread needs to be licensed by the overt neg-particle, cf. Giannakidou (1998).

(43) SpecNegP *niht* + *dehein/kein*:
wan er des **niht** enbern wil von dekeinem menschen
because he that NEG miss wants of no/any man
der ze sinen tagen komen ist
who to his days come is
'because he does not want to miss it of any man who has come to his days'
(Berthold I, 24 (6))

(44) Neg^0 *en/ne* on Vfin + SpecNegP *niht* + *dehein/kein*:
dheyn so gebryset ritter noch so hoch **enist** in diser welt **nicht**
no/any so praised knight nor so high NEG.is in this world NEG
'There is no knight so praised or so high in this world...' (*Prosalancelot* 36, 167)

(45) N-indefinite + *dehein/kein*:
aber sin freude hat niemer mer kein ende
but his joy has never more no/any end
'but his joy will never have an end' (Berthold I, 14 (4))

In contrast to Donhauser's (1998: 297) generalization that NC occurs as Neg-Doubling in OHG, but as Neg-Spread in MHG, one can conclude from these data that Neg-Spread was arguably marginally possible in both, but was never very widespread let alone compulsory with multiple indefinites. Neg-Doubling between the preverbal clitic neg-particle and an n-indefinite was clearly the most common type of NC both in OHG and MHG.

5.2.3 *Early New High German and Modern German*

By the ENHG period, negative concord had become virtually extinct. According to the study by Pensel (1976), there are a few attestations of Neg-Doubling between *en* and an n-indefinite around AD 1500, mostly in Low German (cf. also chapter 6) and partly in West Central German, but none in East Central German. Unfortunately, Pensel did not differentiate between this type of NC and occurrences of the bipartite neg-particle: the two phenomena together make up 4.8% of the negated clauses in his corpus. Very rarely, in 1.8% of negated clauses (especially from the East Central German and West Upper German area), he also found Neg-Doubling between the phrasal neg-particle *nicht* and an n-indefinite. Recall that this type of Neg-Doubling was barely attested in MHG except with the former NPI indefinite *dehein/kein* 'any' > 'no'. The special role of this indefinite is supported by the fact that Pensel's Neg-Doubling data also almost exclusively involve *kein*. Around 1700, Neg-Doubling between *nicht* and an n-indefinite (still almost only *kein*), as in (46), is even rarer with 0.6% of the negated clauses in Pensel's corpus, mostly from East Central and West Upper German.

(46) sie have Keine so wahrhaffte Freude **nicht**

they have no such veritable joy NEG

'They do not have any such real joy.' (II Br. 2,41; after Pensel 1976: 315)

Neg-Doubling including preverbal *en*, however, had died out in all of the German-speaking area around 1700 according to Pensel's data.[41]

In Modern Standard German, NC is ungrammatical. The use of two neg-markers results in highly marked semantic double negation. If an indefinite pronoun, determiner, or adverb occurs in the scope of semantic negation, it generally has to take the form of an n-indefinite. The neg-particle *nicht* is precluded in this case; hence we find *nie* 'never' rather than **nicht je* 'not ever' in (47). Any further indefinites occur as 'normal' indefinites, not as n-indefinites, as in (48). The NC constructions in (47c) and (48b) are fully ungrammatical in standard German but acceptable in some dialects.

(47) a. *Sie hat **nicht** je gesungen.

she has NEG ever sung

b. Sie hat nie gesungen.

she has never sung

c. */#Sie hat nie **nicht** gesungen.

she has never NEG sung

'She has never sung.'

(48) a. Sie hat nie eine Mozartarie gesungen.

she has never a Mozart aria sung

b. */#Sie hat nie keine Mozartarie gesungen.

she has never no Mozart aria sung

'She has never sung a Mozart aria.'

In contrast to the standard language, NC is found in certain Modern German dialects, for instance in Bavarian (cf. Bayer 1990, Weiß 1998):

(49) Koa Mensch is **ned** kema (Weiß 1998: 167)

no human is NEG come

'Nobody came.'

(50) Mia hod neamad koa stikl broud **ned** gschengt (Weiß 1998: 186)

me has nobody no piece bread NEG given

'Nobody gave a piece of bread to me.'

[41] Pensel's (1976) data does not admit any conclusions with respect to Neg-Spread: unfortunately, he excluded negated clauses that did not contain a neg-particle, wrongly taking neg-marking just by n-indefinites to constitute constituent negation rather than sentential negation.

and in some Thuringian dialects:[42]

(51) närjend war kei Schwein ze fingen
nowhere was no pig to find
'No pig was to be found anywhere.' / 'Nobody was around.' (Mansfeld German) (*Thüringisches Wörterbuch*, vol. iv.: 888)

(52) ich bin nerngds **net** hiekumme
I am nowhere NEG get
'I didn't get around.' (Greiz German) (*Thüringisches Wörterbuch*, vol. iv.: 888)

Note that in both Bavarian and in Thuringian, there is evidence for Neg-Doubling with the SpecNegP neg-particle *ned/net/nech* (corresponding to Standard German *nicht*) in (49), (50) and (51), as well as for Neg-Spread in (50) and (51). While a number of Swiss German dialects fail to show NC at all, some Swiss-German dialects have NC of the Neg-Spread type only:

(53) Es cha niemer nüüt defur.
it can nobody nothing there.for
'It's nobody's fault.' (<http://www.medical-info.ch/samwunsch/playlist.php>, July 2006)

The types of NC found in Modern German NC dialects are different from the main type of NC in OHG and MHG: whereas the earlier stages of German mainly showed Neg-Doubling with the Neg^0 neg-particle and co-occurrence of negative XPs was found rarely or not at all, this latter type is the only NC type found in the modern dialects, and the main OHG and MHG NC type is unattested. The dialects thus did not simply preserve the old patterns, but developed new ones.

Dehein/kein arguably played a crucial role in this process: as an original NPI it could of course co-occur with negative XPs. When it changed into an n-indefinite, there were thus instances of an n-indefinite co-occurring with another negative XP in the language. In certain dialects, this syntactic pattern was then extended to the original n-indefinites, so that they too could co-occur with another negative XP and Neg-Doubling with *nicht* as well as Neg-Spread became generally available. Indeed, *kein*, which was virtually the only item allowing these patterns already in MHG and ENHG, is still particularly common in NC constructions in the Modern German NC dialects (cf. Donhauser 1996: 210).

5.2.4 *Analysis*

According to recent syntactic and semantic studies (Zeijlstra 2004, Penka 2010 among others), negative concord may be analysed as an instance of feature checking:

[42] Besides the dialects discussed here, NC is locally attested in Saxon, South-Hessian, Palatine, Brandenburg-Berlin, and Mecklenburg dialects.

n-indefinites are not semantically negative but possess only a formal, uninterpretable negative feature that is checked against the interpretable negative feature residing in NegP. According to this approach, no obligatory movement of n-indefinites to SpecNegP according to the Neg-Criterion (Haegeman and Zanuttini 1991: 244) and ensuing factorization and semantic absorption of negative features need be invoked. Instead, the n-indefinites may be licensed in situ. The OHG example in (54) contains an n-indefinite within a PP. As PPs are generally assumed not to undergo scrambling, the n-indefinite must be in VP-internal *in-situ* position:

(54) (& non respondit ei / ad ullum uerbum)

Inti	**ni**antligita	imo /	zi	noheinigemo	uuorte
and	NEG.answered	him	to	no	word

'and did not answer to a single word' (Tatian 310, 16–17)

Thus, obviously not all n-indefinites move in overt syntax. Furthermore, they can even be shown to be *in situ* at LF in a number of cases: in so-called neg-split constructions, the correct readings can only be derived if the n-indefinite is interpreted *in situ*. Moreover, these constructions demonstrate that, in contrast to the assumptions in Zeijlstra (2004), even n-indefinites in non-NC languages such as Modern Standard German are semantically non-negative (cf. Penka 2010): the n-indefinite needs to be interpreted as a semantically non-negative indefinite while the negation takes scope independently (cf. Jacobs 1980, 1991, Penka and Stechow 2001). This is, for instance, the case with n-indefinites embedded under modals. The sentence in (55) has the preferred reading (i), as well as reading (ii), but not (iii):

(55)

Du	musst	keinen	Schlips	tragen.
you	must.2SG	no	tie	wear.INF

(i) 'It is not necessary that you wear a tie.'
(ii) 'It is necessary that you wear no tie.'
≠ (iii) 'There is no tie for which it is necessary that you wear it.'

In spite of the fact that the n-indefinite, or n-phrase, *keinen Schlips* 'no tie' is the object of *tragen* 'to wear', which is embedded under the modal *musst* 'must, have to', negation can take wide semantic scope over the modal while the indefinite is interpreted *in situ* resulting in the preferred reading (i). The 'negative' and the indefinite part of the n-indefinite are so to speak split up semantically. More precisely, the n-indefinite itself is simply interpreted as a non-negative indefinite. It is licensed by the actual semantic negation in NegP, which may remain abstract, and which may take scope independently. The interpretation of *keinen Schlips* as an inherently negative quantifier, by contrast, would result in the readings (ii) or (iii). The preferred reading (i) could not be derived. A neg-movement analysis for n-indefinites (or quantifier raising at LF) would wrongly result in the unavailable reading (iii) only.

N-indefinites can thus be shown to be semantically non-negative throughout the history of German. Their uninterpretable neg-feature [uNeg] is licensed by the interpretable neg-feature [iNeg] of the (possibly abstract) semantic negation in NegP (see section 5.1.4. above) under c-command (Agree in the sense of Chomsky 2001). The n-indefinite may accordingly remain *in situ* (cf. also Błaszczak 2001). This is illustrated in (56) for the OHG example (31):

(56)

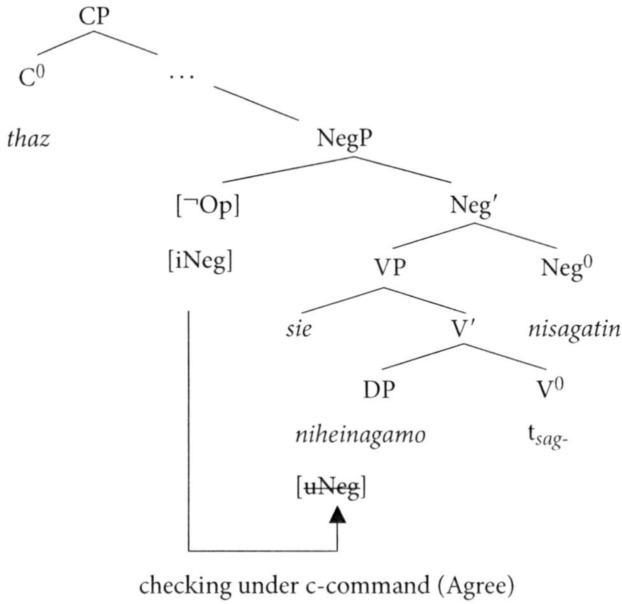

Diachronically, indefinites may gain an uninterpretable neg-feature requiring checking and thus limiting their distribution. Accordingly, they may change from NPIs to n-indefinites as in the case of German *dehein/kein*. Comparable cases of change in polarity type are found in various languages, including also changes towards 'more positive' through the diachronic loss of the relevant uninterpretable feature (see Jäger 2010 for an underspecification-theoretic approach on the basis of the two polarity features ± negative and ± affective).

5.3 Further ways of marking negation

5.3.1 Narrow focus of negation and constituent negation

In OHG, narrow focus of negation on a particular constituent is occasionally expressed just with the usual preverbal neg-particle *ni* and a focus-explicating, contrasting 'but'-phrase that additionally identifies the neg-focus, namely the constituent in square brackets in (57):

(57) Thaz ih **ni** scríbu [thuruh rúam], súntar bi thin lóbduan
that I NEG write through fame but at your praise
'that I do not write for fame, but for your praise' (Otfrid I. 2, 17)

As the neg-particle *ni* in OHG is bound to the finite verb, it is not particularly suitable for marking either narrow focus of negation or constituent negation, especially in written language, where the identification of focus through intonation is not possible. Without a focus-explicating phrase, the focus is not clearly identified. In cases with verbal ellipsis, this kind of construction is altogether impossible. Like other languages in which the usual marker of sentential negation is bound to the verb,[43] OHG therefore posseses a special verb-independent focus-indicating neg-particle *nalles* (also *nalas*, *nals*, compare Old English *nalles/nealles*, not attested in Old Saxon, cf. Mouřek 1903: 37, Behaghel 1918: 229). *Nalles* is composed of negation and an adverbial genitive of *all* 'all' (cf. Braune and Reiffenstein 2004: 233) and thus morphologically corresponds to French *pas du tout* or English *not at all*. Diachronically, a special neg-focus particle is of course particularly expected with languages at Jespersen's cycle stage I.

Just like *ni*, the OHG neg-focus particle *nalles* is used to translate Latin *non* in the translated texts. However, *nalles* is only used in cases with narrow focus of negation, cf. (58) or, more rarely, with narrow semantic scope of negation, i.e. constituent negation ('Konstituentennegation/Sondernegation/Satzgliednegation') cf. (59).[44]

(58) (misericordiam uolo & non sacrificium.)
íh uúili miltida nalles [bluostar]
I want mercy NEG sacrifice
'I want mercy, not sacrifice.' (Tatian 91, 23)

(59) (et non post multos dies / congregatis omnibus / adolescentior filius / peregre profectus est / in regionem longinquam)
nalles [after manegen tagon] / gisamonoten allen / ther iungoro
NEG after many days gathered all the younger
sun / elilentes fuor in uerra lantscaf
son hurrying went in far region
'Not many days later, having gathered everything, the younger son went quickly to a region far away' (Tatian 154, 26–30)

[43] Turkish (cf. van Schaaik 1994: 39 and 42) normally marks negation by means of the verbal neg-affix *-me-*, but uses the special verb-independent neg-particle *değil* for narrow neg-focus/contrastive negation. Japanese (cf. Tanaka 1994: 193–4) employs the topic-particle *wa* after the respective constituent in addition to the usual verbal agglutinating neg-particle *nai*, in order to mark narrow neg-focus.

[44] For example in Tatian, *nalles* marks narrow scope (rather than just narrow focus) of negation in two out of 41 total attestations of *nalles*.

Due to its special function, *nalles* is generally rare: it occurs in around 4% of the negated clauses from Isidor, Tatian, and Otfrid and 11% in Notker. In contrast to *ni*, the neg-focus particle *nalles* is phrasal and its position is verb-independent: it is generally placed immediately to the left of a focused constituent of various types such as NP, AP, PP, V(P), or CP. *Nalles* is therefore syntactically similar to other focus particles. Accordingly, it may be analysed as part of the focused XPs of different categories (cf. the usual analyses for English *even*, *only*, etc.). Alternatively, it could be adverbially adjoined to extended V-projections as Jacobs (1983) and more recently Büring and Hartmann (2001) argue for Modern German focus particles such as *nur* 'only'.

In most cases, *nalles* marks narrow focus rather than narrow scope of negation. The entire proposition is semantically negated. *Nalles* identifies the element that would have to be substituted for one of a contextually restricted set of alternative elements of the same logical type in order to render the proposition true (cf. Jacobs 1991, Rooth 1992).[45] This semantic analysis is supported by the fact that the usual marker for sentential negation, the preverbal clitic neg-particle *ni* may co-occur with *nalles*, as illustrated in (60):

(60) (Non omnis qui dicit mihi / domine domine. Intrabit in regno celorum; / sed qui facit uoluntatem patris mej / qui In caelis est ipse Intrabit / In regno caelorum)
Nalles iogiuuelih ther mir quidit/ truhtin truhtin **ni**gat
NEG anyone who me says lord lord NEG.goes
in himilo rihhi / ouh ther the tuot mines fater uuillon /
in heaven kingdom but he who does my father's will
ther in himile ist hér gát / In himilo rihhi.
who in heaven is he goes in heaven kingdom
'Not everyone who calls to me "Lord, Lord" will go to the kingdom of heaven but he who does the will of my father who is in heaven will go to the kingdom of heaven.' (Tatian 74, 20–4)

For the most part, however, *nalles* occurs in the typical gapping constructions without a finite verb. As narrow focus of negation often comes with a contrastive reading,[46] the focus of negation may, in addition to *nalles*, be identified by a focus-explicating, contrasting 'but'-phrase:

[45] This can be formalized in terms of Jacobs's (1991) INKL-function or Rooth's (1992) normal-semantic vs focus-semantic value based on alternative semantics.

[46] Mostly of the 'not (only) x, but y'-type, in contrast to Wheelock LaBrum's (1982: 215) statement that this contrast type is virtually absent in OHG; besides also in 'x, not y'-type contrasts.

(61) (non quod ego uolo / sed quod tu.)

nalles	[thaz	ich	uuili] /	nibi	thaz	thu.
NEG	that	I	want	but	that	you

'not what I want, but what you want' (Tatian 294, 24–5)

Towards the MHG period, *nalles* became extinct. This does not mean, however, that the language then 'lacked a way of marking narrow focus or scope of negation for a longer period', as Paul (1920: 331) and Qian (1987: 31) state.[47] Instead, the verb-independent and therefore positionally more free neg-particle *niht* took over the function of *nalles*, too, as soon as it was grammaticalized. Consider the use of MHG *niht* in (62) which is strikingly parallel to the use of OHG *nalles* in (61) above:

(62)

als	ér	wil	und	**niht**	[als	wír	wellen].
as	he	wants	and	NEG	as	we	want

'as he wants and not as we want' (Berthold I, 51 (10))

Once the language reached Jespersen's stage II, there was no need for a special focus indicating neg-particle such as *nalles* any more. The new second neg-particle *niht*, being verb-independent in contrast to *ni/ne*, was basically free to be placed immediately before any syntactic constituent to mark narrow focus of negation on that constituent. Right up to the present day, *nicht* is still also used in ways comparable to OHG *nalles*. In the unmarked case, *nicht* occupies SpecNegP (cf. section 5.1.4. above). However, with narrow focus of negation, it may occur sentence-initially together with the constituent that is the focus of negation and also potentially lower than VP, as in (63) and (64) (unless the latter kind of case just involves additional emptying of VP by scrambling). Occasionally, like *nalles*, *nicht* also marks true constituent negation, i.e. narrow semantic scope of negation as in (65):

(63)

Nicht	[Anna]	ist	gekommen,	sondern	Maria.
NEG	Anna	is	come	but	Maria

'It's not Anna who came, but Maria.'

(64)

dass	Anna	nach	Weimar	**nicht**	[fährt],	sondern	wandert
that	Anna	to	Weimar	NEG	goes	but	walks

'that Anna does not GO (i.e. by car/train) but walks to Weimar.'

(65)

In	der	Küche	stehen	mehrere	**nicht**	[geöffnete]	Weinflaschen.
in	the	kitchen	stand	several	NEG	opened	winebottles

'There are several unopened bottles of wine in the kitchen'
≠'It is not the case that there are several open bottles of wine in the kitchen.'

[47] Paul (1920: 331), repeated word-for-word in Qian (1987: 31): *nalles* 'im Mittelhochdeutschen untergegangen [. . .] so dass es längere Zeit an der Möglichkeit zu solcher Verneinung einzelner Satzglieder fehlte' ('*nalles* died out in MHG, such that the language lacked such a possibility for negating individual constituents for a longish period').

5.3.2 *Negative subjunctions and disjunctions*

5.3.2.1 *Subjunctions* In a few cases, OHG *ni* occurs in the left sentence periphery as a negative subjunction or complementizer meaning 'that not'. Partly, it only constitutes expletive negation in a clause dependent on a negated matrix clause. There is one example in the entire OHG Isidor and eight cases among the first 100 negated clauses from Otfrid:

(66) (In qua sententia nemo dubitet secundam esse personam.)
In dhesemu quhide ni bluchisoe eoman, **ni** dhiz sii
in this saying NEG doubt anyone NEG this be
chiuuisso dher ander heit godes, selbo druhtin christ.
certainly the other personality of.god same Lord Christ
'Nobody shall doubt that in this saying it is certainly God's other personality, the same Lord Christ.' (Isidor III, 6)

Furthermore, there is one example in Isidor and two in the entirety of Tatian in which *ni* appears in a left-peripheral position following the question particle *enu/eno/inu*:

(67) (Num angelus qualem cum deo habet imaginem?)
Inu **ni** angil **nist** anaebanchiliih gote?
QU NEG angel NEG.is similar god
'Doesn't the angel look like God?' (Isidor III, 5)

Enu/eno/inu ni translates Latin *num(quid)* or *nonne*, presupposing a positive answer. It is followed by V1 or V2 questions and arguably occupies a position before the prefield (cf. Axel 2007 on *eno/inu*). In none of these cases can *ni* be equated with the standard preverbal Neg^0 neg-particle *ni*, as it optionally co-occurs with it, as in (67).

Another OHG negative complementizer is *nibu/nibi/noba/nub* 'unless'/'if not' (NEG + *ibu* 'if'). It occurs in verb-final clauses that are used as equivalents to clauses with *ibu/oba* 'if' and the usual neg-particle *ni* on the finite verb. In translated texts, *nibu/nibi/noba/nub* stands for Latin *nisi*.[48]

(68) (nisi signa & prodigia uideritis / non creditis.,)
nibi ír zeichan inti uuvntar gisehet / anderuúis nigiloubetír.,
unless you signs and miracles see otherwise NEG.believe
'Unless you see signs and miracles you do not believe' (Tatian 90, 18–19)

[48] *Nibu* also introduces focus-explicating 'but'-phrases, see section 5.3.1. above, and then translates Latin *sed*.

Constructions with negative subjunctions, or complementizers, point to the optional presence of neg-features in CP. Cinque (1999: 201, n. 20) suggests that INFL-projections may generally be doubled within the C-domain. This could accordingly also hold for NegP. Neg-features within CP have also been proposed for Old English by van Kemenade (2000: 74) and for Old Norse by Eythórsson (2002: 216): He assumes that the NegP within the IP-domain is anaphorically dependent on [+ neg] C in these cases. I would suggest, alternatively, that the neg-marker in the C-domain is licensed by the semantic negation in NegP. The possibility of co-occurrence with the usual Neg^0 neg-particle *ni* corroborates the assumption of the presence of NegP in addition to the left-peripheral negative element.

5.3.2.2 Disjunctions In OHG, 'neither…nor' is expressed by the disjunction *noh*. *Noh* may appear in either conjunct or introduce the second conjunct if the first conjunct contains some other neg-marker. If the *noh*-conjunct contains a finite verb, the verbal clitic neg-particle *ni* generally occurs in addition to *noh* in a concordant reading, as in (69) and (70), suggesting that, much like n-indefinites, *noh* is not semantically negative but is licensed by the semantic negation in NegP. This negative-concord pattern is attested in all texts in my corpus.

(69) (nec apostolus dicit nec propheta conperit nec angelus sciuit nec creatura cognouit)
Dhazs ni saget apostolus **noh** forasago ni bifant **noh**
that NEG says apostle nor prophet NEG found.out nor
angil gotes ni uuista **noh** einic chiscaft ni archennida.
angel God's NEG knew nor any creature NEG recognized
'No apostle had said this, nor had any prophet discovered it, nor had any angel of God known it, nor had any creature recognized it.' (Isidor II, 3)

Noh occasionally also co-occurs with an n-indefinite, which constitutes a further instance of negative concord:

(70) (nec / ab ullo potuit curari)
noh / fon iro niheinigemo mohta uuesan giheilit
nor by them no.one could be healed
'nor could be healed by any of them' (Tatian 95, 5–6)

It is even possible that *noh* is used together with *ni* on the finite verb and an n-indefinite so that negation is marked three times in one clause:

(71) (neque patrem quis nouit nisi filius.)
noh then fater niuueiz nioman nibi ther sun
nor the father NEG.knows nobody if.not the son
'nor does anybody know the father but the son' (Tatian 104, 5)

In MHG, the disjunction *noh* occasionally still appears in concordant readings together with the clitic neg-particle as in (72), or with n-indefinites as in (73):[49]

(72) Und da sie zu irselber kam, da ensah sie **noch** enhort ir kint.
and when she to herself came then NEG.saw she nor NEG.heard her child
'And when she recovered, she neither saw nor heard her child.' (*Prosalancelot* 46, 228)

(73) Sô rihtet dem armen als dem rîchen, dem fremden als
so judge the poor as the rich the foreigner as
dem kunden, dem lantman als dem mâge, **weder**
the familiar the countryman as the relative neither
durch liep **noch** durch leit **noch** durch guotes miete
through love nor through harm nor through good bribery
noch durch kein dinc wan nach dem rehten, **noch**
nor through no(/any) thing than after the right nor
nemet von niemen kein guot wan iuwer rehte buoze;
take from nobody no(/any) good than your right due
'So judge the poor man as the rich one, the foreign one as the familiar one, the countryman as the relative, neither on the basis of love nor sorrow nor bribery nor any cause other than the right one, nor take any goods of anyone other than what you are entitled to.' (Berthold I, 84 (14))

As (73) illustrates, a finite verb in the *noh*-conjunct could also appear without the clitic *en/ne* in MHG. In contrast to OHG, the first conjunct of an 'either ... or' construction never contains *noh* in my MHG data, but is negated by other means. The example further shows that there is already evidence for the disjunction *weder ... noch* ('neither ... nor'), which is the standard in Modern German (on *weder noch*, see also Paul 2007: 392, Schmidt 1993: 276). Modern Standard German being a non-NC language, co-occurrence of *noch* with another neg-marker in a concordant reading is excluded today.

5.4 Summary

To sum up the main historical negation patterns and changes, the average negated clause in OHG was negated simply by means of the preverbal negative particle. If indefinites occurred in the scope of negation, these could optionally be neg-marked in addition, producing negative concord. Marking of negation just by means of an n-indefinite is hardly attested at all for OHG. In late OHG, we observe the beginning of the grammaticalization of a second negative particle.

[49] By contrast, the SpecNegP neg-particle *niht* is not attested together with *noh* in my MHG corpus.

In MHG, there are two main syntactic patterns of negation: neg-marking by the newly grammaticalized verb-independent negative particle only or by an n-indefinite only. These two patterns are the only ones still possible in Modern German. In MHG, we also find the preverbal clitic negative particle in combination with either. From OHG to MHG there is a drastic decrease in the frequency of NC related to the loss of the preverbal clitic negative particle in the course of Jespersen's cycle. The preverbal clitic negative particle becomes fully extinct during the ENHG period. NC vanished from the language, too, except for some Modern German dialects which developed new types of NC constructions. As a result of polarity-type changes in individual indefinites, the indefinite system changed over the course of the history of the German language from a three-set system including NPI indefinites to a two-set system with just the opposition between normal indefinites and n-indefinites.

Primary sources

Texts were taken from the TITUS database (<http://titus.fkidg1.uni-frankfurt.de/>) where available.

Berthold von Regensburg, *Vier Predigten Mittelhochdeutsch/Neuhochdeutsch*, ed. Werner Rocke. Stuttgart: Reclam, 1983.

Das Hildebrandslied. In *Althochdeutsches Lesebuch*, ed. Wilhelm Braune and Ernst Ebbinghaus. 17th edn, 84–5. Tübingen: Niemeyer, 1994.

Das Nibelungenlied: Paralleldruck der Handschriften A, B und C nebst Lesarten der übrigen Handschriften, ed. Michael S. Batts. Tübingen: Niemeyer, 1971.

Der althochdeutsche Isidor. Nach der Pariser Handschrift und den Monseer Fragmenten, ed. Hans Eggers. Tübingen: Niemeyer, 1964.

Die lateinisch-althochdeutsche Tatianbilingue Stiftsbibliothek St. Gallen Cod. 56, ed. Achim Masser. Göttingen: Vandenhoeck & Ruprecht, 1994.

Prosalancelot. Nach der Heidelberger Handschrift Cod. Pal. germ. 147, ed. Hans-Hugo Steinhoff and Reinhold Kluge. Frankfurt am Main: Deutscher Klassiker Verlag, 1995.

Notker der Deutsche, *Der Psalter*, ed. Petrus Wilhelmus Tax. Tübingen: Niemeyer, 1979.

Otfrid von Weißenburg, *Evangelienbuch*, ed. Johann Kelle. Regensburg: Manz, 1856.

Otfrid von Weißenburg, *Evangelienbuch*, ed. Wolfgang Kleiber, vol. i. Tübingen: Niemeyer, 2004.

Pariser Gespräche, ed. Wolfgang Haubrichs and Max Pfister. In *"In Francia fui": Studien zu den romanisch-germanischen Interferenzen und zur Grundsprache der althochdeutschen Pariser (Altdeutschen) Gespräche, nebst einer Edition des Textes*. Mainz: Akademie der Wissenschaften und der Literatur, 1989.

Tatian: Lateinisch und altdeutsch, ed. Eduard Sievers, 2nd edn. Paderborn: Schöningh, 1892.

Wessobrunner Predigt. In *Althochdeutsches Lesebuch*, ed. Wilhelm Braune and Ernst Ebbinghaus. 17th edn, 82–3. Tübingen: Niemeyer, 1994.

6

Negation in the history of Low German and Dutch

ANNE BREITBARTH

6.1 Introduction: Low German and Dutch

Low German and Dutch are languages spoken in the northwest of the Continental West Germanic dialect continuum, which includes also High German and Frisian. For Dutch, a standard has developed on the basis of the Hollandish dialect, which is used as the national standard in the Netherlands and the northern part of Belgium. Low German on the other hand is spoken in dialects in the northern part of Germany alongside the High German standard language, but has no written standard of its own. Historically, Dutch dialects belong to the Low Franconian group of West Germanic dialects, while Low German dialects derive from (Low) Saxon dialects. Several dialects in the east of the Netherlands are historically Saxon dialects too (Achterhoek, Drenthe, and Overijssel), while conversely some dialects on German national territory belong to the Low Franconian group.

6.1.1 Low German

Old Low German (Old Saxon) is the language spoken by a group of Germanic tribes calling themselves Saxons and living in what is now northwestern Germany and parts of what are now the Netherlands from *c.*800 to 1200 (cf. Klein 2000: 1245). The first monasteries—and with these the first (mostly Latin) writing—were founded in the area after the so-called Saxon wars (end of 8th c.), though only in the south of the area. Old Low German is only poorly attested textually: the bulk of it is biblical poetry (*Heliand*, *Genesis*) (9th c.), the rest consisting of more minor texts such as verses, ecclesiastical and secular functional prose, and glosses. The *Heliand* (*c.*830) is the largest work by far; at *c.*6000 lines, it makes up around 80% of all Old Low German material (Klein 2000, Sanders 2000).

Middle Low German refers to the dialects spoken in northern Germany between 1200 and 1650 (Stellmacher 1990: 39, Peters 2000b: 1482). In the 14th and 15th centuries, it developed into an international lingua franca in connection with the expansion of the Hanseatic League, spoken all around the North Sea and the Baltic Sea, which led to a certain standardization of the written language incorporating features of different Low German dialects (Härd 1973, 2000, Peters 2000a). Middle Low German was replaced as the written language in the area by (Early New) High German between 1550 and 1650, though Low German continued to exist in spoken dialects.[1]

The scribal dialects of Middle Low German are subdivided into those of the *Altland* (lit. 'old land') and those of the *Neuland* (lit. 'new land') dialects, referring to the colonization history of the area. *Altland* designates the Saxon heartland in the west, *Neuland* the territories east of the river Elbe and along the coast of the Baltic Sea (Peters 2000a). The scribal dialects from west to east in the south of the area are Westphalian, Eastphalian, Elb-Eastphalian, and Brandenburgish, and in the north the North Low Saxon dialects East-Frisian-Oldenburgish, Northalbingian, and East-elbian. The dialects form a continuum bordering Middle Dutch in the west and Middle High German (before 1350), later Early New High German (after 1350), in the south (Peters 1984, 2000a).

6.1.2 *Dutch*

The Old Dutch period is usually assumed to have lasted from *c.*700–1150, although exact delimitation is impossible due to the extremely poor textual attestation. Essentially, only the *Wachtendonck Psalms* (9th or 10th c.) and the *Leiden* (or *Egmond*) *Willeram* (late 11th c.) are recognized as Old Dutch texts, besides smaller fragments and glosses (Quak 1981, Vries et al. 1993, Quak and van der Horst 2002). Both texts are problematic: the *Wachtendonck Psalms* are an interlinear translation of a Latin psalter and the *Leiden Willeram* is a more or less superficial adaptation of a late Old High German text (cf. e.g. Schützeichel and Meinecke 2001), making statements about the syntax of Old Dutch in general, and the expression of negation in particular, very complicated.[2]

The Middle Dutch period (1150–1500/1600) is much more richly attested. Besides a rich literary transmission, mainly from the southern dialects of Flemish and Brabantish, there are also substantial records of charters (Gysseling 1977, van Reenen and Mulder 2000). Because the bulk of the (especially earlier) textual transmission comes from the south, the linguistic literature mainly focuses on southern (Flemish

[1] The present chapter is based on a corpus of charters and other official documents from the 'classical period' (1350–1550; Stellmacher 1990: 39) of Middle Low German. These start being written in Middle Low German around 1325, and the transition to Early Modern German as the language of writing runs to completion between 1525 and 1575.

[2] Cf. Hoeksema (1997: 140) on the impracticality of the *Wachtendonck Psalms* as a witness of Old Dutch syntax.

and Brabantish) texts for this period, though van der Horst and van der Wal (1978) and Burridge (1993) look at texts from both north and south.

The political independence of the northern provinces as a consequence of the Eighty Years' War against Spain (1568–1648) and the economic upturn during the 'Golden Age' (*Gouden Eeuw*, roughly, the 17th century) caused an even greater divergence in the development of the northern and southern dialects. The increased economic importance of the northern provinces from the *Gouden Eeuw* on also shifted the focus of linguistic interest to the textual production of the north, cf. Beheydt (1998) and literature cited there.

This divergence is still apparent in Modern Dutch. Due to the economic superiority of the northern provinces since the *Gouden Eeuw*, the modern standard language, for both the Netherlands and Belgium, is based on the northern Hollandish dialect. Because of the separate history, and the later formation of a Belgian state with Dutch as one of its standard languages, the Dutch dialects in Belgium have preserved a number of archaic features relative to Standard Dutch. This is particularly evident in the expression of negation, as will be seen below.

6.1.3 *The corpus*

The corpus this chapter is based on consists of:

- the *Heliand* epos (*c*.830), the *Genesis* fragments (*c*.840), and the minor texts (verse, ecclesiastic and secular functional prose, glosses; end of 9th–11th century) for Old Low German (696 negative clauses in total);
- the collections of charters and other official texts from ten cities, dioceses, and convents (Barsinghausen, Börstel, Braunschweig, Lübeck, Mariengarten, Oldenburg, Scharnebeck, Steinfurt, Stralsund, and Uelzen) from 1325–1575 for Middle Low German (2829 negative clauses in total); and
- the *Wachtendonck Psalms* and the *Leiden Willeram* for Old Dutch (185 negative clauses in total).

For Middle (and Early Modern) Dutch, the rich existing literature on the development and expression of negation was consulted (van der Horst and van der Wal 1978, de Meersman 1980, de Haan and Weerman 1984, Burridge 1993, Hoeksema 1997, Beheydt 1998, Postma 2002, Postma and Bennis 2006).

6.2 The development of sentential negation in Low German and Dutch

Like all other West Germanic languages, Low German and Dutch underwent the directional development in the expression of negation known as Jespersen's cycle (Dahl 1979). All three steps can be seen in Low German (1) and Dutch (3):

(1) a. Old Low German (Old Saxon): stage I

'**ni**	bium	ic',	quað	he,	'that	barn	godes...'
NEG	am	I	spoke	he	the	child	god.GEN

'"I am not the child of God", he said.' (*Heliand*, 915)

b. Middle Low German: stage II

dar	**en**	sculle	wii	se	**nicht**	ane	hinderen
there	NEG	shall	we	them	NEG	from	bar

'we shall not bar them from it.' (Lübeck 06/01/1450)[3]

c. Modern Low German: stage III

Man	pass	op,	suust	mi	dor	**nich**	daal.
but	watch	out,	scoot	me	there	NEG	down

'But watch out, don't you scoot down there (on me).' (Schleswig-Holstein dialect, <http://inesbarber.de/ib_vertelln.html>)

(2) a. Old Dutch (Old Low Franconian): stage I

ne	farlāt	tu	mi!
NEG	forsake	you	me

'Do not forsake me!' (WP LXX.9.2)

b. Middle Dutch: stage II

Want	ic	**ne**	wille	**niet**,	broeder,	dat	ghi	onwetende	sijt
because	I	NEG	want	NEG	brother	that	you	unknowing	be

'Because I do not want you to be unknowing, brother.' (*Lectionarium Amsterdam* 1348)

c. Modern (Standard) Dutch: stage III

Ik	will	het	**niet**	zien.
I	want	it	NEG	see

'I do not want to see it.'

Thus, at stage I, sentential negation is expressed by means of a single, strictly preverbal element *ni* or *ne*. At stage II, this is done by a bipartite form consisting of the descendant of *ni/ne*, weakened to *en* (or *ne*), and an adverbial element *nicht* or *niet*, which historically derives from a negative indefinite pronoun, *ni-eo-wiht* ('nothing', lit 'NEG.ever.thing'). At stage III, only the adverbial element survives. Low German and Dutch differ in the details of how this development proceeds, as will become clear in the following sections.

In this chapter, the following terminology will be used. Following common practice, albeit potentially confusing in the context of OV languages, I will refer to the newly arising adverbial negator as *postverbal*. An indefinite (pronoun or adverb)

[3] The format of the date in the Middle Low German documents cited is mm/dd/yyyy.

will be called *n-marked* if it shows negative morphology. In Low German and Dutch, such indefinites were formed historically by prefixation of the old negative marker *ni/ne* to an (NPI) indefinite, e.g. *ni-ioman* 'NEG-anyone' > *nioman* 'n.one' or *ni-io* > *nio* 'n.ever'. NPI-indefinites in the scope of negation without such morphological marking will be called *n-free*, e.g. *enig* 'any', *ioman* 'anyone', *io* 'ever'. As a convention, the preverbal marker will be glossed NE, not NEG, when it is no longer clear whether it still expresses sentential negation.

6.2.1 *Low German*

In Old Low German, by far the most common way of expressing sentential negation is by means of the inherited Germanic preverbal marker *ni/ne* alone. More than 99% (617) of the 620 negative clauses in the *Heliand* contain *ni* or *ne*,[4] 582 (94.3%) of them contain no other morphological expression of negation (3). 135 (21.7%) contain an n-free indefinite besides *ni/ne* (4).[5]

(3) **Ni** tharft thu stum uuesan lengron huîla
NEG must you mute be longer while
'You will not be mute any longer.' (*Heliand*, 169–170)

(4) thoh he **ni** mugi **ênig** uuord sprecan
though he NEG can any word speak
'even though he cannot speak a single word.' (*Heliand*, 229)

Of the 38 cases of sentential negation[6] in the minor Old Low German texts, all contain the preverbal marker *ni/ne*. In 32 of these 37 cases, *ni/ne* is the only marker of negation (e.g. (5)), the others contain an n-marked indefinite such as the negative determiner in (6).[7]

[4] The only exception is some negative conjuncts to negative clauses introduced by the disjunction *ni, ne* 'and not, nor', in which the negation marker can be omitted before the finite verb. But even in this type of clause, the preverbal negator is still used in the majority of cases: only in five out of seventeen *ni/ne*-conjuncts is there no additional preverbal negator *ni/ne*.

[5] 401 out of 620 sentences (64.7%) contain only *ni/ne* (that is, without any n-free indefinites or any form of emphasizer), 135 (21.7%) contain an n-free indefinite in addition to *ni/ne*, 74 (11.9%) contain an emphasizer (e.g. 'in this world'), and 35 (5.6%) contain an n-marked indefinite. Two of the clauses with indefinites contain both an n-marked and an n-free indefinite; 23 of the sentences with an n-free indefinite (17%, 3.7% of all negative clauses) and four of the sentences with an n-marked indefinite (11.4%, 0.6% of all negative clauses) contain an emphasizer in addition. For more on indefinites in the scope of negation, see section 6.3 below.

[6] In addition, there are 20 occurrences of the preverbal marker *ni/ne* and two cases of the negative determiner *nian* 'no' which do not occur in full sentences, but in short glosses to Latin texts, mostly only *ne* +verb. Such occurrences are not informative regarding the syntax of sentential negation in Old Low German, as they do not add anything to our knowledge about the existence of regular emphasizers or the interactions of indefinites with negation. These cases are therefore left out of consideration here.

[7] The remaining sentence is a case of an n-marked indefinite occurring without *ni/ne*. Again, on indefinites in the scope of negation, see section 6.3 below.

(5)

ne	uuari	thiu	leccia	heligero	gescriuo
NEG	was	the	lection	holy	written

'the holy lection had not been written.' (Minor OLG texts, PA.15, 16–17)

(6)

tha't	iu	**nian**	scátha	**ni**	uuírthid
that	you	no	damage	NEG	become

'that you do not suffer any damage.' (Minor OLG texts, EsG.53, 31–1)

Given that Modern Low German, like Modern High German, uses *nicht* as the marker of sentential negation, the question arises when and how this entered the language. As is well known, *nicht* derives historically from an n-marked indefinite pronoun *ni-eo-wiht* 'nothing' (lit 'NEG.ever.thing'). As we will see below in section 6.3.1, n-marked pronouns are less common in the scope of the sentential negator *ni* than n-free ones in Old Low German. Therefore, the question of how emphatic strengthening of negation arose in Low German extends to *eo-wiht* 'anything'.

The most frequent type of emphasizer by far in the Old Low German corpus (26 out of 77 negative clauses with emphasizers) are generalizers like 'in this world' or 'in his life' (7).

(7)

That	ni	scal	**an**	**is**	**liƀa**	gio	liðes	anbîtan	uuînes	**an**	**is**	**uueroldi**.
that	NEG	shall	in	his	life	ever	cider	enjoy	wine	in	his	world

'Never in his life will he drink hard cider or wine in this world.' (*Heliand*, 126–7)

Other types of emphasizers include *uuiht* '(any)thing' with a genitive noun or pronoun, (8), and adverbially used *mid uuihti* (lit. 'with anything') meaning 'at all, in the least', (9):

(8)

Ne	sculun	gi	ênigumu	manne	**unrehtes**	**uuiht**	derbies	adêlean.
NEG	shall	you	any	man	injustice.GEN	anything	hostile	give

'You should never pronounce even the slightest detrimental, unjust judgement on any man.' (lit. 'anything of injustice') (*Heliand*, 1691–2)

(9)

that	ic	an	mînumu	hugi	ni	gidar	uuendean	**mid**	**uuihti,**
that	i	in	my	mind	NEG	dare	change	with	anything
of	ic	is	giuualdan	mot.					
if	I	it	affect	could					

'Even if I had the power, I would not dare even to think about changing that in the slightest.' (*Heliand*, 219–20)

In cases like (8), *uuiht* is formally a head noun with a genitive attribute. However, this type of construction is relatively frequent (32.8% of all negative clauses with an emphasizer of some sort in the *Heliand*) and in some cases, *uuiht* can be analysed as an adverbial negation strengthener while the original attributive NP constitutes the actual argument of the verb, occurring in a genitive of negation. This is particularly likely in cases like (10), where the genitive element is a pronoun and occurs separated from its alleged head noun *uuiht*. In this case, the pronoun ('he') is more likely to be the argument of the verb *antkennian* 'to recognize' than *is*... *uuiht* 'anything of him':

(10) sô **is** thea ni mahtun antkenn(i)an **uuiht** the
so he.GEN those NEG could recognize anything/NEG who
thes uuîh[e]s thar uuardon scoldun
the.GEN shrine.GEN there guard should
'they did not recognize him at all who were supposed to guard the shrine.' (lit. 'they did not recognize anything of him...') (*Heliand*, 813)

Emphatic reinforcement with simple *(io)uuiht* as in (11) is much less frequent (six occurrences in the OLG corpus, 0.9%), and *niouuiht* 'nothing', which is the item that eventually grammaticalizes as the postverbal negator in Middle Low German, is hardly attested at all as an adverbial strengthener (three occurrences, 0.4%), (12).[8,9]

(11) ni sculun ûs belgan **uuiht**
NEG shall us anger anything
'They shall not anger us at all.' (*Heliand*, 4895)

(12) Ne ik thi geth ni deriu **(neo)uuiht,** quað he.[10]
and.not I you also NEG harm (no)thing said he
'I will also not harm you at all, either, he said.' (*Heliand*, 3892)

This shows that Old Low German, despite having at its disposal adverbial means of emphasizing sentential negation, had barely begun to establish a regular form that would eventually feed into Jespersen's cycle and become the new neutral sentential negator. When attestation starts again in the 13th century after a gap of about two hundred years, the expression of negation has changed dramatically. In Middle Low German, *nicht* has lost any emphatic value it may have had and has taken over the function of expressing sentential negation.[11]

[8] Cf. Breitbarth, Lucas, and Willis (2013) on cross-linguistically common 'bridging contexts' for negation strengtheners, such as verbs of caring/indifference, or damaging/benefiting, in which the strengtheners initially appear as pseudoarguments expressing extent.

[9] Cf. also the extremely common use of *nichts* 'nothing' in present-day (High) German with the same verb as in (12), *schaden* 'damage, harm':

(i) Das wird dir **nichts** schaden.
that will you nothing harm
'That won't harm you (at all).' lit. 'That will not harm you anything.'

[10] In the older Monacensis manuscript (*c.*850), the form is *neouuiht*, in the Cottonianus manuscript (10th c.), it is *uuiht*.

[11] The Middle Low German corpus used for the present chapter only begins around 1325. The following example is taken from the first volume (1256–1430) of the *Hanserecesse*, the archival records of the Hanseatic League. Note the additional absence of the old preverbal negator:

(13) ... unde dar sole wy en **nicht** hinderlich an wezen
... and there shall we them NEG obstructive to be
'and we shall not bar them from it.' (Steinfurt 12/09/1354)

New emphasizers also enter the picture, but not in a systematic fashion and without starting a new Jespersen's cycle. A common way of emphasizing negation in a large corpus of Middle Low German chancery documents is the use of *yenigh-erleye wiis* 'in any way' as in (14), though this type of reinforcement is even more common with an n-marked form such as *neynerleye wys* 'no way', *in geynerleywyse* 'in no way' or *in nenerley stucke* 'no part', see (15).

(14) dat ik edder myne erven **en**willen noch **en**schullen dem ergenomeden
that i or my heirs EN.want nor EN.shall the aforementioned
heren dem bischupe edder synen nakomelinghen bischupen to Lubeke
sir the bishop or his successors bishops in Lübeck
yenigherleye wiis [...] anklaghen...
any way [...] accuse
'that neither I nor my heirs shall acuse the aformentioned bishop in Lübeck nor his successors in any way.' (Lübeck 10/11/1428)

(15) Und ick und myne erven schollen noch **en**willen den vorgenanten
and I and my heirs shall nor NE.want the aforementioned
heren und oren nhakomelingen in duzen vorschreven jarlickenn
lords and their heirs in these aforementioned yearly
renthen neynen hinder don noch don lathen **nenerleige wis**.
interests no impediment do nor do let no way
'and neither I nor my heirs shall impede the aforementioned lords nor their heirs concerning the aforementioned yearly interest in any way.' (Uelzen 09/19/1505)

With *nicht* clearly having become the standard sentential negator in Middle Low German, *ne/en* on its own becomes very restricted in its occurrence. In the corpus used for this study, it mainly occurs with the new postverbal negator or n-marked indefinites. This is the case in *c.*70% of all clauses with *nicht* or an n-marked indefinite in the period 1325–1375, falling to around 12% in the case of *nicht* and 3.5% in the case of n-marked indefinites by 1575. Single *ne/en* is almost entirely restricted to exceptive clauses in the corpus (16):

(i) Were dat also, dat de koplude an deme hove an jenigeme rechte twivelden,
were that thus that the merchants in the court on any law doubt
dat **nicht** bescreven were...
that NEG laid.down were
'Should it be the case that the merchants in the court doubt any law that is not laid down...' (*Hanserecesse*, Versammlung zu Rostock 14 Oct. 1293)

(16) a. dat **en** sy mit willen der zessen
that NE be.SUBJUNC with permission of.the six
'unless it be with the permission of the six.' (Steinfurt 04/28/1370)

b. vnde dar moste numment yn, he **ne** gheue V mark
and there must n.one in he NE give.SUBJUNC five marks
vp dat minste
on the least
'and no one shall enter there, unless he give/pay at least five marks.' (Stralsund 1392)

In the entire Middle Low German corpus used here (2829 negative clauses), there is only one instance of sentential negation expressed by the preverbal marker alone:[12]

(17) der ik unde myne erven **en**scholed recht warende wesen
of.that I and my heirs NEG.shall law guarding be
'... of which I and my heirs shall not be judges.' (Scharnebeck 05/26/1420)

There are arguments against the preverbal marker actually expressing sentential negation in exceptive clauses. The clauses are neither formally nor functionally equivalent to negative conditionals. Formally, conditionals in Middle Low German are normally either asyndetic with sentence-initial verb placement, or syndetic with an *if*-type complementizer and sentence-final verb placement. Verb-initial asyndetic conditionals always precede their main-clause consequent. Exceptive clauses on the other hand are verb-second clauses which invariably *follow* the clause they depend on. Furthermore, verb-initial asyndetic conditionals are one of the first contexts to *lose* the preverbal marker in Middle Low German (Breitbarth 2008) and Middle Dutch (Burridge 1993), while exceptive clauses are one of the last contexts. They only start losing the preverbal marker once a new element conveying the exceptive meaning, *dan* 'then' > *denn*, is introduced in the course of the 15th century.[13]

[12] In four additional cases, there is a morphologically non-negative NPI indefinite in the scope of negation. In one of them there is additionally the old strengthener *mit ichte* < *mid uuihti* 'at all'. Cf. Postma (2002) for arguments based on Middle Dutch that negative polarity items and negative polarity 'constructions' can also work as licensors of *en*, and the discussion in section 6.2.2.

[13] In Low as well as High German, the preverbal marker is first joined by the adverb *dan* > *denn* 'then', co-occurring with any type of verb at first, and is eventually replaced by the frozen expression *es sei denn* 'unless', lit. 'it be.SUBJUNC then':

(i) a. original exceptive construction
dat **en** sy mit willen der zessen
that NE be.SUBJUNC with declared.intention of.the six
'unless it be with the permission of the six.' (MLG) (Steinfurt 04/28/1370)

b. augmented with *dan*
id **en** sy **dan** myt willen [...] des edelen
it NE be.SUBJUNC then with declared.intention [...] of.the noble
unsers leven juncheren
our dear squire
'unless it be with the permission of our dear noble squire.' (MLG) (Steinfurt 05/07/1486)

Functionally, there is a case for the absence of true sentential (i.e. propositional) negation in this type of clause. While the postverbal marker is firmly established in the entire corpus from the beginning, in none of the 173 exceptive clauses with the preverbal marker do we find a postverbal marker or an n-marked indefinite, even though either would be compatible with *ne/en*. This is concordant with the observation that exceptive clauses with *unless* in English do not license weak NPI indefinites in the absence of an overt clause-internal licenser such as negation:[14]

(18) a. It's no problem if you haven't done anything.

b. It's no problem, unless you have done something / *anything.

The reason for this state of affairs is probably that exceptive clauses do not actually express a negative condition for a consequent to be true, but an underlyingly positive one. They presuppose that the unmarked or expected state of affairs is the opposite, positive statement. It can therefore be assumed that *ne/en* does not indicate the presence of a negative operator in exceptive clauses. The scarcity of examples in the corpus using *ne/en* as sentential negator (see (17)) compared to the number of cases of *en* in the exceptive clauses on the one hand, and the bipartite expression of negation on the other, implies that *ne/en* alone was not generally available any longer as the expression of sentential negation to speakers of Middle Low German.

It can be assumed that *ne/en* was lost from all Low German dialects, in the course of the 17th century at the latest, although this must remain speculative as textual attestation becomes scarce after 1550. This is because High German takes over as the written language in the area. When Low German is used as a written language again in literary texts by authors such as Klaus Groth or Fritz Reuter from the 19th century onwards, there are no longer any traces of *en*, not even together with other negative markers.

Sundquist's (2007) diachronic study of the development of negation in diplomatic letters and chancery texts written in the city of Lübeck between 1320 and 1500 appears to be the only one so far looking at which factors influence the expression of negation in Middle Low German. Sundquist compares the effects of five factors on three types of negation: preverbal only, bipartite, and postverbal only. The factors considered are (1) the clause type (main vs embedded), (2) the verb type (lexical, modal, auxiliary), (3) the position of the verb in the clause (medial or final in embedded clauses), (4) the type of subject (pronominal or full DP), and (5) the date of composition. He applies

c. frozen expression with *denn*
es sei **denn** ...
it be.SUBJUNC then ...
'unless ...' (Modern German)

[14] Note that this only holds for weak NPI indefinites. Free-choice *any*-words are possible in some lexical contexts in *unless*-clauses.

Kroch's (1989) quantitative approach to historical morphosyntactic variation and change in order to see whether the three expressions of negation, in particular the decline of single preverbal negation and the rise of single postverbal negation show a Constant Rate Effect, which would point to them being competing grammatical options. Like Frisch (1997), who made a similar study of the development of the expression of negation in Middle English, Sundquist concludes that the two negation patterns are not mutually exclusive options, but functional doublets, and that bipartite negation is merely the overlap of the former two patterns.

There are a number of problems with Sundquist's approach. First, Lübeck is only one place within a large area of scribal dialects, and one that is known to have undergone a certain amount of dialect levelling (Peters 2000b), owing to its situation in the 'Neuland' colonized by settlers from all over the Low German area and its role as the centre of an international trading network, the Hansa. Therefore, the variation within the Middle Low German dialect area as a whole remains unstudied so far. Second, as we have seen above, it can safely be assumed on the basis of a much larger corpus than Sundquist's that preverbal *ne/en* alone was no longer an expression of sentential negation in Middle Low German.[15] It can therefore not be treated on a par with the other ways of expressing negation (bipartite and postverbal only). All that can be studied is the loss of *ne/en* from the expression of negation with *nicht* (and n-marked indefinites). A third problem with Sundquist's approach is that he only distinguishes 'main' and 'embedded' clauses within the factor 'clause type', but does not consider the position of the verb in them. It is therefore unclear whether he counted e.g. verb-first conditionals as 'embedded clauses', a context where, in Middle Dutch, *en* is lost first (Burridge 1993), as opposed to verb-final clauses, where it is lost last. Furthermore, Sundquist's factor 'position of verb' distinguishes only between final (OV) or medial verb placement (VO) within 'embedded' clauses. He finds this factor to be insignificant for the distribution of negative markers in his corpus. This is not surprising. The Middle Low German data in the corpus on which the present study is based show a certain amount of verb raising and OV leakages, but apparently independently of the distribution of negative markers. Also, whether the subject is a pronoun or a full DP (factor 4) is found to be insignificant by Sundquist, and will therefore be neglected in the following discussion.

The Middle Low German scribal dialects differ significantly in the speed at which they make the transition from stage II to stage III of Jespersen's cycle. The south-western and southern dialects West- and Eastphalian lose *en* more slowly than North

[15] This can also account for Sundquist's observation 'that the decline of Type 1 [single preverbal negation—AB] in the latter half of the 15th century is relatively flat compared to the more drastic increase in Type 3 [single postverbal negation—AB] during this time': the preverbal marker has simply left the negation system.

TABLE 6.1 **Frequency of Middle Low German *en . . . nicht* as a percentage of all cases with *nicht***

	1325–1374	1375–1424	1425–1474	1475–1524	1525–1574
Westphalian	22 (78.6%)	25 (83.3%)	3 (37.5%)	14 (35.9%)	8 (21.1%)
Eastphalian	56 (72.7%)	52 (71.2%)	25 (52.1%)	15 (14.6%)	18 (10.2%)
North Low Saxon	37 (56.1%)	42 (33.1%)	75 (33.0%)	62 (31.2%)	3 (12.0%)
Eastelbian	3 (50.0%)	12 (18.5%)	20 (29.0%)	10 (7.8%)	2 (12.5%)

Low Saxon, and much more slowly than the northeastern dialect Eastelbian.[16] This is shown in Table 6.1.

These patterns can be explained by the different colonization background of the dialect areas in question: the Eastelbian cities of Lübeck and Stralsund in the corpus are in the 'Neuland'; that is, they were founded on formerly Slavonic territory by settlers from the North Low Saxon and Westphalian areas. They therefore constitute a typical urbanization scenario with dialect levelling (Trudgill 1994). Such contexts can lead to simplification, and this factor can therefore account for the accelerated loss of preverbal *en*. Furthermore, Lübeck and Stralsund became centres of the Hanseatic trade at the time, a further likely factor in the levelling of dialect differences and the removal of archaic or marked features. Peters (2000a: 1414) summarizes this development as follows:

> In der Frühzeit Lübecks ist mit einem Nebeneinander verschiedener altländischer Mundarten zu rechnen. Das Zusammenleben in der Stadt führt im Verlauf des 13. Jhs. zu einem innerstädtischen Ausgleich, es entsteht eine städtische Umgangssprache. Es ist anzunehmen, dass sich relativ früh innerhalb der hansischen Gemeinschaft, unter den Fernhandelskaufleuten im Ostseeraum eine lübisch geprägte mündliche Handels- und Verkehrssprache entwickelt hat [. . .][17]

Another significant factor influencing the expression of negation is the position of the verb. As in Burridge's (1993) study of Middle Dutch (see also below, section 6.2.2), verb-first contexts significantly favour the omission of the preverbal marker. The

[16] This is confirmed by a binomial regression analysis in *GoldVarb X*. The factor weights for the factor group 'scribal dialect' are: Westphalian: .769, Eastphalian: .585, North Low Saxon: .475, and Eastelbian .278 (values above .5 favour the bipartite expression of sentential negation).

[17] Translation: 'In the early days of Lübeck, we have to assume a co-existence of different dialects of the Saxon 'Altland'. The collective life in the city leads to an intra-city levelling during the 13th century, to the rise of an urban vernacular. We can assume that, already early on, an oral trade language and lingua franca based on the dialect of Lübeck developed within the Hanseatic community, among the traders around the Baltic Sea.'

TABLE 6.2 Frequency of Middle Low German *en... nicht* in different syntactic contexts

	V1	V2	Vfinal
1325–1374	10 (47.6%)	38 (69.1%)	70 (69.3%)
1375–1424	3 (13.6%)	42 (41.2%)	86 (50.3%)
1425–1474	1 (4.3%)	35 (37.6%)	87 (36.9%)
1475–1524	0/20 (0.0%)	30 (26.1%)	71 (21.3%)
1525–1574	0/18 (0.0%)	3 (4.0%)	28 (17.2%)

difference between verb-second and verb-final contexts, given in Table 6.2, is not statistically significant in the Middle Low German chancery documents.[18]

For High German, it has often been claimed that certain 'high-frequency verbs' like *wissen* 'know' and *tun* 'do', as well as modal verbs, favour single preverbal negation for longer (Behaghel 1918: 230, Paul 2007: 389–90), and also hold on to bipartite negation for longer than other verbs once the postverbal marker arises.[19] In Sundquist's Lübeck data, the factor of verb type seems to be insignificant for single preverbal negation, while bipartite negation slightly prefers lexical verbs, and single postverbal negation *haben* 'have', *sein* 'be', and modal verbs. This is perhaps unexpected given the generalizations in the older literature according to which modal verbs are more conservative. The corpus used for the present study confirms Sundquist's findings; the loss of the preverbal marker is delayed significantly with lexical verbs, and accelerated with auxiliary and modal verbs.[20] Therefore, higher frequency of a verb seems to correlate with (and perhaps trigger) loss of *en/ne*.

6.2.2 *Dutch*

As mentioned at the outset, the extant Old Dutch texts are of limited value for syntactic studies. I shall simply describe the expression of negation in these texts here, but not make generalizations.

In the *Wachtendonck Psalms*, the preverbal marker occurs in 53 of the 57 negative clauses (93%), (19). Of these, 52 do not contain any other negative element, while one contains an n-marked indefinite, (20).[21] Note that the sentential negation particle *ne*

[18] *GoldVarb X* factor weights for the factor group 'position of the finite verb': V1: .138, V2: .561, Vfinal: .520.

[19] Similar results for certain verbs have been found for Middle English, cf. Iyeiri (2001: ch. 5).

[20] *GoldVarb X* factor weights by verb type: auxiliaries: .66, modals: .57, 'special' (lexical) verbs (*wissen, tun*, ...): .41, lexical verbs: .33.

[21] Additionally, there are four negative clauses without *ne*. Two contain the adverbial negator *niuueht* 'not', one contains *niuueht* as a negative determiner 'no, not a single' (for Latin *nequam*) and one contains *fur niuuehte* as an adverbial emphasizer 'at all', lit. 'for nothing'.

is even inserted in (20) against the Latin original, pointing to genuine Old Dutch syntax.

(19) non timebo quid faciat mihi caro
ne sal ic forhtan uuad duo mi flēisc
NEG shall I fear what do me flesh
'I will not fear what flesh can do to me.' (WP LV.5)

(20) nequanto obliuiscantur populi mei
that **nohuuanne** **ne** fargetin folk mīn
that never NEG forget people my
'. . . lest any time my people forget.' (WP LVIII.12.2)

There are already five cases of negation without *ne*, three of them using *niuuiht/niuueht* as constituent negator or even negative determiner (cf. section 3). In two cases, however, *niuueht* is used to render Latin *non*, expressing sentential negation. In (21), the choice of *niuueht* may be attributed to the strict interlinear character of the translation, because it helps maintain the position of the negation before the verb, which in this case has to be non-finite as the Latin form *decidet* can only be rendered by means of a complex verb form in Old Dutch.

(21) et folium eius non decidet
inde lōuff sīn **niuueht** nitheruallan sal
and leaf his NEG down.fall shall
'. . . and his leaf shall not wither.' (WP I.3.3)

In (22) on the other hand, the Latin *non* is rendered as *niuueht* in the first conjunct and as *ne* in the second, pointing at a genuine use of *niuueht* as a negator in this text.

(22) Beatus vir qui non abiit in consilio impiorum, et in uia peccatorum non stetit
Sēlig man ther **niuueht** uuor in gerēde ungonēthero, inde in
blessed man who NEG goes in counsel impious.GEN and in
uueg sundiger **ne** stūnt
way sinners.GEN NEG stands
'Blessed is the man that walketh not in the counsel of the ungodly, nor standeth in the way of sinners.' (WP I.1)

In neither (21) nor (22) can *niuueht* be said to be emphatic (cf. section 6.2.2).

As the Latin original in (23) does not contain a finite verb, the Old Dutch translation has to find an alternative way of rendering *non*, if the strict interlinear character is to be maintained, as *ne* is restricted to finite verbs. This shows that *niuueht* had already established itself as the negator to be used in non-finite contexts, and, as in the case at hand, for constituent negation. Here as well, it is not obvious that this is an emphatic use.

(23) Non sic impii, non sic
Niuueht sō ungonēthe, **niuueht** sō
NEG so impious not so
'The ungodly are not so.' (WP I.4)

Of the 127 occurrences of sentential negation in the *Leiden Willeram*, all contain the preverbal marker, which is mostly cliticized to the finite verb as seen in (24), though in over 80% of the cases, negation is additionally expressed by different elements in the clause, either adverbial *niet* (25) or n-marked indefinites (26).

(24) thaz sie se **ne**wecchan, eer siu selua wolla
that they her NEG.awake before she herself want
'that they may not wake her before she wants to wake herself.' (LW 56.22–3)

(25) Wir **ne**wiilon **niet** uergezzan, thaz...
we NEG.want NEG forget that
'We do not want to forget that...' (LW 42.6)

(26) wanda an hin **nieman** **ni**uindet ieweht unrechtes
because on him n.one NEG.finds anything wrong
'because no one can find anything wrong about him.' (LW 59.4–5)

In 16 of the 25 clauses with single *ne/ni*, the finite verb is the modal *mugan* 'be able', a verb that shows a certain conservative behaviour with respect to keeping single preverbal negation in all West Germanic languages (Behaghel 1918, 1924 and Paul 2007 for High German; Stoett 1923 and van der Horst and van der Wal 1978 for Middle Dutch; and Iyeiri 2001 for Middle English). In addition to the 127 cases of negative sentences, there are six cases of single *ne* in subjunctive verb-second clauses that depend on negative clauses and to which they express an exception (see the discussion of exceptive clauses in Middle Low German above).

(27) wande **nieman** **ni**magh intrare portam regni caelestis, her
because n.one NEG.may enter door kingdom heaven he
newerthe per doctores baptizatus
NE.be by scholars baptized
'because nobody may enter the Kingdom of Heaven, unless he be baptized by the scholars.' (LW 89.22–3)

The old preverbal element *en/ne* is preserved for much longer in Dutch than it is in Low German. Besides the bipartite expression of negation, it continues to be used on its own in a number of other contexts in Middle Dutch (van Helten 1885a, Stoett 1923: 157–9). Two of these contexts, 'paratactic negation'[22] (28) and 'exceptive' clauses (29),

[22] This is the term used by the traditional Dutch literature. Note that Jespersen (1917) uses this term for pleonastic or expletive negation, i.e. non-negative uses of negative markers.

are formally very similar; they are verb-second clauses with *en* preceding the verb. They differ in that clauses with 'paratactic negation' always depend on a clause containing negation or a quasi-negative expression like 'hardly, not long, not far' while their finite verb is in the indicative mood, whereas in 'exceptive' clauses the verb is in the subjunctive. 'Paratactic negation' can fulfil a variety of functions, such as relative, complement, consecutive, and temporal clauses. As (29a) shows, the borders between 'paratactic' and 'exceptive' use of *ne/en* are blurred.

(28) a. Darne was niemen, **hine** was blide
there.NE was no.one he.NE was happy
'There was no one who was not happy.'

b. Het en es niet een dach in die weke, **Hine** es geasselgiert sere
it NE is NEG one day in the week he.NE is attacked badly
'There is not one day in the week on which he is not badly attacked.' (van Helten 1885a: 220–1)

(29) a. Want ik sal keren nemmermere, Ic **en** hebbe vonden
because I shall return never.more I NE have found
mijn gheslacht
my family/lineage
'because I shall never return until/unless I have found my family/lineage.' (van Helten 1885a: 227)

b. Ic salre varen, **in** (ic+**en**) blive doet.
I shall.there go I.NE become dead
'I shall go there unless I die.' (Beheydt 1998: 15)

Exceptive clauses undergo a further development different from that observed for Low German. In Dutch, the common exceptive clause *het en zij/waere* 'it NE be.SUBJUNC/were.SUBJUNC' is reanalysed as a subordinating complementizer *tenzij* 'unless' < *(he)t.en.zij* 'it.NE.be.SUBJUNC' (te Winkel 1901: 171–2), now even triggering sentence-final verb placement. Beheydt argues that *tenzij* and *tenware* were already frozen expressions in southern Dutch in the 15th century and soon grammaticalized as exceptive complementizers with sentence-final verb placement.[23]

Other contexts for single *en/ne* in Middle Dutch include fragment answers with dummy verbs (30), (rhetorical) questions expecting a positive answer (31), and pleonastic or expletive negation in the complement of adversative predicates (such as *doubt, deny, forbid*, etc.), where it is rather rare, however (Burridge 1993: 184–5).[24]

[23] It appears, however, that in the 18th and 19th centuries, *tenzij* was used as a preposition taking a nominal, not clausal, complement, meaning 'except' (Beheydt 1998: 98–9).

[24] In the case of Burridge's example, it is ambiguous with an exceptive clause:

(i) Doe mocht ic hoir qualic weygeren dat(,) ich **en** dede haer begheren
then could I her scarcely deny that I NE did her desire
'Then I could scarcely deny to her that I desired her.' (expletive negation)

(30) Ghi hout u spot. **In** (=ic+**en**) doe
you hold your mockery. I.NE do
'You are mocking. I do not!' (Beheydt 1998: 15)

(31) **En** es dit Floris miin soete lief?
NE is this Floris my sweet lover
'Isn't this Floris, my sweet lover?' (Beheydt 1998: 16)

According to Beheydt (1998), pleonastic or expletive negation in Middle Dutch is more frequent in the standard of comparison.

(32) Ghien moget niet vorder rechten dan u manne **en** wijsen
you.NE can NEG more judge than your men NE tell
'You cannot judge more than your men tell you.' (Beheydt 1998: 16)

Certain verbs are slower to adopt the new adverbial negator. The modals *willen*, *connen*, and *mogen* are reported in the literature as maintaining lone *en/ne* longer (Stoett 1923, van der Horst and van der Wal 1978), as are the verbs *weten* 'know', *roeken* 'care', and *hebben* 'have' when occurring with a *wh*-complement in the function of an indefinite object (33).

(33) a. Si **ne** weten wat best doen
they NE know what best do
'They do not know what would be the best thing to do.'

b. Hem **en** roeket wiet deerde
him NE mattered who.it hurt
'He did not care whom it hurt.'

c. Wi **ne** hebben wat eten
we NE have what eat
'We do not have anything to eat.' (Beheydt 1998: 18)

Another context in which *en* occurs on its own in Middle Dutch involves clauses containing the NPIs *bore* 'much', *meer* '(any)more', and *ander* 'another' (34), where it appears that these elements had been on the way to acquiring a negative value, a development which must have been reversed again later.

(34) a. Dat ghi mi **meer** ne sult sien
that you my anymore NE shall see
'that you shall not see me anymore' (Beheydt 1998: 17)

b. (Het) ne was Persise **bor** leet...
it NE was Persis much sorry
'Persis was not very sorry...' (Postma 2002: 53)

It is not clear how to treat these single uses of *en* in a uniform manner. It appears as though *en* has become ambiguous in Middle Dutch. In some cases it still seems to function as a negation-marking particle, in others it co-occurs with other negative expressions. Postma (2002) argues that it is an NPI 'use' of the argument *wh*-clauses that licenses *ne* in (33).[25] He does not give a formal account of how this licensing should proceed. Normally it is a negative expression (or another overt NPI-licenser such as a question or comparative or conditional operator, or a lexical item like *before* or *without*) that licenses NPIs, not vice versa. Therefore, one might posit that *en* is in fact still a negation particle in constructions like (33). On the other hand, it can occur in non-negative contexts as if it were an NPI itself. Cases like (34) could be analysed either way, with either *bore* and *meer* still being NPIs being licensed by a still-negative *en* or with them having acquired a negative value, licensing 'NPI' *en* by identifying a covert negation scoping over *en*. Given that the *wh*-clauses in (33) fulfil the function of indefinite objects, they might be on the verge of entering a quantifier cycle (see section 6.3 below) as just hypothesized for *bore* and *meer*, that is, on their way to acquiring a negative value. Either way, it is clear that the position of *en* in the system is unstable and it is no longer used as the standard negator on its own.

The loss of *en* from the expression of sentential negation with *niet* is conditioned by a number of factors. Burridge (1993) discusses the factors of dialect and position of the finite verb as very influential. She compares a northern dialect, Hollandish, and a southern one, Brabantish. As in Middle Low German, the type of clause and with it the position of the finite verb is a significant factor in the incidence of *ne/en/n* in Branbantish and Hollandish. In both dialects, *ne/en/n* is dropped more frequently in verb-initial clauses (imperatives, verb-first conditionals, and the like) than in verb-second clauses and especially verb-late clauses. Another factor influencing the loss of *ne/en/n* discussed in the literature is possible haplology, for instance if *en*+Vfin is preceded by the impersonal pronoun *men* (van der Horst and van der Wal 1978, Hoeksema 1997) or an infinitive in *-en* (Burridge 1993). While the southern dialects of Flanders and Brabant significantly lag behind in losing the preverbal marker, northern dialects like Hollandish, from which the modern standard is derived, make the transition to single postverbal negation in the 17th century. Burridge's data from southern Brabantish and northern Hollandish from between 1300 and 1650 show this dialectal split very clearly. Table 6.3 is adapted from her tables 1 and 2.[26]

A sociolinguistic explanation along the lines proposed above for Low German suggests itself here as well. Especially after the independence of the northern provinces (and the end of the Hansa, which had had some influence in Flanders), Holland

[25] Cf. also Hoeksema's (1997: 143–5) speculation that the NPI character of the verbs in question may have something to do with licensing negation with single *ne* in these cases.

[26] Burridge uses the abbreviations MC ('main clause'), SC ('subordinate clause'), and IC ('imperative clause', but including other verb-first contexts) for verb-second, verb-late, and verb-first, respectively. The total numbers (#) refer to the total number of negative sentences with *niet*, the percentages (%) to those without *en/ne/n*.

TABLE 6.3 **Frequency of *ne/en/n*-drop in the context of *niet* in Middle Dutch**

	verb-second				verb-late				verb-first			
	Brabant		Holland		Brabant		Holland		Brabant		Holland	
	%	#	%	#	%	#	%	#	%	#	%	#
1300	0	6	28	39	5	39	8	40	21	17	43	7
1350	2	64	25	12	4	83	36	28	12	17	75	4
1400			11	47			2	66			83	6
1450			17	12			0	22			20	10
1500	9	22	48	29	0	59	28	61	50	12	77	22
1550	12	43			0	59						
1600	5	57	30	67	5	75	8	8	57	7	100	17
1650	9	45	100	50	6	49	98	68	100	14	100	13

(after Burridge 1993: 190–1)

TABLE 6.4 **Frequency of *en* with *nyet/neet* in Drentish Dutch**

	1399–1405 and 1444–7	1488–92
auxiliary verbs	18/150 (12.0%)	30/44 (68.2%)
lexical verbs	17/51 (33.3%)	9/14 (64.3%)

(after Postma and Bennis 2006)

became an international centre of trade and intellectual life, again providing fertile ground for dialect levelling.

The significance of the dialectal split between Holland and Flanders, as well as the difference between the Saxon (Middle Low German) and Low Franconian dialects (Middle Dutch), is confirmed by Postma and Bennis (2006), who, based on a corpus of court proceedings, show that the loss of the preverbal marker was very advanced around 1400 in the northeastern Saxon dialect of Drenthe (see Table 6.4), but that this variety reverts to a more conservative stage around 1490 under the influence of the Hollandish dialect due to the political influence of the chancery of Utrecht.

Beheydt (1998), focusing on the development in letters, travelogues, and chronicles from the southern provinces from the 15th to the 20th century, gives a general overview as presented in Table 6.5 for the development of the expression of negation, for both *niet* and n-marked indefinites.[27]

[27] Looking at the southern dialects individually, East Flemish seems to be first to reduce the use of *en* in bipartite negation, but from the 17th century on, Brabantish becomes the clear forerunner. West Flemish is the slowest dialect in losing *en* (Beheydt 1998: 105–6).

TABLE 6.5 **The development of negation in southern Dutch from the 15th to the 20th century**

	preverbal only	bipartite negation	postverbal only	total
15th c.	4 (2%)	202 (85%)	32 (13%)	238
16th c.	11 (1%)	758 (92%)	54 (7%)	823
17th c.	4 (1%)	717 (80%)	172 (19%)	893
18th c.	1 (0%)	248 (33%)	513 (67%)	762
19th c.	0 (0%)	166 (24%)	531 (76%)	697

(after Beheydt 1998: 105)

It can be concluded from Burridge's figures (see Table 6.3) that, in the northern Dutch provinces, the preverbal marker *en* in combination with *niet* had been lost by 1650. It has been argued that this is the consequence of a ban by prescriptive grammarians and influential writers. Burridge (1993) shows how *en* disappears from the letters of P. C. Hooft (1581–1647) during his lifetime; similarly, van der Wouden (1995: 23) shows how playwright Joost van den Vondel (1587–1679) 'after the model of [...] writers such as P. C. Hooft' deliberately reduces the use of *en* in his plays over the years. However, referring to 'dialectal texts' up to the 19th century, Beheydt (1998: 25) points out that prescriptive pressure is unlikely to have affected the continued use of the preverbal marker in the spoken language. In the southern provinces on the other hand, the loss of the preverbal marker only really accelerated in the 18th century, around 500 years after High German, and still a good 300 years after Middle Low German. Negation with *en* alone had already fallen out of use in the 15th century, and only non-negative or emphatic uses survive, pointing at a reanalysis of *en* as a(n emphatic) polarity marker (Breitbarth and Haegeman forthcoming).

Not only did the southern Dutch dialects take much longer to lose the preverbal marker, some of them still make use of it at the present day, despite the spread of the northern standard language to the south.[28] Beheydt (1998) hints at a possible meaning difference arising in the Early Modern period between the older bipartite and the newer single postverbal expression of negation:

(...) in vraagzinnen (...) blijkt de negatie steeds postverbaal te zijn als de betekenis positief is. Het lijkt heel aannemelijk dat de taalgebruikers vonden dat de tweeledige ontkenning de negatieve betekenis te zeer benadrukte, wat minder het geval was met de postverbale.[29] (Beheydt 1998: 93)

[28] It has been noted that speakers who have preverbal *en* in their native dialects also tend to use it in what is called *tussentaal* ('in-between language'), that is, the dialectally influenced colloquial register of the standard language in Flanders (Lebbe 1997).

[29] Translation: 'In questions, negation seems to be invariably postverbal when the meaning is positive [i.e. when the questions are rhetorical—AB]. It is very likely that language users felt that bipartite negation put too much emphasis on the negative meaning, which was less the case with postverbal negation.'

In fact, in those dialects that have preserved the preverbal marker to this day, it does seem to have developed a certain emphatic value: it signals a contrast between the negative clause it appears in and the discourse context, often with certain emotional overtones such as surprise or irritation.

(35) A: Geef me nen keer Valère zenen telefon.
give me one time Valère his number

B: K' en een-k ik zenen telefon niet.
I EN have=I I his number NEG

A: 'Can you give me Valère's phone number?'

B: 'I don't have Valère's number!' (West Flemish, Haegeman 2002: 180)

Furthermore, single *en* seems to have survived to some extent in non-negative affective contexts (Klima 1964), i.e. where it does not express sentential negation. Weijnen (1956) reports the occasional use of single *en* in 17th-century Dutch in clauses containing *maer* 'only', *nauw* 'near(ly)', and comparatives:

(36) a. En van die eerste jeughd / **en** smaken meestendeel **maer** ouderen de vreughd.
and of the first youth NE taste mostly only old.people the joy
'It's mostly only old people who enjoy the first youth.'

b. hoe wel ter **nauwer** noodt verhaalens waart **en** is
how well to near need telling worth NE is
'although it is hardly worth telling'

c. niet **soeters** [...] als ghy **en** siet
nothing sweeter than you NE are
'nothing sweeter than you are' (Weijnen 1956: 73)

Beheydt (1998) confirms this for her corpus of southern Dutch (15th–20th century), from which examples (37) and (38) are taken:

(37) standard of comparison
ick en bleef aan tafel niet langer, als ik **en** moest
I EN stayed at table not longer than i EN had.to
'I did not stay at the table any longer than I had to.' (17th-c. West Flemish, Maria Petyt, 46)

(38) context of restrictive adverbs
a. ende mijn vaeder **en** leefde **maer** ix maenden nae ons moeder
and my father NE lived only nine months after our mother
'and my father only survived our mother by nine months.' (16th-c. Brabantish, Jan de Pottre, 12)

b. ...dat wij **nauwelijk** **en** derfden spreken in sijne presentie
that we hardly EN dared speak in his presence
'...that we hardly dared speak in his presence.' (17th-c. West Flemish, Maria Petyt, 25)

These non-negative uses of *en* in affective environments can still be found in present-day Flemish dialects. The following examples are from the East Flemish dialect of Ghent and West Flemish Kortrijk:

(39) en **aa't** slecht weer **en** is
and if=it bad weather EN is
'and if the weather is bad' (Ghent) (Leemans 1966: 191)

(40) Ge moet 't zegge **gelijk of** 't **en** is.
you must it say like if it EN is
'You must tell it the way it is.' (Ghent) (Leemans 1966: 191)

(41) **ten** is **maar** een kleintsje
it=EN is only a little.one
'it's only a little one.' (Ghent) (Tavernier 1959: 246)

(42) Je moet niet komen **voordat** ik geschreven **en** heb.
you should not come before I written EN have
'You need not come before I have written.' (Kortrijk) (Barbiers et al. 2008: 60)

On the basis of such examples, Breitbarth and Haegeman (2010) argue that instead of being lost as in other Dutch dialects, the position of *en* was stabilized by its reanalysis as a(n emphatic) polarity marker. As far as is possible to tell from studies on the present-day dialects (Leemans 1966, Vergauts 1971, de Pauw 1973), it appears that the frequency of *en* with *niet* has stabilized at around 10% of the possible cases. This supports arguments for a reanalysis of *en* as an emphatic element; not all negative sentences are emphatic (Breitbarth and Haegeman 2010).[30]

6.2.3 *Summary*

Both Old Low German and Old Dutch start out at stage I of Jespersen's cycle: the standard expression of sentential negation is by means of a preverbal head, *ni* or *ne*, that cliticizes to the finite verb. Already during the oldest period, emphasizers of negative polarity are used. The element that ultimately becomes grammaticalized as a new adverbial negator is derived from an n-marked indefinite pronoun, *niouuiht/niuueht* 'nothing'. In Old Dutch, this element is already used adverbially (22) and as a constituent negator (23), helping to circumvent the finiteness restriction of the old

[30] Cf. also Fonseca-Greber (2007), who argues that the low but stable frequency of the preverbal negator *ne* in spoken Swiss French (around 2.5%) is partly due to its reanalysis as a marker of emphasis.

preverbal negator. In Old Low German, we see the grammaticalization in progress: *niouueht* is used pseudo-argumentally in typical bridging contexts, such as verbs of caring/indifference or damaging, (12).

In the transition to Middle Low German/Middle Dutch, the original preverbal negator changes its status. The new postverbal element *nicht/niet* is clearly the standard expression of sentential negation at this point, although the transition is somewhat more protracted in Middle Dutch. In Middle Low German the preverbal marker has clearly ceased to express negation.

For both Middle Low German and Middle Dutch, there are three factors that significantly influence the loss of the old preverbal marker: dialect, position of the verb, and the type of the verb. While there is clear evidence that the old preverbal marker has lost its negative force in Middle Low German to the new adverbial marker *nicht*—providing a plausible explanation for its demise—things seem to be less clear-cut for Middle Dutch, where *ne/en/n* survives for much longer and is eventually banned from the standard language by what seem to be prescriptive efforts. Throughout the Middle Dutch period, it is still found with *niet* and n-marked indefinites (cf. section 6.3 below), and retains some independent uses, such as in negative clauses with *wh*-complements or certain lexical elements (*meer*, *bore*). It only survives in the Flemish dialects—which for a long time were outside (northern) Dutch political influence—due to reanalysis as a marker of polarity emphasis.

6.3 Indefinites in the scope of negation

Sentential negation in the history of Low German and Dutch is not only marked by negation particles such as *ni/ne/en* or *nicht/niet*. Already at an early date, morphologically n-marked indefinite pronouns, determiners, and adverbs are able to identify sentential negation, whether supported by the presence of one of the sentential negation particles or not (that is, they are n-words in the sense of Giannakidou 2006, see also section 1.8.2). Other indefinite elements licensed in the scope of negation are unable to identify sentential negation by themselves and are therefore to be considered negative polarity items. Diachronically, the licensing conditions of indefinites can change, commonly becoming more restrictive (the 'quantifier cycle', see section 1.9.1). Although such changes happen to individual lexical items, indefinites tend to form series, e.g. for negative or NPI contexts, building up paradigm pressure which causes analogical changes in individual members of a series. This is exactly what characterizes the development of indefinites in the scope of negation in the history of Low German and Dutch. For each language, we will first look at the interaction of indefinites with sentential negation, and then at the developments within the system of indefinites.

6.3.1 *Low German*

Old Low German had one 'neutral' or 'general' series of indefinites, which is not restricted to the scope of *ni*:

(43) **sum** habad iro hardan strîd
some have their hard fights
'some have their hard fights' (*Heliand*, 2493)

Besides this, it had two series of indefinites that could be used in the scope of negation: n-marked ones (such as *niouuiht* 'nothing'), and n-free ones (for example, *(g)iouuiht* 'anything'). The latter was also licensed in non-negative NPI contexts, such as the restriction of a universal quantifier, (44).

(44) allaro barno bezta, thero the **io** giboran uurði
all.GEN children.GEN best those.GEN who ever born was
'the best of all children who was ever born' (*Heliand*, 835)

As seen above (section 6.1.1), the preverbal marker was virtually obligatory in Old Low German negative clauses. This means that there are negative clauses with negative doubling (negation marker + n-marked indefinite) as well as without (negation marker + n-free indefinite). However, the individual Old Low German texts differ with respect to which pattern they prefer (see Table 6.6). While there are no n-marked indefinites used in negative clauses with indefinites in the *Genesis* fragments at all, in the minor texts all of the few clauses with indefinites in the scope of negation use n-marked forms. The language of the *Heliand* disprefers negative doubling—it is only used in around 20% of the possible contexts.[31]

Where n-marked indefinites are used, they co-occur with the sentential negator *ni*, whether they follow or precede it. Old Low German is thus a strict negative concord language (Giannakidou 1998).

TABLE 6.6 **Frequency of indefinites in the scope of negation in Old Low German**

	n-marked	n-free	total neg. clauses
Heliand	35	142	620
Genesis	0	12	37
minor texts	5	0	38
	40	154	695

[31] Note that the number of indefinites in the scope of negation in the *Heliand* is higher (177) than the number of negative clauses containing indefinites (169). This has to do with the fact that four clauses contain both an n-marked and an n-free indefinite and six contain two or more n-free indefinites.

(45) a. **Ni** scal **neoman** lioht, the it haƀad, liudiun dernean
NEG shall nobody light who it has people.DAT conceal
'No one who has light should hide it from people.' (*Heliand*, 1405–6)

b. **Neo** endi **ni** kumid, thes uuîdon rîkeas giuuand
never end NEG comes the.GEN wide.GEN kingdom.GEN end
'The broad kingdom will never end.' (*Heliand*, 267–9)

Remarkably, also n-free indefinites are able to precede the negative marker, (46c), in violation of Jespersen's 'Neg-First Principle' (Haspelmath 1997, Mazzon 2004).

(46) a. **ne** dragu ic **ênig** drugi thing.
NEG carry I any deceptive thing
'I am not bringing any kind of trick/deception.' (*Heliand*, 264)

b. Sia **ni** namon is tho **niam** (=**niaman**)
they NEG took it then nobody
'They did not take it from anyone.' (GG.63,10–11)

c. sô is **io** endi **ni** cumit
thus it.GEN ever end NEG comes
'thus the end (of it) will never come.' (*Heliand*, 1324)

Old Low German does not have negative spread. If more than one indefinite occurs in the scope of negation, at most one of these is n-marked:

(47) a. Nis thes tueho **ênig** gumono **nigiênumu**
NEG=is the.GEN.SG doubt any men.GEN.PL none.DAT.PL
'None of the men have any doubt about it.' (lit. 'there is not any doubt about it to none of the men') (*Heliand*, 3190–1)

b. it ni mag iu te **ênigoro** frumu huuergin uuerðan
it NEG can you to any benefit at.all redound
te **ênigumu** uuilleon.
to any happiness
'It is not able to do you any good at all, nor bring you any happiness.' (*Heliand*, 1854–5)

The three subcorpora indicate a diachronic development between the 9th and 11th centuries, where the oldest stage (represented by the *Genesis* fragments) had no n-marking on indefinites in the scope of negation, the middle stage (the *Heliand*) had optional (and still dispreferred) n-marking, and the final stage (the minor texts) had obligatory n-marking.[32] As we saw above, the preverbal negator essentially ceases to

[32] Despite the low numbers in the *Genesis* fragments and the minor texts, the *Heliand*, with 620 negative clauses, 169 of which contain a total of 177 indefinites, forms a solid body of data and can be used to show

express sentential negation in Middle Low German. Therefore, this shift may indicate the weakening of preverbal *ni* already in the 10th and 11th centuries.

Like Old Low German, Middle Low German distinguishes two series of indefinites which can occur in the scope of negation. N-marked indefinites can co-occur with the old preverbal marker (now *ne/en*), but, as we saw in section 6.2.1, there are arguments that this was no longer the negative marker in this period. They cannot, apparently, co-occur with the new postverbal negator *nicht*. There are two exceptions to this in the corpus out of 1,263 sentences with n-marked indefinites, that is, only 0.1%. In both cases, they are extraposed from the negated clause. In (48), *nynerleye wijs* is an adjunct and *nyman van unser weghene* could be interpreted as an afterthought to the coordinated subject, that is, 'we, our heirs, *nor anyone on our behalf*', but as both phrases occur outside the clause as a kind of afterthought, one may assume that they received the n-marking to indicate their still being in the scope of the sentential negation.

(48) ...dar wy ofte unse ervende unsen vader **nicht** an engen ofte
...to.which we or our heirs our father NEG in obstruct or
hinderen **ne**solen nynerleye wijs ofte nyman van unser weghene
hinder NE.shall no way or no.one of our behalf
'...in which neither we nor our heirs nor anyone on our behalf shall obstruct or hinder our father in any way.' (Steinfurt 01/07/1355)

In (49), the extraposed complement PP cannot be seen as an afterthought; rather, it appears to be focused by its right-peripheral position. Again, however, one may assume that the extra-clausal position of the n-marked indefinite DP made the scribe use n-marking to indicate the connection to the earlier sentential negation here as well.

(49) dat ze sik **nycht** **en**scholen vorbynden tjegen **nyne** **heren**
that they REFL NEG NE.shall ally against no masters
ofte landesheren
or territorial.lords
'that they would not form any alliance against any masters nor territorial lords.' (Oldenburg 05/01/1436)

that the three subcorpora represent different diachronic stages. The probability that the 12 instances of n-free indefinites in the *Genesis* fragments happen to belong to the 80% share of n-free indefinites in the *Heliand* and that the 20% of n-marked indefinites are accidentally unattested is 0.13. While this is not statistically significant at the 0.05 level, it does indicate a tendency, as there is still an 87% chance that the distribution of indefinites is different in these two texts. The probability that the five instances of n-marked indefinites in the minor texts happen to belong to the 20% of n-marked indefinites in the *Heliand*-grammar and that the 80% share of n-free indefinites is accidentally unattested is 0.0004, in other words, highly improbable.

In any event, 0.1% is an extremely low frequency, indicating that this pattern is essentially ungrammatical.

Incidentally, the co-occurrence of *nicht* and n-free indefinites is also rare in the corpus (3; 0.2%): the bulk are co-occurrences of one n-marked indefinite with one or more n-free ones, (50), or of two or more n-marked indefinites, (51) and (52). This means that the form of negative concord preferred by Middle Low German was negative spread, provided our above analysis that the old preverbal marker was no longer the expression of sentential negation is correct.

(50) dat we **nemende**, dem we schuldich synt to dessem jare [...]
that we no.one who we due are to this year
ichtes gheven konnen
anything give can
'that we cannot give anything to anyone we are indebted to this year.' (Uelzen 08/13/1396)

(51) dar en willen wy **nemande** **nyner** **helpe** uop plichtich wezen
there NE will we no.one.DAT no.GEN help up obliged be.INF
'we will not be obliged to give any help to anyone in this regard.' (Steinfurt 08/28/1354)

(52) Ock so **enschal** **nymant** **nyn** **buwholt** over de slachte schepen
also thus NE.shall no.one no timber over the piling ship
by vorluße des gudes
under loss the.GEN good
'Equally, nobody shall ship any timber across the piling, under punishment of losing the good.' (Oldenburg 1500)

As can be seen in the following examples, indefinites of the old n-free series, which was preferred over the n-marked series in the scope of negation in Old Low German, are now mainly used outside the scope of negation, in weak NPI contexts. Examples (53) and (54) are polar (yes/no) questions; in (55) *yergen* 'anywhere' occurs in the complement of a superlative, in (56) in a conditional.[33]

(53) Is dar **iemant** deper wen gij?
is there anyone brave than you
'Is there anyone as brave as you?' (*Alexander* 2 24: Seel., 2: 53, 2)

(54) Vader, mach men juw myt **yenigen** dingen helpen?
father may one you with any things help
'Father, can we help you with anything?' (*Alexander* 1 13: Seel., 1: 23, 14)

[33] As weak NPI contexts other than conditionals, such as yes/no questions, are naturally very rare in official documents, the corpus was complemented by a literary source (*Alexander*).

(55)

de	wiseste	fruwe,	de	**yergen**	wesen	mochte
the	wisest	lady	who	anywhere	be	might

'the wisest lady who may be found anywhere' (*Alexander* 6 14e: Seel., 6: 214, 7)

(56)

vnde	kumpt	he	dij	**yerghen**	to,	wise	ene	na	my	to	Paris
and	comes	he	you	anywhere	to	direct	him	to	me	to	Paris

'Should you meet him anywhere, direct him to me in Paris.' (*Alexander* 8 4: Seel., 8: 229, 33)

The indefinite pronoun *icht* 'something/anything' is infrequently used as an adverbial negator besides *nicht* in Middle Low German.

(57)

De	brutscho	scolen	ok	**icht**	betere	wesen	wan	v	sol.
the	wooing	shall	also	NEG	better	be	than	five	shilling

'Bride negotiations shall also not be more expensive than five shillings.' (Braunschweig 1349)

All in all, it seems that Low German developed from a language without negative doubling via one with optional and finally obligatory strict negative doubling (negation particle + n-word) to a language with negative spread (n-word + n-word(s)). This can probably be attributed to the weakening of the old preverbal negation marker *ni/ne*: it was first sufficient to identify sentential negation, not requiring indefinites in its scope to be n-marked as well. With its increasing weakening, n-marking became more and more common in indefinites in the scope of negation, leading first to obligatory doubling with *ni*, and then to common negative spread.

Turning now to the developments within the indefinite system of Low German, we have seen that Old Low German distinguishes three series, a 'neutral' one, a series of indefinites licensed in NPI contexts, and a series of indefinites restricted to the scope of negation. Among the NPI contexts, licensing the second series, we find questions (58), the complement of universal quantifiers (59), the complement of superlatives (60), and indirect negation (61), in all of which the n-free indefinites appear, here exemplified with *io* 'ever'.

(58)

Huan	uuas	thi	[**io**]	manno	tharf	[...]?
When	was	you	ever	man	need	

'When did you ever need any man?' (*Heliand*, 4433)

(59)

endi	cumad	[alle]	tesamne	liudi,	the	**io**	thit	lioht	gisâun,
and	come	all	together	people	who	ever	this	light	saw

'and all the people who ever saw this light come together.' (*Heliand*, 2596–7)

(60)

allaro	barno	bezt,	thero	the	**io**	[giboren]	uurði
all.GEN	children	best	of.those	who	ever	born	were

'the best of all children who were ever born.' (*Heliand*, 5267)

(61) that **ni** haƀit ênigan gigadon huergin, / thiu uuord an thesaro uueroldi,
that NEG have any of.their.kind anywhere the words in this world
that io uualdand mêr, drohtin diurie
that ever ruler more Lord dear
'There is nothing like it in words anywhere in this world that could glorify the ruler, our dear Lord, more!' (*Heliand*, 25–6)

In Middle Low German, the old 'neutral' indefinite *sum* 'some(one)' is lost, while the old item *wat* < *hwat* 'something' remains a 'neutral' indefinite (62), besides potentially the newly formed *ichteswat* 'something/anything', which is also available in weak NPI contexts and as a free-choice item (63).

(62) suwar de rede to langk is [...], dar wil ik **wat** affbreken.
whenever the speech too long is there will I something off.break
'Whenever the speech is too long, I shall make it a bit shorter.' (*Alexander*, 0 6: Seel., 0: 4, 10)

(63) De wile, dat de moder heft **ichteswat** to geuene, so hebben se
the time that the mother has something to give so have her
de kindere sere leff.
the children very dear
'While/as long as the mother has something/anything to give, the children are very fond of her.' (*Alexander*, 4 14 : Seel., 4: 137, 33)

The 'neutral' uses of *yerghen* < *hwergin* 'some-/anywhere' (< Gmc. *hwer 'where' + *gen 'any') have been lost in favour of newly formed *ichteswo*, in parallel formation to *ichteswat*.

(64) Hijrumme, dat du dat hilge cruce **ichteswo** geeret heuest,
because.of.this that you the holy cross somewhere venerated have
des schaltu geneten.
of.this shall.you enjoy
'Because you have venerated the holy cross in some place you shall enjoy this.' (*Alexander*, 2 33 : Seel., 2: 66, 37)

There do not appear to be any great changes in the licensing conditions on the old weak NPI indefinites and the n-marked indefinites, apart from the conditions governing their co-occurrence with the sentential negation marker and with each other as discussed above. N-marked indefinites continue to be restricted to the scope of negation, while the n-free series are licensed in all NPI contexts, weak and strong. This is shown in Table 6.7, where the left-hand columns represent Old Low German, and the right-hand ones Middle Low German.

TABLE 6.7 **Changes in the indefinite system between Old and Middle Low German**

	OLG	MLG	OLG	MLG
	determiner		'entity'	
'neutral'	*sum* *ên*	*eyn, een, ein*	*hwat*	*wat,* *ichteswat*
NPI	*ênig, hwergin*	*enich, jenigh*	*iouuiht,* *eowiht*	*icht(es)*
n-word	*neg(h)en* *nigên* *nênig*	*n(e)yn(e)* *nen(e)* *ny(n)erley/* *nenerley* *geen*	*niouuiht* *neouuiht*	*nicht(s)*
	'person'		'time'	
'neutral'	*sum, hwê, man*	*(eyn man)*	?	*up een tid*
NPI	*ioman* *man*	*iemand, jemant*	*io*	*ye*
n-word	*nioman*	*ne(y)man* *n(e)ymand*	*nio* *neo*	*nie, nummer,* *to nenen* *tiden*
	'place'			
'neutral'		*ichteswo*		
NPI	*hwergin*	*yerg(h)en,* *iergen*		
n-word	?[a]	*nergen(d),* *newerlte*		

Notes [a] The expected element *nio-hwergin* is not attested.

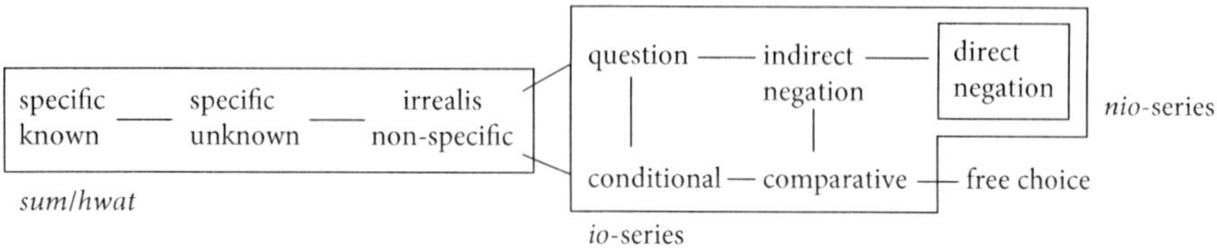

FIGURE 6.1 The system of indefinites in Old Low German

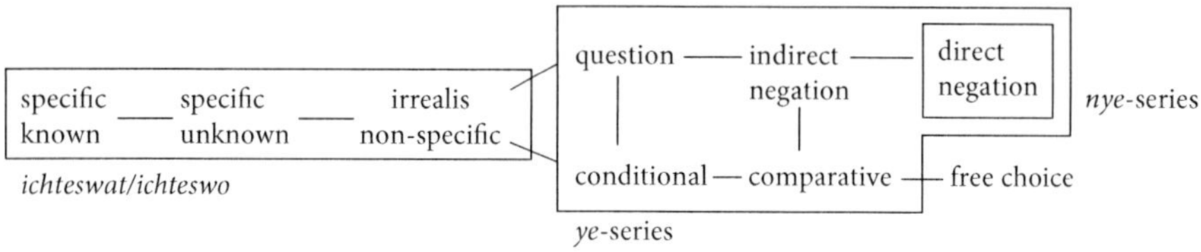

FIGURE 6.2 The system of indefinites in Middle Low German

The major aspects of the development of indefinite pronouns are summarized in Figures 6.1 and 6.2 using Haspelmath's (1997: 63–4) implicational map of indefinite-pronoun functions.

It seems that the indefinites in the scope of negation in Low German are affected by a certain cyclical development. In Old Low German, the n-free indefinites combine with the negation particle *ni* to form the n-marked series. In Modern Low German, a similar development seems to be incipient with the formation of emphatic multiple negative expressions.

6.3.2 *Dutch*

We saw above that in the Old Dutch of the *Wachtendonck Psalms*, *niuueht* is occasionally used to render Latin negative elements, mostly in order to maintain the word order of the Vulgate text in the interlinear translation. In (65), *nieuuiht* appears to be used as a negative determiner to *uuort* 'word' (i.e. 'no word', lit. 'nothing word'). *Niuueht* may have been chosen here to convey emphasis.

(65) Firmauerunt sibi sermonem nequam.
Gefestoda sig uuort **nieuuiht**
assert.PAST.3PL REFL word nothing
'They asserted themselves with no word (at all?).' (WP LXIII.5.2)

In the *Leiden Willeram*, there are 35 clauses with n-marked indefinites, all with the preverbal marker; that is, the text shows strict negative doubling.

(66) Thich **neminnot** **nieman**, her nesii recht
you NEG.loves n.one he NEG.be righteous
'Nobody loves you who is not righteous.' (LW 42.12–13)

(67) so **newillon** ouch ich **negheyn** arbeyd thurgh sinan willan scuwan
so NEG.will also I no effort through his will spare
'thus I shall spare no effort by his will.' (LW 56.10–11)

Besides negative doubling, negative spread appears to be possible (68), but does not seem obligatory (69):

(68) thaz sie **nietemer** **neheine** uirtutem **ni**mugan hauen
that they n.to.more no virtue NEG.can have
nisi tantum per me
unless only through me
'that they cannot (just as little) have any virtue, unless it be through me.' (LW 88.11–12)

(69) wanda an hin **nieman** **ni**uindet **ieweht** unrechtes
because in him n.one NEG.finds any thing unrighteous.GEN
'because no one finds anything unrighteous in him' (LW 59.4–5)

Here it looks as though indefinites preceding the finite verb negated by *ni* are n-marked to indicate that they are in the scope of negation, while indefinites following it do not need this marking as their being in the scope of sentential negation is sufficiently clear. If true, the system employed in the *Leiden Willeram* is similar to the Old High German one (Donhauser 1998: 289),[34] perhaps not surprising given the essentially Old High German syntax of the text, cf. section 6.1.2.

In Middle Dutch, n-marked indefinites continue to co-occur with the preverbal marker, now mostly weakened to *en*. Negative spread is available too, (70)–(71).

(70) Gode ne sach noyt gheen man[35]
God ne saw never no man
'God never saw any man.' (*Lectionarium Amsterdam*, 1348)

(71) **Nieman** **en** sijt ghi **niet** sculdich . . .
nobody NE are you nothing due
'You do not owe anything to anybody . . .' (New Testament, North Dutch translation, 1399)

[34] 'Die *n*-losen Indefinita des Althochdeutschen dagegen finden sich ausschließlich nach der Stellungsgruppe *ni*+V$_{fin}$ . . .' ('The n-free indefinites of Old High German, however, occur exclusively *after* the group *ni*+V$_{fin}$').

[35] Note that *gheen* 'no' is counted as an n-word here. It never seems to be used outside the scope of negation in Middle Dutch, unlike its Middle High German cognate *dehein*, cf. also n. 37.

Negative doubling with the new negator *niet* does not seem to have been available immediately;[36] the 65 n-marked indefinites[37] in the official documents 1200–1280 in the Corpus Gysseling (Gysseling 1977) do not co-occur with *niet*, only with preverbal *en*, and in fact in 25% of the cases even express sentential negation on their own as in (72), cf. Breitbarth (2009: 103).

(72)

Dat	**niemen**	vortane	hem	sal	onderwinden	moghen	der
that	nobody	henceforth	him	shall	begin	may	of.the

bruederscap...
fraternity...
'that no one henceforth may become part of the fraternity...' (CG 14:63, 8–10; Mechelen 1254)

Negative doubling with *niet* only becomes available very late in Middle Dutch, potentially indicating a change in the status of n-marked indefinites, and remains rare.[38] In (73) *niet* in fact has two readings, as an indefinite pronoun or as a sentential negator.

(73)

Maeldegijs	seide:	"Ic	**en**	liet	**niemant**	**niet.**"
Maeldegijs	said	I	NE	left	nobody	NEG/nothing

'Maeldegijs said: "I did not leave anyone"' or
'Maeldegijs said: "I left nothing to anyone"' (*Vier Heemskinderen* 1508)

A special case involves the temporal n-marked indefinites *nie/noyt/nemmer* 'never (more)', which can occasionally be used instead of the n-free equivalent *oyt* 'always, ever' in weak NPI contexts. Van Helten (1885b) gives examples with *noyt* occurring in restrictive relative clauses after superlatives (74), in comparatives (75), in *before*-clauses (76), in questions (77) and in the restriction of a universal quantifier (78).

(74)

Du	best	die	scoenste	creature,	Die	ic	met	oghen	**nie**	ghesach
you	are	the	most.beautiful	creature	that	I	with	eyes	n.ever	saw

'You are the most beautiful creature that I ever saw with my eyes.' (Fr. 7195) (van Helten 1885b: 236)

[36] In (71) *niet* is a pronoun ('nothing').

[37] This figure in fact contains 19 occurrences of *gheen* 'no(one)', which seems to behave in exactly parallel fashion to n-marked *engheen*, unlike its High German counterpart *dehein*, which starts out as a weak NPI indefinite and retains NPI characteristics for a long time (Jäger 2008: 260–6).

[38] Hoeksema (1997: 141) gives a made-up 'Middle Dutch' example (his (3)), based on the fact that negative doubling between n-words and *niet* is possible in Present-day West Flemish. Emphatically, his example is not from any Middle Dutch source or text. This also shows that despite its apparent conservatism, neither West Flemish nor any other Flemish dialect is a simple preservation of the Middle Dutch state of affairs.

(75) Dat hi sochter sliep… Dan **noit** up sijn bedde te voren
that he more.softly slept… than n.ever on his bed to before
'that he slept more softly than ever before on his bed' (Esm. 281) (van Helten 1885b: 235)

(76) Het moet al weder ghekeert zijn, eer de sonden
it must already back turned be before the sins
nemmermeer vergheven selen werden
n.ever.more forgiven shall be
'It has to have returned before the sins shall be forgiven.' (La. 4,8128) (van Helten 1885b: 239)

(77) Wie horde **noyt** secgen… Dat des gelijcs yet gesciede?
who heard n.ever say… that of.it same something happened
'Who ever heard tell that something like that happened?' (Ve. 3,33,40) (van Helten 1885b: 240)

(78) Al tfolc dat **nie** was geboren
all the.folk that n.ever was born
'all the people who were ever born' (Vdl. 2817) (van Helten 1885b: 243)

Such 'pleonastic' use of *nooit* is furthermore attested in the complement of certain adversative predicates:

(79) mi rouwet dat ic **noit** was geboren
me regrets that I n.ever was born
'I regret that I was ever born.' (Ovl.G. 3,110,118) (van Helten 1885b: 241)

Besides such NPI uses, *noyt* can also be used as an emphatic negator,[39] and even as a (possibly emphatic) negative determiner, (80).

(80) Die hertoge sweech al stille, ende en antwoirde **noyt** woort
the duke be.still already silent and NE answered never word
'The duke fell silent and didn't answer a single word.' (Grimb. I 790, *Middelnederlandsch Woordenboek*, entry on *nooit*)

The n-free counterpart of *noyt*, *oy(e)(n)t*, is itself special in that it undergoes a cross-linguistically typical 'quantifier cycle' from a positive element to a 'more negative' element (cf. section 1.9.1), while most of the Middle Dutch indefinite system is simply a continuation of the Old Dutch one (in so far as can be ascertained). Originally meaning 'always' (81),[40] it becomes a weak NPI indefinite in Middle Dutch (82).

[39] This is a cross-linguistically common strategy of emphatic denial, cf. Lucas and Willis (2012).

[40] The *Middelnederlandsch Woordenboek* mentions Frisian *ô* + *yet* < *eo* 'ever' + *ieuueht* 'anything' as the possible etymology of *oyt*, however.

(81) semper veritas odio fuit
oyt is de waerheyt behaet ghewest
always is the truth hated been
'the truth has always been hated.' (Kil., *Middelnederlandsch Woordenboek*, entry on *ooit*)

(82) die lelicste mans figure die **oint** ieman hadde gesien
the ugliest man's figure which ever anyone had seen
'the ugliest man that anyone had ever seen' (*Franc.* 4112, *Middelnederlandsch Woordenboek*, entry on *ooit*)

A possible scenario for this development might be that after the loss of *iuwerlte*, there was no temporal NPI indefinite ('ever') and that both *oyt* 'always' and *noyt* 'never' were used to fill this gap in the system. The use of *oyt* can be accounted for by its underspecification: it is not contextually restricted, and can therefore be used both in neutral and NPI contexts. The use of *noyt* is subject to the Elsewhere Condition (Kiparsky 1973): *noyt* is more specific than *oyt* in that it is specified for, or restricted to, downward-entailing contexts. It is of course overspecified by additionally being restricted to a subset of the downward-entailing contexts, namely negative contexts. Ultimately, *oyt/ooit* wins this competition for the weak NPI slot, but loses its positive uses (seen in (81)).[41] Hoeksema (1998, 1999) reports only finding NPI *ooit* in his corpus of 19th-century Dutch. *Noyt* retreats to its original licensing context, (direct) negation.

In so far as they have been possible to reconstruct, the developments in the system of indefinites from Old to Middle Dutch can be summarized as in Table 6.8.

The major aspects of the development of indefinite pronouns (as reconstructed) are summarized in Figures 6.3 and 6.4 using Haspelmath's (1997: 63–4) implicational map of indefinite pronoun functions.

On the way to Modern (Standard) Dutch, further shifts have occurred in the system of indefinites. All n-marked indefinites are now (again) restricted to the scope of negation:

(83) dat hij zachter sliep dan (*n)ooit tevoren
that he more.softly slept than (n.)ever before
'that he slept more softly than ever before'

[41] Hoeksema (1999: 159) mentions relics of positive uses in the Brabantish dialect of Dutch, where it can mean 'sometimes, occasionally', as in (i):

(i) ... roggebrood, dat wel **ooit** bij spek gegeten
[...] rye.bread that well sometimes along.with bacon eaten
maar meestal voor de honden en de paarden bestemd werd
but mostly for the dogs and the horses meant was
'rye bread, which was occasionally eaten with bacon, but mostly made to feed the dogs and horses'

TABLE 6.8 Changes in the indefinite system between Old and Middle Dutch

	Old Dutch	Middle Dutch	Old Dutch	Middle Dutch
	determiner		'entity'	
'neutral'		*een*		
NPI		*enig*	*ieueht*	*iet(s)*
n-word	*ne(g)hein, chein,* (*nieuuiht*)(cf. (68))	*(ne)gheen*	*niauuiht*	*niet(s)* *nyet*
	'person'		'time'	
'neutral'				*oy(e)(n)t*
NPI	*ieman*	*yemant*	*iuwerl(t)e* *ietemer*	*ye(t)* *iet*
n-word	*niaman*	*nyemant* *niemand*	*nohuuanne*	*noy(e)(n)t* *nie* *nemmer*

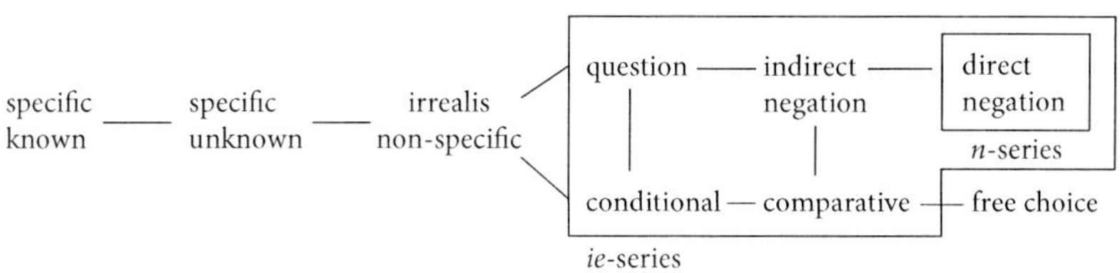

FIGURE 6.3 The system of indefinites in Old Dutch

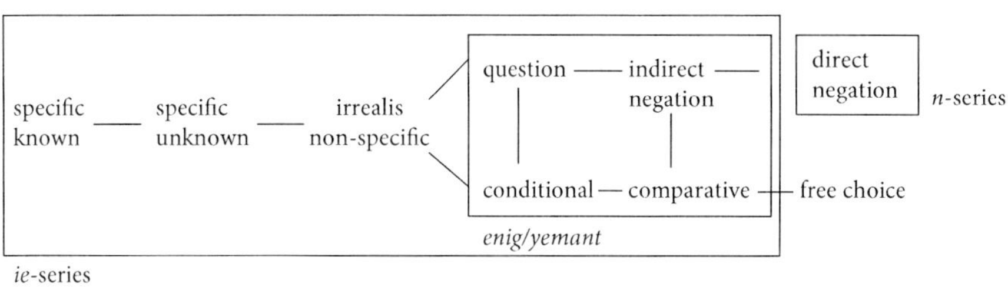

FIGURE 6.4 The system of indefinites in Middle Dutch

Ooit has undergone a lexical split into an NPI indefinite 'ever' and a PPI element 'once' (Hoeksema 1998, 1999). NPI *ooit* needs to appear in the same intonational phrase as the expression of negation (84a), while PPI *ooit* is infelicitous with clausemate negation, as seen in (84b), where *ooit* appears in a separate intonational phrase. The determiner *enig* 'some, any' when used with singular count nouns has become a weak NPI indefinite (Hoeksema 2007a), (85).

(84) a. Niemand was ooit blij ('ever', NPI)
n.one was ever happy
'No one was ever happy.'

b. #Niemand was blij, ooit. ('once', PPI)
n.one was happy ever/once
'Once (upon a time), no one was happy' (marginal, hence the #) (after Hoeksema 1999: 154)

(85) a. Geen van hen heeft enig dier geslacht.
none of them has any animal slaughtered
'None of them has slaughtered any animal.'

b. Heeft u ooit enig voorstel verworpen?
have you ever any proposal rejected
'Have you ever rejected any proposal?'

c. Als ik enig voorstel verwerp, wordt hij boos.
when I any proposal reject, becomes he mad
'When I reject any proposal, he gets mad.'

d. Hij was langer dan / zo lang als einige andere speler.
he was taller than / as tall as any other player
'He was taller than/as tall as any other player.' (Hoeksema 2007a: 10)

6.3.3 *Summary*

The diachronic developments concerning indefinites in the scope of negation in Low German and Dutch can be summarized as follows.

In Old Low German we see a rise of n-marked indefinites in the scope of negation. While the expression of indefinite quantification in negative clauses was arguably initially *ni*... n-free indefinite (as in the *Genesis* fragments), the use of n-marked indefinites becomes obligatory over the period; it is still optional and dispreferred in the *Heliand* epos, but exceptionless in the minor texts. In Middle Low German, n-marked indefinites remain the main expression of indefinite quantification in negative clauses. Unlike in Old Low German, n-marked indefinites can now co-occur with each other. While n-marked indefinites are compatible with the old preverbal particle *ne/en*, the new sentential negator *nicht* does not seem to co-occur with them, *ne/en* no longer being the standard expression of sentential negation at this stage.

As argued in section 6.2, this means that Low German lost negative doubling between Old and Middle Low German, but gained negative spread. The system of indefinites in historical Low German went from one with a 'positive' and an 'NPI' series to one with a 'positive', an 'NPI', and a 'negative' series. Initially, the latter two are both available in the scope of negation. Towards the Middle Low German period, the NPI series retracts to non-negative NPI contexts while the (newer) n-marked series becomes the only one available in negative clauses ('bagel'-distribution; cf. Pereltsvaig 2006).

The oldest Dutch texts express indefinites in the scope of negation by means of n-marked forms (*ne*... n-marked indefinite), where the preverbal marker may already be missing. Negative spread—that is, the co-occurrence of n-marked indefinites—is also already attested, apparently optionally. It becomes the rule in Middle Dutch, much like in Middle Low German, and, as there, n-marked indefinites may co-occur with the disappearing preverbal marker (*en*). Later in Middle Dutch, negative doubling with the new postverbal element *niet* becomes available. As far as we have been able to reconstruct the system of indefinites, the main changes seem to be the widening of *iouueht* > *iet*(*s*) 'anything' > 'anything, something' and the competition between two temporal adverbs for the weak NPI slot. After a competition between *nooit* 'never' and *ooit* 'always' > 'ever', the latter wins out.

In Low German, the type of negative concord available at each stage can easily be correlated with the relative 'strength' of the sentential negation markers *ni* > *ne*/*en* and *nicht*. The use of n-marked and n-free indefinites follows from this; while both n-free and n-marked indefinites were used in negative clauses in Old Low German, with a preference for n-free ones, the weakening of the old preverbal marker meant that n-marked indefinites became the exclusive expression of indefinite quantification in negative clauses, while n-free indefinites became restricted to non-negative NPI contexts.

The rise of negative spread can be accounted for if one assumes that the n-marked indefinites were originally semantically negative when they first arose in Old Low German, arguably by univerbation of *ni* with n-free indefinites, and later on became semantically non-negative, as will be argued in the next section.

6.4 The development of negation

In order to account for the Low German and Dutch developments, I will adopt an adapted version of Zeijlstra's (2004) account of Jespersen's cycle and negative concord. According to Zeijlstra (2004), languages can express negation either syntactically or semantically. In a system of the latter type, every overt particle or negative indefinite contributes semantic negation, leading to double negation (logical affirmation) in cases where two such elements co-occur. In a language expressing negation syntactically, 'negative elements mark the presence of a (c)overt negative operator' (Zeijlstra 2004: 244) by bearing syntactic (formal) negation features. Only one element carries an interpretable negation feature [iNeg] which licenses the

TABLE 6.9 **Types of negative-concord language (Zeijlstra 2004, 2009)**

	negator	n-word
non-strict NC	[iNeg]	[uNeg]
strict NC	[uNeg]	[uNeg]
NegSpread only	$\neg(\exists)$	[uNeg]
no NC	$\neg(\exists)$	$\neg\exists$

uninterpretable negation features [uNeg] of the other elements. Language variation is the result of variation in which element carries the interpretable feature, potentially a covert element. In languages expressing negation syntactically, which by Zeijlstra's assumptions are necessarily negative-concord languages, all n-words are endowed with an uninterpretable negation feature [uNeg], and therefore need to be licensed by an element carrying [iNeg]. This can be either the sentential negator, deriving the non-strict negative concord (doubling) languages, or a covert negation operator OP¬, deriving the strict negative concord languages. In non-strict negative-concord languages, the sentential negator licenses all indefinites in its scope and a covert OP¬ is only inserted as a last resort in cases where an indefinite occurs outside its scope (for example a preverbal subject). In strict negative-concord languages, the overt sentential negator is always uninterpretable and merely indicates the presence of a covert negation operator, which licenses all [uNeg] elements in its scope. Languages in which the sentential negator is a syntactic head are always negative-concord languages, according to Zeijlstra, and therefore always express negation syntactically.

An additional assumption has to be made for languages like French, in which negative spread is possible, but n-words cannot co-occur with the sentential negator *pas* (that is, there is no negative doubling). Zeijlstra (2009) proposes that in such languages the sentential negator is semantically negative ($\neg(\exists)$),[42] and would therefore clash, that is, lead to double negation, with the covert [iNeg] operator projected to license the [uNeg] n-words. Taken together, Zeijlstra predicts there to be three types of negative-concord language, as in Table 6.9.

6.4.1 *Low German*

We saw above that, in those Old Low German texts that use n-marked indefinites in the scope of negation, these can precede the preverbal marker. Under Zeijlstra's approach, Old Low German is hence to be classed as a strict NC language, with a

[42] The parentheses indicate the optionality of existential closure induced by the negative operator. This only happens where there is an open variable (Zeijlstra 2004: 247).

[uNeg] feature on the sentential negator *ni* that is able to identify the presence of an interpretable covert negation operator OP¬ with an [iNeg] feature in a scope position. This could also explain the grammaticality of the order *io*...*ni*-V in (46c), repeated here as (86).

(86) OP¬[iNeg] sô is **io** endi ni[uNeg] cumit

sô	is	**io**	endi	**ni**	cumit
thus	it.GEN	ever	end	NEG	comes

'thus the end (of it) will never come.' (*Heliand*, 1324)

Here, an n-free (NPI) indefinite *io* 'ever' precedes the expression of sentential negation, in an apparent violation of both the Neg-First Principle (Haspelmath 1997, Mazzon 2004), according to which sentential negation has to be marked at the earliest opportunity in a clause, and the generalization that NPIs are licensed only within the scope of negation. Under the assumption that the negation feature of *ni* is uninterpretable and the covert OP¬ identified by *ni* is in a position c-commanding *io*, this is not a problem. Under an approach such as the one proposed by Penka (2010: 61), for instance, OP¬ is not confined to the specifier of a postulated functional projection NegP (as assumed by Zeijlstra), but can adjoin to any propositional node (semantic type *t*), minimally above VP, but potentially higher, too.

There are a number of problems with applying Zeijlstra's approach to historical Low German as is. First, the negator *ni/ne* is arguably a syntactic head, predicting negative concord in Old Low German to be exceptionless. This is not what is found in the main texts of Old Low German, the *Heliand* and especially the *Genesis*. Secondly, in Zeijlstra's account all negative indefinites (n-words) in negative-concord languages bear a formal [uNeg] feature. Unless Zeijlstra's operation of Multiple Agree, by which one interpretable feature is able to license all uninterpretable features in its scope, is parameterized (an option not mentioned by Zeijlstra), we expect Old Low German to have negative spread, contrary to fact.[43]

A likely scenario, which is still able to make use of Zeijlstra's main insights, is the following. The Old Low German negator *ni* is indeed [uNeg], as witnessed by the availability of preverbal NPI-indefinites, even in the *Genesis* fragments, where n-marked indefinites are not used together with *ni* (87).

[43] However, see Haegeman and Lohndal (2010) for arguments against Zeijlstra's Multiple Agree account of negative concord in West Flemish. Depending on further properties of the n-marked elements involved (indefinites and negators), namely other formal features besides [u/iNeg] such as quantificational features, different types of negative concord may possibly be derived in languages other than West Flemish. We will not explore this option here.

(87) that is **ênig** seg **ni** ginas
that of.it any man NEG was.saved
'that no man was saved from it.' (*Genesis*, 322)

The newly forming n-marked indefinites, on the other hand, bear an interpretable syntactic negation feature, [iNeg]. As they are already available in the earliest text, the *Heliand* epos, we have no indication as to when exactly they arose. Assuming that a principle such as van Gelderen's (2008: 297) Feature Economy, (88), is operative in language change, we can hypothesize that the preverbal negation marker *ni* originally carried an [iNeg] feature. We can therefore reconstruct the emergence of n-marked indefinites in Old Low German by univerbation of this preverbal negation marker with indefinites of the NPI-series (for instance, *ni+ioman* > *nioman* 'no one'). Under Feature Economy, *ni* changed to [uNeg] before the beginning of textual attestation, while the n-marked indefinites are still [iNeg] in the *Heliand* on this scenario.

(88) Feature Economy
Minimize the semantic and interpretable features in the derivation, e.g.

VP-adverbial		CP-adverbial		C-head
semantic	>	[iF]	>	[uF]

(van Gelderen 2008: 297)

The question of course arises why postulating an [iNeg] feature on an n-word (negative quantifier) does not lead to double negation with the covert [iNeg] operator projected by the [uNeg] feature on the sentential negator *ni*. The position assumed in the present chapter is that covert [iF] operators should only be projected as a matter of last resort in order to ensure syntactic licensing of all uninterpretable features [uF] (cf. also Zeijlstra 2004: 246). Assuming, as is common within the Minimalist framework, that syntactic derivations proceed bottom-up, the [uNeg]-feature on *ni* is already licensed before any covert operator is merged (in SpecNegP under Zeijlstra's approach), at the level of vP. This is true for both object and subject indefinites, assuming subjects have their base position within *v*P.[44] Even adverbial n-indefinites, in the present corpus only *nio* 'never', if analysed as adjoined to *v*P, are unproblematic. In all cases, [iNeg] c-commands [uNeg] and thus ensures the syntactic licensing. We illustrate this here for an object indefinite:

(89) a. **ne** antuuordida **niouuiht** uuid iro uuretun uuord
NEG answered nothing against their hostile words
'He didn't reply anything to their hostile words.' (*Heliand*, 5382–3)

[44] The VP-internal subject hypothesis goes back to Zagona (1982) and Koopman and Sportiche (1985, 1991) and is now standard in Minimalist approaches.

b.

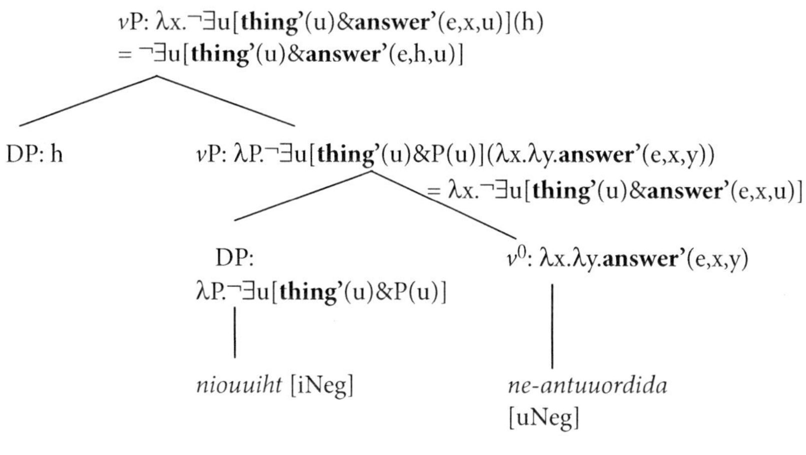

The rise of negative doubling in Old Low German can be accounted for as follows: once they are available in the scope of negation, n-marked indefinites are preferred in this environment due to either:

(a) the fact that they are more specifically designated to appear there than n-free indefinites, which are licensed in all NPI contexts (weak and strong), by being restricted to the scope of negation and by being morphologically marked for this restriction (a form of the Elsewhere Condition; Kiparsky 1973); or perhaps
(b) a universal functional preference for marking negation as clearly as possible (Ramat 2006b).

Once n-marked indefinites are the standard way of expressing indefinite quantification in the scope of negation, a reanalysis of the system as a 'standard' strict negative-doubling language as described by Zeijlstra, that is, one with a [uNeg] negator (*ni*) and [uNeg] n-words, is possible. Once the n-marked indefinites are reanalysed as carrying a [uNeg] feature (by Feature Economy, (88)), the rise of negative spread is possible, accounting for the change towards Middle Low German. Independently, the rise of a semantically negative adverbial negator (*nicht*) leads to the reanalysis of the old preverbal marker.[45] Essentially, it seems that the old preverbal negator comes to behave as a weak NPI at stage II of Jespersen's cycle in Middle Low German—it is licensed in the scope of negation as well as in certain non-negative contexts.[46] However, becoming a weak NPI cannot immediately account for the loss of the preverbal marker during the Middle Low German period. Breitbarth

[45] Cf. Breitbarth (2013) for more arguments for the independence of Jespersen's cycle and negative concord in historical Low German.

[46] Zeijlstra (2009) proposes such an NPI-analysis for present-day French *ne*.

(2009) has proposed that it is reanalysed as a marker of affective polarity. The preverbal marker is lost before this reanalysis can fully actualize (Timberlake 1977), that is, spread to other non-negative affective contexts besides exceptive clauses in Middle Low German, which would independently confirm its new status to a language learner (sentential negation already entails and thus indicates the affective polarity of a clause).[47] The fact that the new negator is semantically negative, but not syntactically, while n-marked indefinites carry a syntactic [uNeg] feature means that Middle Low German develops negative concord of the French type (Zeijlstra 2009): negative spread without negative doubling.

6.4.2 *Dutch*

The standard sentential negator in Old Dutch is *ne*. It is even inserted against the Latin original, as seen in (20). However, at the same time, there are already unemphatic uses of adverbial *niuueht* in Old Dutch. In Middle Dutch, *niet/nyet* appears to be the neutral sentential negator, but there are still a number of contexts in which *ne/en* can be used on its own to express sentential negation (as in paratactic negation or the context of certain verbs). Furthermore, the old preverbal marker eventually seems to undergo, at least in the southern dialects, a reanalysis which definitively removes it from the immediate expression of sentential negation.

Negative doubling seems to be the rule in Old Dutch, and negative spread is at least optionally available, becoming standard in Middle Dutch. Considering the above analysis of Old Low German, we could say that the preverbal negator in Old Dutch had a formal [uNeg] feature, while the arising postverbal negator was semantically negative, initially preventing negative doubling between n-words and *niuueht* > *niet* well into the Middle Dutch period, parallel to Middle Low German (cf. above) or French (Zeijlstra 2009). Unlike their Old Low German cognates, the n-marked indefinites had a [uNeg] feature in Old Dutch. Depending on one's analysis of the status of the old preverbal marker and whether one wants to argue that it is still a negative marker or not, Middle Dutch either has or does not have negative doubling

[47] It may be objected that affective polarity does not have to be marked in non-negative contexts in most languages; however, the option may arise as a by-product of Jespersen's cycle. It has long been acknowledged that old preverbal *ne* in French has ceased to be a negation marker proper but as it were 'switches off' the affirmative concept (cf. Tesnière 1959: 224–5), sentential negation being expressed by a *forclusif* (*pas* or n-words), and thus acts as a polarity marker. Biberauer (2008) argues that one of the two elements of the bipartite expression of negation in Afrikaans, *nie*$_2$, is not a negation marker, but the realization of a CP-related polarity head and gives examples in which it can be used in non-negative affective contexts:

(i) Ek kan my nouliks/skaars inhou **nie**$_2$.
I can me barely in.hold NEG
'I can barely contain myself', i.e. 'I'm very excited.' (Biberauer 2008: 116)

(ii) Ek weier om saam te kom **nie**$_2$.
I refuse COMP.INF together to come NEG
'I refuse to come along.' (Biberauer 2008: 116)

in addition to negative spread. Later in Middle Dutch, by van Gelderen's (2008) Feature Economy (89), postverbal *niet* seems to change from being semantically negative to syntactically negative carrying a formal [iNeg] feature, entering into agreement with [uNeg] n-words.

The rise of pleonastic uses of *nooit* '(n)ever' (examples (75)–(80)) does not seem to affect the other n-marked indefinites (*nyet(s)* 'nothing', *niemand* 'no one', *nergens* 'nowhere', etc.). Reversing Haspelmath's (1997: 210) generalization, reproduced here in (90), we can say that an n-marked indefinite is open to the loss of its [uNeg] feature and subsequently to become an NPI by reanalysis in cases where it continues to occur with the verbal negator.

(90) If a negative indefinite never co-occurs with verbal negation, it has only the direct-negation function. (Haspelmath 1997: 210)

Let us assume then that in Dutch too, the old preverbal marker *ne* becomes ambiguous between a negation marker and a polarity marker in the transition between Old Dutch and Middle Dutch, as argued by Breitbarth (2009). The reanalysis is further corroborated by the emerging use of *ne/en* in weak (non-negative) NPI-contexts like (32), a sign that the change is beginning to actualize. On the other hand, the continued ability to license certain negative polarity constructions (Postma 2002) as seen in (33) may point at its continued ability to express sentential negation.[48] Later, the ambiguity is resolved in favour of the reanalysis of *en* as a polarity marker, spreading after the Middle Dutch period to more non-negative contexts in those dialects that maintain it (cf. (36)–(38)), while it is lost in others.

6.4.3 *The loss of the preverbal marker*

The old preverbal marker *en/ne* is lost at different speeds in the different continental West Germanic languages. In High German (cf. Jäger 2008, this volume), *en* is lost around 1300, in the Low German dialects between 1400 and 1500 (cf. also Breitbarth 2008), and in Dutch in the 17th century (Burridge 1993), in southern dialects even later (Beheydt 1998). It therefore looks as though this innovation spread from the south to the north and northwest. Furthermore, in both Low German and Dutch, dialect is a significant factor, with the loss being more advanced in areas with population mixture and resulting dialect levelling, such as the northeastern *Neuland* east of the river Elbe. Some Low German dialects adjacent to the progressive High German area on the other hand are slower in their transition than some dialects further away. That is, the geographical diffusion of the innovation is to a significant

[48] As indicated above (section 6.2.2), not all of the cases Postma counts as NPI constructions need to be analysed as such. *Bore*, *twint*, and *meer* could also have been on their way to becoming negative indefinites specified as [uNeg], following the grammaticalization path outlined in Haspelmath (1997: section 8.3), but reverted or were lost before the actualization of the reanalysis was complete.

degree influenced by the patterns of migration in the area. While rural areas are more conservative due to their close-knit social networks (Trudgill 2004), areas (especially cities) with increased in-migration from disparate areas of origin and looser social networks tend to be more innovative. Often, this goes hand in hand with structural simplification, as in the case of the loss of *en/ne*.[49]

As a particular factor behind the delay of the loss of *en* in the southern Dutch dialects we have furthermore identified its reanalysis as an emphatic (polarity) marker (Breitbarth and Haegeman 2010, forthcoming). This functional distinction from the expression of negation led to its prolonged maintenance in Flanders.

Besides the geographical factor, two grammatical factors influencing the loss of *en* have been identified, the position of the finite verb and the type of the verb. In both Low German and Dutch, the preverbal marker is lost earlier in verb-first contexts, and last in verb-final contexts. Lexical verbs are most conservative in maintaining *en*, (temporal) auxiliaries most innovative.

For Dutch, Burridge (1993), considers two possible scenarios for the fate of *en/ne*. First, after verb placement became rigidly bound to clause type, *en* is deleted in contexts where it would violate or interfere with these word-order constraints, namely in verb-first and verb-second contexts.[50] Second, Burridge considers the option of *en* being reanalysed as a part of the verb (clitic or affix), which is, she argues, what has happened in Flemish. Problematic for extending the first scenario to Low German is that we have seen above that, in Middle Low German, verb-second and verb-final contexts do not show a significant difference in influencing the loss of *en*; only verb-first contexts clearly favour the loss. Furthermore, correlating the type of verb with its position does not yield a unified account of the loss of the preverbal marker in Middle Low German and Middle Dutch. Postma and Bennis's (2006) account of the system of negation in the Saxon dialect of Drenthe around 1400 in terms of such a systematic correlation[51] does not extend to other Middle Dutch or Middle Low German dialects. Postma and Bennis furthermore show that the system assumed for Drentish around 1400 was rather short-lived and is lost under contact with Hollandish around 1500.

[49] The same can be observed in the High German dialects at the same time. As Pensel's (1976) study reveals, of the four High German dialects he considers, West Upper German, East Upper German, West Central German, and East Central German, only the former three show some vestiges of the preverbal marker in the period of 1470–1530. East Central German, the result of dialect mixture due to the colonization history of the area (which is also the reason that this group of dialects became the basis of the modern standard language), does not show any traces of it at all in the relevant period.

[50] This presupposes counting *en* as an independent element, potentially problematic in the light of de Haan and Weerman's (1984) observation that *en* is the only element capable of intervening between a separable prefix and its verb.

[51] Postma and Bennis argue that, in Middle Drentish around 1400, the preverbal marker *en* appears on auxiliaries in C (V1- and V2-contexts) but not in final position in clauses without verb movement to C. For lexical verbs, the inverse correlation is argued to hold, *en* appears on them in final position, but not on lexical verbs in C.

Breitbarth (2009) essentially spells out a version of the second scenario. She proposes that the preverbal marker ceases to function as the expression of sentential negation and is reanalysed as a bound morpheme spelling out the formal features of a left-peripheral polarity head on the finite verb.[52] This polarity head is assumed to be situated in the CP layer, reflecting the polarity of the clause [±affective] at its interface. The reasons why the preverbal marker is lost from most West Germanic dialects are that the reanalysis fails to fully actualize (spread to more, and in particular, non-negative affective contexts) in most dialects, and that [+affective] is entailed by the presence of an overt marker of negation and therefore does not need to be spelled out separately on the finite verb. Jespersen (1917) argues that the reason why the unaccented preverbal marker is lost first from sentence-initial position may have to be sought in the phonological weakness of this position.[53] As argued above, the preverbal marker is not entirely lost in some southern Dutch dialects due to another reanalysis, this time as an emphatic element. Under such a scenario, where the preverbal marker is analysed as a verbal affix, the fact that auxiliaries and modals are so much more progressive in losing the preverbal marker than lexical verbs can be explained uncontroversially by the common tendency for high-frequency items to undergo deflection earlier than lower-frequency items.

6.5 Conclusion

Regular bipartite negation only seems to establish itself in the attestation gap between the 'Old' and 'Middle' periods of Low German and Dutch, although the late Old Dutch *Leiden Willeram* already used the bipartite expression of negation quite regularly.[54] When attestation resumes in the 13th century, the preverbal marker already seems optional to some extent, as seen above, and is in fact restricted to specific (not always negative) contexts, while the adverbial negator acts as the standard, non-emphatic negator in all respects. The inherited preverbal marker is beginning to be reanalysed as a polarity marker, leading to a period of ambiguity especially in the Dutch dialects, where we saw continued uses of preverbal *en* in certain grammatical and lexical contexts. This reanalysis as a polarity marker is never fully actualized in most dialects and the preverbal marker is eventually lost, especially in contexts of dialect mixture due to population migration, which tends to lead to structural simplification—in the present case, of the expression of negation. Only in

[52] By Alternative Realization (Emonds 1987, 2000), according to which the features of a higher functional head are realized on the lexical head of a sister of that head. Emonds argues that this mechanism is at work in a variety of agreement scenarios.

[53] For arguments from first language acquisition regarding the omission of unstressed syllables from prosodically difficult positions, see Gerken (1994).

[54] To the extent that its syntax can be taken to be representative at all, following the late Old High German original so closely.

TABLE 6.10 Changing features of lexical items involved in the expression of negation in historical Low German and Dutch

	Old Low German	Middle Low German
preverbal marker	[uNeg]	[Pol]
postverbal marker	n/a	$\neg(\exists)$
n-marked indefinites	[iNeg]	[uNeg]
n-free indefinites	strong NPI	weak NPI
	Old Dutch	**Middle Dutch**
preverbal marker	([uNeg])	[uNeg] > [Pol]
postverbal marker	($\neg(\exists)$)	$\neg(\exists)$ > [iNeg]
n-marked indefinites	([uNeg])	[uNeg]
n-free indefinites	(individual changes)	

certain southern Dutch dialects can the reanalysis gain some hold and form the input to a further reanalysis, this time as a marker of emphasis (on polarity) (Breitbarth and Haegeman 2010, forthcoming).

Regarding the developments in the system of indefinites in the scope of negation and the availability of negative concord, both Low German and Dutch witness the rise of negative spread towards their 'Middle' periods. The Low German evidence suggests that the use of n-marked indefinites in the scope of negation is an innovative development within Old Low German. One development within the Dutch system of indefinites is noteworthy; the originally positive (PPI) temporal adverb *oy(n)t* 'always' and its n-marked equivalent *noy(n)t* 'never', originally restricted to the scope of negation, compete for the weak NPI slot ('ever') vacated by Old Dutch *niuwerlte* during the Middle Dutch period until *ooit* 'ever' < *oy(n)t* wins out.

The historical developments in the expression of negation in Low German and Dutch, analysed using Zeijlstra's (2004, 2009) approach, are summarized in Table 6.10.

Primary sources

Old Low German

Heliand, Genesis

Heliand und Genesis. Ed. Burkhard Taeger. Tübingen: Niemeyer, 1996.

Minor OLG texts

Kleinere altsächsische sprachdenkmäler. Mit anmerkungen und glossar, ed. Elis Wadstein. Norden/Leipzig: D. Soltau's Verlag, 1899.

Old Dutch

Wachtendonck Psalms, WP

Die altostniederfränkischen Psalmenfragmente, die Lipsius'schen Glossen und die altsüdmittelfränkischen Psalmenfragmente, ed. W. L. van Helten. Groningen: J. B. Wolters, 1902.

Leiden Willeram, LW

(Expositio) Willerammi Ebenspergensis Abbatis in Cantico Canticorvm. Die Leidener Handschrift, ed. Willy Sanders. Munich: Wilhelm Fink Verlag, 1971.

Middle Low German

Alexander

Der große Seelentrost, ed. M. Schmitt. Köln/Graz: Böhlau, 1959. Available at <http://titus.uni-frankfurt.de/texte/etcs/germ/mnd/a_seelen/a_see.htm>.

Barsinghausen

Urkundenbuch des Klosters Barsinghausen, ed. A. Bonk. Hanover: Hahn, 1996.

Börstel

Urkundenbuch des Stifts Börstel, ed. R. Rölker und W. Delbanco. Osnabrück: Selbstverlag des Vereins für Geschichte und Landeskunde von Osnabrück, 1996.

Braunschweig

Urkundenbuch der Stadt Braunschweig, ed. L. Hänselmann, H. Mack, M. R. W. Garzmann, and J. Dolle. Osnabrück: Wenner, 1975.

Hanserecesse

Hanserecesse. Die Recesse und andere Akten der Hansetage 1256–1430, vol. 1, ed. Karl Koppmann. Leipzig: Verlag von Duncker und Humblot, 1870.

Lübeck

Urkundenbuch der Diözese Lübeck, vol. 1, ed. W. Leverkus (1856); vols. 2–5, ed. W. Prange. Neumünster: Wachholtz, 1994–7.

Mariengarten

Urkundenbuch des Klosters Mariengarten, ed. Manfred von Boetticher. Hildesheim: Lax, 1987.

Oldenburg

Urkundenbuch der Stadt Oldenburg, ed. D. Kohl. Münster: Aschendorff, 1914. [Inventare der nichtstaatlichen Archive Westfalens, Vol. 6]

Scharnebeck

Urkundenbuch des Klosters Scharnebeck: 1243–1531, ed. Dieter Brosius. Hildesheim: Lax, 1979.

Steinfurt

Inventar des Fürstlichen Archivs zu Burgsteinfurt, ed. A. Bruns and W. Kohl. Münster Westf.: Aschendorff, 1971–1983.

Stralsund

Der Stralsunder liber memorialis, ed. Horst-Diether Schroeder. (Veröffentlichungen des Stadtarchivs Stralsund. Hg. Herbert Ewe).

Part 1: Fol. 1–60, 1320–1410. Schwerin: Petermänken-Verlag, 1964.

Part 2: Fol. 61–120, 1410–1422. Weimar: Hermann Böhlaus Nachfolger, 1969.

Part 3: Fol. 121–186, 1423–1440. Weimar: Hermann Böhlaus Nachfolger, 1972.
Part 4: Fol. 187–240, 1366–1426. Rostock: Hinstorff Verlag, 1966.
Part 5: Fol. 241–300, 1426–1471. Weimar: Hermann Böhlaus Nachfolger, 1982.
Part 6: Fol. 301–344, 1471–1525. Weimar: Hermann Böhlaus Nachfolger, 1988.

Uelzen

Urkundenbuch der Stadt Uelzen, ed. T. Vogtherr. Hildesheim: Lax, 1988.

Middle Dutch

Corpus Gysseling

Corpus van Middelnederlandse teksten (tot en met het jaar 1300), ed. Maurits Gysseling. The Hague/Leiden: Nijhoff, 1977.

Lectionarium Amsterdam

Het Amsterdamse Lectionarium, ed. Cebus Cornelis de Bruin. Leiden: Brill, 1970. In Instituut voor Nederlandse Lexicologie (ed.), CD-ROM *Middelnederlands*. The Hague/Antwerp: Sdu Uitgevers/Standaard Uitgeverij, 1998.

New Testament, North Dutch translation

Het Nieuwe Testament van de Moderne Devotie, ed. Cebus Cornelis de Bruin. Leiden: Brill, 1979.

Vier Heemskinderen

De historie van den vier Heemskinderen, ed. G. S. Overdiep. Groningen: Wolters, 1931. In Instituut voor Nederlandse Lexicologie (ed.), CD-ROM *Middelnederlands*. The Hague/Antwerp: Sdu Uitgevers/Standaard Uitgeverij, 1998.

7

Negation in the history of the Brythonic Celtic languages

DAVID WILLIS

7.1 Introduction

The Brythonic Celtic languages, Welsh, Breton, and Cornish, are particularly interesting from the point of view of negation in that two of them, Welsh and Breton, have independently undergone Jespersen's cycle. Apart from some residue in the form of initial consonant mutations (see section 7.3.1 below), present-day spoken Welsh, in (2), has replaced the Middle Welsh preverbal negative marker *ni(t)* in (1) with a postverbal marker *ddim* in main clauses, reaching stage III of Jespersen's cycle (V–Neg) today (Morris-Jones 1931, Poppe 1995: 154–5):

(1) . . . **ny** wnn i pwy wyt-ti.
NEG know.PRES.1SG I who be.PRES.2SG-you
'. . . I don't know who you are.' (*Pedeir Keinc y Mabinogi* 2.22–3, Middle Welsh)

(2) Wn i **ddim** pwy wyt ti.
know.PRES.1SG I NEG who be.PRES.2SG you
'I don't know who you are.' (Present-day Welsh)

Literary Welsh still allows stage I (Neg–V) preverbal negation, although, increasingly, formal writing too is switching to stage III. The new negative marker *ddim* derives from an indefinite pronoun *dim* 'thing, anything, nothing'. This item has given rise to a series of homophonous or nearly homophonous items in Present-day Welsh. There are now six synchronically and diachronically related but distinct items: argument *dim*, adverb/pseudoargument *ddim*, quantifier *dim*, pseudoquantifier *dim o*, sentence-final adverbial *dim*, and focus-negating/constituent-negation *dim* (Borsley and Jones 2000, 2005). Adverb/pseudoargument *ddim* is the main sentential negator of Present-day Welsh, as illustrated above in (2). Argument *dim* is an indefinite pronoun 'nothing, anything', illustrated in (3), while quantifier *dim* 'any, no' is illustrated in (4).

(3)

Dyw	Dafydd	ddim	wedi	dweud	**dim**.
be.PRES.3SG	Dafydd	NEG	PRF	say.INF	nothing

'Dafydd has said nothing.' (argument *dim*, Present-day Welsh)

(4)

Dyw	Dafydd	ddim	wedi	cael	[**dim**	anrheg].
be.PRES.3SG	Dafydd	NEG	PRF	get.INF	no	present

'Dafydd hasn't received a (any) present.' (quantifier *dim*, Present-day Welsh)

Pseudoquantifier *dim o*, normally shortened to *mo*, marks a definite object in the scope of negation:

(5)

Welodd	Dafydd	**(ddi)m**	**o**	'r	ffilm.
see.PAST.3SG	Dafydd	nothing	of	the	film

'Dafydd didn't see the film.' (pseudoquantifier *dim o*, Present-day Welsh)

Sentence-final adverbial *dim* reinforces already present sentential negation:

(6)

Dw	i	ddim	wedi	cysgu	**dim**.
be.PRES.1SG	I	NEG	PRF	sleep.INF	NEG

'I haven't slept at all.' (Borsley and Jones 2005: 142) (Present-day Welsh)

Finally, *dim* is used (alongside more formal *nid* and colloquial *ddim*) as a constituent and focus negator with the constituent being negated obligatorily fronting to initial focus position:

(7)

Dim	**Dafydd**	enillodd	y	râs.
NEG	Dafydd	win.PAST.3SG	the	race

'It wasn't Dafydd who won the race.' (Present-day Welsh)

Breton has also undergone Jespersen's cycle. It has replaced an inherited stage I pattern with preverbal *ni(t)*, illustrated for Old Breton in (8), by stage II (Neg–V–Neg) bipartite negation *ne ... ket*, illustrated for Present-day Breton in (9).

(8)

Ni	gus	coucant.
NEG	know.PRES.IMPERS	certain

'It is not known exactly.' (Fleuriot 1964a: 268) (Old Breton)

(9)

N'	ouzhon	**ket**	piv	out.
NEG	know.PRES.1SG	NEG	who	be.PRES.2SG

'I don't know who you are.' (Present-day Breton)

The bipartite pattern is already dominant, although not compulsory, by the time of the earliest Middle Breton texts (1450), and becomes compulsory over the next two centuries. More recently, many dialects have begun to omit the preverbal marker *ne*, leading to a stage III negative pattern with postverbal *ket* alone.

This chapter will begin by tracing these developments in some detail. It will then move on to examine the main developments in the system of indefinites, paying particular attention to indefinites found in negative and non-assertive (negative polarity) contexts. Here, a number of items have become increasingly negative in the historical period, leading to the cyclic creation of new series of pronouns for non-negative contexts. Finally, various other contexts for negation will be considered, for instance, negative non-finite clauses and negative imperatives, all contexts where there has been considerable change over the history of the languages.

7.2 Textual and linguistic background

The Brythonic languages are descended from the Brythonic (British, Brittonic) parent language, the language spoken by the Britons during the Roman and pre-Roman periods. Another branch of Celtic, Goidelic, was spoken in Ireland, subsequently spreading to Scotland and the Isle of Man, giving rise to Irish, Scottish Gaelic, and Manx. Brythonic and Goidelic are traditionally classified as Insular Celtic languages, in contradistinction to the Continental Celtic languages (Gaulish, Celtiberian, Lepontic, Galatian) once spoken over much of continental Western Europe. Another distinction commonly made is that between P-Celtic, those languages where the reflex of Indo-European $^*/k^w/$ is /p/, namely Brythonic and Gaulish (Welsh *pedwar* 'four') and Q-Celtic, those languages where it is /k/, namely Goidelic and Celtiberian (Irish *ceathair* 'four').

Despite certain gaps, there is substantial textual attestation of all the Brythonic Celtic languages in the historical period. Welsh is conventionally divided into Old Welsh (800–1150), Middle Welsh (1150–1500), and Modern Welsh (1500 to the present day). Within the modern period it is useful to refer to Early Modern Welsh (1500–1700) and to Present-day Welsh. Old Welsh is attested in a modest corpus of glosses on Latin texts, charters, and similar texts in contemporary manuscripts, and a larger body of poetry preserved largely in later manuscripts. Middle Welsh is preserved in an extensive corpus of texts including both native and translated tales and romances, fixed-metre poetry, law codes, chronicles, saints' lives and other religious texts, and scientific and medical works. For the modern period, attention here will focus heavily on the evidence of sources that give a reliable reflection of spoken practice: dialect plays, ballads, court case depositions, personal letters by less educated individuals, and the representation of speech in novels.

Breton is conventionally divided into Old Breton (800–1100), Middle Breton (1100–1650), and Modern Breton (1650 to the present day). Old Breton is attested only in glosses and similar documents giving only isolated words and phrases. There is a substantial gap in attestation between Old Breton and the earliest Middle Breton texts, which appear from 1450. This makes it difficult to trace the development of the language. Middle Breton boasts a moderately sized corpus of verse drama on religious topics (saints' lives, mystery plays, etc.), carols, and various prose works, including homilies and travellers' phrase books. A continuous tradition of printed and manuscript sources exists since then.

Apart from Old Cornish glosses, Cornish is attested from religious plays dating from the fifteenth and sixteenth centuries onwards, as well as a collection of sixteenth-century prose homilies and various other prose pieces from the seventeenth and eighteenth centuries. Discussion here focuses on the language of the canonical Middle Cornish verse plays, with some reference to developments in Late Cornish (1575–1800), where these can be identified.

The standard grammars are those of Evans (1964) for Middle Welsh, Hemon (1975) for Middle Breton, and Lewis (1946, 1990) for Middle Cornish. General overviews of the languages can be found in MacAulay (1992), Russell (1995), and Ball and Müller (2009), while there is an introduction to the historical syntax of Welsh in Borsley, Tallerman, and Willis (2007: 286–337).

Contact with English and French has been a feature of all the Brythonic languages since the early Middle Ages. In the case of Cornish, this led to complete language shift to English and the death of the language in the eighteenth century. In Wales, contact with English culminated in the emergence of mass bilingualism in the nineteenth century and rapid language shift to English in many areas. In Brittany, mass bilingualism developed in the late nineteenth century, followed by even more rapid shift to French. Both Breton and, especially, Welsh are the focus of lively revitalization efforts today, and this may affect the grammatical structure of both languages.

7.3 The Welsh Jespersen cycle

7.3.1 Preverbal negative markers in Middle Welsh

In both Old Welsh (800–1150) and Middle Welsh (1150–1500), sentential negation is expressed using a particle in immediately preverbal position. In declarative main clauses, this is *ny* /ni/, *nyt* /nɪd/ before a vowel (spelled *ni*(*d*), but also pronounced /nɪd/ in Modern Welsh). Word order in negative clauses is predominantly verb-initial, although the language as a whole is verb-second in main clauses (Willis 1998), so this results in the negative marker being clause-initial with negation–verb–subject–object order:

(10)	. . . **ny**	cheffy	di	varch	gennyf	i.
	NEG	get.PRES.2SG	you	horse	with.1SG	me

'. . . you will not get a horse from me.' (*Ystoryaeu Seint Greal*, l. 1940, Middle Welsh)

The present tense of *bot* 'be' in the third person singular, whether existential or a copula, may be expressed by the special form *nyt* (also the source of the focus-negation marker, section 7.11 below):

(11)

...nyt	kyfle	yni	trigyaw	yma.
NEG.be.PRES.3SG	chance	to.us	live.INF	here

'...there is no chance for us to live here.' (*Pedeir Keinc y Mabinogi* 57.28–58.1, Middle Welsh)

Regular re-formations, with the ordinary negative particle plus the usual present-tense verb forms, are also found, and come to replace this in later Welsh (cf. Croft's cycle, section 1.6).

In embedded clauses, a different particle, *na*(*t*) /naːd/ (spelled *na*(*d*), but also pronounced /naːd/ in Modern Welsh), is used:

(12)

Pann	welas	Kicua...	**nat**	oed	yn	y	llys	namyn
when	see.PAST.3SG	Cigfa	NEG.COMP	be.IMPF.3SG	in	the	court	except
hi	a	Manawydan...						
she	and	Manawydan						

'When Cigfa saw that there was no one in the court except her and Manawydan...' (*Pedeir Keinc y Mabinogi* 57.13–14, Middle Welsh)

A third particle, *na*(*c*) /naːg/, is used to negate imperatives, as in (13), subjunctives (optatives), as in (14), and to give a negative response to a question (responsives), as in (15). Note that, before a consonant, the two particles /naːd/ and /naːg/ are homophonous. Only before a vowel can their form be distinguished.

(13)

...**nac**	**arch**	dim	namyn	lloneit	y	got	o	uwyt.
NEG	ask.IMP.2SG	anything	except	fill	the	bag	of	food

'...don't ask for anything except for the fill of the bag of food.' (*Pedeir Keinc y Mabinogi* 15.4–5, Middle Welsh)

(14)

na	**ueidych**	ditheu	dangos	dy	wyneb	lliw	dyd	byth
NEG	dare.PRES.SUBJ.2SG	you	show.INF	2SG	face	colour	day	ever

'...may you not dare to show your face in daylight ever again...' (*Pedeir Keinc y Mabinogi* 91.10, Middle Welsh)

(15)

...a	doy	ti	y	dangos	imi	yr	aneueil	hwnnw?
QU	come.PRES.2SG	you	to	show.INF	to.me	the	animal	that
Nac	**af**							.
NEG	go.PRES.1SG							

'Will you come to show me that animal?' 'No.' (lit. 'I will not go.') (*Peredur* 68.15, Middle Welsh)

All of these particles trigger morphophonological alternations on the following verb, part of the wider system of initial consonant mutations found at all stages of Welsh. The negative particles uniquely trigger a mixture of two mutation patterns: aspirate mutation, /p t k/ > /f θ x/, if the verb begins with /p t k/, and soft mutation, /b d m r^h ɬ/ > /v ð v r l/ and /g/ > zero, otherwise. Aspirate mutation occurs in example (10) above (*keffy* /k/ > *cheffy* /x/), and soft mutation occurs in example (14) (*beidych* /b/ > *ueidych* /v/).

7.3.2 *The emergence of negative reinforcement*

In the course of Middle Welsh, a new adverb emerges to reinforce negation, based on a reanalysis of the indefinite pronoun *dim* 'anything, nothing'. In early Middle Welsh, this pronoun is used exclusively as a verbal argument (see section 7.6.2 below), typically as subject or direct object. Use as a direct object, found at all stages of Welsh, is illustrated in (16).

(16) Ac ny mynnwys ef **dim**.
and NEG want.PAST.3SG he anything
'And he didn't want anything.' (*Pedeir Keinc y Mabinogi* 27.10–11, Middle Welsh)

It is also used as an 'extent argument', that is, as complement to a verb of succeeding, a verb of caring or indifference, or a verb of compensation (*tygyaw* 'work, help', *diwygyaw* 'compensate, put right', etc.), expressing the extent to which something succeeds, the extent to which it is (un)important, or the extent to which compensation is made:

(17) ac ny thygyawd ydunt **dym**...
and NEG avail.PAST.3SG to.3PL anything
'and it didn't help them at all...' (*Brenhinedd y Saesson* 192.13–14, Middle Welsh)

However, a new pattern emerges in texts from the second half of the thirteenth century onwards. In this pattern, *dim* has no selectional relationship with the verb; that is, it is not an argument of the verb, but rather reinforces the negative polarity of the clause (Willis 2006, Poppe 2009):

(18) A gwedy gwelet o 'r brenhin hynny ny chyffroas arnaw **dim**...
and after see.INF of the king that NEG agitate.PAST.3SG on.3MSG at.all
'And after the king saw that, he did not become agitated (in the least)...' (*Brenhinedd y Saesson* 70.9–10, Middle Welsh)

(19) A phan weles ynteu daruot llad y varch, ny
and when see.PAST.3SG he PRF.INF kill.INF 3MSG horse NEG
lidiawd **dim** yr hynny...
become.angry.PAST.3SG at.all despite that.NEUT
'And when he saw that his horse had been killed, he still didn't get angry (in the least).' (*Ystoryaeu Seint Greal* 2874, Middle Welsh)

This reanalysis represents a split of one item, argument *dim*, into two items: one continues argument *dim* 'anything', the other represents a new emphatic marker. The conservative pattern is clearly continued today as the indefinite pronoun 'argument *dim*' (see section 7.6.2 below).

This item differs in several ways from the Present-day Welsh negative marker *ddim*. First, Middle Welsh pseudoargument *dim* is actually a negative polarity adverb rather than a negator in its own right. This can be seen from the fact that, like the indefinite pronoun *dim*, it occasionally occurs in non-assertive environments other than negation, for instance in the embedded interrogative in (20). Effectively, then, its distribution is equivalent to that of a negative polarity adverb such as English *at all*.

(20) Ac yno y wylyaw a orugant y edrych a allei
and then 3MSG watch.INF PRT do.PAST.3PL to look.INF PRT can.IMPF.3SG
ymdidan **dim** ac wynt.
converse.INF at.all with them
'And then they watched him to see whether he could converse with them at all.' (*Ystoryaeu Seint Greal* 5213, Middle Welsh)

Secondly, it was emphatic. This is partly demonstrated by its low frequency (under 5% of negative clauses) even in those Middle Welsh texts where it is found. Its emphatic character is clear in example (20), where it seems to mean 'in any way'; that is, they watched him to see whether he could converse with them by means of any language, or by signs, or by any other means imaginable. This is also demonstrated by sixteenth-century texts translated from or influenced by English, where it normally corresponds to some emphatic element in English, as in (21), where *ddim* corresponds to *in no wise* in the English King James Bible, with which the Welsh Bible was brought into line:

(21) a 'r hwn a ddêl attaf-i, **ni** 's bwriaf
and the DEM PRT come.PRES.SUBJ.3SG to.1SG-me NEG ACC.3 cast.PRES.1SG
ef allan **ddim**
him out NEG
'and anyone who may come to me, I shall not cast him out.' (cf. King James Bible *and him that commeth to me, I will in no wise cast out.*) (*Y Bibl cyssegr-lan*, John 6: 37, 1620)

From a syntactic perspective too, it differs from Present-day Welsh *ddim* in occupying a late clausal position. In particular, it may follow prepositional-phrase complements of verbs, as with *arnunt* '(waiting) for them' in (22) (Willis 2011a: 105–6), and may follow non-finite verbs in periphrastic tenses, as with the continuous *oeddynt yn mendio* 'were mending' in (23) (Willis 2010: 123–4).

(22) Ac wynteu... nyt arhoyssant [arnunt] **dim**...
and they NEG wait.PAST.3PL on.3PL at.all
'And they didn't wait for them...' (*Ystoryaeu Seint Greal* 1919, Middle Welsh)

(23) Eithr **nid** oeddynt yn mendio **dim.**
but NEG be.IMPF.3SG PROG mend.INF at.all
'But they did not mend their ways at all.' (*Darn o'r Ffestival* 106.6, mid-sixteenth century)

In the sixteenth and seventeenth centuries, modern-like properties emerge. The form of the sentential negator becomes fixed as *ddim*, rather than alternating between *dim* and *ddim* according to syntactic environment. It loses its emphatic semantics, appearing in translated texts in contexts where there is no corresponding emphatic element in the English:

(24) ac o.herwydd hynny... ni ddeuaf i **ddim** i mewn.
and because this NEG come.PRES.1SG I NEG to in
'...and because of this I will not come in.' (Translation of *and therefore [I will not accept of it] nor come in*) (Edward Fisher, *Madruddyn y difinyddiaeth diweddaraf*, p. 147, 1651)

It also appears in an earlier sentence position, coming to occupy the position immediately following the subject, as it does today. This means that it comes to appear consistently before a prepositional-phrase complement of the verb, as illustrated in (25), where the negator precedes *i'r Cyngor* 'to the council', rather than following it.

(25) Hefyd ni ddoe Escobion y Dwyrain **ddim** i 'r Cyngor
also NEG come.IMPF.3SG bishops the east NEG to the council
yn Syrmium
in Sirmium
'Also the eastern bishops did not come to the Council of Sirmium...' (John Jewel, *Deffynniad ffydd Eglwys Loegr*, p. 81, 1595)

It also appears in the middle of an aspectual periphrasis, illustrated with *oedd hi... yn pechu* 'she was sinning' in (26). The contrast in both these cases with examples (22) and (23) above shows clearly the shift to an earlier clausal position.

(26) Ac velly pawb a wybu nad oedd hi **ddim**
and thus everyone PRT know.PAST.3SG NEG.COMP be.IMPF.3SG she NEG
yn pechu.
PROG sin.INF
'And thus everyone realized that she was not sinning.' (*Teithie Syr Sion Mandefyl*, Peniarth 218, ll. 127–8, 1605–10)

We thus reach the system of Present-day Welsh, where pseudoargument *ddim* in (27) occupies a position after the auxiliary and the subject but before the aspectual particle and conveys pragmatically neutral sentential negation:

(27) Doedden nhw **ddim** yn gwella.
NEG.be.IMPF.3PL they NEG PROG improve.INF
'They weren't improving.' (Present-day Welsh)

A residue of the earlier system is the maintenance of an emphatic marker of negation in Present-day Welsh, limited to clause-final (or at least clause-late) position:

(28) Doedden nhw ddim yn gwella **dim**.
NEG.be.IMPF.3PL they NEG PROG improve.INF at.all
'They weren't improving at all.' (Present-day Welsh)

Note that this can co-occur with the plain sentential negator *ddim* in (28), demonstrating that the two are different items today.

Once again, then, we have a split, this time in the sixteenth century. Middle Welsh pseudoargument *dim* /ðɪm, dɪm/ split into two items. One item, clause-late *dim*, represents continuity in acquisition of the inherited system. The other item results from a reanalysis based on sentences where the syntactic position of *dim* (in the form /ðɪm/) is not self-evident. A new item, *ddim*, was created in immediately post-subject position. Clause-late *dim* remained in a prosodically prominent position favourable to it retaining its emphatic character, whereas the move to an earlier, prosodically weaker position for the new *ddim* encouraged the loss of its emphatic force.

As we saw above, Middle Welsh pseudoargument *dim* was a weak negative polarity adverb and could occur in a variety of non-negative environments meaning 'at all'. Today, however, it is inherently negative and cannot occur in interrogative and conditional clauses without giving rise to a negative interpretation. This development is difficult to date as non-negative uses were never particularly common anyway. Some possible historical examples in non-negative contexts from the sixteenth century are given in examples (29) (interrogative) and (30) (conditional):

(29) Ai tybied, y lleihaant hwy **ddim** ar eu balchder...?
QU wonder.INF PRT reduce.PRES.3PL they at.all on GEN.3PL pride
'I wonder, will they reduce their pride at all...?' (translates *Will these men abate their ambition and pride?*) (John Jewel, *Deffynniad ffydd Eglwys Loegr* 80, 1595)

(30)

pe	y	baei	r	iaith	yn	talu	**dim**
if	PRT	be.IMPF.SUBJ.3SG	the	language	PROG	count.INF	at.all

'if the language mattered at all' (*Drych Cristianogawl* A4v.28–9, 1585)

The pattern in (29), with *ddim* in interrogative clauses, has been reinterpreted as expressing a negative question in Present-day Welsh:

(31)

Wyt	ti	ddim	yn	gwybod	lle	mae	Mair?
be.PRES.2SG	you	NEG	PROG	know.INF	where	be.PRES.3SG	Mair

'Don't you know where Mair is?' (Present-day Welsh)

Negative questions may also be expressed using the negative question particle *on'd* (< Middle Welsh *pony*(*t*)). Note that this is the first context in which *ddim* appears alone without *ni*(*d*) with a negative interpretation, a phenomenon also observed in French (Price 1978, 1993, Muller 1991: 225, Ayres-Bennett 1994: 66–7, Martineau and Mougeon 2003: 120).

7.3.3 The loss of preverbal negation

Ddim has gone further in becoming the main marker of sentential negation in speech in northern dialects (on southern dialects, see section 7.10.1). Today, in speech, *ddim* is obligatory in negative main clauses containing no n-word (such as *neb* 'anyone, no one'), and *ni*(*d*) has disappeared entirely from main clauses. Thus, (32) has been replaced by (33), and (34) has been replaced by (35). The forms in (32) and (34) survive in literary written Welsh but are never found in spontaneous speech.

(32)

Ni	chysgais	i.
NEG	sleep.PAST.1SG	I

'I didn't sleep.'

(33)

Chysgais	i	ddim.
sleep.PAST.1SG	I	NEG

'I didn't sleep.'

(34)

Nid	ydwyf	i	wedi	cysgu.
NEG	be.PRES.1SG	I	PRF	sleep.INF

'I haven't slept.'

(35)

Dydw	i	ddim	wedi	cysgu.
NEG.be.PRES.1SG	I	NEG	PRF	sleep.INF

'I haven't slept.'

The mutation effect of the preverbal particle *ni*(*d*) (mixed soft and aspirate mutation) remains on the verb to some extent. In (33), negation is partially marked by the aspirate mutation of the verb from *cysgais* to *chysgais*. The verb *bod* 'be' prefixes a *d-* in the negative, the remains of the final consonant of *nid*, as can be seen in (35) (cf. negative *dydw* with (*y*)*dw* 'I am'). However, many varieties have generalized soft mutation to negatives. In (36), the verb (radical form *clyw*(*a*)*ist* 'heard') undergoes soft mutation to *glywist*, rather than more traditional aspirate mutation *chlywaist*:

(36)

Glywist	ti	ddim!
SM.hear.PAST.2SG	you	NEG

'You didn't hear!' (Angharad Jones, *Y dylluan wen*, p. 142, 1995)

In many, but not all, cases, therefore, the verb appears in a distinct form in negative clauses: affirmative *cysgais* vs negative *chysgais* in (33), and affirmative *dw* vs negative *dydw* in (35). Even in (36), there is a distinction between obligatory soft or aspirate mutation in the negative, contrasting with variability between no mutation and soft mutation in the affirmative.

The extension of *ddim* in speech to all negative main clauses containing no n-word belongs to the period after the emergence of a standard language in the sixteenth century, and this makes dating it difficult against the background of literary texts where *ddim* is not and has not become obligatory. The most likely scenario seems to be that the frequency of pseudoargument *ddim* rose sharply in speech in the late eighteenth century, perhaps becoming obligatory in negative main clauses lacking an n-word by the mid-nineteenth century. In late eighteenth-century colloquial texts, we find all three stages of Jespersen's cycle (stage I *ni*(*d*) verb, stage II *ni*(*d*) verb *ddim*, and stage III verb *ddim*) represented with lexical verbs from the 1770s. However, stage I patterns dominate, perhaps even until as late as the second quarter of the nineteenth century. The three patterns are illustrated below:

(37)

Ni	'mâd	â	hi,	bydd	siwr	o	'i	chalyn.
NEG	leave.PRES.3SG	with	her	be.FUT.3SG	sure	of	GEN.3FSG	follow.INF

'He won't leave her; he'll be sure to follow her.' (Hugh Jones, *Protestant a neilltuwr* 19.8, 1783)

(38)

nid	ydwi	**ddim**	'ch	dynabod
NEG	be.PRES.1SG	NEG	GEN.2PL	recognize.INF

'I don't recognize you.' (Ellis Roberts, *Gras a natur* 19.24, 1769)

(39)

Mi	roedd	hi	yn	discwyl	iddo	fo	ei	chymeryd	hi
PRT	be.IMPF.3SG	she	PROG	expect.INF	to.3MSG	him	GEN.3SF	take.INF	her

ac	wnei[ff]	o	**ddim**.
and	do.FUT.3SG	he	NEG

'She was expecting him to take her but he won't.' (*Welsh defamation suits*, Caernarfon Sessions, 1778)

Stage III patterns emerge with lexical verbs in the 1770s. With the verb *bod* 'be', reduction of *ni(d)* to an initial *d-* is well attested somewhat earlier, with a high frequency in colloquial texts by the mid-eighteenth century:

(40) **doedd** mo 'i fath am fyta oddyma i fon
NEG.be.IMPF.3SG NEG 3MSG kind for eat.INF from.here to Anglesey
'There was no one like him for eating from here to Anglesey.' (*Brenin Llur* 230, *c.*1700–50)

A few other common verbs beginning with /a/ are also advanced with respect to the innovation of a Stage III pattern. This principally concerns inflected forms of *mynd* 'go' and *gallu* 'be able', which are found with *d*-initial forms in colloquial texts in the mid-eighteenth century. This is illustrated for *mynd* in (41) (*ei* '(you) go') and for *gallu* in (42) (*(g)all* 'is able').

(41) 'deidi buth i 'r ne ni dwaunost mo 'r llwubre
NEG.go.PRES.2SG.you ever to the heaven NEG know.PRES.2SG NEG the ways
'You'll never get to heaven; you don't know the way.' (*Brenin Llur* 677, *c.*1700–50)

(42) 'Dall neb o.ran gwaith salach
NEG.be.able.PRES.3SG anyone on.account.of work wretched.COMP
swilio.
be.ashamed.INF
'No one can be ashamed on account of more wretched work.' (*Barn ar egwyddorion y llywodraeth* 46.14, 1784)

The stage II pattern illustrated in (38) is never particularly common, and the transition from stage I to stage III seems to run to completion in perhaps just over half a century from around 1770 to around 1850. It is thus a characteristic feature of the Welsh Jespersen cycle, unlike, for instance, Breton, Dutch, or French, that stage II is highly unstable and disappears very quickly.

7.3.4 Summary of the Welsh Jespersen's cycle

The chronological development of the Welsh Jespersen's cycle is summarized below:

- Middle Welsh had a stage I preverbal negative marker *ny(t)*.
- In late Middle Welsh *dim* 'anything' was reanalysed to create a new negative polarity adverb in a clause-late position, the ancestor of Present-day Welsh sentence-final *dim* 'not at all'.
- In Early Modern Welsh *dim* split into two items: the new one, ancestor of Present-day Welsh pseudo-argument *ddim*, occupied the immediately post-subject position and was non-emphatic, while the old one, the ancestor of Present-day Welsh clause-late *dim*, continued the distribution and pragmatics found in Middle Welsh.

- In Early Modern Welsh before forms of *bod* 'be', the final /d/ of *nyt* was reanalysed as part of the verb, effectively creating a distinct non-assertive paradigm for this verb.
- *Ni*(*d*) became optional, first with forms of the verb *bod* 'be' beginning with a vowel, with this pattern spreading to other verbs towards the end of the eighteenth century.
- *Ddim* was introduced in all cases where negation was not marked by other means (an n-word), becoming compulsory in such environments in the nineteenth century.
- *Ni*(*d*) disappeared from speech in the early nineteenth century, leaving only a stage III pattern with verb + *ddim*.

7.4 The Breton Jespersen cycle

Like Welsh, Breton has undergone Jespersen's cycle. While Old Breton had preverbal negation markers like Old and Middle Welsh, a new item, Middle Breton *quet*, Modern Breton *ket*, develops and spreads in the Middle Breton period, giving rise to bipartite negation, which, as *ne*... *ket*, is now the norm in literary Breton. Colloquial Breton and many of the dialects, particularly the southeastern Vannes dialect, have gone further towards eliminating the preverbal particle altogether (Le Roux 1924–63, maps 206, 235, 242, 250, and 251).

Middle Breton has two preverbal markers of negation *ne*(*nd*) and *na*(*c*). *Ne*(*nd*) is used in main clauses, where Middle Welsh would use *ny*(*t*), and *na*(*c*) is used in all embedded contexts (including relative clauses) and in imperatives, that is, a range of contexts where Middle Welsh would use *na*(*t*), *na*(*c*), or relative *ny*(*t*). Thus, in main clauses, negation is expressed using the preverbal marker *ne*:

(43) **Ne** mem caffaff da vezaff quen sauant;
NEG 1SG.ACC.REFL find.PRES.1SG to be.INF so wise
'I do not find myself to be so wise.' (*Le mystère de sainte Barbe* 87, 1557)

While, unlike in Welsh, no consonant is systematically added before a vowel, a special form *nenn d-* is used before those forms of the highly irregular verbs *bout* 'be' and *monet* 'go' that begin with a vowel:

(44) Maz aff araucq **nenn d-** aff adreff
if go.PRES.1SG forward NEG go.PRES.1SG back
'If I go forward, I do not go back.' (*Trois poèmes en moyen-breton* 42, 1530)

In embedded clauses, the negative marker *na* is used:

(45) Hoz tat eu /A archas dimp... /**Na** rasemp ny muy...
your father be.PRES.3SG REL ask.PAST.3SG to.1PL NEG do.COND.PAST.1PL we more
'It is your father who asked us not to make more...' (*Le mystère de sainte Barbe*, stanza 250, 1557)

This is also found with imperatives, in (46), and in *wh*-questions and parallel contexts (relative clauses etc.), in (47):

(46) Ach autrou doe **na**-m ancoffet
oh lord God NEG-1SG.ACC forget.IMP.3PL
'Oh, Lord God, do not forget me.' (*Trois poèmes en moyen-breton*, stanza 28, 1530)

(47) Perac **na** rechuy... /Quichen e quichen tri frenest...
why NEG do.PRES.2PL.you side in side three window
'Why don't you make three windows side by side...?' (*Le mystère de sainte Barbe*, stanza 248, 1557)

Before a future form of the verb, *na* gives optative meaning, whereas *ne* gives ordinary future meaning.

While Middle Welsh distinguishes between *na*(*t*) in embedded complement clauses, *ni*(*t*) in *wh*-questions and relative clauses, and *na*(*c*) in imperatives, optatives, and responsives, neither Middle Breton nor Middle Cornish shows such a distinction, all these clauses being treated alike. The Breton–Cornish system may reflect the ancestral state more faithfully, with Welsh having innovated the distinctions (Willis 2011b: 440–1). Furthermore, while the Welsh preverbal negative markers trigger a mixture of soft and aspirate mutation on the following verb, depending on the nature of the initial consonant, the Breton markers always trigger soft mutation. This is not marked in Middle Breton orthography, but can be inferred from patterns of alliteration and from the modern language. Many modern Breton dialects have given up the distinction between *ne* and *na*, generalizing *ne* (with schwa) to all contexts (Hemon 1975: 283–4), while Late Cornish generalized *na* (see section 7.8 below).

While all the Middle Breton examples given so far instantiate stage I of Jespersen's cycle, containing only a preverbal marker of negation, the language was already well advanced along the cycle. Stage II patterns, using an innovated postverbal marker *quet* (Modern Breton *ket*) are well represented in the texts, although they are far from compulsory:

(48) hac euitce ne scuyzont quet.
and yet NEG tire.PRES.3PL NEG
'and yet they do not tire.' (*La vie de sainte Catherine* 80.9, 1576)

In Present-day Breton, *ket* is compulsory except in the presence of another n-word, and *ne* may be elided, as in French.

Such patterns, while attested extensively in even the most conservative Middle Breton texts, are not found in Old Breton, where main-clause negation is marked using *ni*(*t*), as in Old Welsh, or occasionally *ne*, as in Middle Breton, and there is no sign of *ket* (Fleuriot 1964b: 275–7). It can therefore be concluded that *ket* emerged during the lengthy gap in attestation between Old Breton (ninth to eleventh centuries) and Middle Breton (from 1450).

The origin of *ket* is unclear. Fleuriot (1964b: 283) suggests that it is 'not impossible' that *ket* developed from the Old Breton complementizer *cet* 'although' (Middle Welsh *kyt*). However, Hemon (1975: 284), while noting that Middle Breton *quet* sometimes seems to mean 'indeed' (but see below), ultimately rejects the connection. It is indeed difficult to conceive of a convincing means to connect the two. More recently, Schapansky (1994) has defended this etymology. Building on Hemon's comments, she argues that Old Breton *cet* 'although' was shifted to other positions in the sentence, taking on the meaning of an adverb 'indeed', from where it came to reinforce negation. The role of Middle Breton *quet* in non-negative environments is certainly crucial to understanding its development. However, the proposal that a complementizer 'although' was reanalysed as an adverb 'indeed' and thence as a marker of negation seems a priori improbable, and does not conform with what we know of the more general paths of development of new markers of negation, where we know that minimizers (French *pas* < '(not even) a step') and indefinites (English *not* < 'nothing') are overwhelmingly the dominant sources.

More promising, given general pathways of grammaticalization, is an etymological connection with Welsh *cadach* 'rag' (a loan from Irish *cadach* 'calico') or with Welsh *ceden* 'coarse hair, nap, shag, cotton' (cognate with Irish *catán* 'nap of shagged cloth'). While the bare-root formation *cet* is not found independently in Breton, it is found as part of a compound noun in Old Breton *guelcet* 'festival clothing' (< *guel* 'festival' + *cet* 'clothing') and possibly in Old Welsh *bronnced* 'breast veil' (< *bronn* 'breast' + *ced* 'clothing'), although the latter is itself dubious, cf. Falileyev (2000: 19). In this case, *quet* would have developed from use as a minimizer ('he didn't eat a scrap' > 'he didn't eat at all'). This can be compared to the development of English *scrap*, which has also grammaticalized as a weak negative polarity item (including as a quantifier e.g. *The police didn't have a scrap of evidence* but not **The police had a scrap of evidence*). Although this may be a promising line of inquiry, the connection between these items and Middle Breton *quet* is not phonologically straightforward, and more research is needed to evaluate this hypothesis.

Middle Breton *quet* occurs freely in a variety of weak negative polarity environments, in addition to its use as a reinforcer of the preverbal negative markers. Examples are given below, where (49) shows its use in a conditional; (50) and (51) show it in main and embedded yes–no interrogatives; and (52) shows it inside the scope of universal quantification.

(49) an tat han mam, á dle blam pep amser, /Ho
the father and.the mother PRT should.3SG reproach.INF every time their
buguel mar bez **quet**, coezet en pechet scler
child if be.HAB.3SG *quet* fall.PP in sin clear
'the father and the mother should always reproach their child if he/she has fallen into sin' (*Le mirouer de la mort*, ll. 1039–40, 1519/75)

(50)

Ac	eff	so	**quet**...	den	sauant...
QU	he	be.PRES.REL	*quet*	man	wise

'Is he (at all) a wise man...?' (*Le mystère de Saine Barbe*, stanza 133, 1557)

(51)

me	ya...	da	gouzout	a	den	an	menez	en	quarter	se	he
I	go.PRES.3SG	to	know.INF	QU	man	the	mountain	in	district	this	3FSG.ACC

guelse	quet...
see.IMPF.3SG	*quet*

'I shall go... to find out whether a mountain dweller in this district has seen her (at all)...' (*Le mystère de sainte Barbe*, stanza 365, 1557)

(52)

ancouffnez...	ho	holl	poan	hodeues	quet	gouzaffet...	a-raint
forget.INF	their	all	pain	have.PRES.3PL	*quet*	suffer.PP	PRT-do.FUT.3PL

'They will forget all the pain that they have ever suffered.' (*Le mirouer de la mort*, ll. 3312–4, 1519/75)

It is not possible in any of these contexts in Present-day Breton, where it has become purely a marker of negation.

Middle Breton *quet* differs from a fully grammaticalized negation marker in a number of other ways. First, its word order is freer than in Present-day Breton. It may be fronted to clause-initial position, as in (53), and it may occur in a clause-late position, separated from the verb by complements or adjuncts, as in (54).

(53)

quet	n-en	gry
quet	NEG-3SG.ACC	do.FUT.3SG

'He won't do it.' (*Le grand mystère de Jésus* 52a, 1530)

(54)

En	caffou	na	saouzan	n-oz	lesyf	aman	**quet**
in	bereavement	nor	surprise	NEG-2PL.ACC	leave.FUT.1SG	here	*quet*

'I shall not leave you here in bereavement and astonishment.' (*L'ancien mystère de Saint-Gwénolé* 97, 1580)

Neither of these word orders is particularly common, and their presence may be due to the needs of (internal and external) rhyme in poetic texts. It is possible though that they represent real phenomena, since they have analogues in other languages at the early stages of stage II of Jespersen's cycle. In Old French, for instance, *pas* is sometimes found in fronted position (Buridant 2000, Detges 2003: 214), and, as we saw above, the emerging Welsh negator *ddim* occurs for a while in a clause-late position. These phenomena are, therefore, not unexpected in the history of Breton.

Secondly, *ne* and *quet* may occur in different clauses, with *ne* occurring in the main clause, and *quet* in a subordinate clause:

(55) Muy **ne** guelaf ez duhen **quet**
more NEG see.PRES.1SG PRT come.COND.1SG *quet*
'I do not see that I should return again at all.' (*L'ancien mystère de Saint-Gwénolé* 69, 1580)

This suggests that *ne* and *quet* were once semantically independent from one another, having different scopes. In (55), for instance, 'at all, to any extent' modifies the embedded verb 'return', while the negation has scope over the entire proposition, including the main clause verb 'see'.

Quet also has various uses as a quantifier. With the following preposition *a* 'of', we find *quet a*, sometimes used as an adnominal quantifier 'any':

(56) ...hoguen nemeus **quet a couff** pe en lech bonamant
but NEG.have.PRES.1SG *quet* of memory which in place exactly
'...but I do not remember (I have no memory) where exactly.' (Guillaume Quiquer, *Dictionnaire et colloques françois et breton*, *Chrestomathie bretonne*, 309.8–9, 1633)

More commonly it is found without a preposition, in which case it is hard to know whether it actually forms a constituent with the following noun phrase. In common, perhaps fossilized, phrases such as *hep quet gou* 'without any lie' or *hep quet sy* 'without any doubt', it seems clear that *quet* forms a constituent with the following noun, and is therefore a quantifier inside the noun phrase.

Finally, in Modern Breton, *ket* can be used as a pronoun:

(57) N' eo ket hep ket.
NEG be.PRES.3SG NEG without anything
'He's not without anything.' (Favereau 1997: 283) (Present-day Breton)

In terms of its distribution in Middle Breton, *quet* shows a number of features found in other cases of emergent negative markers: an initial period where it occurs in non-assertive contexts as well as with negation, a greater freedom of positioning than later, and co-existence of adverbial and adnominal (quantifier) uses. All of these are features of *ne... point* in the French of the same period, and, to a lesser extent, of the other emerging French negators *ne... pas* and *ne... mie* (Catalani 2001: 100). Since they are found in other cases where a new negative marker emerges from a noun, the existence of these phenomena all point to a nominal original for *quet*, rather than a connection with 'although'.

While in Welsh the new marker of negation *ddim* ultimately comes to occupy a syntactic position immediately after the subject, in Breton, *ket* occupies a position immediately before the subject (Borsley and Stephens 1989: 413–15, Borsley and Roberts 1996: 22):

(58) Ne sell ket james Marijo (*ket james) ouzh an dud war ar blasenn.
NEG look NEG never Marijo NEG never at the people on the square
'Marijo never looks at the people on the square.' (Jouitteau 2005: 156)

While it is possible that this reflects a difference in the properties of the negation markers, it has generally been interpreted as being due to a difference in the position to which subjects move in the two languages, subjects raising higher in Welsh than in Breton (Borsley and Roberts 1996: 46).

7.5 Sentential negation in Cornish

Cornish maintains the inherited stage I system of sentential negation, and participates in Jespersen's cycle only to a very limited degree. In Middle Cornish, the preverbal negative particle is *ny* in main clauses (*nyns*, the regular phonetic correspondence of Middle Breton *nend*, before the verbs *bos* 'be' and *mones* 'go') (Lewis 1946: 48–9):

(59) **Ny** won a raf.
NEG know.PRES.1SG REL do.PRES.1SG
'I don't know what I shall do.' (*Bewnans Ke*, l. 331, Middle Cornish)

To negate imperatives, embedded clauses (including relatives), and responsives, *na* (*g*) is used:

(60) **Na** vith ydyot.
NEG be.IMP.2SG idiot
'Don't be a fool!' (*Bewnans Ke*, l. 268, Middle Cornish)

(61) Rys ew gwelas orth an wel/ **nag** ota ge mowas lows
necessary is see.INF by the work NEG.COMP be.PRES.2SG you girl loose
'We must see by the work that you are not a slack girl.' (*Bewnans Ke*, ll. 1115–16, Middle Cornish)

(62) **Nag of,** ou arluth, defry...
NEG be.PRES.1SG my lord indeed
[in response to (61)] 'Indeed, my lord, I am not...' (*Bewnans Ke*, l. 1117, Middle Cornish)

While Cornish does not undergo Jespersen's cycle, it does undergo change: in Late Cornish, *na*(*g*) is extended to main clauses and replaces *ny* entirely (Wmffre 1998: 57), perhaps via reinterpretation of responsives such as (62) as not being triggered by a question. An example of the new pattern is given in (63).

(63) Nages travith dale talues an bees...
NEG.be.PRES.3SG anything should.PRES.3SG value.INF the world
'There is nothing that the world should value...' (*The Cornish writings of the Boson family*, p. 8, *c*.1660–1700)

The development is paralleled in some southern Welsh dialects.

The Middle Cornish system, then, is identical to what Breton must have had before embarking upon Jespersen's cycle. Despite its general conservatism with respect to Jespersen's cycle, Middle Cornish does use a number of originally nominal items adverbially to reinforce negation. Price (1996) has argued that use of these amounts to a discontinuous (stage II) construction. He identifies 19 examples of negative reinforcement in Middle Cornish texts, using the items *cam(an)* 'step', *banna* 'drop', *tam* 'bit', and *poynt* 'point', all cognate with items used in this way in Breton. Middle Cornish uses these items at a much lower frequency than Middle Breton uses its postverbal negator *quet*, and individual texts differ according to which reinforcer is preferred, suggesting a lower degree of grammaticalization of these items than in Breton. The fact that Cornish and Breton use identical lexical items suggests a connection, although it is difficult to determine whether this is through genetic inheritance (southwestern Brythonic providing the basis for conventionalization of these items), through medieval contact between Breton and Cornish (which was extensive) or through external influence on both. Price considers the possibility of French or Anglo-Norman influence on Cornish in this respect, but this seems unlikely.

7.6 Welsh indefinites

Similar tendencies to those found in the Welsh and Breton Jespersen cycles, including a move towards increasingly negative meaning, are seen also with indefinite pronouns and adverbs. On the one hand, emphatic forms lose their emphatic quality, and new forms develop to renew the old items. On the other hand, forms previously found in all weak negative polarity contexts become inherently negative and restricted to negative contexts. This section examines these developments in Welsh, while section 7.7 considers parallel independent developments in Breton, and section 7.8 looks at Cornish.

Middle Welsh has two series of indefinite pronouns, a fully grammaticalized series found in negative polarity contexts, (64) (henceforth the *neb*-series), and a semi-grammaticalized series based on generic nouns found predominantly in affirmative contexts, (65) (henceforth the generic-noun series). Some minor items or items that grammaticalize during the course of Middle Welsh are omitted from these lists.

(64) *neb*-series (Middle Welsh)

person	*neb* 'anyone'
thing	*dim* 'anything'
quantity	*dim* + noun / *un* + noun 'any'
quality	*neb* + noun 'any'

(65) generic-noun series (Middle Welsh)

person	*dyn* 'a person, anyone' (= *dyn* 'person')
thing	*peth* 'a thing, anything' (= *peth* 'thing')
quantity	*peth o* (mass noun) / *rei o* (count noun) 'some'
quality	*ryw* 'some' (< *ryw* 'kind, type')

Items for 'any (amount of)', 'any (kind of)', 'anyone', and 'anything' are distinguished. The items expressing 'any (amount of)' distinguish mass from count nouns consistently. The *neb*-series shows no synchronic morphological motivation: each member is monomorphemic and there is no particular series marker. Such lack of transparency is uncommon cross-linguistically (Haspelmath 1997: 21–4). The items in the generic-noun series, with the exception of *rei o*, are all homophonous with ordinary indefinite noun phrases ('a person', 'a thing', etc.), hence this is a poorly defined series. Other ontological categories (place, cause, reason, manner) are expressed using generic nouns such as *lle* 'place' or *mod* 'manner' in both negative polarity and affirmative contexts.

The general developments are as follows: the *neb*-series develops inherently negative meaning, as in (66); the generic-noun series is reformed and given transparent morphological motivation by the creation of new items based on *ryw* 'some', giving rise to the Present-day Welsh *rhyw*-series in (67); and a new series based on the innovation *unrhyw* (< *un* 'one, any' + *rhyw* 'kind') in (68) has more recently been created. While initially the non-assertive negative polarity functions of the *neb*-series were taken over by the *rhyw*-series, more recently the *unrhyw*-series has become specialized for this use. The three main Present-day Welsh series of indefinites are thus as listed in (66) (the *neb*-series), (67) (the *rhyw*-series), and (68) (the *unrhyw*-series). The time adverbs *byth* and *erioed* have special distributions that do not conform to the general patterns (Borsley and Jones 2005: 109–12) and, in a sense therefore, lie outside of these series.

(66) *neb*-series (Present-day Welsh)

person	*neb* 'no one'
thing	*dim byd* 'nothing'
quantity	*dim* (mass or count nouns) / *'run* (< *yr un* 'the one') (count nouns) 'no, none'
place	*nunlle* / *unman* / *lle'm byd* (dialectally variable) 'nowhere'
time	*byth* (generic or future-oriented) / *erioed* (past-oriented)

(67) *rhyw*-series (Present-day Welsh)

person	*rhywun* 'someone'
thing	*rhywbeth* 'something'
quantity	*peth o* / *rhai* (*o*) / *rhywfaint o* 'some'
quality	*rhyw* 'some (kind of)'
place	*rhywle* 'somewhere'
time	*rhywbryd* 'sometime'
manner	*rhywsut* / *rhywfodd* 'somehow'

(68) *unrhyw*-series (Present-day Welsh)

person	*unrhyw un* 'anyone'
thing	*unrhyw beth* 'anything'
quality	*unrhyw* 'any'
place	*unrhyw le* 'anywhere'
time	*unrhyw bryd / byth / erioed* 'ever'
manner	*unrhyw sut* 'any way'

7.6.1 'Anyone, no one'

Middle Welsh expresses 'anyone, no one' using the inherited pronoun *neb* in all negative polarity contexts: in negative clauses in (69) and (70), in interrogatives in (71), in conditionals in (72), and in comparatives in (73).

(69)

A	**neb**	ny	dieghis	odyna	namyn	ef	a	'e	wreic.
and	anyone	NEG	escape.PAST.3SG	from.there	except	he	and	GEN.3MSG	wife

'And no one escaped from there except him and his wife.' (*Pedeir Keinc y Mabinogi* 36.21–2, Middle Welsh)

(70)

ny	welynt	**neb.**
NEG	see.IMPF.3PL	anyone

'... they saw no one.' (*Pedeir Keinc y Mabinogi* 52.7, Middle Welsh)

(71)

...a	weleisti	**neb**	o	'r	llys	yn	dyuot	y'm hol i?
QU	see.PAST.3SG	anyone	from	the	court	PROG	come.INF	after.me

'... have you seen anyone from the court coming after me?' (*Peredur* 14.6–7, Middle Welsh)

(72)

...pei	kymerwn	i	**neb**	y	'm	kanlyn,	mi	a	'th
if	take.COND.1PL	I	anyone	to	GEN.1SG	follow.INF	I	PRT	ACC.2SG

gymerwn	ditheu.
take.COND.1SG	you

'... if I took anyone to accompany me, I would take you.' (*Ystoryaeu Seint Greal* 731, Middle Welsh)

(73)

...canys	mvy	y	carei	ef	Eudaf	no	**neb**...
for	more	PRT	love.IMPF.3SG	he	Euddaf	than	anyone

'... for he loved Euddaf more than anyone.' (*Brut Dingestow* p. 71.24 = ms. 99.12–13, Middle Welsh)

Occasionally, the generic nouns *dyn* 'person', either alone or as *un dyn* 'any person', and *gwr* 'man' are used in this function too, although they are more usual in affirmative contexts:

(74) Ny lafasswys **dyn** vynet y 'r forest ys blwydyn.
NEG dare.PAST.3SG person go.INF to the forest since year
'No one has dared to go to the forest for a year.' (*Peredur* 68.15–16, Middle Welsh)

The combination *neb un* 'any one' occurs occasionally too in negative polarity environments (see section 7.9.1 below), as does *un* 'one' alone, to mean 'anyone'. Of these miscellaneous items, only *undyn* 'one, any man' seems to have grammaticalized as a fixed item, becoming moderately common in Modern Welsh, where it bears a single stress on the first syllable rather than two stresses, and which seems historically to have followed *neb* in terms of its distribution.

Neb has come to be inherently negative. When it occurs in interrogative and conditional clauses, it is now interpreted as negative. In such clauses, an affirmative meaning is now conveyed either by *rhywun* or the recently innovated item *unrhyw un* 'any one' (< *unrhyw* 'any' + *un* 'one').

7.6.2 *'Anything, nothing'*

In Middle Welsh 'anything' in negative polarity contexts is expressed using *dim*, a grammaticalized form of the noun of the same form meaning 'thing'. While the generic noun exists to a certain extent in Middle Welsh, its use had been seriously curtailed and it was limited to some fixed expressions and certain narrowly defined constructions. Middle Welsh use of *dim* to express 'anything' in a negative context is illustrated in (75). Its use in other weak negative polarity contexts is illustrated in (76) (interrogative) and (77) (comparative).

(75) Ac ny mynnwys ef **dim**.
and NEG want.PAST.3SG he anything
'And he didn't want anything.' (*Pedeir Keinc y Mabinogi* 27.10–11, Middle Welsh)

(76) A wdom ninheu **dim** y wrth hynny?
QU know.PRES.1PL we anything about that
'Do we know anything about that?' (*Pedeir Keinc y Mabinogi* 53.16, Middle Welsh)

(77) . . . rac ouyn colli y wreic, yr hon a garei ynteu y wuy
for fear lose.INF the woman the DEM PRT love.IMPF.3SG he PRED more
no **dim** **daearavl**.
than anything earthly
'. . . lest he should lose the woman that he loved more than anything on earth.' (*Brut Dingestow*, p. 136.26–7 = ms. 211.5, Middle Welsh)

Middle Welsh already has a range of emphasizing prepositional phrases based on 'in the world'. The exact form varies slightly, as does the position of the prepositional phrase with relation to the indefinite pronoun. The main forms found are *yn y byt* 'in the world' and *o'r byt* 'of the world' or *o'r holl uyt* 'of the whole world'. This seems to be a way of producing an emphatic negative with an indefinite pronoun. Examples with modification of *dyn* 'person, anyone' are given in (78)–(80). Note that the form of the prepositional phrase varies between *yn y byt* in (78) and (80), and *o'r byt* in (79); and that the *byd*-element may follow, as in the first two examples, or precede, as in (80).

(78) ...ny adwn ni drwc arnam ny hunein yr **dyn** **yn y** **byt**.
NEG allow.PRES.1PL we harm on.1PL us REFL for person in the world
'...we shall not allow ourselves to be harmed for anyone's sake (in the world).'
(*Pedeir Keinc y Mabinogi* 21.4–6, Middle Welsh)

(79) ...ac nyt **dyn** bydawl **o 'r byt** a 'e lliwyassei wyntwy.
and NEG person earthly of the world PRT 3PL colour.PLUPRF.3SG them
'...and it was no earthly person (in the world) who had coloured them.'
(*Ystoryaeu Seint Greal* 4255–6, Middle Welsh)

(80) ...nyt oes **yn y** **byt** **dyn** uwy y galar no
NEG be.PRES.3SG in the world person greater 3FSG grief than
hi yn y ol.
she after.him
'...there is no one (in the world) whose grief for him is greater than hers.'
(*Pedeir Keinc y Mabinogi* 26.22–4, Middle Welsh)

These prepositional phrases appear to have played a similar emphasizing role with *dim*: as illustrated by examples from Middle Welsh and from the sixteenth century in (81) and (82).

(81) Yr hynny ual kynt ny wneuthum j **dim** **drwc** o 'r byt.
despite that as before NEG do.PAST.1SG I anything bad of the world
'Nevertheless as before I didn't do anything bad at all (in the world).' (*Ystoryaeu Seint Greal* 2122, Middle Welsh)

(82) ...am nad oedd ef yn kasav **dim** **yn y** **byd**
for NEG.COMP be.IMPF.3SG he PROG hate.INF anything in the world
yn gymaint a medd-dod.
PRED so.much as drunkenness
'...since he hated nothing in the world (nothing at all) as much as drunkenness.' (*Gesta Romanorum* 1889, sixteenth century)

Today, *dim yn y byd* has contracted to *dim byd* 'anything, nothing'. In doing so, it undergoes a loss of its emphatic quality: speakers today do not perceive *dim byd* to be an emphatic version of *dim*, but merely a variant. Furthermore, it undergoes phonetic reduction with the loss of the preposition *yn* and the article, such that its meaning can no longer be derived compositionally. Phonological reduction of *dim yn y byd* to *dim byd* had occurred by the mid-nineteenth century:

(83) fel na cheis i ddim amser i syfenu **dim byd.**
so.that NEG.COMP get.PAST.1SG I any time to write.INF anything
'…so that I didn't get any time to write anything.' (William Rees, *Llythyrau 'Rhen Ffarmwr* 5.3–4, 1847)

7.6.3 'Any, no'

Middle Welsh expresses 'any, no' using a range of quantifiers. It is expressed by *dim* (< *dim* 'thing') if the head noun is a mass noun:

(84) A guedy nat oed **dim bvyt** gan y Saesson…
and after NEG.COMP be.IMPF.3SG any food with the English
'And once the English didn't have any food (left)…' (*Brut Dingestow* p. 147.17 = ms. 228.12, Middle Welsh)

If the head noun is a count noun, *un* 'one' is the most frequent option:

(85) a diamheu yw gennym na welsam eiroet uilwraeth yn **un**
and doubtless is with.1PL NEG.COMP see.PAST.1PL ever valour in any
wreic kymeint ac ynot ti.
woman so.much as in.2SG you
'…and we have no doubt that we have never seen as much valour in any woman as in you.' (*Pedeir Keinc y Mabinogi* 20.26–8, Middle Welsh)

We also find *neb* 'any' (< *neb* 'anyone'):

(86) Ny byd kylus **neb brawdwr** yr rodi a datganu
NEG be.FUT.3SG faulty any judge for give.INF and announce.INF
brawt o awdurdawt yscriuennedic…
judgement from authority written
'No judge is at fault for giving and announcing a judgement on the basis of written authority…' (*Llyfr Blegywryd* 102.5–6, Middle Welsh)

Neb tends to be used with animate count nouns, as in (86), but it is not entirely restricted in this way, and less frequent cases with inanimate count nouns or inanimate mass nouns do occur:

(87) mal na chaffvn y.ganthunt vynteu weithyon **neb amdiffin**.
so NEG.COMP get.COND.1SG from.3PL them now any defence
'... so that we could not get any defence from them now.' (*Brut Dingestow* p. 85.6–7 = ms. 125.6–7, Middle Welsh)

Where 'any' means 'any member of a contextually salient group', *yr un* 'the one' is used:

(88) Ac yn hynny tyuu kedymdeithas y rydunt yll pedwar,
and in that grow.PAST.3SG companionship between.3PL all four
hyt na mynnei **yr un** uot heb y gilid na
until NEG.COMP want.IMPF.3SG the any be.INF without 3MSG RECIP neither
dyd na nos.
day nor night
'And thereby companionship grew between all four of them, such that none wanted to be without the others day or night.' (*Pedeir Keinc y Mabinogi* 51.9–11, Middle Welsh)

These distinctions became blurred with time, and the three items *dim*, *neb*, and *un* were evidently more or less equivalent by the sixteenth century. The 1567 New Testament translation, which gives synonyms designed to bridge dialect differences, for instance offers the following, glossing *dim* as *neb* or *vn*:

(89) val y gallom ddiddanu yr ei 'sy mewn
as PRT can.PRES.SUBJ.1PL comfort.INF the ones be.PRES.REL in
* dim [:- * neb, vn] gorthrymder
any affliction
'that we may be able to comfort them which are in any trouble' (King James Version) (*Testament Newydd*, 2 Cor. 1: 4, 1567)

Neb and *un* die out as adnominal quantifiers. Quantifier *dim* has spread to occur before any noun, not just mass nouns. We find free use of *dim* before plural nouns for instance in the sixteenth century:

(90) Ag nid oedd yddo ef ddim plant, namyn vn verch...
and NEG be.IMPF.3SG to.3MSG him any children except one daughter
'And he didn't have any children, except for one daughter...' (*Gesta Romanorum* 425, sixteenth century)

Such usage forms the basis for Present-day Welsh, where *dim* is the usual quantifier for 'any, no'.

While Middle Welsh *un* has died out as a means of expressing 'any', the related form *yr un* has survived, in speech normally in the reduced form *'run*. This represents an extension of its use, since in Middle Welsh it is found only to refer to any member of a previously definite group. That is, while in Middle Welsh the definite article *yr* contributes a definite interpretation (requiring a previously defined, hence definite, group), this requirement has been lost in Modern Welsh and there is no longer any semantic connection with the definite article. Examples where no connection with a previously defined group is evident appear already in late Middle Welsh:

(91)

Ac	ony	ellwch	chwi…	y	dwyn	hi,	myui	a	'e
and	unless	can.PRES.2PL	you	GEN.3FSG	take.INF	it	I	PRT	GEN.3FSG

dygaf,	kan	nyt	oes	im	**yr**	**vn**.
take.PRES.1SG	since	NEG	be.PRES.3SG	to.me	the	one

'And if you can't take it, I will take it, since I haven't got one (any) (shield).' (*Ystoryaeu Seint Greal* 538–9, Middle Welsh)

Here, a magic shield has been discussed, but the speaker is saying that he has no shield of any kind, not simply that he does not have the shield just discussed. Phonetic reduction to *'run*, with loss of the initial schwa of the definite article, had occurred by the seventeenth century at the latest. It is now essentially synonymous with *dim* 'no', and has undergone the same shift to being inherently negative. However, it retains the requirement that the following noun should be singular.

An ultimately unsuccessful grammaticalization occurs in later Middle Welsh: a new form for 'any' arises, namely *neb ryw*. Originally this meant 'any kind of… (at all)' ('generalizer any'), from *neb* plus *ryw* 'kind, type', but it seems to be used as a simple synonym for the other items by late Middle Welsh:

(92)

heb	dywedut	vn	geir	wrth	**neb**	**ryw**	Gristyawn
without	say.INF	one	word	to	any	kind	Christian

'without saying a word to any Christian' (*Peredur* 40.15–16, Middle Welsh)

While very widespread in late Middle Welsh, this died out completely in Early Modern Welsh.

7.6.4 Other neb-*series indefinites*

Middle Welsh expresses 'ever' using either *ermoet* (< *er'm oet* 'in my life') / *eiroet* (< *er y oet* 'in his life') (Modern Welsh *erioed*) or *byth* 'ever'. The former is past-oriented, while the latter must refer to a generic event or to an event in the future. These are not actually negative polarity items, and, while they fill slots in the negative system, they have affirmative uses too:

(93)

direidwreic	uuost	**eiroet**…
evil.woman	be.PAST.2SG	ever

'you have always been an evil woman…' (*Pedeir Keinc y Mabinogi* 83.14, Middle Welsh)

As might be expected from the fact that they are somewhat detached from the negative system, *erioed* and *byth* do not participate in the general drift for *neb*-series items to become inherently negative, and are still possible in non-assertive and even affirmative contexts today:

(94)	Mae	o	wedi	sgwennu	'n	dda	erioed.
	be.PRES.3SG	he	PRF	write.INF	PRED	good	ever

'He has always written well.' (Borsley and Jones 2005: 160) (Present-day Welsh)

There is no conventionalized indefinite referring to place in Middle Welsh. Instead various circumlocutions using the generic noun *lle* 'place' are used. In many varieties, particularly northern ones, *yn un lle* 'in any place' conventionalizes and gives rise to a new indefinite *nunlle* 'anywhere, nowhere', which joins the *neb*-series, and which has joined the drift to become inherently negative. In the south, *unman* (< *un man* 'any place') undergoes the same development. A third item, *lle'm byd*, derives from *lle* 'place' plus the emphatic use of *yn y byd* 'in the world', as discussed above, paralleling the development of *dim byd* 'nothing'.

7.6.5 The Welsh quantifier cycle

All of the items discussed in the previous section, with the exception of the 'ever' items, *erioed* and *byth*, undergo a shift towards being inherently negative. We saw above, (69)–(73), that *neb* occurred in a range of non-assertive negative polarity environments in addition to negative clauses in Middle Welsh. The same is true of all the other items mentioned in this section. In Middle Welsh, their negative interpretation depends on their co-occurrence with a marker of sentential negation such as *ni*(*t*) or *na*(*t*). Two changes affect the status of these items. First, they come to express negative meanings in the absence of a marker of sentential negation. Secondly, they lose the ability to appear in non-assertive environments with non-negative meanings.

Even in Middle Welsh, *neb*-series items may have negative interpretations where the scope of negation is limited to the item itself, and where the overall proposition is not negative. Thus, in (95), there is narrow scope negation over *dim* 'nothing' alone.

(95)	Ac	y velly	e	dielws	ryuyc	y	Freinc	hyt ar	**dim**.
	and	thus		PRT avenge.PAST.3SG	arrogance	the	French	as.far.as	nothing

'And thus he avenged the arrogance of the French down to nothing.' (*Historia Gruffud vab Kenan* 23.2, Middle Welsh)

Negative interpretations of *neb*-series items in sentence fragments (typically answers to questions) are found at least as early as the seventeenth century:

(96) Scot: Pa ryw newydd, noble Crwmel?
Scot what kind news noble Cromwell
Crwmel: **Dim** ond darfod cwrs y rhyfel.
Cromwell nothing but finish.INF course the war
'Scot: What news, noble Cromwell?
Cromwell: Nothing except (only) that the course of the war has ended.' (*Rhyfel cartrefol*, ll. 810–11, after 1660)

In the nineteenth century, negative interpretations appear in certain non-elliptical syntactic environments. Initially, irrealis conditional clauses, as in (97), and absolute clauses, as in (98), seem most favourable to negative interpretations.

(97) tase **dim** arath i 'ch atal chi
be.COND.3SG nothing other to 2PL stop.INF you
'if there was nothing else to stop you' (William Rees, *Llythyrau 'Rhen Ffarmwr* 62.15–16, 1870)

(98) . . . yr oedd y pin ysgrifennu wedi mynd ar goll
PRT be.IMPF.3SG the pen write.INF PRF go.INF on lose
. . . a **dim** sgrap o bapur gwyn yn y tŷ.
and no scrap of paper white in the house
'the writing pen had become lost . . . and not a scrap of white paper in the house.' (William Rees, *Helyntion bywyd hen deiliwr* 52.1–3, 1877)

Absolute clauses containing *neb*-series items were once interpreted non-negatively, as witnessed by the following example from the 1588 Bible translation (retained in the 1620 Bible) where *dim cîg noeth byw*, with quantifier *dim*, is interpreted as 'any live raw flesh' rather than 'no live raw flesh':

(99) Ac edryched yr offeiriad, yna, os chŵydd gwynn [a fydd] yn
and look.IMPERS the priest then if swelling white PRT be.FUT.3SG in
y croen, a hwnnw wedi troi y blewyn yn wynn, a **dim cîg**
the skin and that PRF turn.INF the hair PRED white and any flesh
noeth byw yn y chŵydd;
naked live in the swelling
'And let the priest look, then, if [there will be] a white swelling in the skin, and it has turned the hair white and there is any naked live flesh in the swelling . . .' (Tyndale Bible: 'and let the preast se him. Yf the rysinge apeare white in the skynne ad haue also made the heer white, ad there be rawe flesh in the sore also') (Leviticus 13: 10, 1588)

There are also nineteenth-century examples of inherently negative indefinites in tenseless complement clauses.

Conversely, *neb*-series items have disappeared (or have come to be interpreted as negative) in interrogatives and in conditionals, the two major non-assertive environments where they were once possible. In the nineteenth century and into the twentieth century, *neb*-series items were possible in these contexts, although they were already being replaced by *rhyw*-series items (see below). The *neb*-series items *dim* 'any' and *'run* 'any' (cf. (66) above) are shown in interrogative and conditional contexts in (100) and (101) respectively.

(100) a. A oes **dim** **rhew** ac eira yn Awstralia?
QU be.PRES.3SG any frost and snow in Australia
'Is there any frost and snow in Australia?' (John Williams, *Awstralia a'r cloddfeydd aur*, 1852)

b. ous gynoch chi **run** ci arall
be.PRES.3SG with.2PL you the.one dog other
'Have you got any other dog?' (William Rees, *Llythyrau 'Rhen Ffarmwr* 30.6, 1850)

(101) a. tae **dim** **synwyr** yn dy goryn di
be.COND.3SG any sense in 2SG skull you
'if you had any sense in your skull' (William Rees, *Llythyrau 'Rhen Ffarmwr* 45.11, 1851)

b. mi 'rydw' i wedi cael troad os cafodd **yr un dyn** erioed dro
PRT be.PRES.1SG I PRF get.INF conversion if get.PAST.3SG the one man ever conversion
'I've had a conversion if any man ever had a conversion.' (William Rees, *Helyntion bywyd hen deiliwr* 84, 1877)

Middle Welsh had already grammaticalized *ryw* 'kind, type' as an adnominal quantifier 'some (kind of)'. When followed by a generic noun it often fulfilled the same function as the generic-noun series. Collocations such as *ryw beth* 'some thing' and *rhyw le* 'some place' conventionalized and came to be used in place of generic nouns functioning as indefinites. There is also a semantic shift from 'such a thing' or 'something such that it has the property...' etc. to the indefinite meaning that these items have today. Specific indefinite readings seem to emerge in late Middle Welsh, with non-specific readings slightly later. An example where *ryw beth* seems to function as a specific indefinite pronoun 'something' in late Middle Welsh is given in (102).

(102) 'Nac ef, y rof a Duw,' heb yr iarll, 'ef a vu **ryw** **beth** yn ymdidan a thi.'
no between.1SG and God said the earl it PRT be.PAST.3SG some thing PROG converse.INF with you
'"No, between me and God," said the earl, "there was something talking to you."' (*Kedymdeithyas Amlyn ac Amic*, ll. 536–7, Middle Welsh)

In these cases, the *rhyw*-items seem to be competing with generic nouns, ultimately replacing them to form a morphologically uniform series of indefinites.

From the point of view of negation, however, the most significant shift comes rather later, when the *rhyw*-series pronouns start being used freely in non-assertive contexts. In the nineteenth century (at the latest), we find the *rhyw*-series being used in conditionals and interrogatives where there is no presupposition of the existence of a referent for the pronoun:

(103) a oes gennych chwi **rywbeth** arall a recomendwch
QU be.PRES.3SG with.3PL you something else REL recommend.PRES.2PL
i mi?
to me
'Have you anything / something else that you recommend to me?' (David Owen, *Wil Brydydd y Coed*, p. 4, 1863–5)

(104) a. os bydd **rhywbeth** ynddo.
if be.FUT.3SG something in.3MSG
'...if there is anything / something in it.' (David Owen, *Wil Brydydd y Coed*, p. 7, 1863–5)

b. os oes **ryw** ystyr iddo
if be.PRES.3SG some meaning to.3MSG
'if it has some / any meaning' (William Rees, *Helyntion bywyd hen deiliwr* 62.14, 1877)

This amounts to encroachment on the previous territory of the *neb*-series: in both (103) and (104), we might have expected to find *dim* 'anything, any' at an earlier date. It is thus symptomatic of the ongoing narrowing of the *neb*-series to negative contexts at this period, a process which reached its full expression in the twentieth century. The disappearance of *neb*-series items in interrogatives and conditionals probably dates to the last hundred years, and its course may have varied from item to item. Fynes-Clinton gives no non-negative uses for quantifier *dim* 'no, any' in his comprehensive 1913 description of the Bangor dialect, while he does give non-negative uses for *neb* 'anyone, no one' (Fynes-Clinton 1913: i.88–9).

In the nineteenth century, the *neb*-series lost ground in non-assertive negative polarity contexts to the *rhyw*-series. Later, in the twentieth century, when the *neb*-series was ousted completely from interrogatives and conditionals, a different series, namely the *unrhyw*-series, also played a significant role. This series is a relatively recent innovation. Although a Middle Welsh word *unryw* is attested in the meaning 'same' (< *un* 'one, same' + *rhyw* 'kind'), this does not seem to be the source of the modern item. Rather, modern *unrhyw* is a new creation dating from the sixteenth century, when it first appears with the meaning 'any kind of, any' (that is, based on *un* 'one, any' + *rhyw* 'kind'):

(105) llymach nac vnrryw gleddau-daufinioc
sharper than any sword-two.edged
'sharper than any (kind of) two-edged sword' (*Testament Newydd* 330b, Hebrews 4: 12, 1567)

The pronouns *unrhyw beth* 'anything' and *unrhyw un* 'anyone' are of more recent provenance, however, being first attested in 1711 and 1852 respectively according to the University of Wales Dictionary. *Unrhyw*-series items are quite rare until the second half of the nineteenth century. However, they become common by the late twentieth century, appearing extensively in negative, interrogative, conditional, and comparative clauses, as well as being used as free-choice items. Transfer from English is apparent here, as *unrhyw*-series items have come to be identified as translation equivalents of English *any*-series items, and have adopted a syntactic distribution to match (Willis 2008).

The changes in the patterns of distribution of indefinites across the history of Welsh are summarized in Figures 7.1 and 7.2, using Haspelmath's (1997: 63–4) implicational map of indefinite pronoun functions (cf. also section 1.9).

7.6.6 *Negative concord in Welsh*

Present-day Welsh has two forms of negative concord. First, the result of the loss of the preverbal negative marker *ni(d)* is that verbs have special negative forms. Either the *d*- of *nid* has prefixed to the verb (mainly restricted today to the verb *bod* 'be'),

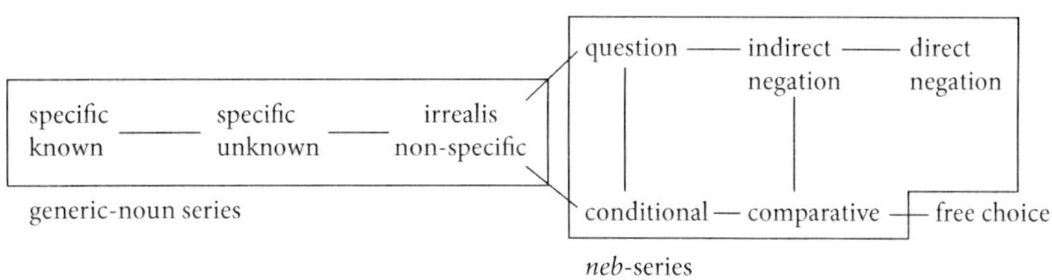

FIGURE 7.1 Expression of indefinites in Middle Welsh

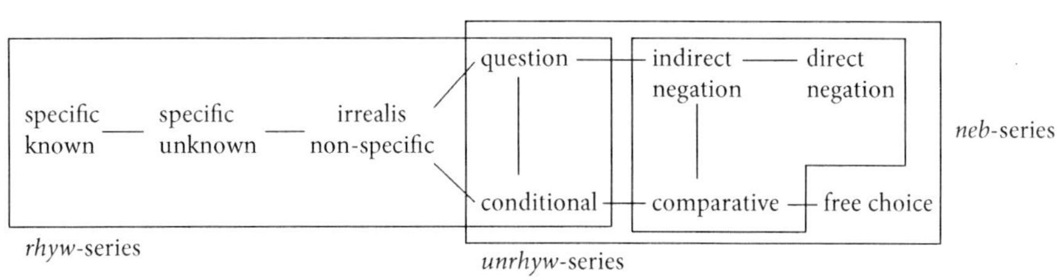

FIGURE 7.2 Expression of indefinites in Present-day Welsh

or else the mixed soft–aspirate mutation triggered by *ni(d)* remains on verbs in negative clauses. If a clause contains an n-word (from the *neb*-series) and a finite verb, for most speakers, the finite verb must take on the special negative form if it can, hence negative *does* rather than affirmative *mae* in the following examples:

(106) Does neb yn ennill.
NEG.be.PRES.3SG no.one PROG win.INF

(107) *Mae neb yn ennill.
be.PRES.3SG no.one PROG win.INF
'No one is winning.' (Present-day Welsh)

Secondly, there is limited negative concord (negative doubling) between the postverbal negative marker *ddim* and an n-word. This concord occurs in some environments but not in others. Where the n-word is the object of a verb in a compound tense, such as the periphrastic perfect in (108), *ddim* is mostly compulsory. Omission of *ddim* would lead to ungrammaticality here.

(108) Dyw Dafydd ddim wedi gweld neb.
NEG.be.PRES.3SG Dafydd NEG PRF see.INF no.one
'Dafydd hasn't seen anyone.' (Present-day Welsh)

Where the n-word is in subject position or the object in a synthetic clause, then negative concord with *ddim* is generally impossible:

(109) Welodd neb (*ddim) Dafydd.
see.PAST.3SG no.one NEG Dafydd
'No one saw Dafydd.' (Present-day Welsh)

(110) Welodd Dafydd (*ddim) ddim byd.
see.PAST.3SG Dafydd NEG nothing
'Dafydd saw nothing.' (Present-day Welsh)

There is variability when an n-word is within a prepositional phrase, negative concord with *ddim* being optional:

(111) Dydy Dafydd (ddim) yn siarad am ddim byd ond ei hun.
NEG.be.PRES.3SG Dafydd NEG PROG talk.INF about anything but 3MSG.REFL
'Dafydd doesn't talk about anything except himself.' (Present-day Welsh)

Finally, an n-word in a position that does not require negative doubling with *ddim* licenses further instances of n-words (negative spread):

(112) Does **neb** yn gwneud **dim byd**.
NEG.be.PRES.3SG no.one PROG do.INF nothing
'No one is doing anything.' (Also 'No one is doing nothing.') (Present-day Welsh)

Whether *ddim* is also present, giving negative doubling, depends on the rules given above. If the highest n-word is a position that requires negative concord, *ddim* may co-occur with two (or more) n-words, giving negative doubling (*ddim* ... *dim byd*) and negative spread (*dim byd* ... *neb*) at the same time:

(113) Dyw Dafydd **ddim** wedi dweud **dim byd** wrth **neb**.
NEG.be.PRES.3SG Dafydd NEG PRF say.INF nothing to no.one
'Dafydd hasn't said anything to anyone.' (Present-day Welsh)

Where two n-words co-occur, both a negative spread (single-negation) interpretation and a double-negation interpretation are possible (but with different intonations), hence the ambiguity of (112) above.

Historically, this system is of very recent provenance. In Middle Welsh, *neb*-series items cannot convey negation on their own, and they can be analysed as weak negative polarity items. At some point in the Early Modern Welsh period, they became inherently negative, disappearing from non-negative environments and sufficing to convey negation on their own in some contexts. Since *ni*(*d*) continued to co-occur with *neb*-series items, Early Modern Welsh became a strict negative-concord language with *ni*(*d*) ... *neb* 'not ... no one' as result of this change in status of the *neb*-series items. As Jespersen's cycle progressed, *ni*(*d*) was dropped, leaving only a residue in special negative verb forms, and *ddim* was introduced. Negative concord between the special verb forms and the *neb*-series was maintained, and Welsh remained a negative-concord language with respect to the special negative verb forms.

The spread of *ddim* adds a new dimension to the picture, since it raises the question of whether negative concord holds between *ddim* and the *neb*-series. In the nineteenth century (and before), there is only limited negative concord between *ddim* and the *neb*-series. Examples are found where a *neb*-series n-word acting as object of a verb in a compound tense is not doubled by an instance of *ddim* after the subject (contrast (108) and (113) above). Note that, in (114) and (115), the object is argument *dim* 'nothing' in its soft-mutated form (not the marker of sentential negator *ddim*).

(114) tydi hi wedi gwneud **fawr ddim** ond fy synu
NEG.be.PRES.3SG she PRF do.INF much nothing but 1SG surprise.INF
'She hasn't done anything except surprise me.' (Beriah Gwynfe Evans, *Dafydd Dafis*, p. 326, 1898)

(115) dydi ysbrydoliaeth wedi deyd **dim** am dano fo
NEG.be.PRES.3SG inspiration PRF say.INF nothing about.3MSG him
'Inspiration hasn't said anything about him.' (Annie Harriet Hughes, *Plant y gorthrwm*, p. 26, 1908)

The spread of negative concord seems to occur earlier with *neb* 'no one' than with *dim* (*byd*) 'nothing'.

The result of these developments taken together is that Present-day Welsh is a strict negative-concord language with respect to the relationship between special negative verb forms and n-words, while it shows a variant of non-strict negative concord with respect to the relationship between *ddim* and n-words.

A prominent analysis of negation in Present-day Welsh is that of Borsley and Jones (2005), who develop an HPSG model that imposes distributional constraints on the various elements involved. The relevant question in our context is whether this analysis can be extended to provide an insightful interpretation of the historical developments.

Borsley and Jones divide verbal forms into weak, strong, and extra-strong negative verbs, given in slightly simplified form below:

(116) **weak negative verbs** (special negative verbal forms)
d-forms (e.g. *oes* be.PRES.3SG > *does* NEG.be.PRES.3SG)
mixed soft or aspirate mutation on verbs (*cafodd* get.PAST.3SG > *chafodd* NEG.get.PAST.3SG)

strong negative verbs
embedded negative particle *na*(*d*) + verb (*bydd* be.FUT.3SG > *na fydd* NEG.COMP be.FUT.3SG)
main-clause negative particle *na*(*g*)
negative infinitive marker *peidio* (see section 7.10.2 below)

extra-strong negative verbs
negative imperative marker *paid* / *peidiwch* (see section 7.12 below)
s-forms of the negative auxiliary (*sa*, *so*, *s*(*i*)*mo*, etc.) (see section 7.10.1 below)

Weak negative verbs require a negative dependent (either *ddim* or an n-word in a relevant syntactic position) (the Negative Dependent Constraint); strong negative verbs do not require a negative dependent but are compatible with them; and extra-strong negative verbs allow a dependent n-word, but cannot co-occur with *ddim*.

For n-words, Borsley and Jones invoke a Negative Context Requirement, which states that Welsh n-words must appear in a negative context. They treat Welsh n-words as semantically negative (negative quantifiers), but argue that the quantifier can only be retrieved from storage at various nodes with a clausal interpretation. The contexts in which n-words may appear are then defined as those that allow the quantifier to be retrieved from storage. This includes both straightforward negative-concord contexts, and a range of contexts where n-words can have negative interpretations in the absence of another element.

The Negative Dependent Constraint is formalized as a requirement that weak negative heads have a complement (one member of their COMPS list) marked

[NEG +]. In HPSG analyses of Welsh, subjects and objects in synthetic VSO structures are all complements of the finite verb, hence an n-word in either of these positions can fulfil this requirement. Hence, in (117), the COMPS list of the weak negative verb *welodd* 'saw' contains two items: the subject *Dafydd* and the object *neb*. Since *neb* is marked [NEG +], *welodd* fulfils the Negative Dependent Constraint.

(117) [$_{S}$ [$_{V}$ Welodd] [$_{NP}$ Dafydd] [$_{NP}$ neb]]
see.PAST.3SG Dafydd no.one
[POL *weak-neg*] [NEG +]
'Dafydd saw no one.' (Present-day Welsh)

In the periphrastic structure in (118), on the other hand, the n-word is part of a larger verb phrase *gweld neb* 'see no one', itself embedded within an aspectual phrase (AspP). Only the n-word itself bears the feature [NEG +], hence the weak negative verb *dyw* 'is (not)' does not have a complement marked [NEG +]. The only way to resolve this is to have *ddim* as an additional complement of *dyw*, as is done in (119).

(118) *[$_{S}$ [$_{V}$ Dyw] [$_{NP}$ Dafydd] [$_{AspP}$ wedi [$_{VP}$ gweld neb]]]
NEG.be.PRES.3SG Dafydd PRF see.INF no.one
[POL *weak-neg*] [NEG +]
'Dafydd hasn't seen anyone.'

(119) [$_{S}$ [$_{V}$ Dyw] [$_{NP}$ Dafydd] [$_{Adv}$ ddim] [$_{AspP}$ wedi [$_{VP}$ gweld neb]]]
NEG.be.PRES.3SG Dafydd NEG PRF see.INF no.one
[POL *weak-neg*] [NEG +] [NEG +]
'Dafydd hasn't seen anyone.'

On this analysis, negative concord between n-words and special verbal forms is a real phenomenon, mediated by the Negative Context Requirement. Negative concord between n-words and *ddim*, on the other hand, is a by-product of the fact that weak negative verbs cannot use n-words embedded within other phrases to satisfy the Negative Dependent Constraint.

How would this type of analysis deal with the historical developments? First consider the spread of negative concord. We have seen that sentences like (114) and (115), which are, in very broad structural terms, parallel to (118), were grammatical in the nineteenth century. To accommodate this, it would be necessary to say that, in nineteenth-century Welsh, non-finite verbs and aspectual heads shared their value for NEG with their complements—this is the approach that Borsley and Jones adopt for prepositional phrases, as in (113) above. This mechanism has declined since then. While this could capture the facts, it hardly offers a satisfying explanation of the change. The introduction of *ddim* in the first place could be dealt with by positing a shift in the polarity specification of special negative verbs from strong negative verbs to weak negative verbs as *ni(d)* was eroded.

Next, consider changes in the distribution of Welsh n-words. There are various non-assertive contexts (interrogatives, conditionals) where n-words were once found but where they are no longer found. There has also been an increase in the range of environments where n-words are grammatical, with a negative meaning, in contexts where there is no other marker of negation at all. Borsley and Jones leave open the question of how to account for use of n-words today in comparatives and 'before'-clauses, noting that either a mechanism could be proposed to eliminate the inherently negative meaning of the n-word, or else n-words could be analysed as ambiguous between a negative and non-negative item (Borsley and Jones 2005: 89). Changes in the range of environments where negative interpretation of n-words are allowed in the absence of any other negative marker could be achieved by allowing the list of contexts in which a negative quantifier may be retrieved from storage to change over time. While this can deal with the historical facts, it does little to explain why the list of such contexts should have been continually expanding.

7.7 Breton indefinites

In Middle Breton, the main indefinite pronouns form a single series found in both negative and other non-assertive contexts:

(120) *negun*-series

person	*negun, nigun* 'anyone'
	den 'anyone'
thing	*netra* 'anything'
quantity	*nep* 'any' (also *quet* (*a*), as in (56) above)
time	*bizhuyquen* (generic or future-oriented) 'ever'
	bezcoaz (past-oriented) 'ever'
	nepret (< *nep pred* 'any time') 'ever'
place	*en nep lech* 'anywhere' (lit. 'in any place')

Den is homophonous with a generic noun *den* 'person', which is found in non-negative-polarity environments. *Negun* is a loan from Latin *nec unus* or a Romance descendant of it (Hemon 1975: 156). In general, in Middle Breton, *negun* tends to be used in negative clauses, while *den* tends to be used in other non-assertive contexts, but this is not an absolute rule.

Netra is derived historically from *nep tra* 'any thing'. The generic noun *tra* 'thing' is found in non-negative-polarity environments. It sometimes occurs in place of *netra* even in negative and other negative polarity contexts.

While negative concord with *ne* is compulsory if a negative interpretation is to arise, negative doubling with the postverbal negative marker *quet* is possible but avoided. This seems to be the case irrespective of the relative positions of the items. Examples (121) and (122) show the normal pattern, with an n-word and no *quet*.

(121) Necun ne deu a dref.
no.one NEG come.FUT.3SG back
'No one will come back.' (*L'ancien mystère de Saint-Gwénolé* 477, 1580)

(122) Eno ne guelo den.
there NEG.3FSG.ACC see.FUT.3SG no.one
'No one will see her there.' (*Le mystère de sainte Barbe*, stanza 36, 1557)

Doubling of *quet* and an n-word is illustrated in (123).

(123) a. Rac se nepret da monet de metou / Ne lesiff **quet** **den** **en** **bet**...
for this ever to go.INF among.3FSG NEG let.FUT.1SG QUET anyone in.the world
'For this reason, I shall never allow anyone ever to go near her...' (*Le mystère de sainte Barbe*, stanza 37, 1557)

b. Na nemeux **quet** **tra** **en** **bet** nement huy
and NEG.have.PRES.1SG QUET anything in.the world except you
'Nor do I have anything at all except you.' (*Le mystère de sainte Barbe*, stanza 101, 1557)

This series is renewed by the addition of various new items. The noun *mann* 'sign, trace' (cf. Welsh *man* 'blemish') has given rise to an indefinite pronoun 'nothing' (also in Cornish) via a minimizer use in southwest Brythonic. Another noun, *seurt* 'sort', has also given rise to a similar pronoun. Here the likely historical development is 'a sort of X' > 'such an X' > 'such (a thing)' > (not) 'such a thing' > 'anything, nothing'. As is typical in such cases of grammaticalization, this item still survives in other uses, as a noun meaning 'sort, kind' and as an adjective 'such (a)', and the existence of these other uses supports the proposed historical development. Their present-day use is illustrated in (124).

(124) Ne gomprenan **seurt** (ebet) / **mann** (ebet).
NEG understand.PRES.1SG anything (at all)
'I don't understand anything.' (Present-day Breton)

In some dialects, Middle Breton *heny* 'one' has grammaticalized as a negative indefinite. The southeastern Vannes dialect has a pronoun *hañni* 'no one' derived in this way. French *jamais* 'never' has also been borrowed to give a new item *james* 'never'.

While the basic membership of the Middle Breton series in (120) has largely remained intact in present-day Breton, its distribution has shifted. In Middle Breton, *negun*-series items are found in a variety of non-assertive negative-polarity contexts, including interrogatives, conditionals, and comparatives, as illustrated for *netra* 'anything' in (125) (interrogative), (126) (conditional), and (127) (comparative).

(125) Huy ó eus **nettra** da dibriff?
you have.PRES.2PL anything to eat.INF
'Do you have anything to eat?' (Parallel French text: 'Avez-vous quelque chose à manger?') (Guillaume Quiquer, *Dictionnaire et colloques françois et breton, Chrestomathie bretonne* 305.10–11, 1633)

(126) Mar comandet **netra** a gement a allen
if order.PRES.2PL anything of everything REL be.able.COND.1SG
'if you order anything that is within my power' (*La vie de Saint Patrice* 255, eighteenth century)

(127) An Barnn á vezo... da vezaff douget, meurbet muyguet **netra**
the judgement PRT be.FUT.3SG to be.INF feared much more.than anything
'The judgement will be... to be feared much more than anything' (*Le mirouer de la mort*, ll. 643–4, 1519/1575)

In Present-day Breton, this series has disappeared from all of these contexts except comparatives; contrast the ungrammatical interrogative in (128) and conditional in (129) with the grammatical comparatives in (130).

(128) *Daoust ha **den** a zo amañ?
QU QU anyone PRT be.PRES.3SG here
'Is there anyone here?' (grammatical as 'Is there no one here?') (Present-day Breton)

(129) *ma'z eus **den** amañ
if is anyone here
'if there's anyone here' (grammatical as 'If there's no one here') (Present-day Breton)

(130) a. Gouzout a rez gwelloc'h eget **den.**
know.INF PRT do.PRES.2SG better than anyone
'You know better than anyone.' (Present-day Breton)

b. Anavezout a ra ar vro-mañ gwelloc'h eget **nikun.**
know.INF PRT do.PRES.3SG the country-this better than anyone
'He knows this country better than anyone.' (Cornillet 2008: 73) (Present-day Breton)

As in Welsh, an emphatic prepositional phrase 'in the world', Middle Breton *en bet*, provides the basis for new grammaticalization in the system of negative indefinites. Already by Middle Breton, it seems that *en bet* had lost its compositional meaning. For instance, the meaning of *en bet* in (131) does not seem to be derivable from 'a window in the world':

(131) Memoa dit... gourchemennet /... na grases
have.IMPF.1SG to.you ordered NEG.COMP make.COND.PAST.2SG
quet **/Prenest en bet**... nemet dou
NEG window in.the world except two
'I had ordered you... that you should not make any window... except two (... that you should make only two windows)' (*Le mystère de sainte Barbe*, stanza 286, 1557)

Phonological reduction of *en bet* results in the Present-day Breton form *ebet*. In accordance with its historical origin as a prepositional phrase, *ebet* follows its head-noun. This is strange for a determiner in Breton, a generally rigidly head-initial language. *Ebet* joins the *negun*-series, and, as with other members of the series, it occurs in weak negative polarity contexts in Middle Breton. Today, however, *ebet* is restricted to direct negative contexts only (Hendrick 2011: 99–101). Present-day Breton thus has the following series of n-words descended from the Middle Breton *negun*-series:

(132) *den*-series
person *den (ebet)* / *nikun* 'no one'
thing *tra (ebet)* / *netra* 'nothing'
quantity N *ebet* / *nep* N 'no'
[time *gwech ebet* / *james* / *morse* 'never'
birviken/biken (generic or future-oriented) / *biskoazh* (past-oriented) 'ever']
place *neblec'h* 'nowhere'

In interrogatives and conditionals, items from the *den*-series have been replaced by items from a new series, innovated only in Breton, namely the *un... bennak*-series:

(133) *un... bennak*-series
person *unan bennak* / *un den bennak* 'someone, anyone' (plur. *ur re bennak*)
thing *un dra bennak* 'something, anything'
quantity *un... bennak* 'some, any'
[time *ur wech bennak* 'once, ever'
birviken/biken (generic or future-oriented) / *biskoazh* (past-oriented) 'ever']
place *ul lec'h bennak* / *un tu bennak* 'somewhere, anywhere'

As in Welsh, the items relating to time do not belong straightforwardly to either series. *Birviken* / *biken* and *biskoazh* 'ever' may appear in both negative contexts, in (134), and in non-negative contexts, an interrogative in (135) and a superlative in (136):

(134) **Biskoazh** n' en deus graet an dra-se.
never NEG have.PRES.3SG done the thing-that
'He never did that.' (Cornillet 2008: 103) (Present-day Breton)

(135) Daoust hag ho peus gwelet kig rostet war ma zaol **biskoaz**?
QU QU have.PRES.2PL seen meat roast on my table ever
'Have you ever seen roast meat on my table?' (Per-Jakez Heliaz, *An dachen piz-bihan*, 1953)

(136) Hennezh eo kaerañ levr a-m eus lennet **biskoazh**
that be.PRES.3SG nicest book REL-have.PRES.1SG read.PP ever
hag a lennin **biken**.
and REL read.FUT.1SG ever
'That's the nicest book that I've ever read or will ever read.' (Cornillet 2008: 103) (Present-day Breton)

For further discussion of the Present-day Breton system of negative indefinites, see Schapansky (2000).

All Brythonic languages have a cognate of Breton *bennak* (Welsh *bynnag*, Cornish *penag*) used to form free relatives (Willis 2011b). While all the languages allow this item to follow a *wh*-word, giving Welsh *pwy bynnag*, Breton *piv bennak*, and Cornish *pyv penagh* 'whoever', Breton alone has extended its use, combining it with the indefinite article to form the indefinites listed in (133). With an ordinary singular count noun, *un*... *bennak* is an indefiniteness marker meaning 'some, approximately, an X or so', as in *ur gudenn bennak* 'some problem, a problem' or *un dek vloaz bennak* 'some ten years'. These items are found in affirmative contexts and in weak negative polarity contexts (interrogatives and conditionals), but are marginal in negative contexts unless used to refer to a specific entity outside of the scope of negation:

(137) Bepred e vez **un dra bennak** dedennus da gavout eno.
always PRT be.HAB.3SG something interesting to have.INF there
'There's always something interesting to be had there.' (Present-day Breton)

(138) ma z' eus **un dra bennak** dedennus
if PRT be.PRES.3SG something interesting
'if there's anything interesting' (Present-day Breton)

(139) Bez' hoc'h eus **un dra bennak** da zebriñ?
PRT have.PRES.2PL something to eat.INF
'Do you have anything to eat? (Present-day Breton)

(140) *N' eus ket **un dra bennak** da welet.
NEG be.PRES.3SG NEG something to see.INF
'There isn't something to see.' (Present-day Breton)

The element *bennak* derives historically from *py*, the unstressed form of a general interrogative pronoun 'who, what', plus a negative element *na*(*g*). The original context for its use must have been something like the free-relative construction that survives in Middle Cornish:

(141)

panak	vo	age	deses
whatever	be.PRES.SUBJ.3SG	their	disease

'whatever their disease may be' (*Life of Saint Meriasek*, l. 3104) (Lewis 1946: 46)

Here the negative must once have been interpreted as an instance of expletive negation, cf. French *Je doute, qu'il ne soit là* 'I doubt that he'll be there' (Rowlett 1998: 26–7), or German *Was es nicht alles gibt!* '(look) what is(n't) there'.

Outside of free relatives, *bennak* remains rare in Middle Breton, although is sometimes found in its modern indefinite sense in affirmative clauses:

(142)

Ret	eu	diff	gouzout...	/Diouz	**un re pennac**...	/Vn	tra...
necessary	be.PRES.3SG	to.me	know.INF	from	someone	a	thing

'I must learn a thing from somebody.' (*Le mystère de sainte Barbe*, stanza 107, 1557)

The pathway by which this use arises may be from free relative ('Bring whoever you like') to free-choice indefinite ('Bring whoever' with omission of the relative clause) and thence to ordinary specific indefinite. The spread to non-assertive contexts seems to postdate the Middle Breton period. The relevant shifts seem to parallel the slightly earlier development of Old French *quel... que* from free relative 'whatever' along the same pathway to free-choice indefinite marker, attested in the fourteenth and fifteenth centuries, illustrated in (143), to the Modern French *quelque*-series (*quelque* 'some', *quelqu'un* 'someone', *quelque chose* 'something', etc.) (Foulet 1919). This seems likely then to be a case of transfer from French.

(143)

Qui	femme	prend,	de	quelque	taille,	/Il	ne	puet	faillir	a	bataille.
who	wife	take.PRES.3SG	of	any	size	he	NEG	can.PRES.3SG	lack.INF	at	battle

'Anyone who takes a wife, of whatever / any size, he cannot be short of battles.' (Jean le Fèvre, *Les lamentations de Matheolus* l. ii, v. 3817–18, *c*.1371) (Foulet 1919: 227)

The full set of changes in the indefinite system from Middle to Present-day Breton is summarized in Figures 7.3 and 7.4.

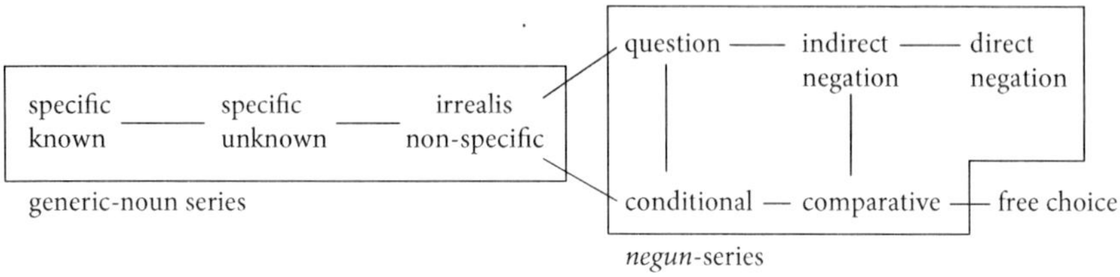

FIGURE 7.3 Expression of indefinites in Middle Breton

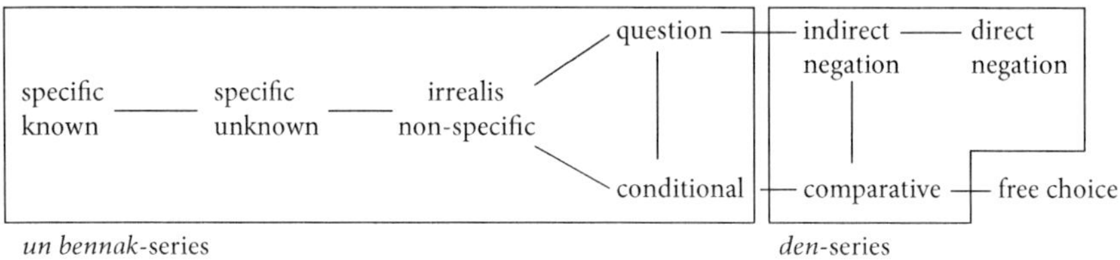

FIGURE 7.4 Expression of indefinites in Present-day Breton

7.8 Cornish indefinites

Middle Cornish indefinites effectively form a single series with a distribution that is not sensitive to negative polarity; that is, all items are found in negative, non-assertive and affirmative contexts. The system is thus as in Figure 7.5. The forms themselves are given in (144).

(144) *nep*-series

person	*den* (*vyth*) / *nep* (*onon*) 'someone, anyone'
thing	*nep peyth* / *nep tra* / *tra* (*vyth*) 'something, anything'
quantity	*nep* N 'some, any' (also N *vyth*)
time	*neffra* / *bythqueth* / *byth* / *vyth* / *nep preys* 'ever, always'
place	(*in*) *nep pow* / *nep le* / *nep tu* 'somewhere, anywhere'

The generic nouns *den* 'a person' and *tra* 'a thing' are used alone as indefinites as in Middle Breton. The inherited quantifier *nep* 'some, any' may also be used with various generic nouns to create indefinites such as *nep peyth* 'some thing' or *nep*

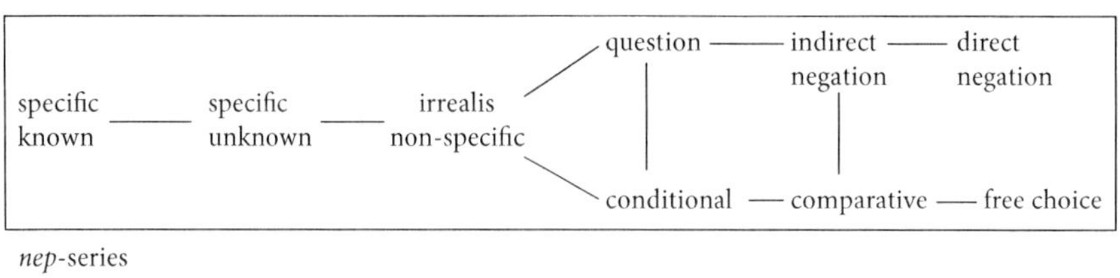

FIGURE 7.5 Expression of indefinites in Middle Cornish

preys 'some time'. This strategy is used to create various place indefinites with the generic nouns *pow* 'country', *le* 'place', and *tu* 'side'. Of these, *nep pow* seems to show the greatest degree of grammaticalization and greatest degree of semantic separation from its etymological source.

Examples in (145) and (146) show lack of sensitivity to negative polarity: *nep le* 'somewhere, anywhere' is used indiscriminately in an affirmative context in (145) and in a negative context in (146).

(145)

Hy	re	gafes	dyhogel	/dor	dyseghys	**yn**	**nep**	**le.**
she	PRF	get.PAST.3SG	certainly	earth	dried.out	in	some	place

'She has certainly found dry land somewhere.' (*The Ancient Cornish Drama, Origo mundi*, ll. 1143–4, Middle Cornish)

(146)

rag bytqueth	my	ny	welys	/benen	thy'm	a	wel	plekye
for ever	I	NEG	see.PAST.1SG	woman	to-me	REL	better	please.IMPF.3SG

/wheth	**yn**	**nep**	**le**
yet	in	any	place

'For never have I seen a woman who pleased me more in any place.' (*The Ancient Cornish Drama, Origo mundi*, ll. 2107–9, Middle Cornish)

Strikingly, this patterning is even extended to the English loanword *neffra* (< Old or Early Middle English *næfra* 'never'), which adopts the distribution of its Middle Cornish equivalent *bythqueth*, and is therefore found in both negative and affirmative contexts. Example (147) shows its unexpected affirmative use to mean 'always'.

(147)

ha	**neffra**	me	a	'th	vynyk.
and	ever	I	PRT	you	bless.PRES.3SG

'and I shall always bless you.' (*Bewnans Ke*, l. 791, Middle Cornish)

The only sensitivity to negative polarity seems to be introduced by the item *vyth* 'ever', which may optionally be added to a generic noun to create a weak negative polarity item:

(148)

Mars	ues	**den**	**vith**	a	vyn	cows	/py	gul
if	be.PRES.3SG	person	ever	REL	want.PRES.3SG	speak.INF	or	do.INF

erbyn	both	e	vres…
against	will	his	heart

'If there's anyone who speaks or acts against the wishes of his heart.' (*Bewnans Ke*, ll. 1496–7, Middle Cornish)

In such cases, the temporal sense of *vyth* 'ever' seems to have bleached, and it typically follows the generic noun immediately, which is suggestive of grammaticalization. It is possible that Late Cornish developed this more fully to give rise to polarity sensitivity, encouraged by contact with English, with *nabonnen* (< Middle

Cornish *nep onon* 'some one') corresponding in distribution to English *someone* and *denveeth* (< Middle Cornish *den vyth* 'person ever') corresponding to *anyone* (but see also Wmffre 1998: 24, 39):

(149) ha na ore **den veeth**...
and NEG know.PRES.3SG anyone
'and no one knows...' (*The Cornish writings of the Boson family*, p. 29, *c*.1660–1700)

Pervasive lack of sensitivity to negative polarity in indefinites is unique to Middle Cornish among the Brythonic languages. It has no parallel in English—indeed, the later emergence of sensitivity to negative polarity is likely to be due to contact with English—but has parallels in Old Irish (cf. examples (157) and (158) below), and hence is a good candidate for a feature that reflects the syntax of the parent language (see section 7.9.1 below).

7.9 Issues in the reconstruction of Brythonic indefinites

7.9.1 Neb *in Common Brythonic*

Brythonic Celtic languages make extensive use of indefinites arising via grammaticalizations based on generic nouns. In most cases, the earliest items based on generic nouns are different enough to suggest that the Brythonic parent language possessed a productive pool of forms, with the daughter languages only later fixing on particular items. For instance, in the case of indefinites for things, Middle Welsh *dim*, Middle Breton *nep tra* > *netra* and *tra* and Cornish *neb peyth* and *neb tra* 'anything, something' are grammaticalizations of the same general type, all derived from generic nouns meaning 'thing', but are based on different lexical items. We can conclude that Brythonic made extensive use of generic nouns for indefinites, but that particular items had not yet conventionalized or else had conventionalized differently in different areas. Some similarity of patterning in grammaticalization in Cornish and Breton suggests this latter option to some extent.

All medieval Brythonic languages share a quantifier *neb, nep* 'any'. Cornish allows it freely across affirmative and negative contexts, while Middle Welsh and Middle Breton show more complex patterning. In addition to using *neb* as a weak negative polarity item, both allow *neb* as the antecedent of a free relative. This is illustrated for Middle Welsh in (150).

(150) A **'r** **neb** a dodes hut ar y wlat, a beris
and the anyone PRT put.PAST.3SG magic on the land PRT cause.PAST.3SG
bot y gaer yma.
be.INF the fortress here
'And whoever bewitched the land caused the fortress to be here.'
(*Pedeir Keinc y Mabinogi* 56.4–5, Middle Welsh)

Furthermore, *neb* is also an element within the specific unknown indefinite quantifier *nebun* 'any' (< *neb* + *un* 'one'), which occurs freely in both negative polarity and non-negative-polarity environments. Its use in an affirmative environment is given in (151).

(151) ...e wynvydedic wyry a emdangosses y **nebun** **yscolheic**
the blessed virgin PRT appear.PAST.3SG to some scholar
a dywedut urthav...
and say.INF to.3MSG
'... the Blessed Virgin appeared to some (a certain) scholar and said to him ...'
(*Gwyrthyeu e wynvydedic Veir*, Peniarth 14, p. 19, ll. 5–6, Middle Welsh)

Both these uses are archaic today.

These uses of *neb* in Middle Welsh and Middle Breton are surprising, since these languages do not otherwise allow *neb* in affirmative contexts. When compared to more general use of *neb* in affirmative contexts in Cornish, the Welsh and Breton use looks like the fossilized relic of an earlier more productive system. For instance, it looks as though Middle Welsh *nebun* was created as an item at a point when Welsh did allow affirmative uses of *neb*. This suggests that the Cornish pattern, with *neb* freely available in affirmative, non-assertive, and negative contexts, is the one that should be reconstructed for the Brythonic parent language.

Middle Welsh and Cornish use *neb* also as an animate indefinite pronoun. Middle Breton, apart from its use as an antecedent to free relatives, does not use it as a pronoun. However, such use is attested in Old Breton and should therefore be reconstructed for the whole of the Brythonic parent language:

(152) Na dimicit nep.
NEG despise.IMP.2PL anyone
'Do not despise anyone.' (Fleuriot 1964a: 262) (Old Breton)

We can conclude that the Brythonic ancestor of *neb* was both a pronoun and a quantifier, and was freely available in all environments, both affirmative and negative.

The Old Irish quantifier *nach/na* 'some, any' (the 'dependent' form of *nech/ní* 'anyone/anything') is also cognate. This item is evidently pronominal in origin (< Common Celtic *ne-k^{w}os NEG + 'who') (see section 7.9.2 below), so the use of *neb* as an adnominal quantifier, in (86) above, is an innovation, based on the abductive reanalysis given in (153).

(153) [$_{DP}$ *neb*] > [$_{DP}$ [$_{Q}$ *neb*] [$_{NP}$ Ø]]
anyone any

In (153), *neb* is hypothesized to contain a null head noun and therefore to be an adnominal quantifier rather than, or perhaps in addition to, being a pronoun. Once this reanalysed structure becomes entrenched, it is manifested by the emergence of examples such as (86). Given that use of *neb* as a quantifier is paralleled by the syntax of the cognate items in Old Irish and Middle Breton, we could posit that this reanalysis took place early on in the development of the Celtic languages; however, this reanalysis is so common cross-linguistically that independent innovation in Brythonic and Goidelic cannot be ruled out. The original morphological formation of *neb* is based on the animate form of the interrogative pronoun. This is reflected in Middle Welsh by the fact that pronominal *neb* is animate ('anyone' rather than 'anything'), and that the free-relative antecedent *y neb* is also restricted to animate uses ('anyone who, whoever' rather than 'anything that, whatever'). On the other hand, quantifier *neb* is possible with a following inanimate, as in (87) above. Historically, this must therefore represent an extension in the environments in which it occurs. It is made possible by the fact that Brythonic does not continue the neuter forms of the pronoun (found in Old Irish as *ní* 'anything', *na* 'any'). The evidence of Breton and Cornish, where quantifier *nep* is used irrespective of animacy, suggests that either: (i) Welsh *neb* was once used more widely for inanimates, and that its rarity with inanimate nouns is due to competition from the innovative quantifier *dim*; or (more economically) that Middle Welsh is conservative and maintains an original restriction to use with an animate head noun.

Quantifier *dim*, on the other hand, is a Welsh innovation, as it is not found in any other Celtic language. It is based on the same form of reanalysis as posited for *neb*, only based on the indefinite pronoun *dim* 'anything'. That is, *dim* 'anything' is (abductively) hypothesized to contain a null or elided head noun:

(154) $[_{\mathrm{DP}}$ *dim*$]$ > $[_{\mathrm{DP}} [_{\mathrm{Q}}$ *dim*$] [_{\mathrm{NP}}\ \emptyset\ ]\ ]$
anything any

Once this hypothesis is accepted, a new item, quantifier *dim* is posited and phrases like that in (84) become possible.

The Welsh quantifier *yr un*, *'run* 'any, no' has parallels in other Celtic languages, principally in Irish. Although Middle Breton and Cornish have some examples where *un/unan* can be interpreted as meaning 'any', they are fairly rare, suggesting that this use had not become grammaticalized in these languages. The development of *un* into an indefinite article in Middle Breton (but not in Welsh or Cornish) may have precluded the development of the quantifier. Irish *aon* 'one' shows a greater tendency to develop into an 'any'-word, but this development is so common cross-linguistically (cf. English *any* < Old English *ǣnig* derivative of 'one') that the development in Irish is likely to be independent.

7.9.2 *Common Celtic and the historical development of Old Irish* nech, *Middle Welsh* nep

In the documented history of Welsh, it is clear that, in very general terms, *neb* becomes more negative, as was seen in section 7.6.5 above. In Middle Welsh, it has non-negative non-assertive uses and appears in contexts entirely unconnected with negation. However, in Present-day Welsh, it is an inherently negative item. This seems like a straightforward unidirectional development. However, problems arise when we turn to internal and comparative reconstruction.

Brythonic *nep* is cognate with the Old Irish indefinite pronoun *nech* (masculine and feminine), *ní* (neuter). In Old Irish, the 'independent' forms *nech* and *ní* are used as pronouns 'anyone' and 'anything' respectively. Parallel 'dependent' forms exist, *nach* (masculine and feminine) and *na* (neuter). These are used as adnominal quantifiers 'any'. These forms are found in negative polarity contexts, whether negative, as in (155), or other non-assertive environments, as in (156).

(155) ním-raib ní
NEG.1SG-be.PRES.3SG anything
'may I not have anything' (*Zeitschrift für Celtische Philologie* 7: 308, §1) (*Dictionary of the Irish language*, s.v. 1 *ní*)

(156) cech duine shirfess ni fort
every man seek.FUT.REL anything on.2SG
'every man who (whoever) shall ask anything of you' (*Leabhar Breac* 462) (*Dictionary of the Irish language*, s.v. 1 *ní*)

They have non-negative uses rather more extensively than their Middle Welsh cognates. *Nach* etc. is used in affirmative environments to mean 'some' and 'something':

(157) itá nách cumachta fora.cul na nén-sa
be.PRES.3SG some power behind the birds-DEM
'there is some power behind these birds' (*Serglige Conculain* 7) (*Dictionary of the Irish language*, s.v. 1 *nach*)

(158) ní do thabairt do neuch
something.ACC to give.INF to someone.DAT
'to give something to someone' (glossing *aliquid proferre*) (*Milan glosses* 98.a.4) (*Dictionary of the Irish language*, s.v. 1 *ní*)

Old Irish *nech* may serve as the antecedent to a free relative, although in contrast to Middle Welsh usage in (150), it is not preceded by a definite article in a free-relative construction:

In the varieties in question, *smo* has spread to be the negative of the present tense of the verb 'be' in all contexts, including, for instance, periphrastic tenses that use auxiliary 'be', as with the present progressive in (172).

(172) Smo fi 'n gweitho heddi.
NEG.AUX.1SG I PROG work.INF today
'I'm not working today.' (Present-day southern Welsh)

Another pattern of contraction from *nid oes mohono* 'he isn't' gives another variant of the same auxiliary, *sano* (Morris 1910).

Many southern varieties have also undergone another non-Jespersen development. In these varieties, the negative reponsive particle *na(g)* has been generalized to become a possible marker of negation in any clause:

(173) Nag 'yn ni 'n prynu bara o 'r siop, fel pobl normal.
NEG be.PRES.1PL we PROG buy.INF bread from the shop like people normal
'We don't buy bread from the shop like normal people.' (<http://tadarmab. wordpress.com>)

Effectively, the pragmatic limitation on *na(g)* as occurring only in responses to yes–no questions failed to be acquired in the history of these dialects. Generalization of *na(g)* to embedded clauses in place of *na(d)* is also very common, again predominantly in the south.

7.10.2 Negation in other non-finite contexts

In general, non-finite verbs are not negated directly in Middle Welsh. If they need to be negated independently of the finite verb on which they depend, then the clause is made finite (either indicative or subjunctive) and this finite clause is negated:

(174) Ni allaf i na chyscwyf.
NEG can.PRES.1SG I NEG.COM sleep.PRES.SUBJ.1SG
'I cannot not sleep.' (*Gereint*, *White Book Mabinogion* 427.23) (Richards 1948: 377)

In Present-day Welsh, non-finite verbs can be negated directly using the negative auxiliary *peidio*. While this is also an ordinary lexical verb meaning 'stop, cease' (see also section 7.12 on negative imperatives below), it is used with non-finite verbs with purely grammatical meaning:

(175) Dwi 'n methu peidio cysgu.
be.PRES.1SG PROG be.unable.INF NEG sleep.INF
'I cannot not sleep.' (Present-day Welsh)

There are very limited environments where this is possible in Middle Welsh. It is found only in bipartite embedded non-finite yes–no focus questions where the non-finite verb is elided in the second clause, that is, the pattern in (176).

(176) Guedy treiglav o Ioachym yn y vedul beth a vnelei
after turn.INF of Ioachim in 3MSG mind what PRT do.IMPF.SUBJ.3SG
ae ymchuelut ae **peidyav**...
QU.FOC return.INF QU.FOC NEG
'After Ioachim had considered in his mind what he should do, whether to return or not...' (*Mabinogi Iesu Grist*, Peniarth 5, folio 15r, ll. 19–20, Middle Welsh)

Even here, it is not the only possibility, and a finite clause without ellipsis is possible instead:

(177) dewis di ae kerdet ae **na cherdych.**
choose.IMP.2SG you QU.FOC walk.INF QU.FOC NEG walk.PRES.SUBJ.2SG
'Choose whether to walk or not to walk.' (*Breudwyt Ronabwy*, Jesus 111, folio 136r, col. 561, l. 27, Middle Welsh)

From this context, use of *peidio* to negate a non-finite verb seems to spread to other contexts, reaching the dominant position that it has today.

7.11 Constituent and focus negation

In Middle Welsh, both constituent and focus negation are expressed using the particle *nyt* (for definitions, see section 1.1). Note that this differs from the sentential negation marker *ny*(*t*) in that the final /d/ (orthographic <t>) is always present, irrespective of whether the following word begins with a vowel. Constituent negation is illustrated in (178).

(178) Yna y kychwynnawd Lawnslot y wrth Wenhwyuar, ac nyt heb
then PRT start.PAST.3SG Lancelot PRT by Guinovere and NEG without
dolur a thristit.
pain and sadness
'Then Lancelot started out away from Guinovere, and not without pain and sadness.' (*Ystoryaeu Seint Greal* 478–9, Middle Welsh)

The element in the scope of focus negation normally fronts:

(179) **nyt** teilygdawt uy anryded a 'm etteil am hynny
NEG worthiness 1SG honour PRT 1SG.ACC prevent.PRES.3SG on this
'It is not the honour of my rank that prevents me from [doing] this.' (*Pedeir Keinc y Mabinogi* 2.10–11, Middle Welsh)

In embedded clauses, if there is fronting of the element in the scope of negation, then the subordinate constituent negation marker *nat*, is used. Again, this is distinguished from the marker of sentential negation in subordinate clauses by obligatory presence of the final /d/:

(180) A ryued oed genhyf, **nat** kyn rodi morwyn
and strange be.IMPF.3SG with.1SG NEG.COMP.FOC before give.INF maiden
gystal a honno ym, y gwneit y gwaradwyd a
as.good as that to.me PRT do.IMPF.IMPERS the disgrace PRT
wnelit ym.
do.IMPF.SUBJ.IMPERS to.me
'And I found it strange that it was not before giving a maiden as good as that to me that the disgrace that was perpetrated against me was done.' (*Pedeir Keinc y Mabinogi* 33.1–3, Middle Welsh)

While this remains a possible standard pattern today (in the form of Present-day Welsh *nid* in main clauses and *nad* in embedded clauses), a number of dialect or colloquial variants have very wide currency. For main-clause constituent negation, we also find *dim*, *ddim*, and *nage*. The first two represent the extension of different forms of the sentential negation marker *ddim* to mark constituent negation. The third looks superficially like an extension of the anaphoric negator *nage* 'no', used in response to focus questions, to be a marker of constituent negation in its own right, although it is unclear if that is actually what happened.

In embedded clauses, affirmative focus markers have tended to spread, giving rise to combinations such as *mai ddim* to mark embedded focus, where *mai* is the affirmative embedded focus marker and *ddim* is the negative focus marker:

(181) Dwi 'n sicr mai (d)dim ni oedd e.
be.PRES.1SG PRED sure FOC NEG us be.IMPF.3SG it
'I'm sure it wasn't *us*.' (lit. 'I'm sure that it's not us that it was.') (Present-day Welsh)

This replaces the more traditional pattern, still current, in which embedded negative focus clauses are marked using *nad*, an embedded counterpart for the main-clause focus marker *nid*:

(182) Dwi 'n sicr nad ni oedd e.
be.PRES.1SG PRED sure FOC.NEG us be.IMPF.3SG it
'I'm sure it wasn't *us*.' (lit. 'I'm sure that it's not us that it was.') (Present-day Welsh)

7.12 Negative imperatives

Middle Welsh negated imperatives using the preverbal marker *na(c)* (Modern Welsh *na(g)*), which was illustrated above in section 7.3.1, example (13). In Present-day spoken Welsh, this is no longer possible, and in place of a true negative imperative we find a negative auxiliary, singular *paid*, plural *peidiwch*, followed by the preposition *â* 'with' (now optional) and a non-finite verb:

(183) Paid (â) gadael!
NEG.2SG with leave.INF
'Don't leave!' (Present-day Welsh)

Furthermore, the expected output of Jespersen's cycle does not arise. In main clauses postverbal *ddim* ultimately replaces preverbal *ni(d)*. We might expect a parallel development in imperatives, with postverbal *ddim* replacing preverbal *na(g)*. While *na... ddim* is occasionally found, postverbal *ddim* is not generalized with imperatives and is now ungrammatical:

(184) *Ad ddim! *Dere ddim!
leave.IMP.2SG NEG come.IMP.2SG NEG
'Don't leave!' 'Don't come!' (Present-day Welsh)

Paid is the imperative of the verb *peidio* 'cease, stop', a verb that remains in the language, but, in (183), it functions simply in place of a negative imperative. It is not limited to inhibitive readings. It does not imply that a leaving event has begun or that there has been a previous leaving event, and so does not have the expected compositional meaning 'Stop leaving!' There is little evidence that *paid* was used for ordinary negative imperatives in Middle Welsh. Middle Welsh examples such as (185) are open to a purely compositional interpretation with the 'stop' meaning:

(185) Yna y dyvat Maxen vrth y vyr, 'Kymeruch Porffir heb ohir a dyguch y gantav y eneit, a'e aelodeu.' Porffir a gyuodes y ar y veigc, ac a gymerth ysgaul yn y lav, a phedeir mil o wyr Maxen a ladaud ef rac bronn Maxen. Ac yna yd ofynnes Maxen, ac y crynavd rac ofyn o tebygu y lledit ynteu. A phedeir mil ereill a vrathaud Porffir. Ac yna Catrin a velas hynny ac a dyvat vrth Porffir, '**Peit a'e llad** a choffa dioddeiueint Duv yn harglvyd ny Iessu Grist.'
'Then Maxen said to his men, "Capture Porffir without delay and take away his soul and his limbs." Porffir got up from his bench, and took a ladder in his hand, and he killed four thousand of Maxen's men in front of Maxen. And then Maxen took fright, and trembled with fear thinking that he too would be killed. And Porffir wounded another four thousand. And then Catrin saw this and said to Porffir, "**Stop killing them** and remember the suffering of our Lord God Jesus Christ."' (*Buched Catrin Sant*, Peniarth 5, folio 22v, ll. 30–6, Middle Welsh)

By the eighteenth century, however, *paid* had spread to ordinary negative imperatives, and had therefore become a prohibitive marker. Example (186), for instance, seems to mean 'don't kill' and not 'stop killing'.

(186) **paid** tithe a lladd yr henddun
NEG.2SG you.CONJ with kill.INF the old.man
'Don't kill the old man.' (*Brenin Llur*, l. 1192, 1700–50)

The two coexist in the eighteenth century, the former pattern with preverbal *na*(*g*) still being found productively in colloquial texts, alongside the new pattern with *paid*:

(187) **Na** ddewch ddim i mewn yma.
NEG come.IMP.2PL NEG to in here
'Don't come in here.' (*Welsh defamation suits*, Archdeaconry of Brecon, 1771)

(188) **Paid** ai fissio...
NEG.2SG with.2MSG miss.INF
'Don't miss it.' (Edward Thomas, *Cwymp dyn* 45.14, 1767)

However, negative imperatives with *na*(*g*) seem to have died out in speech soon afterwards, in the nineteenth century. The loss of true negative imperatives occurs at about the same time as the loss of the preverbal negative marker *ni*(*d*) in main clauses, and it is tempting to connect it with Jespersen's cycle. In both cases, a preverbal negative particle is renewed by formerly lexical material, but, by adopting a different strategy, the imperative retains preverbal marking of negation. As Horn (1989: 449–50) notes, the Neg-First Principle is particularly strong in imperatives, and the introduction of *paid*, rather than the expected diffusion of postverbal *ddim* in imperatives (to give the pattern in (184) above), may have been preferred because it allows the negative content of imperatives to be accessed immediately.

Other accounts of why languages often lack true negative imperatives have linked this phenomenon to a variety of factors. Postma and van der Wurff (2007) link absence of true negative imperatives to ambiguity between the anaphoric negator and the negative particle. That is, languages with the potential for confusion between 'No, go!' and 'Don't go!' disallow negative imperatives. While this is difficult to evaluate in Welsh, which does not have and never has had a single word for 'no', the basic intuition here does not seem promising: contrary to the prediction of this approach, no significant changes are observed in the anaphoric negator at the time when true negatives disappear.

More promising generally is Zeijlstra's (2004, 2006) proposal that whether a language has true negative imperatives depends on properties of its negative marker. Zeijlstra argues, following Han (2001), that the imperative operator must outscope the negative operator by moving to c-command it. Languages with negative heads bearing interpretable negation [iNeg] (for instance, Italian) lack true negative imperatives, because the negative head blocks movement of the imperative verb to a position above the interpretable negation. Conversely, a language where the negative marker is phrasal will always allow true negative imperatives because a marker in a phrasal position cannot prevent head movement of the imperative verb to a position c-commanding negation. These claims are summed up in Zeijlstra's (2006: 414) two generalizations:

(189) Generalization 1: Every language with an overt negative marker X^0 that carries [iNeg] bans true negative imperatives.
Generalization 2: Every language that bans true negative imperatives exhibits an overt negative marker X^0.

The Middle Welsh pattern with *na*(*g*) can be successfully accounted for using this system. Although Middle Welsh negators are all negative heads, there are no syntactic differences between preverbal and postverbal n-words, and hence, in Zeijlstra's system, the negative particles could be uninterpretable, with negation taking scope from an interpretable abstract negative operator in Spec, NegP. While both *na*(*g*) and the imperative verb would move to C, only the imperative verb would bear an operator feature, and hence the imperative operator in C would outscope the abstract negative operator in Spec, NegP. The appearance of *na*(*g*) as the negative particle rather than the indicative particle *ni*(*d*) would simply mean that the Middle Welsh negative particles were sensitive to the realis–irrealis distinction.

However, problems arise when we consider the shift to the Present-day Welsh system. Jespersen's cycle results in the creation of a new phrasal negator *ddim*, which for Zeijlstra would be located in Spec, NegP, bearing an interpretable [iNeg] feature. Occupying a specifier position, *ddim* cannot block movement of the verb from Neg to C. Thus, the configuration in (190), in which the imperative operator legitimately c-commands the negative operator, should be available in Present-day Welsh.

(190) *Dere ddim!
come.IMP.2SG NEG
[CP [C dere] [NegP ddim [Neg ~~dere~~] [VP ~~dere~~]]]
[iImp] [iNeg]

This approach then also ultimately fails to account for the loss of true negative imperatives in Welsh.

In Breton, Jespersen's cycle spread the bipartite pattern to imperatives, giving a pattern parallel to French:

(191) N' a ket re vuan!
NEG go.IMP.2SG NEG too fast
'Don't go too fast!'

However, while the pattern in (191) survives, there is an increasing tendency to avoid true negative imperatives, subsitituting either the present indicative, as in (192), or a negative imperative marker *arabat* (from *arabat da* X *ober* Y 'it is folly for X to do Y') plus a non-finite verb, as in (193) (Hemon 1975: 258, Favereau 1997: 201, 252–3).

(192) N' ez ket re vuan!
NEG go.PRES.2SG NEG too fast
'Don't go too fast!'

(193) Arabat mont re vuan!
NEG go.INF too fast
'Don't go too fast!'

The reasons for the ongoing loss of true negative imperatives in Breton are not clear, but it seems unlikely to be related to a change in the status of the ordinary sentential negator.

7.13 Conclusion

This chapter has surveyed the extensive changes that Brythonic Celtic negation has undergone in almost all areas over the past thousand years. We have seen how independent Jespersen cycles have arisen in both Welsh and Breton, yielding new postverbal markers of negation in both languages. In the indefinite system, the general trend has been towards the specialization of former negative polarity items as purely negative items, with various new items being created from various sources, including free-choice items, to fill the gap left behind. Comparison of the daughter languages suggests a parent language with relatively little sensitivity to negative polarity, such sensitivity developing over the history of the languages. Of the various minor developments outside of these areas, perhaps the most significant is the loss of negative imperatives, above all in Welsh. In all of these areas, the Brythonic Celtic developments have the potential to provide an important source of comparative data to help us understand the ways in which many aspects of negative systems develop.

Acknowledgements

I would like to express my thanks to Perynn Bleunven, Anders Richardt Jørgensen, Oliver Padel, and Janig Stephens for discussion of material dealt with in this chapter. Naturally, they are not responsible for the uses to which I have put their comments and suggestions.

Primary sources

An analysis and calendar of early modern Welsh defamation suits, ed. Richard F. Suggett. SSRC Final Report (HR 6979).
L'ancien mystère de Saint-Gwénolé, ed. Émile Ernault. Rennes: Plihon, n.d.
The Ancient Cornish Drama. ed. Edwin Norris. Oxford: Oxford University Press, 1859.

Yr anterliwt goll: Barn ar egwyddorion llywodraeth, ed. Emyr Wyn Jones. Aberystwyth: Llyfrgell Genedlaethol Cymru, 1984.

Bewnans Ke: The life of St Kea. ed. Graham C. G. Thomas and Nicholas Williams. Exeter: University of Exeter Press, 2007.

Y Bibl Cyssegr-lan. London: B. Norton a I. Bill, 1620.

Brenhinedd y Saesson or the Kings of the Saxons, ed. Thomas Jones. Cardiff: University of Wales Press, 1971.

Breudwyt Ronabwy, Jesus 111, from *Rhyddiaith Gymraeg 1350–1425*, <http://www.rhyddiaithganoloesol.caerdydd.ac.uk>.

Brut Dingestow. ed. Henry Lewis. Llandysul: J. D. Lewis a'i Feibion, 1942.

Buched Catrin Sant. ed. J. E. Caerwyn Williams. *Bulletin of the Board of Celtic Studies* 25, 247–68, 1973.

Chrestomathie bretonne, ed. Joseph Loth. Paris: E. Bouillon, 1890.

The Cornish writings of the Bosun family, ed. Oliver J. Padel. Redruth: Institute of Cornish Studies, 1975.

Cyfreithiau Hywel Dda yn ôl Llyfr Blegywryd, ed. Stephen J. Williams and J. Enoch Powell. Cardiff: Gwasg Prifysgol Cymru, 1942.

Darn o'r Ffestival (Liber Ffestialis), ed. Henry Lewis, Supplement to the Transactions of the Honourable Society of Cymmrodorion (Session 1923–24). London: Cymdeithas y Cymmrodorion, 1925.

Evans, Beriah Gwynfe. 1898. *Dafydd Dafis.* Wrexham: Hughes a'i Fab.

Gesta Romanorum, ed. Patricia Williams. Cardiff: Gwasg Prifysgol Cymru, 2000.

Le grand mystère de Jésus, ed. Hersart de la Villemarqué. Paris: Didier, 1865.

Gwyrthyeu e wynvydedic Veir, Peniarth 14. In *Rhyddiaith Gymraeg o lawysgrifau'r 13eg ganrif: Testun cyflawn*, ed. Simon Rodway and Graham R. Isaac, <http://hdl.handle.net/2160/1207>.

Heliaz, Per-Jakez. 1953. *An dachen piz-bihan.* <http://embann.an.hirwaz.online.fr>.

Historia Gruffud vab Kenan, ed. D. Simon Evans. Cardiff: Gwasg Prifysgol Cymru, 1977.

Historia Peredur vab Efrawc, ed. Glenys Witchard Goetinck. Cardiff: Gwasg Prifysgol Cymru, 1976.

Hughes, Annie Harriet (Gwyneth Vaughan). 1908. *Plant y gorthrwm.* Cardiff: Cwmni Cyhoeddiadau Addysgol.

Jones, Angharad. 1995. *Y dylluan wen.* Llandysul: Gomer.

Jones, Hugh (Llangwm). 1783. *Enterlute newydd; ar ddull ymddiddan rhwng protestant a neilltuwr.* Shrewsbury: T. Wood.

Kedymdeithyas Amlyn ac Amic, ed. Patricia Williams. Cardiff: Gwasg Prifysgol Cymru, 1982.

Mabinogi Iesu Grist, Peniarth 5, from *Rhyddiaith Gymraeg 1350–1425*, <http://www.rhyddiaithganoloesol.caerdydd.ac.uk>.

Le mirouer de la mort, ed. Émile Ernault. Paris: H. Champion, 1914.

Le mystère de sainte Barbe, ed. Émile Ernault. Paris: Libraire du Collège de France, 1888.

Owen, David (Brutus). [1863]. *Wil Brydydd y Coed*, ed. Thomas Jones. Cardiff: Gwasg Prifysgol Cymru, 1949.

Pedeir Keinc y Mabinogi, ed. Ifor Williams. Cardiff: Gwasg Prifysgol Cymru, 1930.

Rees, William (Gwilym Hiraethog). [1877]. *Helyntion Bywyd Hen Deiliwr*, ed. Dafydd Jenkins. Aberystwyth: Clwb Llyfrau Cymreig, 1940.

Rees, William (Gwilym Hiraethog). *Llythyrau 'Rhen Ffarmwr*, ed. Edward Morgan Humphreys. Cardiff: Gwasg Prifysgol Cymru, 1939.

Rhyddiaith Gymraeg, vol. ii: 1547–1618, ed. Thomas Jones. Cardiff: Gwasg Prifysgol Cymru, 1956.

Roberts, Ellis. 1769. *Llyfr enterlute newydd wedi gosod mewn dull ymddiddanion rhwng Gras a Natur*. Warrington: William Eyres.

Thomas, Edward (of Rhydwen, Flintshire). n.d. [*c*.1767–8]. *Cwymp dyn*. Chester: Read a Huxley.

Trois poèmes en moyen-breton, ed. Roparz Hemon. Dublin: Dublin Institute for Advanced Studies, 1982.

La vie de saint Patrice: Mystère breton en trois actes, ed. Joseph Dunn. Paris: Honoré Champion, 1909.

La vie de sainte Catherine, ed. Émile Ernault. *Revue Celtique* 8, 76–95, 1887.

Williams, John (Glanmor). 1852. *Awstralia a'r cloddfeydd aur*. Denbigh: Thomas Gee.

Ystoryaeu Seint Greal, ed. Thomas Jones. Cardiff: Gwasg Prifysgol Cymru, 1992.

The following texts are cited from the *Historical Corpus of the Welsh Language 1500–1850* (ed. Ingo Mittendorf and David Willis), <http://people.pwf.cam.ac.uk/dwew2/hcwl/hafan.htm>: *Brenin Llur*; Teithie Syr Sion Mandefyl; Huw Morus, *Rhyfel Cartrefol*; 1588 Bible.

Remaining printed texts are cited from contemporary printed editions in Early English Texts Online (<http://eebo.chadwyck.com/home>) or Eighteenth Century Collections Online.

8

Negation in the history of Greek

JO WILLMOTT

8.1 Introduction

Throughout the history of the language, the Greek system of negation is interesting for its complexity, and for its interaction with the independently complex and developing system of modality. This chapter cannot go into great depth about all the research that has been done in this area. Instead it will touch on a few aspects, tying certain key articles to matters of current theoretical interest. Considering data from Homeric, Classical, and Standard Modern Greek (SMG), it will focus particularly on those areas which benefit from a historical approach.

To give a broad outline of the relevant developments of the language, in the ancient language (comprising Homeric Greek, the result of a long oral tradition, finishing in around the 8th century BC, and Classical Greek, a collection of dialects centring around the language of Athens in the 5th century BC), there are two negators, *ou* and *mē*. The existence of two negators continues through the history of the language: in SMG these are *δen* and *min*. In form the negators at each stage are related, although the relationship of *ou* to *δen* is not straightforward, and will be discussed in more detail in section 8.2. The nature of the meaning of, and relationship between, the two negators is one of the most interesting questions and will be discussed in sections 8.3 and 8.4. In the final two sections, issues of theoretical interest connected to each of the negators will be investigated, namely negative concord and the expression of negative imperatives.

8.2 The development of *ou* to *δen*: Jespersen's cycle at work?

8.2.1 *Introduction*

SMG *δen* is clearly functionally equivalent to Ancient Greek *ou*: both are the markers of 'standard' negation, as defined by Payne, namely they both negate declarative main clauses (Payne 1985: 198–201, with further discussion by Miestamo 2007). In this section I will consider the formal relationship between the two negators. Although the development is fairly transparent in outline, I will argue that previous analyses can be improved through a closer consideration of the data. Although Jespersen's cycle has been invoked to describe the developments, there are certain important differences between Greek and the other languages which are used as prototypical examples of the cycle.

8.2.2 The development

In outline, the Classical Greek negator *ou* comes to be replaced by *ouden* (Jannaris 1897: §1796–800). Used in Classical Greek to mean 'nothing', and apparently therefore a negative quantifier, in form *ouden* is made up of the negator itself (*ou*), a particle (*de*), and the neuter form of the word for 'one' (*hen*). It may be found either on its own (1), or strengthening the negator (2).[1]

(1)

οὐδὲν	διοίσεις	Χαιρεφῶντος	τὴν	φύσιν
ūden	dioiseis	k^{h}airep^{h}ōntos	tēn	p^{h}usin
OUDEN	differ.2.FUT	Chairephon.GEN	the	nature.ACC

'you will not in any way differ in nature from Chairephon' (Aristophanes *Clouds* 503)

(2)

οὐ	γὰρ	ᾦζυρὲ	τούτων	ἐπιθυμῶ	μανθάνειν	**οὐδέν**
ū	gar	ōsdure	tūtōn	epit^{h}umō	mant^{h}anein	ūden
OU	PTCL	miserable.VOC	those.GEN.PL	want.1SG	learn.INF	nothing.ACC

'You miserable man, I don't want to learn about anything of those' (Aristophanes *Clouds* 656)

Incidentally, the above evidence reveals that Classical Greek is a negative-concord language, of the type in which a negated verb may be accompanied by a negative indefinite (post verbally), but where the negative indefinite may function as the sole negative element in the sentence, similar to Italian. To use Haspelmath's terminology, it may be described as a (N)V–NI language (Haspelmath 1997: 201) (cf. also section 1.8.2). I will return to the issue of negative concord in section 8.5.

To return to the development from *ou* to *δen*, post-classically *ouden* is found more and more regularly as the standard negator (3). The form soon develops through aphaeresis (loss of first unstressed vowel) to *δen* (4). This is the form of the standard negator in SMG (5).

(3)

Ἰουδαίους	**οὐδεν**	ἠδίκησα
Iūdaiūs	ūden	idikisa
Jews.ACC	OUDEN	wronged.1SG

'I have not wronged the Jews.' (Acts 25.10; from Jannaris 1897: §1798)

[1] In addition to a gloss and rough translation, all examples are accompanied by a phonetic transcription. This is meant merely to help the reader and does not claim to be fully accurate.

(4)

δὲν	ἠξεύρεις...	τὸ	φῶς	τῶν	ὀφθαλμῶν	μου
ðen	iksevris	to	fos	ton	ofθalmon	mu
ΔEN	find.2SG.PAST	the	light	the.GEN.PL	eyes.GEN.PL	me.GEN.SG

'You did not find the light of my eyes.' (Digenes Acritas, Escorial manuscript line 859 *TLG*, 12th century)

(5)

Δεν	το	έγραψα	για	σένα
ðen	to	éγrapsa	ja	séna
ΔEN	it	write.1SG.PAST	for	you.ACC

'I did not write it for you.' (SMG) (Holton et al. 1997: 204)

In spite of the apparent simplicity of this development, a closer analysis reveals that it is more complicated (for further discussion, see Landsman 1988: 20–4). As ever in Greek, dating the development is difficult, due to a persistent drive towards Atticism and an almost constant diglossia of the written and spoken language (Horrocks 1997). This helps to explain how Jannaris can describe *ou* as only becoming obsolete in the nineteenth century (Jannaris 1897: §1797) while Horrocks claims that *ouden* starts to replace *ou* from the sixth century on (Horrocks 1997: 208).

A search of the canon of *Thesaurus Linguae Graecae* (*TLG*) reveals that *δen* is regularly found after the ninth century. Unsurprisingly, there appears to have been a period of variation. For example, the *Chronicle of the Morea* (early fourteenth century 'vernacular' verse, see Horrocks (1997: §12.3.3) for description of the text and other linguistic features) has 331 instances of *ouden* and 117 of *ou*. This contrasts to a ratio in the texts of Plato (5th/4th-century prose) of 1222 examples of *ouden* to 3262 of *ou*, almost exactly the opposite proportion. We may therefore conclude that *ou* is still in use as late as the fourteenth century, although *ouden* is clearly the more common form.

As well as being difficult to date precisely, we may see that the development did not take place at all in some dialects of Greek. For example, in Tsakonian, the negator is *o*, directly from *ou* (Joseph 2001: 350), while in standard Pontic, the negator is *ki* from *ouki* (Horrocks 1997: 312). It would appear that from the variation seen in the medieval texts, some dialects generalized one negator while others generalized another (Io Manolessou, p.c.).

8.2.3 Mechanism of the development

Although the exact details of the timing and localization of the development are not entirely straightforward, we may nonetheless assert that SMG *δen* is the reduced form of *ouden*, which has replaced Ancient Greek *ou* as clausal negator. This has been generally described as an example of Jespersen's cycle, as a strengthened form of the negator (*ouden*) has developed into the basic form of the negator. For example, Roussou claims that *δen* may be compared with French, where *pas*, originally a positive strengthener to the negator, has developed (in colloquial spoken French at any rate) to carrying the negative force by itself (Roussou

2007: 21). Certainly *ðen* appears to continue the 'extra', 'non-negative' elements in the Classical Greek *ouden* (particle *de* + *hen* 'one'). However, the development deserves to be looked at more closely, as various previous analyses may be criticized.

Roberts and Roussou explain the development in terms of their Minimalist theory of grammaticalization and claim that *ouden* moves to a higher position in the clause, from the DP where it modifies the noun, to the CP where it modifies the sentence (Roberts and Roussou 2003: 157–60). They use the following two examples from Classical and post-Classical (Koine) Greek (4th century AD) to exemplify the separate stages in their proposed development (I reproduce their translations):

(6) **οὐδεν** *αὐτῶν* *ἀτιμάσεις*
ūden autōn atimaseis
nothing them.GEN.PL undervalue.2SG
'You will undervalue none of them.' (Plato, *Parmenides* 130e) (Roberts and Roussou 2003: 158)

(7) *ὅτι* **[ο]ὐδὲν** *ἔχωμεν* *μάρτυρων*
oti ūðen exōmen martiron
that none.NEUT.ACC.SG have.1PL witnesses.GEN.PL
'... that we have no witnesses.' (P. Oxy. 1683) (Roberts and Roussou 2003: 158)

Roberts and Roussou argue that in the second example *ouden* has already developed from the negative (adnominal) quantifier they claim is seen in the first example, as it is stranded from its noun. However, these examples are in fact rather problematic. As *ouden* is neuter the translations given are questionable. We would expect to find the masculine *oudena* to express 'not one of them', 'not one witness'. In both of these examples *ouden* instead appears to be functioning adverbially: 'you will not dishonour them (at all)' (with the verb *atimaseis* taking the genitive of the pronoun); 'we do not have (any) witnesses'.

It is true that there are some clear examples of *ouden* functioning as a negative quantifier in the Classical period, for example:

(8) *ὁ δὲ Ἕλιξος καὶ ὁ Κοιρατάδας* **οὐδὲν** *τούτων εἰδότες ἐβοήθουν*
ho de Heliksos kai ho Koiratadas ūden tūtōn eidotes eboēt^{h}ūn
the PTCL Helixus and the Coeratadas n.thing these.GEN.PL knowing hurried
'Helixus and Coeratadas, knowing nothing about what was going on, hurried. . .' (Xenophon *Hellenica* 1.3.21)

Thus, although the examples used by Roberts and Roussou are rather problematic, we might still be able to agree with their account of the development.

However, more problematically still, there are several examples in Classical Greek where *ouden* appears to be adverbial, modifying the verb rather than a noun, as we saw above in example (1), repeated here for ease:

(9) **οὐδὲν** διοίσεις Χαιρεφῶντος τὴν φύσιν
ūden dioiseis k^h^airep^h^ōntos tēn p^h^usin
OUDEN differ.2.FUT Chairephon.GEN the nature.ACC
'You will not in any way differ in nature from Chairephon.' (Aristophanes, *Clouds* 503)

This use of *ouden* may not be analysed so easily as belonging to a determiner phrase. The fact that *ouden* could already be used adverbially in Classical Greek surely contributed to its development to clausal negator and therefore causes difficulties for the account proposed by Roberts and Roussou.

Furthermore, we may see that there are also some difficulties with describing the development from *ou* to *δen* in terms of Jespersen's cycle, or at least of comparing the development with that seen in French. Firstly, it is not clear that in the early stage of the language *oude hen* may be described in the same way as French *ne…pas*.

It is true that the expression *oude hen* is found in Ancient Greek. For example (the reading is guaranteed metrically):

(10) σαφὲς δ' ἂν εἶπεν **οὐδὲ** ἕν—
safes d an eipen ūde hen
clear PTCL PTCL(MODAL) said NEG one
'(lit.) The clear (thing) he would say was not one.' (= 'Nothing he said was clear.') (Aristophanes, *Frogs* 927)

However, it is difficult to ascertain the exact structure of this sentence: is this sentential negation ('he did not say anything clear') or constituent negation ('he said not one clear thing')? In any case, it is a very restricted construction (only twenty-seven examples in a search of *TLG*).

The evidence that *ouden* developed from a construction where *hen* had a positive meaning is therefore doubtful, unlike in French. Certainly, in Classical Greek *ouden* is already an inherently negative quantifier (Roussou 2007: §2.4). When preceded by the negator this may be used to emphasize the negation, as in example (2) above. In this sense Classical Greek is a language like French, where elements are used to strengthen simple negation. But *ouden* is also used alone as a negative adverb and the only negative element in the sentence already in the early stages of the language, as in example (9) above. Word order suggests that it is from this construction that its use as sole negator has developed. Unlike French, as well as several other languages that undergo Jespersen's cycle, therefore, we do not end up with a postverbal negator (cf. Haspelmath 1997: section 8.2.3.1).

Finally, it should be mentioned that the fact that the 'preserved element' (*δen*) was originally non-negative, which again suggests a similarity between this development and the French one, appears to be purely coincidental: the Ancient Greek *ouden* developed to *δen* by a process of aphaeresis that widely affected unstressed initial vowels (Horrocks 1997: 207).

There are thus clear differences between Greek and French. What about other prototypical Jespersen languages? On the basis of the developments in English, Zeijlstra has drawn up certain phases for the development of negation in prototypical Jespersen languages, as laid out in Table 8.1 (Zeijlstra 2004: 56).[2]

It is perhaps tempting to see Greek as an example of a Jespersen language, with *ou* being used first on its own (phase I), then optionally strengthened by *ouden* (phase II?), with *ouden* finally being used as the only negative marker (phase V), presumably having gone through phases III and IV. However, even ignoring the lack of evidence for the intervening phases, we can see various important differences between the developments in Greek and the developments in English. Firstly, we have no evidence that driving force behind the strengthening of *ou* with *ouden* is the phonological weakening of *ou*. Although phonological weakening plays a part in the process from *ou* to *den*, it is only after *ouden* is used as sole negator. In fact, the role of phonological weakening (as opposed to inflationary effects) in the cycle for other Jespersen languages has been questioned elsewhere (e.g. Detges and Waltereit 2002). Secondly, however plausible its existence, there is no clear evidence for a stage when *ou* was used alone as the negator, and we have no clear evidence that the final negative marker (*δen*) has developed from the strengthening element used to reinforce the negator rather than the free-standing negative adverb. Instead we have

TABLE 8.1 **Phases of the development of negation in Jespersen languages**

Phase	Description
Phase I	Negation is only expressed by a single negative marker that is attached to the finite verb.
Phase II	The negative marker that is attached to the finite verb becomes phonologically too weak to express negation by itself and a second negative adverb becomes optionally available.
Phase III	Sentential negation is obligatorily expressed by the negative marker that is attached to the finite verb and the adverbial negative marker.
Phase IV	The negative adverb is the obligatory marker for negation and the use of the negative marker that is attached to the finite verb becomes optional.
Phase V	The negative adverb is the only available negative marker. The negative marker that is attached to the finite verb is no longer available.
Phase VI	The negative marker is available in two forms: it can appear either as a negative adverb or as a negative marker that is attached on the finite verb, though sometimes simultaneously.
Phase VII=I	Negation is only expressed by a single negative marker that is attached to the finite verb.

[2] Joseph (p.c.) points out that several languages have phonologically weak negators, such as Albanian (written in standard Albanian *s'*) thus throwing some doubt on the validity of phase II.

TABLE 8.2 **Phases of the development of negation in Greek**

Phase	Description
Phase I (Classical)	a) Negator (*ou*) is used alone or strengthened by *ouden* b) *ouden* can be used alone as negative adverb
Phase II (Post-classical)	a) *ouden* replaces *ou* as sole negative marker b) aphaeresis affects *ouden* (develops to *den*)

the phases of development given in Table 8.2, each made up of two 'sub-phases', the relative chronology of which is unclear.

Jespersen's cycle has been invoked in another description of the development of negation in Greek. Kiparsky and Condoravdi claim that, throughout the history of the language, a distinction is made between 'plain' and 'emphatic' negation and that the development of the system may be described as several iterations of Jespersen's cycle (Kiparsky and Condoravdi 2006). However, we may again see difficulties with their account.

They outline the developments as laid out in Table 8.3 (a few accents have been corrected and the glosses are mine).

We need to consider carefully what Kiparsky and Condoravdi mean by 'plain' and 'emphatic' negation since they appear to use the terms in different ways over the course of their paper. Since they describe *ou(k)* as a marker of plain negation the following pair of sentences would appear to express what they mean by the terms:

(11) John didn't eat. (= plain negation)

(12) John didn't eat a thing. (= emphatic negation)

However, when explaining the table given above, they assert that it displays the 'plain and emphatic versions of "nothing", "not any" of the modern Cretan dialect and three of its antecedent stages' (Kiparsky and Condoravdi 2006: 1). Thus the sentences should correspond more to the following English sentences:

(13) I didn't eat anything. (= plain negation)

(14) I didn't eat a crumb. (= emphatic negation)

Whether or not it may be shown that the n-words in Greek versions of example (13) above develop into the n-words in the equivalent of example (14), this is important as it reveals that this is rather different to the parallels cited with, for example, English *not* and French *ne*, which are sentential negators developed from 'plain' examples of 'nothing' (English *not* develops from *nawiht*, French *ne* from Latin *nōn*, previously **ne oenum*, see Jespersen 1917: 14–16 and section 1.4). We are thus dealing with developments in the quantifier cycle rather than the negator cycle described as the standard Jespersen cycle.

TABLE 8.3 **Developments in plain and emphatic negation in Greek (Kiparsky and Condoravdi)**

		PLAIN	EMPHATIC
I	Ancient Greek	*οὐ . . . τι* [u . . . ti] 'not . . . anything'	*οὐδὲ . . . ἕν* [ude . . . hen] 'not . . . one'
II	Early medieval Greek	*(οὐ)δέν . . . τι* [(u)den . . . ti]'not . . . anything'	*δέν . . . τίποτε* [den . . . tipote] 'not . . . anything at all'
III	Greek dialects	*δέν . . . τίποτε* [ðen . . . tipote]'not . . . anything'	*δέν . . . τίποτε* [ðen . . . tipote] *δέν . . . πρâμα* [ðen . . . prama] . . . 'not . . . anything at all/a thing'
IV	Cretan	*δέν . . . πρâμα* [ðen . . . prama] 'not . . . anything'	*δέν . . . δροσά* [ðen . . . ðrosa] *δέν . . . ἀπαντοχή* [ðen . . . apandoxi] . . . 'not . . . a drop/a hope'

We may then look in more detail at some of the stages outlined above, and will see some problems with the claims made by Kiparsky and Condoravdi. First, we may consider the development from stage III to IV. In stage III, according to their analysis, *ðén . . . típote* is used to express plain 'nothing', while *ðén . . . práma* is used as emphatic 'nothing'. An example of the second use is given in the following sentence (from Kiparsky and Condoravdi 2006: 10):

(15) *βρίσκει μιὰν κοπέλλα . . . ποὺ* ***δὲν*** *ἤξερε* ***πρâμα***
vríski mian kopélla . . . pú ðén íksere práma
find.3SG one girl who NEG knew thing
'finds a girl who has no clue' (Theran, emphatic)

This is the usage expected from the etymology of the 'strengthening elements': *típote* consists of the elements *ti* 'something/anything' and *pote* 'ever', while *práma* was originally a positive noun, meaning 'thing'.

In stage IV, *ðén . . . práma* is said to have developed to express plain negation, while other items, such as *apandoxi* ('hope') and *drosa* ('drop') express emphatic negation:

(16) *Ἐδώκασί σου πρâμα; —**Ἀπαντοχή**!*
eðókasí su práma apandoxí
give.3PL.PAST you.DAT thing hope
'Did they give you anything? Nothing! ('not a hope!')' (Cretan, plain and emphatic)

(17) *'Έφαες πρᾶμα; —Δροσά!*
efaes prama ðrosa
eat.2SG thing dewdrop
'Did you eat anything? Nothing! ('not a dewdrop!')' (Cretan, plain and emphatic)

However, unfortunately the examples given are not sufficient for the claims about the stages of the development, since they do not come from different periods of the language but rather from different dialects. What we may certainly see from these examples is that *práma* is a weak negative polarity item in Cretan (at the date of the examples), while it appears to be an emphatic negator in Theran. However, no examples are given to show that *práma* is used to express 'plain nothing' in Cretan. What these data therefore show is that, in the case of *práma*, a word which is used in one dialect as an emphatic indefinite form is used in another as a weak polarity item. If this dialectal diversity does in fact reflect a diachronic development within one dialect, we have already seen that it is a case of a weakening of the quantifier, rather than an example of Jespersen's cycle proper, where the original strengthener comes to take over the functions of the negator alone.

8.2.4 Conclusion

The development of *ou* to *δen* in Greek provides an interesting example of the development of expressions of sentential negation. What originally appears to be a negative quantifier develops into a sentential negator, which in subsequent periods of the language is strengthened by various different elements. The developments have been described in various different accounts as examples of Jespersen's cycle, but I have here argued that a closer analysis shows that such a claim is not straightforward. In the case of the examples considered by Kiparsky and Condoravdi, they illustrate developments in quantifiers rather than developments in the expression of sentential negation. That is, we find merely a serial weakening and replacement of the strengthening element. And although the replacement of *ou* by *ouden* shares certain features with Jespersen's cycle it does not go through the phases noted for the prototypical languages and is different in important ways from the developments seen in French and English.

This development will be considered again in section 8.5.4, where I will argue that the detailed history of the negator is important for understanding its use at various stages, and may indeed account for some of the differences we may note between the modern and ancient stages of the language.

8.3 The development of *min* from *mē*: a constellational approach?

8.3.1 *Introduction*

Formally there are no significant problems in the development of *mē* to *min*. The change in vowel quality occurred in the later Roman or early Byzantine period (Horrocks 1997: 109). The only area of uncertainty is the presence of the [n] (which only occurs in certain phonological/morphological contexts, as we will see in further detail below). One explanation is that it arose by analogy with the ending of *δen* (Janda and Joseph 1999: 347). Whatever the explanation for this element, it is clear that *mē* and *min* are formally closely related.

The more interesting question is the meaning of the negator, and how it changes over time. For this negator we will consider the later period first, since there is a recent article dedicated to the subject. Janda and Joseph have argued that SMG *mi(n)* should be understood as a 'morphological constellation' of ten different elements rather than a single entity (Janda and Joseph 1999. Some of the data and argumentation are reproduced in Joseph 2002). By the term 'morphological constellation' they mean a 'group of elements which share at least one characteristic property of form but are distinguished by individual idiosyncrasies—of both form and function—that prevent their being collapsed with one another' (Janda and Joseph 1999: 343). In this case, the ten elements are argued to share a common formal core [*mi*] and a functional core relating in some way to negation.

I will argue that, although they make an important point in realizing that we should not treat the negator as a monolithic entity, the number of different elements argued for by Janda and Joseph is too large. I will then examine the negator in Ancient Greek and argue that here too we may distinguish more than one use, and that these uses will help to explain the uses found in SMG.

8.3.2 *Examples in SMG*

The ten different elements argued for by Janda and Joseph are described and exemplified as follows (Janda and Joseph 1999: 344–7; I have reproduced their terminology, spelling, and glosses):

a. negator of subjunctive clauses

(18) *μπορεί να μην έχουν κοιμηθεί*
borí na min éxun kimiθí
can.3SG SUBJUNC MI have.3PL slept
'It is possible that they haven't gone to bed yet.'

(19) *ας μην έρθει τώρα ο Γιάννης*
as min érθi tóra o jánis
SUBJUNC MI come.3SG now the John.NOM
'Let John not come now.' or 'John should not come now.'

b. negator of active participles

(20)

μην	έχοντας	ιδέα	για	όλα	αυτά,
min	éxondas	iðéa	ja	óla	aftá,
MI	have.ACC.PTCP	idea.ACC	about	all	these
ο	Γιάννης	την	παντρέυτηκε		
o	jánis	tin	pandréftike		
the	John.NOM	her.ACC	married.3SG		

'Not having any idea about all these things, John married her.'

c. pleonastic negator in clausal complements of verbs with negative force (e.g. preventatives)

(21)

φοβάμαι	να	μην	έρθει
fováme	na	min	érθi
fear.1SG	SUBJUNC	MI	come.3SG

'I am afraid that he may come.' (NOT 'I am afraid he may not come.')

(22)

δε	σε	εμποδίζω	να	μην	μιλάς
ðe	se	emboðízo	na	min	milás
NEG	you.ACC	prevent.1SG	SUBJUNC	MI	speak.2SG

'I do not prevent you from speaking.' (NOT 'I do not prevent you from not speaking.')

d. negator of imperatives and hortatives (i.e. introducer of prohibitives)

(23)

μην	το	πετάξεις
min	to	petáksis!
MI	it.ACC	throw.2SG

'Don't throw it out!'

(24)

μην	ξεχνάμε	πως	ο	Γιάννης	είναι	ακόμα	έκει
min	ksexnáme	pos	o	jánis	íne	akóma	ekí
MI	forget.1PL	that	the	John.NOM	is.3SG	still	there

'Let's not forget that John is still there!'

e. introducer of negatively evaluated clausal complements to verbs and nouns of fearing (with variant [*mípos*/*míbos*])

(25)

το	έσκασε	από	φόβο	μην	τον	χτυπήσουν
to	éskase	apó	fóvo	min	ton	xtipísun
it.ACC	burst.3SG	from	fear.ACC	MI	him.ACC	beat.3PL

'He ran off for fear that they might beat him.'

(26)	*φοβάμαι*	*μην* / *μήπως*	*έρθει*
	fováme	min / mípos	érθi
	fear.1SG	MI	come.3SG

'I fear that he might come.'

f. introducer of tentative main-clause questions (with variant [*mípos*/*míbos*])

(27)	*μην* / *μήπως*	*είδες*	*το*	*παιδί;*
	min / mípos	íðes	to	peðí?
	MI	saw.2SG	the	child.ACC

'Did you perhaps (happen to) see the child?'

g. independent utterance expressing negative actions (i.e. prohibitions)

(28) *μη!*
mi
MI
'Don't!'

h. negator of lexical items (ones that are not fully verbs)

(29)	*γύρισε*	*δυο*	*μη*	*εμπορικά*	*φιλμ*	*μαζί*	*του*
	jírise	ðjo	mi	emboriká	film	mazí	tu
	turned.3SG	two	MI	commercial	films.N.ACC	with	him

'He shot two non-commercial films with him.'

i. negator of ellipted (i.e. 'understood') elements

(30)	*παρκαρισμένα*	*και*	*μη*	*αυτοκίνητα*	*ήταν*	*παντού*
	parkarizména	ke	mi	aftokínita	ítan	pandú
	parked.NPL	and	MI	automobiles.N	were	everywhere

'Parked and unparked cars (i.e. 'cars that are parked and (ones that are) not (parked)') were everywhere.'

(31)	*μη*	*τα*	*χέρια*	*σου*	*έξω*
	mi	ta	xérja	su	ékso
	MI	the	hands.ACC	your	outside

'Don't (put) your hands out!'

j. negative combining-element in isolated derivational word-formations

(32) *μήτε* (*míte*) 'not even; neither' (cf. *ούτε* (*úte*) 'not even; neither')

(33) *μηδέν* (*miðén*) 'nought, zero'

(34) *μήπως* (*mípos*) 'lest', 'perhaps' (cf. complementizer (*πως* (*pos*) 'that').

Janda and Joseph isolate five different 'formal' features, of which their different elements share a certain subset, as follows:

i. whether a final (assimilating) *-n* is allowed[3]
ii. whether the element is a bound or a free form
iii. whether the element occurs syntactically in COMP (the complementizer-node)
iv. whether the element occurs primarily with verbs or instead (regularly) with other word-classes
v. whether the element has a semantic force that is strongly negative or instead only weakly so or even only indirectly associated with negativity.

The features and their distribution amongst the ten elements isolated by Janda and Joseph are represented in a table given here as Table 8.4.

8.3.3 *Discussion of claims for SMG*

The methodology used to derive this 'constellation' may be questioned on various levels. We may first ask the status of the 'derivational' uses of *min* (j). While these are clearly developed from the non-derivational use it is not clear that they should be considered as a single type (the fact that two of the 'formal features' may not be stated in this instance would appear to confirm this). In any case, they would appear to have a rather different status from the other uses.

We may then consider the formal differences. It is not in fact clear in what sense the fifth (strength of negation) counts as formal at all. We may also question its

TABLE 8.4 **Functional elements and formal features of *min* in Standard Modern Greek (Janda and Joseph 1999)**

		final n	bound	C^0	pre verbal	strong
a.	subjunctive	+	+	–	+	+
b.	participial	±	+	–	+	+
c.	pleonastic	+	+	–	+	–
d.	imperative	+	+	+	+	+
e.	complementizer	+	+	+	+	±
f.	interrogative	+	+	+	+	–
g.	prohibitive	–	–	+	–	+
h.	lexical	–	+	–	–	+
i.	elliptical	–	–	±	–	+
j.	derivational	±	+	–	–	±

[3] When allowed, this *-n* appears regularly before vowels (see (19) above), and variably before nasals and fricatives (it is 'generally omitted in fast speech but is possible in more careful articulation' (Janda and Joseph 1999: 347), so that μην μιλάς (as in sentence (22)) above can surface as [*mi milás*]). It shows various effects before stops (for further information see Janda and Joseph 1999: n. 10).

use as a diagnostic feature on semantic terms: while it appears straightforward to state that the negative force of an imperative (d) is 'stronger' than that of fear clauses (e) and pleonastic constructions (c), both constructions are common environments in which to find negators cross-linguistically, for example in French *je crains qu'il ne vienne* (now usually known as 'expletive negation', though I will retain the term 'pleonastic' here, see Rowlett 1998: 27–8). Given that this difference in 'strength' is the only way in which the pleonastic element (c) can be argued to differ from the subjunctive element (a), the distinction between them may therefore be questioned.

Similarly, it is unclear whether the interrogative element (f) may be clearly separated from the imperative element (d), from which it again only differs in the 'strength' of the negative force. The difference between the two types appears clear when considering the translations used by Janda and Joseph ('Don't throw it out' vs 'Did you perhaps (happen to) see the child'). In the interrogative sentence there seems to be no element of negation at all. However, in English too we can use a negative element in such a sentence, for example 'You didn't see the child, did you?'. While this negative element is being questioned, and therefore could be described as weaker, the difference could be explained in pragmatic rather than strictly semantic terms. In any case, this does not appear to be a clear formal criterion for distinguishing between different uses.

Moving on to the other four formal features, while the first (final *-n*) is fairly objective and verifiable, the other three are more questionable. For example, as a one-word construction the 'prohibitive' element (g) is necessarily unbound and not preverbal. If we take these features out of the equation, the difference between this and the imperative use (d) is reduced to the presence or absence of final *-n*. Even this difference (and therefore the justification for claiming a separate meaning) may be questioned: given that the prohibitive is necessarily followed by a phonological pause, we do not expect to find the assimilating *-n* as found in the imperative construction, as its presence depends on the nature of the following element.

A similar point may be made for the elliptical element (i). Necessarily not bound and not preverbal in its surface form, its analysis depends on the theoretical analysis of ellipsis used. Again the lack of final *-n* in this use could be explained if we posited a phonological pause in the place of the ellipsis. It would seem more reasonable to explain elliptical uses of the negator by considering what element precisely has been elided.

The claim that one or other of the elements resides in the C node depends again on one's theoretical point of view. The structure of the left periphery in SMG is a matter of debate, and several different models have been put forward (for a summary of these models, see Roussou 2000). It is notable that in each of these models the negator has a different relationship with the complementizer node. Given the unresolved nature of the claims about the complementizer phrase and the relative position of the negator in Greek, Janda and Joseph's claim about the position in C^0

remains rather subjective. It would seem advisable to change this formal feature to whether or not the element is clause-initial.

I therefore conclude that the criteria used by Janda and Joseph to distinguish ten different uses for *min* are not secure. I believe that it is not shown conclusively that we should separate the derivational, elliptical, prohibitive, interrogative, and pleonastic uses, for reasons given above. I have thus reduced the ten uses argued for by Janda and Joseph to five, namely the imperative, complementizer, subjunctive, participial, and lexical uses. (We might query further the distinction between the participial and lexical uses since in both cases this appears to be an instance of 'constituent negation', the difference only depending on the nature of the constituent. However, given the formal difference I have left the two uses separate.)

Given the description of a morphological constellation as a group of elements which are 'distinguished by individual idiosyncrasies' it is perhaps more significant for my critique of Janda and Joseph's account that I have argued against the validity of using the features 'bound', 'preverbal', and 'strong'. This leaves two formal criteria: the presence or absence of assimilating *-n* (which I only consider apparent in cases where the negator is followed by another word), and whether or not the element is clause-initial. The table of formal and functional elements may therefore be redrawn as Table 8.5.

In this account the five uses are no longer distinguished by formal criteria. This throws some doubt on the claim that 'constellation' may be described as being 'morphological'. Nonetheless, I agree with Janda and Joseph in saying that we may distinguish different uses of *min* in SMG, if only on semantic and in some cases syntactic grounds. Furthermore, as we will see, these different uses have interesting counterparts and origins in Ancient Greek.

8.3.4 *The situation in Ancient Greek*

Janda and Joseph explain the constellation of uses of *min* that they see in SMG as creations by speakers 'out of earlier more unified situations' (Janda and Joseph 1999:

TABLE 8.5 Functional elements and formal features of *min* in Standard Modern Greek, redrawn

		final -n	clause-initial
a/c	subjunctive/pleonastic	+	–
b	participial	±	–
d/g/f	imperative/prohibitive/questions	+	+
e	complementizer	+	+
h	lexical	–	–

350). However, I will show that each of the uses they observe in SMG *min* has counterparts in Ancient Greek *mē* and that in fact there are even further uses. These uses may not be mapped across to clear differences in surface form, and there are therefore again no grounds to describe *mē* as a 'morphological constellation'. Even on functional grounds alone it is not desirable to claim that each of these uses necessitates a separate element. So, just as we have seen for SMG, I will argue that the distinguishable elements are fewer than Janda and Joseph might claim. Nonetheless, the complexity of the negator deserves to be examined further. Indeed, a comparison between the two stages of the language would appear to help explain the multiplicity of the uses in SMG.

8.3.5 Uses of mē *which correlate to uses of* min

We may begin with those uses of the negators in Ancient Greek which correlate to those in SMG, summarized in Table 8.6.

The following are examples of each of these uses.

a. negator of 'subjunctive' clauses (final clauses (a1), and wishes with the optative (a2))

(35) *ἄγε δὴ σὺ φράσον ἐμοὶ σαφῶς πρὸς τουτονί, ἵνα*
age dē su p^{h}rason emoi sap^{h}ōs pros tutoni hina
come.2.IMP PTCL you say.2SG.IMP me.DAT clearly to this so.that
μή σε βάψω βάμμα Σαρδιανικόν·
mē se bapsō bamma sardianikon
MĒ you.ACC dip.1SG.SUBJUNC dip Sardian
'Come now, answer me clearly on this question, so that I do not dip you in Sardian dip.' (Aristophanes, *Acharnians* 110)

(36) *εἰ γὰρ μὴ νύμφαι γε θεαὶ Βάκιν ἐξαπάτασκον*
ei gar mē nump^{h}ai ge t^{h}eai bakin exapataskon
if PTCL MĒ nymphs PTCL gods Bacis fooled.3.PL
'if only the nymphs had not fooled Bacis' (Aristophanes, *Peace* 1070)

b. negator of active participles

(37) *οὐκ ἂν δύναιο μὴ καμὼν εὐδαιμονεῖν*
ūk an dunaio mē kamōn eudaimonein
OU PTCL can.2.OPT MĒ toiling.MASC.NOM be.happy
'You would not be able to be happy if you hadn't toiled.' (Xenophon, *Anabasis* 1.2.22)

TABLE 8.6 **Uses of the negator in Ancient Greek with their counterparts in Standard Modern Greek**

	Environment	Form
a.	negator of subjunctive clauses	a1. *Mē* is found in final clauses (introduced by complementizer *hina*, the ultimate origin for the SMG subjunctive marker *na*). a2. *Mē* is also found in wishes, though these are expressed by the optative or indicative in Ancient Greek.
b.	negator of active participles	both *mē* and *ou* are found negating participles (for more on their distribution, see section 8.3.6)
c.	pleonastic negator in clausal complements of verbs with negative force	also found in Ancient Greek
d.	negator of imperatives and hortatives	also found in Ancient Greek
e.	introducer of negatively evaluated clausal complements to verbs and nouns of fearing	also found in Ancient Greek
f.	introducer of tentative main-clause questions	also found in Ancient Greek
g.	independent utterance expressing negative actions (i.e. prohibitions)	also found in Ancient Greek (rare)
h.	negator of lexical items	also found in Ancient Greek, but may also be negated with *ou* (see section 8.3.6)
i.	negator of understood elements ellipted	also found in Ancient Greek
j.	negative combining element in isolated derivational word formations	More common than in SMG, in Ancient Greek, a quick search of the standard dictionary revealed at least sixteen of these compound forms. For example, μήτε (mēte) 'and not'; μηδείς (mēdeis) 'no-one'; μήποτε (mēpote) 'never'; μηδαμῶς (mēdamōs) 'in no way'; μήτις (mētis) 'no-one'.

c. pleonastic negator in clausal complements of verbs with negative force

(38) *οὐ φυλάξεσθ' ὅπως μὴ… δεσπότην εὕρητε;*
ū p^hulaxest^h hopōs mē… despotēn heurēte
OU guard.2PL.FUT so.that MĒ master.ACC find.2PL.SUBJUNC
'Beware! You might find a master!' (Demosthenes, *Speeches* 6.25)

use of negators with participles and nouns and is thus arguably less unified rather than more unified as Janda and Joseph argue.

In the protasis of conditional clauses the usual negator in Ancient Greek is *mē* (Smyth 1956: §2702). This contrasts with SMG, where the usual negator is *δen* (Holton et al. 1997: 457). For example:

(49) *εἰ μὴ βούλεται Πρωταγόρας ἀποκρίνεσθαι, οὗτος μὲν ἐρωτάτω,*
ei mē bouletai prōtagoras apokrinest^{h}ai, houtos men erōtatō,
if MĒ wants Protagoras answer.INF he PTCL ask.3.IMP
ἐγὼ δὲ ἀποκρινοῦμαι,
egō de apokrinoumai
I PTCL answer.1.FUT
'If Protagoras is not willing to answer, let him put the questions, and I shall answer.' (Plato, *Protagoras* 338c7)

There are some examples of *ou* used in the context of conditional protases in Ancient Greek. In the majority of instances *ou* appears to be being used as a constituent negator rather than as a sentential negator. Found particularly frequently with verbs of saying, thinking, or wanting, 'adherescent *οὐ*' in Smyth's terms gives the opposite meaning of the verb (Smyth 1956: §2691–2, and see further Landsman (1988: 18). For example:

(50) *εἰ δ' ἀποστῆναι Ἀθηναίων* ***οὐκ*** *ἠθελήσαμεν…, οὐκ ἠδικοῦμεν*
ēi d apostēnai at^{h}naiōn ūk ēt^{h}elēsamen… ūk ēdikūmen
if PTCL revolt.INF Athenians OU want.1PL OU do.wrong.1PL
'but if we refused to revolt from the Athenians, we were not doing wrong'
NOT 'but if we didn't want to revolt…' (Thucydides, *Histories* 3.55)

However, there are also examples in conditional protases where *ou* is clearly sentential. Often these are 'quotes' of what could be strongly asserted (Smyth 1956: 2698b). For example:

(51) *εἰ, ὡς νῦν φήσει,* ***οὐ*** *παρεσκεύαστο*
ei hōs nūn p^{h}ēsei ū pareskeuasto
if as now say.FUT.3SG OU made.preparations.3SG
'if, as he will soon assert, he had not made preparations' (Demosthenes, *Speeches* 54.29)

There are rather more examples of *ou* in conditional protases in Homer, usually with the indicative. Smyth notes two explanations which have been given for this phenomenon, firstly as a 'retention of the original use' of distributing the negators according to mood, and secondly as '*οὐ* went with the predicate, whereas *μή* was closely attached to *εἰ*' (Smyth 1956: 2699a). I have argued elsewhere that in origin *mē* may have been used because of its wider scope, being often used in 'unless' conditionals in Homeric Greek (Willmott 2007: 209–10). Basset has also investigated the opposition between the two negators in conditional protases, concluding that they are used with different 'conditions de vérité' (Basset 1989: 57). While the exact difference between them remains rather elusive, what is clear is that the modern language has changed from the ancient language in this respect.

Another difference between the two stages of the language may be found in relative clauses. In Ancient Greek there is a choice of negator. According to Smyth, *ou* is used when the antecedent is definite, and *mē* when it is generic (Smyth 1956: 2705d and g). For example:

(52)	προσημαίνουσιν	ἅ	τε	χρὴ	ποιεῖν	καὶ	ἃ	**οὐ**	χρή
	prosēmainūsin	ha	te	k^{h}rē	poiēn	kai	ha	ū	k^{h}rē
	signify.beforehand.3.PL	REL	PTCL	must	do.INF	and	REL	OU	must

'They signify beforehand what must be done, and what must not.' (Xenophon, *Cyropaedia* 1.6.46)

(53)	ἃ	**μὴ**	οἶδα	οὐδὲ	οἴομαι	εἰδέναι
	ha	mē	oida	ūde	oiomai	ēdenai
	REL	MĒ	know.1SG	OU	think	know.INF

'What I do not know, I do not even think I know.' (Plato, *Apology* 21d.)

In SMG, too, relative clauses may be negated with *δen* or *min*, but the 'modality' is made explicit by the presence or absence of *na* (Holton et al. 1997: 447).

With the infinitive the usual negator in Ancient Greek is *mē*, the regular exception being in indirect speech, where the infinitive is 'representing' the original indicative (Smyth 1956: 2711–27), for example:

(54)	εἰκὸς	μέντοι	σοφὸν	ἄνδρα	**μὴ**	ληρεῖν
	eikos	mentoi	sop^{h}on	andra	mē	lērein
	proper	PTCL	wise	man	MĒ	talk.idly.INF

'It is proper for a wise man not to talk idly.' (Plato, *Theaetetus* 152b)

(55)	ἐμοὶ	δὲ	δοκοῦσιν	οὗτοι	**οὐ**	τὸ	αἴτιον	αἰτιᾶσθαι
	emoi	de	dokūsin	hūtoi	ū	to	aition	aitast^{h}ai
	me.DAT	PTCL	seem.3.PL	these	OU	the	cause	blame

'I think that these people do not blame the real cause.' (Plato, *Republic* 329b)

As the infinitive is lost from the language during the post-classical period (Horrocks 1997: 4.6, Joseph 1983), we have no comparable usage in SMG.

As we have seen, active participles are negated with *mē* in SMG. Again, the use in Ancient Greek appears rather different. Firstly, there we also find middle and passive participles which may be negated with *mē*. For example:

(56)	οἱ	μὴ	δυνάμενοι
	hoi	mē	dunamenoi
	the.MASC.PL	MĒ	can.PTCP.PRES.MASC.PL

'any who are not able' (Xenophon *Anabasis* 4.5.11; from Smyth 1956: §2734)

Secondly, in Ancient Greek we may find *ou* and not *mē* negating participles, for example:

(57)	*Φαρνάβαζος,*	**οὐ**	*δυνάμενος*	*συμμεῖξαι*	*πρὸς*	*τὸν*	*Ἱπποκράτην*
	P^{h}arnabasdos,	ū	dunamenos	summeixai	pros	ton	Hippokratēn
	Pharnabasus,	OU	can.PTCP.PRES.MASC.SG	agree.INF	to	the	Hippocrates

'Pharnabazus, unable to come to terms with Hippocrates...' (Xenophon, *Hellenica* 1.3.6)

The difference is apparently related to uses that *ou* and *mē* have in other constructions. Smyth claims that the negator used with a participle is '*ou* when it states a fact, *mē* when it states a condition' (Smyth 1956: 2728). We could therefore explain this split as a kind of contamination from the constructions that the participle is 'standing for'. We could rewrite the participle from the first sentence above (51) as a conditional clause ('if any are not able'), in which case the negator *mē* would be required. The second participle, on the other hand, could be rewritten as a finite causal clause ('since he was not able').

Finally, we may compare the use of negators with nouns. In SMG, nouns may not be negated with *ðen* but only *min* (or *oxi*) (Thanasis Giannaris, Dimitris Michelioudakis, p.c.), as we have seen exemplified as the lexical use in example (29) above. In contrast, Smyth has shown that both negators negate nouns in Ancient Greek. He claims that generic nouns are negated with *mē*, while non-generic nouns are negated with *ou*, pointing out that here *mē* is more common than *ou*, giving the following examples (Smyth 1956: 2735):

(58)	*ἡ*	*τῶν*	*γεφυρῶν...*	**οὐ**	*διάλυσις*
	hē	tōn	gep^{h}urōn...	ū	dialusis
	the.SG	the.GEN.PL	bridge.GEN.PL	OU	destruction

'the non-destruction of the bridges' (Thucydides, *Histories* 1.137)

(59)	*ὁ*	**μὴ**	*ἰατρός*
	ho	mē	iatros
	the	MĒ	doctor

'a non-doctor' (Plato, *Gorgias* 459b)

The distribution of negators with nouns therefore compares with their distribution with participles. These additional uses are summed up in Table 8.7.

TABLE 8.7 **Differences between Ancient Greek and Standard Modern Greek in uses of *mē***

		clause-initial	notes	SMG
b2	all participles	–	Comparable to other constructions, e.g. relative, conditional. Can use *ou*.	only active negated by *mē*
h	nouns	–	Can use *mē* or *ou* depending on meaning.	only *min*
k	conditional protases	–	Usually *mē*, though occasionally *ou*.	only *en*
l	infinitive	–		infinitive does not survive
m	relative	–	Can use *mē* or *ou* for generic vs specific.	*en/min* depending on *na*

8.3.7 Discussion of uses in Ancient Greek

Although I have discussed an additional four uses for *mē*, as well as showing that the ten uses of *min* found by Janda and Joseph are also expressed in Ancient Greek, I will again argue that it is unwarranted to claim that all fourteen of these correspond to separate 'elements'. For the same reasons given above I will claim that there is no need to distinguish the derivational, elliptical, prohibitive, interrogative, and pleonastic uses. This then leaves us with nine uses of the negator in Ancient Greek, as summed up in Table 8.8.

TABLE 8.8 **Uses of *mē* in Ancient Greek**

	construction	clause-initial
a1	final clauses	–
b2	all participles	–
d/g/f	imperative/prohibitive/questions	+
a2	wish	+
e	complementizer	+
h	lexical	–
k	conditional protases	–
l	infinitive	?
m	relative	–

In Table 8.8 I have noted whether or not the element is found clause-initially (the same distinction as drawn in SMG). The status of the infinitive construction is questionable as it depends whether infinitives are interpreted as belonging to the same clause as the main verb: if not, *mē* could be explained as being clause-initial in this use.

This feature leads me to claim that these nine uses may be grouped into three broad categories. Of the clause-initial uses, it seems plausible to draw a distinction between the complementizer (and possibly infinitive) use and the others (imperative/prohibitive/questions and wishes). In the latter the negator is not only clause-initial but sentence-initial, and appears to be connected with the specific speech act in question (see further section 8.4).

The use of the negator as a complementizer has been argued to have derived from its use in imperatives, with fear clauses and negative purpose clauses originally paratactic (Chantraine 1948: §397, also see Willmott 2007: 156). Some examples of this type of clause in Homer may still be interpreted in this way. For example:

(60)

ἀλλ'	ἄγε	νῦν	ἐλέαιρε	καὶ	αὐτοῦ	μίμν'	ἐπὶ	πύργῳ,	μὴ
all	age	nun	eleaire	kai	autu	mimn	epi	purgō	me
but	go.2SG.IMP	now	pity.IMP	and	here	stay	on	wall	MĒ

παῖδ'	ὀρφανικὸν	θήῃς	χήρην	τε	γυναῖκα
paid	orp^{h}anikon	t^{h}ēēs	k^{h}ērēn	te	gunaika
child	orphan	make.2SG.SUBJUNC	widow	PTCL	wife

'Wait, take pity and stay here on the wall, so you don't make an orphan of your child and a widow of your wife', or, 'wait... do not make your child an orphan' (Homer, *Iliad* 6.432)

However, already at this stage of the language we have examples in which the negator can only be interpreted as a complementizer, for example:

(61)

αἰεί	μιν ποτὶ	νῆας ἀπὸ	στρατόφι	προτιειλεῖν	ἔγχει	ἐπαΐσσων,
aiē	min poti	nēas apo	stratop^{h}i	protiēlēn	egk^{h}ei	epaissōn,
always	him against	ships from	army	press	spear.DAT	rush.PTCP

μή	πως	προτὶ	ἄστυ	ἀλύξῃ
mē	pōs	proti	astu	aluxē
MĒ	at.all	to	city	escape.3SG.SUBJUNC

'Keep pressing him against the ships, away from his forces, and rush at him with your spear so he can't escape back to town.' (Homer, *Iliad* 10.348)

These first two sets of uses are therefore possibly historically related, although the exact details of the development remain obscure.

The remaining uses (i.e. non-clause-initial) may seem to form a rather more nebulous set. Nonetheless, they do share a certain feature, namely that in all of these contexts it is possible (if only rarely in some instances) to find *ou* as well as *mē*.

In section 8.4 I will return to the difference between the two negators in more detail and will argue that there is a semantic and possibly syntactic difference between the two, in some way connected to the level of 'irrealis' in the clause. For the time being, we may note that these uses do share a common feature and are not just 'the rest'.

We are thus left with three different 'sets' of usages of *mē* in Ancient Greek:

α: speech act
β: complementizer
γ: contrast with *ou*

Just as the different uses distinguished for *min* by Janda and Joseph, these all share certain features. In addition to the negative meaning, we may describe all of them as in some way 'irrealis': in addition to the (*γ*) category, speech acts are by their nature irrealis, and *mē* is only used as the complementizer of 'modal' subordinate clauses (non-modal complementizers being *hoti* or *hōs*). But although they share similarities there is a clear distinction between the three uses which may be described in semantic or syntactic terms. It therefore seems that *mē* is syntactically and semantically complex just as *min* is.

It will be instructive to compare these three uses with the five I distinguished for SMG. For ease of comparison I have restated the five uses in SMG here in Table 8.9.

The first category in SMG (A), would appear to continue the imperative category in Ancient Greek (*α*) unproblematically. The second (B) again appears to map fairly closely to its Ancient Greek counterpart (*β*). The third (C) appears to spring from one of the uses where there is a contrast with *ou* (*γ*), in particular the use in subordinate clauses. This is now rather more 'grammaticalized', and clearly connected with the modality of the clause, as *mē* is always required after *na*, and not possible in other subordinate contexts: there is therefore no longer strictly a choice between the two negators in this context. The fourth (D) and fifth (E) might be best described as 'relics' of the ancient use. Again, however, there is no longer a choice of which negator to use. Instead a usage in ancient Greek which was explicable in semantic terms appears to have been generalized according to syntactic context. The particular origin of *δen* may also be relevant in this context (see section 8.2.2): its development

TABLE 8.9 Uses of *min* in Standard Modern Greek restated

A	Imperative
B	Complementizer
C	Subjunctive
D	Participial
E	Lexical

from negative quantifier might explain why it is not used with nouns and why therefore *min* has been generalized in this use.

8.3.8 Conclusion

Janda and Joseph distinguish a large number of different uses for *min* in Modern Greek, arguing that they form a 'morphological constellation'. I have argued that the methodology behind their conclusions may be questioned in various important respects. Nevertheless, using syntactic and semantic criteria it is clearly necessary to distinguish more than one use for both *min* and *mē*. I have argued that we can see nine different uses of *mē* in Ancient Greek (which may be grouped into three sets), and five uses of *min* in SMG. I have also argued that a consideration of the complex situation in the ancient language may shed some light on the nature of the different uses and their relationship in the modern language. We now need to consider the meaning of *mē/min* with respect to the other negator.

8.4 The difference between the two negators

As previously stated, one of the key issues in considering Greek negation is establishing the difference between the two negators, *ou* and *mē* in Ancient Greek, and *δen* and *min* in SMG. Now we have seen how these develop individually over the course of the language we are in a better position to consider their relative meanings.

An explanation of their meaning might begin by considering their etymology. The two negators are compared with the two negators reconstructed for Proto-Indo-European, namely *ne and *mē (Moorhouse 1959: 12). In the proto-language the former is said to have been used to negate statements, and the latter prohibitions. However, while Greek *mē* appears directly cognate with Sanskrit *mā́* and Armenian *mi* and is formally the direct descendant of the Proto-Indo-European 'prohibitive' marker *mē, the etymology of *ou* is more difficult. It is found in several forms:

οὐ (ū)—before a consonant
οὐκ (ūk)—before an unaspirated vowel
οὐχ (ūk^{h})—before an aspirated vowel
οὐκι (ūki)—emphatic form

It is thus clearly not a direct descendant of *ne. However, there is one account which claims that *ne played a part in the formation of *ou*. Cowgill (1960) proposed a phrasal origin for the negator. On the basis of comparison with Armenian *oč*, he claims that the [k] element is part of the basic form. He then derives this *ouk* from a 'pre-Greek phrase' *ne oiu kwid, made up of the elements *ne (negator) + *oiu ('life, age') + k^{w}id ('something'), meaning something like 'not ever in my life'. Joseph (2005) supports this etymology, claiming the use of the modern Albanian word *jetë* 'life' in the phrase *përjetë* 'forever' uses the same metaphor. This would apparently be an example of Jespersen's cycle, where a sentence negator is strengthened and the strengthening element then takes on the negative meaning, with the original negator itself finally dropping out of use.

However, this derivation is not universally accepted. Lehmann states that the origin of *ou* is unknown (Lehmann 1974: 4.3.3), and Landsman claims that it is 'etymologically puzzling' (Landsman 1988: 15). Clackson argues that Greek *ou* and Armenian *oč* must be explained in different ways (Clackson 1994: 158). It should also be noted that there have been other explanations for the [k] element in this form. For example, Ruijgh (1992) claims that it has arisen due to a missegmentation between the negator and the modal particle *kán* (seen in Ionic *an*) (see further Chantraine 1948: §503 and Willmott 2007: 200). Such an explanation for the modal particles would not be possible if the etymology of *ou* proposed by Cowgill was correct.

In spite of the disagreements about the actual etymology of the forms, semantically the two Greek negators are said to preserve the distinction claimed for Proto-Indo-European, namely a negator of statements as against a negator of prohibitions. So, for example, Lehmann says that 'despite the difference in surface form, ... the functions of the negative pair correspond closely with those in Sanskrit' (Lehmann 1974: §4.3.3). However, such claims do not correlate with the data: we have already seen that Ancient Greek *mē* does much more than negate imperatives. Thus the continuance of a binary distinction from PIE to Ancient Greek does not mean that the distribution of each negator has stayed the same.

I will argue that the same may be said about the development from Ancient to Modern Greek. This is in spite of the fact that at first glance the distinction may seem to remain rather similar, even if it is not the declarative versus imperative distinction claimed by Lehmann. After all, in the modern language *min* is found in *na* clauses, and *δen* elsewhere; in Homeric and Classical Greek, *mē* negates imperatives, wishes, and purpose clauses, while *ou* negates assertions. We might then conclude that there is little change in the function of the two negators through the history of Greek, and that one negator is used for +modal contexts and one negator for −modal contexts. This indeed is suggested by Zeijlstra, who claims that *δen* is marked as [−IRR] and *min* as [+IRR] (Zeijlstra 2006: 419). However, a closer look at the data will show again that it is not so straightforward.

Firstly, in both Homeric and Classical Greek it is not the case that a particular negator is associated with a particular mood. For example *mē* is used with the future indicative in purpose clauses (Philippaki-Warburton 2004: 794). In Homeric Greek we also find *ou* with the subjunctive (Smyth 1956: 2707a). These uses could lead one to the conclusion that the ±modal semantic environment does not overlap straightforwardly with particular inflectional moods. This would not appear to be overly controversial. It has, for example, often been argued that the indicative may be used with 'modal' meaning, particularly in the future tense (eg. Fleischman 1982, Tsangalidis 1999, Palmer 2000: 105, Willmott 2007: 56–60).

However, there are also environments in Classical Greek which appear to be *semantically* modal in which the 'wrong' negator is found. For example, the negator *ou* is the usual negator of the optative in its potential use in (future unreal) conditional consequents. The negator *ou* is also used in conditional consequents with the indicative in so-called 'counterfactual' conditionals. These two uses of the negator are well established in the grammar books and uncontroversially contradict any claim that *mē* is semantically +modal and *ou* is −modal. It is unsurprising, then, that two recent papers should have argued for a different relationship between the negators and type of modality.

Philippaki-Warburton and Spyropoulos claim that, throughout the history of the language, *mē* (later *min*) is associated with deontic modality and *ou* (later *δen*) is associated with epistemic modality (Philippaki-Warburton 2004). The distinction between deontic and epistemic modality is of course fundamental in studies of modality, being usually exemplified with English modal verbs:

(62) You must hand in work on time or else. = Deontic

(63) You must be Fran. = Epistemic

However, I have recently argued that their claim does not stand (Willmott 2009). In brief, in SMG not all *na* clauses are deontic, *min* is used with gerunds, and *δen* is used in conditional clauses, which are not clearly an epistemic environment. In Classical Greek too, *mē* is also found in some non-deontic circumstances, namely in conditional sentences, with participles with a conditional meaning, and as the complement to certain verbs (e.g. verbs of expecting and swearing etc., see Goodwin 1889: §685). In Homeric Greek I showed elsewhere that the optative is used with a 'dynamic' meaning, where it is again negated with *ou* (Willmott 2008). For example:

(64) τὸν δ' **οὐ̃** κε δύ' ἀνέρε δήμου ἀρίστω ῥηϊδίως ἐπ' ἄμαξαν
ton d ū ke du anere dēmū aristō r^{h}idiōs ep amaksan
that PTCL OU PTCL two men region best easily on wagon
ἀπ' οὔδεος ὀχλίσσειαν οἷοι νυν βροτοί εἰσ'
ap ūdeos ok^{h}lisseian hoioi nūn brotoi eis
from floor lift.OPT.3PL as now mortals are
'Two men, the best from the region, would not be able to lift it easily from the floor to the wagon, such as men now are.' (Homer, *Iliad* 12.448)

The uses of the negators in Homeric Greek may therefore be summed up as in Table 8.10 (this does not include the use of *mē* as complementizer, discussed in sections 8.3.5 and 8.3.7 above).

As I pointed out in Willmott (2009), the fine-grained nature of the distribution of the negators bears some similarities to the finely-grained functional category of modality argued for by Cinque (1999, 2004). Based on the relative order of a range

TABLE 8.10 **The uses of the negators in Homeric Greek**

Mood	Construction	Negator
Imp/subj	Directives	*mē*
Opt	Wishes	*mē*
Opt/subj/indic	Most conditional antecedents	*mē*
Opt/subj/indic	Purpose clauses	*mē*
Opt/indic	Conditional consequents	*ou*
Opt	Statements of obligation	*ou*
Opt	Statements of ability	*ou*
Indic/subj	Assertions	*ou*

$\text{MoodP}_{\text{speechact}} > \text{MoodP}_{\text{evaluative}} > \text{MoodP}_{\text{evidential}} > \text{MoodP}_{\text{epistemic}} > \text{TP(Past)} > \text{TP (Future)} > \text{MoodP}_{\text{irrealis}} >$

$\text{ModP}_{\text{alethic}} > \text{AspP}_{\text{habitual}} > \text{AspP}_{\text{repetitive(I)}} > \text{AspP}_{\text{frequentative(I)}} > \text{ModP}_{\text{volitional}} > \text{AspP}_{\text{celerative(I)}} > \text{TP(Anterior)} >$

$\text{AspP}_{\text{terminative}} > \text{AspP}_{\text{continuative}} \geq \text{AspP}_{\text{retrospective}} > \text{AspP}_{\text{proximative}} > \text{AspP}_{\text{durative}} > \text{AspP}_{\text{generic/progressive}} > \text{AspP}_{\text{prospective}} >$

$\text{ModP}_{\text{obligation}} > \text{ModP}_{\text{permission/ability}} > \text{AspP}_{\text{completive}} > \text{VoiceP} > \text{AspP}_{\text{celerative(II)}} > \text{AspP}_{\text{repetitive(II)}} > \text{AspP}_{\text{frequentative(II)}}$

FIGURE 8.1 Model of the IP (Cinque 2004: 133)

of different adverbs and other verbal elements in various Romance languages, he proposed a universal hierarchy of functional projections (Cinque 2004). A subset of this hierarchy is shown in Figure 8.1.

This model of the clause structure has already been related to different negative markers in different Romance dialects (Zanuttini 1997: 101), and would appear to correlate well to the environments for the different negators distinguished above in Homeric Greek. The modality of imperatives and wishes appears to compare semantically to Cinque's $\text{MoodP}_{\text{speechact}}$, while statements of obligation appear to compare to his $\text{MoodP}_{\text{obligation}}$ and statements of dynamic modality could be compared to his $\text{ModP}_{\text{permission/ability}}$. It is then tempting to correlate the modality of purpose clauses and conditional clauses with another of his types, perhaps $\text{MoodP}_{\text{irrealis}}$. We may thus redraw the list of uses as in Table 8.11.

According to this correlation, *mē* negates the types of modality higher in the hierarchy, while *ou* is found lower down. Needless to say, the different uses of *mē* in Homeric Greek do not map onto Cinque's model in a completely straightforward manner. One problem is the use of *mē* in (most) conditional antecedents and *ou* in conditional consequents. Both of these are 'irrealis' contexts and thus have been correlated with Cinque's $\text{MoodP}_{\text{irrealis}}$ category. The fact that *mē* is found in one context and *ou* in the other, however, suggests either that the choice of negator is not

TABLE 8.11 **Uses of the negators in Homeric Greek with Cinque's functional categories**

Construction	Negator	Functional category
Directives	*mē*	$\text{MoodP}_{\text{speechact}}$
Wishes	*mē*	$\text{MoodP}_{\text{speechact}}$
Purpose clauses	*mē*	$\text{MoodP}_{\text{irrealis}}$
Most conditional antecedents	*mē*	$\text{MoodP}_{\text{irrealis}}$
Conditional consequents	*ou*	$\text{MoodP}_{\text{irrealis}}$
Statements of obligation	*ou*	$\text{ModP}_{\text{obligation}}$
Statements of ability	*ou*	$\text{ModP}_{\text{permission/ability}}$

dependent purely on semantic grounds, with syntactic context instead playing some part, or else that there needs to be further division between two different types of 'irrealis' modality.

A detailed consideration of the constructions in which the two negators are found in Homeric Greek thus suggests that the negators may not be differentiated according to any simple division such as declarative versus prohibitive or deontic versus epistemic. The distribution supports Cinque's view of a fine-grained model of modality. It also suggests that each individual negator may be complex, operating at various different positions in the sentence, thus supporting the claims made in section 8.3 that the negators should not be analysed as monolithic entities.

More work clearly remains to be done on this issue, particularly on how the distribution of the negators changes over the course of the history of the language. For example, it is interesting to note that, one of the first post-classical developments is an encroaching of the domain of *ou* by *mē* (Gildersleeve 1880), even though *ou*, or its successor *δen*, eventually comes to have a wider distribution in SMG (for example in conditional clauses).

8.5 Negative concord

8.5.1 Introduction

From the Classical period onwards, Greek is one of several languages which display what is now usually called 'negative concord', and which was discussed in detail by Jespersen as a form of 'double negation' (Jespersen 1917: 62–80, and see Haspelmath 1997: 201–3 for a discussion of why the term 'double negation' is problematic). The topic of negative concord has been of considerable theoretical interest with respect to various modern languages, including Modern Greek. Here I will argue that the particular nature of negative concord in Classical Greek is no less interesting, and may shed some light on the later developments.[4]

8.5.2 *Definition and examples*

Negative concord may be briefly described as follows: in sentences with two (or more) apparently negative elements (normally a negator and one or more so-called 'n-words'), they do not 'cancel out' the negative meaning (as they do in English), but rather, the sentence has a negative meaning. We may see examples of this in Classical Greek in (65) and (66) below (from Smyth 1956: §2760), and for Modern Greek in (67) and (68) (adapted from Holton et al. 1997: 322 and 421):

(65)	**οὐ**	*γάρ*	*με*	*νῦν*	*χέζοντά*	*γ'*	**οὐδεὶς**	*ὄψεται.*
	ū	gar	me	nun	k^h^esdonta	g	ūdeis	opsetai
	OU	PRT	me	now	shitting	PTCL	n.person	see.3SG.FUT

'Now no one will see me shitting.' (Aristophanes, *Ecclesiazusae* 322)

(66)	**οὐ**	*γὰρ*	*ᾤζυρὲ*	*τούτων*	*ἐπιθυμῶ*	*μανθάνειν*	**οὐδέν**
	ū	gar	ōsdure	tūtōn	epit^h^umō	mant^h^anein	ūden
	OU	PRT	miserable.VOC	those.GEN.PL	want.1SG	learn.INF	n.thing

'You miserable man, I don't want to learn about anything of those.' (Aristophanes, *Clouds* 656)

(67)	**δε**	*θέλω*	**τίποτα**
	ðe	θélō	típota
	ðEN	want.1SG	n.thing

'I don't want anything.' (Modern Greek)

(68)	***Κανένα***	*του*	*βιβλίο*	**δεν**	*άξιζε*
	Kanéna	tu	vivlío	ðen	áksize
	n.thing	his	book.NOM	ðEN	value.3SG.PAST

'No book of his was worth anything.' (Modern Greek)

In Classical Greek the n-words appear to be inherently negative, being made up of the negator in compound with another element (e.g. *oudeis* 'no one', *ouden* 'nothing', *oudepote* 'never' etc.). In SMG, on the other hand, n-words are formally (positive) indefinite pronouns. As Horrocks has shown, *tipota* derives from Classical Greek *ti* 'something' and *pote* 'ever', and *kanénas* from Classical Greek *kai* 'and' and *heis* 'one' (Horrocks 1997: 223–4 and 274–5). And indeed, in subjunctive, conditional, and

[4] In a recent paper (Willmott 2011) I have argued that the situation is rather different in Homeric Greek. See further there for discussion of the apparently embryonic stage of the construction in that period. Due to the limited amount of data from the earliest period of the language, in this section I will be discussing the evidence from Classical Greek onwards.

imperative contexts they do not have a negative reading at all. For example (adapted from Holton et al. 1997: 321):

(69) *Ήρθε* **κανένας** *στο γραφείο μου;*
írθe kanénas sto grafío mu
came KANENAS to.the office my
'Did anyone come to my office?'

As Haspelmath has shown, it is common for indefinite pronouns that are used in negative contexts to also have other uses (Haspelmath 1997: section 8.1, also see his Figure 4.4 for an implicational map of the functions of indefinite pronouns). It is thus often difficult to determine whether an element is negative and thus that the language has negative concord. In SMG, however, these elements are usually described as n-words (and SMG is thus usually described as a negative-concord language), since they can be used in fragmentary answers with negative meaning (Giannakidou 2006, see also section 1.8.2, and, for further discussion on the use of this context to determine 'negativeness', Haspelmath 1997: section 8.1.2), for example:

(70) *Τι θελείς;* ***Τίποτα***
ti θelís? típota
what want.2SG n.thing
'What do you want? Nothing.' (Modern Greek) (Holton et al. 1997: 322)

Before going on to examine the differences between the different stages of the language, it should be noted that, in Classical Greek, negative concord may be found with *mē* as well as *ou*. The negator and negative indefinite pronoun in such cases always share the same negative element (e.g. *mē*... *mē*- or *ou*... *ou*-), for example:

(71) ***μή*** *νυν πρότερον* ***μηδεὶς*** *ὑμῶν ἀντείπῃ*
mē nun proteron mēdeis humōn anteipē
MĒ now first no.one of.you contradict.3SG.SUBJUNC
'Let no one contradict nor interrupt me.' (Aristophanes, *Ecclesiazusae* 590)

Given that negative concord is generally discussed with reference to an 'assertive' sentential negator and only (as far as I am aware) with reference to *δen* in SMG, I will, however, only discuss the use of *ou* in negative-concord environments in Classical Greek.

8.5.3 Differences between Classical Greek and Modern Greek

Although both Classical Greek and SMG may be said to display negative concord, the construction is rather different in the two periods. Namely, in SMG the n-word must be accompanied by the negator, which may come before or after it (see examples (67) and (68) above). It is thus described as a strict NC language, or in

Haspelmath's terms an NV–NI language (Giannakidou 2000: 462, Zeijlstra 2006: 411–12, Haspelmath 1997: 201).

In Classical Greek, on the other hand, the n-words may stand by themselves and generate a negative reading, for example:

(72) *οὐδεὶς* *αὐτῶν* *ἅψεται*
ūdeis autōn hapsetai
n.person them.GEN.PL touch.3PL.FUT
'No one will touch them.' (Lysias 1.36)

Furthermore, negative concord only occurs when the negator is followed by the n-word, as in examples (65), (66) (and (71)) above, or when there are two (or more) n-words in the sentence (for further examples of several accumulating negatives, see Kühner and Gerth 1898: 203), for example:

(73) *καὶ* ***οὐδενὶ*** *πώποτε* ***οὔτε*** *ἡμεῖς* ***οὔτε*** *ἐκεῖνος* *δίκην*
kai ūdeni pōpote ūte hēmeis ūte ekeinos dikēn
and n.person.DAT ever neither we.NOM.PL neither he.NOM.SG case.ACC
οὔτε *ἐδικασάμεθα* ***οὔτε*** *ἐφύγομεν*
ūte edikasamet^{h}a ūte ep^{h}ugomen
neither prosecute.1PL.PAST neither defend.1PL.PAST
'Neither we nor he either ever prosecuted or defended a case against anyone.' (Lysias 12.4)

If the n-word comes first and is followed by a negator (cf. (68) in SMG), a negative-concord reading does not result: in this situation the negations cancel each other out (see Smyth 1956: §2760).

Greek is thus a non-strict NC language, in Zeijlstra's terms, or an (N)V–NI language in Haspelmath's terms (Zeijlstra 2006: 411, Haspelmath 1997: 201). (Haspelmath 1997: 224 mistakenly claims that Classical Greek *oudeis* is V–NI and only develops to (N)V–NI in New Testament Greek. Examples such as (65) and (73) above prove otherwise.)

Perhaps it is this 'mixed' behaviour which leads Giannakidou to claim that Classical Greek is not an NC language (Giannakidou 2000: 487). In fact, Classical Greek patterns like Italian, which *is* usually described as displaying negative concord. For example:

(74) Ieri nessuno (*non) ha telefonato.
Yesterday n.person (*NEG) has called
'Yesterday nobody called.' (Italian)

Greek is therefore interesting as an example of a language which has developed from a non-strict to a strict negative-concord language (cf. section 3.8 and Zeijlstra 2006: 421–2 for comparable developments in Italian). The direction of this development in Greek is expected since NV–NI (or strict) is the preferred option cross-linguistically (Haspelmath 1997: section 8.2).

8.5.4 *Explaining the construction*

The negative-concord construction has been explained in many different ways, using many different models. In this section I will show that the historical development of Greek appears to support one model of the phenomenon. However, as I will show in the following section, a closer examination of the evidence casts some doubt on this explanation.

Giannakidou has explained the phenomenon in SMG semantically. She claims that we must generally distinguish between emphatic and non-emphatic n-words, pointing out that in Greek it is only emphatic n-words which are licensed to appear before the negative marker and in fragmentary answers. She argues that emphatic n-words (at least in SMG) are semantically not negative and claims that they are instead universal quantifiers (Giannakidou 2000: section 2). Even in fragmentary answers (the context in which the negative meaning is said to prove that the n-words are negative) she argues that they are not really negative, but rather that we should understand them as having a negative element in ellipsis.

However, subsequent scholars have pointed out problems with the 'ellipsis' argument, arguing that the existence of the fragmentary answers prove the negative semantics of the words. They have thus argued that other explanations are needed for the presence of two negative elements resulting in one semantic negation. For example, Zeijlstra (2004) has put forward a syntactic explanation of the construction, using a Minimalist framework. In brief, he argues that the various different types of negative concord (strict or non-strict) should be described in different ways but in general as an example of 'syntactic agreement'. He distinguishes between negative operators that carry an 'interpretable' negative feature (iNeg) and elements that carry 'uninterpretable' negative features (uNeg).

Zeijlstra's approach may be illustrated with the following sentences from the non-strict language Italian:

(75) a. Gianni non ha telefonato a nessuno.
Gianni NEG have.PRES.3SG phone.PP to n.one
'Gianni didn't phone anyone.'

b. *Nessuno non ha telefonato.
n.one NEG have.PRES.3SG phone.PP
'No one phoned.'

c. Nessuno ha telefonato.
n.one have.PRES.3SG phone.PP
'No one phoned.' (Italian)

The negative operator *non* is said to be iNeg while the n-words (including *nessuno*) are said to be uNeg. In example a) the uNeg feature of *nessuno* is said to be 'checked against' the iNeg feature of *non*. As the negative operator must c-command the n-word it follows that the negative operator must precede the n-word, explaining why example (75b) is impossible. Of course, the n-words may appear without the negative marker, as in (75c). In such examples the negative reading of the sentence is explained in the following way: the uNeg feature on *nessuno* is said to trigger an abstract negative operator (without a phonological realization) which provides the negative force.

In strict NC languages like SMG, however, the negative marker itself is said to carry uNeg rather than iNeg. In these languages the negative force of the sentence is always provided by an abstract operator rather than a phonological realization. It is this that allows them to have the negators and n-words in either order.

The development of the negators in Greek would appear to support this account. If Classical Greek should be analysed in the same way as Italian, with an 'interpretably negative' negative marker (*ou*) going together with 'uninterpretably negative' n-words such as *oudeis, ouden*, etc. we may have an explanation of why SMG is an example of a strict NC language. After all, as described in section 8.2.2 above, the standard negator is said to derive from an original n-word, which is uNeg in a non-strict language like Classical Greek. Because *δen* has developed from this n-word to the standard negator, we would expect the modern language to be a strict NC language.

8.5.5 *Discussion*

Although the theory appears to fit the data well, there are several issues with this analysis. Firstly, the claim that the n-words in Classical Greek are 'uninterpretably negative' and thus in some way 'not negative' might seem to be rather troubling given their transparent relationship to the negator. Just as in English, where the use of transparently negative words such as *nothing* and *never* has been argued to prevent negative concord (Giannakidou 2000: 487, see also Haspelmath 2011 for the general trend), we might expect these to be just as 'interpretably' negative as the negative marker.

Zeijlstra's approach also forces a rather questionable explanation of sentences such as (75c) above or an equivalent Classical Greek sentence, such as (9) above, reproduced here for ease:

(76)
οὐδὲν	*διοίσεις*	*Χαιρεφῶντος*	*τὴν*	*φύσιν*
ūden	dioiseis	k^{h}airep^{h}ōntos	tēn	p^{h}usin
OUDEN	differ.2.FUT	Chairephon.GEN	the	nature-ACC

'You will not in any way differ in nature from Chairephon.' (Aristophanes, *Clouds* 503)

As we have seen, the negative force of such sentences is explained by Zeijlstra as being accounted for by an 'abstract negative operator', which allows the uNeg force of the n-word to be 'checked', rather than by anything negative in the word itself. However, the status of these 'abstract negative operators' is doubtful. We might ask why they are only invoked to explain the n-words and not the negators themselves. Proposing their existence to explain why apparently negative elements do not cancel out the negative meaning of the negator is weak without independent evidence of their existence.

Nonetheless, the existence of negative concord in both Italian and Classical Greek shows that negative concord is possible even when n-words belong to the paradigm of negative quantifiers. A claim that negators and n-words 'express their negative quality' in different ways is plausible, for both languages.

Leaving aside the theoretical issues with Zeijlstra's argument, there is further evidence which is problematic for the claim that SMG is a 'strict' negative language because its negative marker is derived from a 'non-negative' n-word. Namely, the data from the period of development from Classical Greek to SMG. It has been shown that the system of 'strict' negative concord like that found in SMG is found already in the twelfth century (Horrocks 1997: 275). But the example used to support this demonstration uses *ou* not *δen*:

(77) *τίποτε οὐ λογίζεται*
tipote u lojízete
n.thing NEG think.PRES.3SG
'He thinks of nothing.' (Digenes Acritas (E), 706)

Thus, it appears that the appearance of strict negative concord in Greek predates the replacement of *ou* by *δen*. The development from a non-strict to a strict negative-concord language therefore may not be straightforwardly explained in terms of the etymology of the negative marker, and further research is needed to account for the details of the development.

8.6 Negative imperatives

8.6.1 Introduction

In the expression of negative commands, Ancient Greek again demonstrates an important difference from SMG. The difference has important ramifications for the interpretation of negative commands and their theoretical explanation, both from a syntactic and a semantic point of view.

8.6.2 SMG: a common pattern

In SMG a morphologically inflected imperative form is used in positive commands, for example:

(78) Μίλησε με το Γιάννη γί αυτό
mílise me to jánni ji aftó
speak.2SG.IMP with the John about this
'Speak with John about this.' (Modern Greek) (Holton et al. 1997: 411)

However, in negative commands the negative *min* is used not with the imperative but with the normal, indicative form of the verb. The 'subjunctive' particle *na* is optional before *min*, for example:

(79) (να) μην του το πεις
(na) min tu to peis
(NA) MIN him it say.2SG.IND
'You should not tell him./Don't tell him.' (Modern Greek) (Holton et al. 1997: 420)

The avoidance of the imperative in negative commands is very common cross-linguistically. Most commonly, languages avoid the imperative with the usual negator, instead tending either to use a different negative particle or a different form of the verb (van der Auwera and Lejeune 2011). SMG is rather unusual in that it has a different negative particle for prohibitions but still avoids using the imperative.

There has been much work done on the general tendency to avoid negating imperatives from a syntactic point of view. Zeijlstra (2006) offers an explanation for SMG which is connected to his general theories on the 'interpretable' nature of negation (see section 8.5.4). He notices that *min* interacts with 'n-words' differently from *δen*: unlike with *δen* where n-words may come either before or after the negator (the sign of a strict NC language), n-words may only follow, not precede *min* (Zeijlstra 2006: 419). For example:

(80) *Thelo KANENAS na **mi** fiji
want.1SG n.one SUBJUNC MIN leave.3SG
'I want nobody to leave.'

(81) Thelo na mi fiji KANENAS
want.1SG SUBJUNC MIN leave.3SG n.one
'I want nobody to leave.'

Zeijlstra claims on the basis of this behaviour with n-words that *min* carries 'interpretable negative features' (iNeg, see above). It is this feature of *min* that is said to rule out the possibility of 'true negative imperatives' or TNIs in SMG. He claims that the negative force of the interpretably negative negator would negate the illocutionary force of the imperative verb form.

The claim has often been made also from a semantic point of view that in negated commands the negation must lie within the scope of the illocutionary force, and therefore that negative elements may not 'outscope' imperatives (see for example Han 2001, Horn 1989). Just as we may describe negated declaratives as positive assertions of a negative proposition, we may describe negated commands as a positive command of a negative proposition. For example:

(82) The sun does not shine in July.
= I assert the sun NOT shine in July

(83) Don't shout.
= I command that you NOT shout

This analysis partly stems from the interpretation of a command as being made up of deontic modality affecting the proposition, for example:

(84) Shout.
$\approx$ You must shout/It is necessary that you shout

It is certainly the case that in the case of negative commands the negation is within the scope of this obligation rather than without it, for example:

(85) Don't shout.
$\approx$ It is necessary that you NOT shout
$\neq$ It is NOT necessary that you shout

The observation that the negation should lie without the scope of the illocutionary force of the sentence is supposed to have syntactic implications too. For example, (Zeijlstra 2006: 416) claims: 'V_{imp} must raise to C^0 and as the negative marker Neg^0 must be attached to V^0, this negative marker c-commands [IMP]' and 'since the negative head adjoins to V_{imp} and V_{imp} must raise to C^0, Op_{IMP} cannot avoid being outscoped by negation'. Zeijlstra therefore comes to the conclusion that 'every language with an overt negative marker X^0 that carries [iNeg] bans TNIs' (Zeijlstra 2006: 416). As SMG *min* is shown to be [iNeg], the lack of TNIs in the language is therefore apparently explained.

8.6.3 Ancient Greek: a problem for the Minimalist account

However, the data from Ancient Greek presents some difficulties for this analysis. Just as with *min*, n-words may follow *mē* but may not precede it. So for example:

(86) μή νυν πρότερον μηδεὶς ὑμῶν ἀντείπῃ μηδ'
mē nun proteron mēdeis humōn anteipē mēd
MĒ now first no.one of.you contradict.3SG.SUBJUNC and.not
ὑποκρούσῃ,
hupokrusē
interrupt.3SG.SUBJUNC
'Let no one contradict nor interrupt me.' (Aristophanes, *Ecclesiazusae* 590)

(87) *μηδεὶς νυν πρότερον **μὴ** ὑμῶν ἀντείπῃ μηδ' ὑποκρούσῃ
mēdeis nun proteron mē humōn anteipē mēd hupokrusē
no.one now first MĒ of.you contradict and.not interrupt
'Let no one contradict nor interrupt me.'

Thus, according to the criterion Zeijlstra used with SMG, in Ancient Greek *mē* must carry iNeg features and we would not expect it to be followed by the imperative in negative commands. This is not problematic for the examples above, as the subjunctive is found. However, elsewhere *mē is* found with the imperative in Ancient Greek. For example:

(88) μὴ θορυβεῖτε
mē t^{h}orubeite
NEG cause.disturbance.IMP.PRES.2PL
'Don't make a disturbance.' (Plato, *Apology* 21a)

The acceptability of the imperative appears to depend on the aspect of the verb form: present (imperfective) imperatives are possible, while aorist (perfective) ones are not, as detailed in Table 8.12.

In perfective cases, as in sentence (86) above and in the following from just a few lines before the imperative example in (88), we find the aorist subjunctive:

(89) μὴ θορυβήσητε
mē t^{h}orubēsēte
NEG cause.disturbance.SUBJUNC.AOR.2PL
'Do not make a disturbance.' (Plato, *Apology* 20e)

If *mē* is 'interpretably negative', as its behaviour with n-words suggests, we therefore do not have an explanation of how it can be used with the imperative, apparently outscoping the illocutionary force of the imperative. These data support van der Auwera's argument that the cross-linguistic preference for a dedicated marker for prohibitions stems from something other than the scope argument (van der Auwera 2010b: section 3). More work clearly needs to be done on the syntactic status of *mē* in Ancient Greek, and the acceptability of TNIs more generally.

TABLE 8.12 Aspect and mood in Ancient Greek commands

Positive	Negative
present imperative	μή + present imperative
aorist imperative	μή + aorist subjunctive

8.6.4 Semantic explanation of Ancient Greek

Instead of answering the syntactic conundrum, I have considered the semantic nature of the use of the imperative and subjunctive in this construction, trying to explain why the two different moods interact in this way with the aspectual system (Willmott 2010).

While previous explanations for this phenomenon have argued that the two constructions differ purely in aspectual terms and that there is no significance to the change of mood (e.g. McKay 1986), I argued that the difference in mood suggests that two different constructions ought to be distinguished. I compared languages in which scholars have distinguished what are termed 'preventive' from 'prohibitive' constructions such as Russian, Aleut, Tatar, Even, and Armenian (see Birjulin and Xrakovskij 2001).

Broadly speaking, the 'prohibitive' is used to prohibit *controllable* actions (e.g. 'don't paint', 'don't read') while the 'preventive' is used to prevent *uncontrollable* actions (e.g. 'don't break the glass', 'don't fall') (Birjulin and Xrakovskij 2001: 34). Putting aside the difficulties of using the language of the Homeric poems as a linguistic database and the fact that it is a very limited selection (only thirteen aorist subjunctives in negative commands), I claimed that the constructions match the meanings of the two constructions observed in the other languages, with the aorist subjunctive resembling a preventive marker, and the present imperative a prohibitive marker.

Six of the thirteen examples of the aorist subjunctive in negative directives in Homer are found preventing emotions. In Armenian, 'emotive' verbs would normally be prevented with the preventive, and therefore appear to be seen as 'uncontrollable' (Kozintseva 2001: 257). Those which are not found with verbs that may be thought of as more 'controllable' may be described as either cautions/warnings or strong prohibitions, just as the 'preventive' in Armenian (for examples, see the original paper).

The overwhelming majority of the present imperatives (over 80%), on the other hand, are found with controllable verbs. Most of the exceptions to the above tendency are found correcting adverse emotions, just like the prohibitive markers in the languages analysed. There are also a few examples which could be described as strong prescriptions to control the situation, marked with the prohibitive in Armenian etc. Of course, the analysis is extremely subjective. Given the difficulty of ascertaining whether certain actions are really 'controllable' or 'uncontrollable', and given that there is rather an overlap between the two categories where they are grammatically distinguished, it could be said that it would be all too easy to describe the Ancient Greek data in a similar way. The inconclusiveness of a semantic analysis is particularly marked for the Homeric data, where there are so few examples of the aorist subjunctive.

However, I pointed out that there is also a non-semantic way of comparing the constructions. In Homeric Greek, the present imperative construction is much more common than the aorist subjunctive one. Out of 167 examples of the second person negative directive, only 13 of them are in the aorist subjunctive, while the rest are in the present imperative, therefore under 10% of the total (Willmott 2010: 537–8). A similar skewing is observed in languages with a prohibitive and preventive distinction. For example, one Russian novel (Goncharov's *Oblomov*) had 126 negative directives, of which 114 (90%) were prohibitive and only 12 preventive (10%). In a selection of Armenian texts, the percentages were 79% prohibitive vs 21% preventive (Kozintseva 2001: 259). These are similar ratios to that noticed between the aorist subjunctive and present imperative in Homeric Greek. This more 'objective' comparison therefore supports the semantic analysis.

8.6.5 Conclusion

The avoidance of the imperative with *min* in Modern Greek reflects a general linguistic tendency, apparently caused by a syntactic interplay between the illocutionary force and the scope of the negator. However, a comparison with Ancient Greek reveals a more complex situation. The acceptability of *mē* with the imperative causes problems for modern analyses, and the choice between imperative and subjunctive in the negative command construction appears to have a semantic basis. There is again more work to be done on the correct analysis of this construction, both in general and specifically for Greek.

8.7 Conclusions

In this survey of some of the literature on certain key aspects of negation in Greek I have focused on those aspects which show that the historical approach helps to explain the synchronic data. For instance, understanding SMG *ðen* is helped by a consideration of its development from a negative quantifier in Ancient Greek. Similarly, the complex nature of the uses of *min* may be simplified by a comparison with the similarly complex, but importantly different *mē*. And although the SMG avoidance of *min* with the imperative appears entirely unsurprising, a comparison with the ancient period of the language reveals a more complicated situation.

I also hope to have shown that current theoretical advances and a detailed empirical study of the language can mutually illuminate each other. Greek is an (unusual) example of a language which develops from a non-strict to a strict negative-concord language. The consideration of the development of the sentential negator in SMG may go some way to explaining why the change might take place. The complex distribution of the negators in both Ancient and Modern Greek provide support for fine-grained models of modality and approaches to grammatical elements which acknowledge the variety of their uses. Finally, the acceptability of *mē* and an imperative form in Ancient Greek presents troubling evidence for theoretical accounts of the avoidance of negators and imperative forms cross-linguistically.

In sum, the tale of two negators in Greek is one of interesting and complex developments which are belied by the preserved binary distinction and which have important theoretical consequences. In spite of the considerable work done on this area, particularly in Modern Greek, there are significant questions remaining.

Acknowledgements

This chapter has benefited significantly from several discusssions with those in the negation project, particularly David Willis, Chris Lucas, and Sten Vikner. I am also very grateful to Brian Joseph who kindly read a draft and made very useful comments on the work in progress.

Primary sources

Unless otherwise indicated, examples from Ancient Greek texts come directly from the *Perseus Digital Library Project*, accessed on different dates during the writing of this chapter:

Perseus Digital Library Project, ed. Gregory R. Crane. Tufts University. <http://www.perseus.tufts.edu>.

Example (4) is from TLG:

Thesaurus linguae Graecae. Irvine: University of California, Irvine. Columbia URL: <http://www.columbia.edu/cgi-bin/cul/resolve?AUV9026>.

9

Negation in the history of the Slavonic languages

DAVID WILLIS

9.1 Introduction

While the Slavonic languages have been fairly conservative with respect to the expression of sentential negation and have not experienced the processes of renewal found across much of Western Europe, other aspects of the syntax of negation have undergone considerable change over the last thousand years. Use of the genitive to mark a direct object in the scope of negation is a characteristic feature that goes back to Common Slavonic, but this system has been remodelled, lost, or extended in various different ways in the daughter languages. The core set of negative indefinites has remained fairly stable, and strict negative concord has become an increasingly entrenched feature of the grammars of all the Slavonic languages. Indefinites in other negative polarity environments have, however, been historically highly unstable, with multiple instances of renewal. The basic division between one set of indefinites found in negative clauses and a different set or sets found in other negative polarity environments has generally remained intact, despite innovations in the way that non-negative indefinites are expressed.

This chapter will give an overview of these developments, setting the historical data against the backdrop of rather more well-developed analyses of synchronic comparative Slavonic syntax. After setting out the textual and linguistic background (section 9.2), we begin by looking at the expression of the main types of (chiefly) sentential negation (section 9.3). Section 9.4 considers the development of the genitive of negation, the marking of objects and some subjects within the scope of negation with genitive case. Section 9.5 focuses on indefinites, especially those in negative and negative polarity contexts. Inevitably, attention will be focused on those languages and varieties which have been the subject of the most in-depth research;

however, an attempt will be made to give an overall sense of changes as they have occurred across the group.

9.2 Textual and linguistic background

The modern Slavonic languages are all descended from the reconstructed Common Slavonic language spoken up to the sixth century CE, centred on an area covering what is now western Ukraine and southern Belarus. Slavonic speakers spread out across the whole of Eastern Europe in the fifth and sixth centuries, with the first reconstructable dialect differences appearing soon afterwards. Accumulation of these differences led to the split of the parent language into three branches: West Slavonic (Czech, Polish, Slovak, Upper Sorbian, Lower Sorbian, etc.), East Slavonic (Belarusian, Russian, and Ukrainian), and South Slavonic (Bulgarian, Croatian, Macedonian, Serbian, Slovene, etc.). Some phonological innovations (loss of or vocalization of the reduced vowels, palatalizations of velar stops) are shared or largely shared across the entire group up until the tenth century; thereafter, innovations affect only part of the Slavonic speech area, but often spread between neighbouring languages within either the North Slavonic area (covering West and East Slavonic collectively) or the South Slavonic area. From the start of the ninth century, the South Slavs were separated from other Slavs by the arrival of the Magyars in Hungary, leading to a clearer division between the South and West Slavonic languages and the elimination of the dialect continuum that once existed between them (via Slovene and Slovak).

While the languages have an amply attested historical tradition, it is not always easily interpretable owing to various conservative aspects of the written record. Old Church Slavonic (OCS) is the earliest variety of Slavonic for which we possess substantial written records. Written in the two earliest Slavonic alphabets, the older Glagolitic and the newer Greek-based Cyrillic, it forms a body of religious texts found in manuscripts thought to date from the tenth and eleventh centuries, mostly translations from Greek of gospels, prayers, and saints' lives. Although originally based on the Slavonic varieties used around the Greek city of Thessaloniki around 850 to 900, it came to be used, in various local redactions, as a literary language in all the Orthodox Slavonic countries and in Romania. Old Church Slavonic is already recognizably South Slavonic, sharing various phonological features with modern South Slavonic languages, in particular, Bulgarian and Macedonian. However, it is early enough to reflect many linguistic features of Common Slavonic.

In the Orthodox parts of the South Slavonic area, the Church Slavonic tradition remained strong throughout the Middle Ages, leading to a highly conservative written tradition through which developments in the spoken languages are only partially visible. Middle Bulgarian, attested primarily through a rich body of religious literature, partially reflects the extensive changes that must have been occurring in the spoken language, such as the erosion of the case system, emergence of a definite article, and loss of the infinitive. More vernacular traits are found in documents from Romanian-speaking Wallachia, where Bulgarian was used as a chancery language.

The South Slavonic languages collectively form a single dialect continuum. Their division into distinct languages results from social and political factors which led to separate standardization of Bulgarian and Macedonian in the nineteenth century, the unification of Serbian and Croatian as Serbo-Croatian in the late nineteenth and early twentieth centuries, and its subsequent and ongoing disintegration into a number of standard varieties today. In the west, the Church Slavonic tradition (gospels, saints' lives, missals, etc.) remained strong even in Catholic Croatia, where use of the Glagolitic alphabet survived for religious purposes until the eighteenth century. Historical vernacular Croatian is better reflected in chancery documents from fourteenth- and fifteenth-century Dubrovnik and in the humanist literary tradition that flourished in sixteenth-century Dalmatia. Tracing historical change in Slovene is more difficult. It is attested only via fragmentary prayers (Mikhailov 1998) until the Reformation, which saw the production of a substantial body of Protestant literary texts in the second half of the sixteenth century. However, with the Counter-reformation, use of Slovene in writing declined until the nineteenth century.

Records from the medieval East Slavonic lands show a continuum of usage from vernacular Old East Slavonic (OES) right up to religious literature composed in the local form of Church Slavonic. Vernacular usage is best reflected in the short but numerous birchbark documents discovered during excavations in Novgorod since the 1950s and in chancery documents throughout the East Slavonic area, with chronicle redactions also providing some evidence of vernacular features. Dialect variation is in evidence from the start, with northern texts from Novgorod and Pskov already showing distinct phonological developments in the eleventh century (Zaliznjak 2004). Distinctly southern Russian and Ukrainian developments are well established in chancery documents by the fifteenth and sixteenth centuries (Kotkov 1990).

In the West Slavonic area, Old Polish (mid 12th to end 15th century) is attested in a range of religious literature, with continuous texts from the fourteenth century. Old Czech is attested in quantity from the fourteenth century, with a variety of courtly literature, Bible translations, and other religious literature, as well as chancery documents from the fifteenth century. The development of a standard language incorporating various conservative features dating back to the sixteenth century has meant an increasing gulf between written and spoken Czech which often hides linguistic change. Chancery documents from Slovakia appear from the fifteenth century alongside other literature in a mixed Czech–Slovak form, with distinctively Slovak texts emerging only in the seventeenth century. Upper Sorbian, spoken around Bautzen in Saxony, and Lower Sorbian, spoken around Cottbus in Brandenburg, are the two surviving Slavonic languages of Germany. Polabian, spoken above all in Lower Saxony, became extinct in the mid eighteenth century. A Sorbian written

FIGURE 9.1 Family tree for Slavonic

tradition arose in the sixteenth century linked to the Reformation, with printed devotional books and a manuscript translation of the New Testament from this period.

A full family tree for Slavonic is given in Figure 9.1. For more detailed overviews of the languages, see Comrie and Corbett (1993) and Sussex and Cubberley (2006).

9.3 Sentential negation

We begin by looking at the expression of standard sentential negation, considering also special cases, such as the negative of imperatives, negative auxiliaries, and negative existentials, all of which have distinctive properties in some or other variety of Slavonic.

In all Slavonic languages, the primary marker of sentential negation is the inherited negative marker *ne (Russian *ne*, Polish *nie*, Bulgarian *ne*, etc.). This appears in immediately preverbal position:

(1) Marija **ne** ožidala takoj rezkoj reakcii.
Maria NEG expect.PAST.FSG [such sharp reaction].GEN
'Maria didn't expect such a sharp reaction.' (Russian)

(2) Marija **ne** očakvaše takava ostra reakcija.
Maria NEG expect.IMPF.3SG such sharp reaction
'Maria didn't expect such a sharp reaction.' (Bulgarian)

(3) Maria **nie** spodziewała się takiej ostrej reakcji.
Maria NEG expect.PAST.3FSG REFL [such sharp reaction].GEN
'Maria didn't expect such a sharp reaction.' (Polish)

It is used in all types of clauses, whether main or subordinate, or indicative, subjunctive, or imperative. All stages of the Slavonic languages thus have true negative imperatives, as in the following Russian example:

(4) Trogajte menja! Ne trogajte menja!
touch.IMP.2PL me NEG touch.IMP.2PL me
'Touch me!' ~ 'Don't touch me!' (Russian)

Alongside the ordinary negative, various South Slavonic languages have also developed a special negative imperative auxiliary. In Serbian, Croatian, and Macedonian, this is *nemoj* (< *ne* 'not' + imperative of *moći* 'be able') (Hansen 2004, van der Auwera 2010b), which is followed by an infinitive or a *da*-clause:

(5) **Nemoj** da izlaziš kasno!
NEG.IMP.2SG that leave.PRES.2SG late
'Don't go out late!' (Serbian) (Hammond 2005: 89)

In Bulgarian, it is *nedej*, followed by a bare verb stem (a historically truncated infinitive) or a *da*-clause (cf. affirmative infinitive *Piši* 'Write!'):

(6) Nedej pisa!
NEG.IMP.2SG write.Ø
'Don't write!' (Bulgarian) (Lindstedt 2010: 412)

The negative particle generally forms a phonological unit with the following verb, and may in some cases therefore attract stress onto itself, according to the phonological rules of the language in question, for instance, Polish *nie wiem* 'I don't know' with stress on *nie*. Accordingly, the negative marker has generally been treated as either a clitic or a verbal prefix. In historically perfect auxiliary structures of the form 'be' + past participle, there is some variation between languages in the element to which the negative marker attaches. In Czech and Slovak, (7) and (8), negation prefixes to the (first) past participle and is preceded by finite auxiliaries, while in South Slavonic, illustrated here by Bulgarian in (9) and (10), negation precedes any auxiliary:

(7) Ja som nenapísal.
I be.PRES.1SG NEG.write.PP.MSG
'I have not written.' (Slovak) (Rivero 1991: 344)

(8) *Ja ne som napísal.
I NEG be.PRES.1SG write.PP.MSG
'I have not written.' (Slovak)

(9) Az ne săm go viždal.
I NEG be.PRES.1SG him.OCL see.PP.MSG
'I haven't seen him.' (Bulgarian)

(10) *Az săm go ne viždal.
I be.PRES.1SG him.OCL NEG see.PP.MSG
'I haven't seen him.' (Bulgarian)

This seems to suggest that the negative head occupies a lower position (for instance, with the order T–Neg–V) in Czech and Slovak than in South Slavonic (where the order would be Neg–T–V) (Rivero 1991). Historically, Old East Slavonic also followed the Czech–Slovak pattern, as demonstrated in (11), but evidence of this has since been lost with the disappearance of the perfect-tense auxiliary in these languages.

(11) a knjazju **esme** zla **ne** stvorili nikotorago že
and prince.DAT be.PRES.1PL evil.GEN NEG do.PP.PL no.GEN PRT
'...and we have done no harm at all to the prince.' (OES) (*Novgorodskaja pervaja letopis'*, p. 67, l. 34)

The modern South Slavonic pattern continues what is found in Old Church Slavonic, where negation almost always preceded the perfect auxiliary (and merged with it phonologically) (Večerka 1989: 33–5):

(12) **Něstŭ** umrŭla děvica.
NEG.be.PRES.3SG die.PP.FSG girl
'The girl has not died.' (OCS) (*Codex Marianus*, Luke 8:52)

Negation generally precedes future and conditional auxiliaries (rather than the participle) throughout the entire group. The order negation–future/conditional auxiliary seems to go back to Common Slavonic, as witnessed by its presence in Old Church Slavonic (Večerka 1989: 35–7):

(13) Ašte biste slěpi byli, **ne biste** iměli grěxa.
if be.COND.2PL blind be.PP.PL NEG be.COND.2PL have.PP.PL sin.GEN
'If you had been blind, you would not have had sin.' (OCS) (*Codex Marianus*, John 9:41)

The contrast between future/conditional auxiliaries and the perfect auxiliary discussed above probably reflects an original difference in the status of the two sets of auxiliary verb, rather than a property of the negative particle itself (see Willis 2000).

Expletive negation occurs in Old Church Slavonic with verbs such as *(u)bojati sę* 'fear', *bljusti sę* 'be careful', *vŭzbranjati/-iti* 'prevent, impede', *xraniti sę* 'guard against', *vŭnĭmati (si/sebě)* 'guard against', *sǫmĭneti sę* 'guard against, avoid', *straxŭ jestŭ* 'it is to be feared that...' (Večerka 1995: 518–19, 1996: 140–1). A subordinate clause following these verbs is introduced by either *(j)eda* 'lest' with no negative particle, or else (rarer) with the expletive-negation option with *da ne* 'that not'. If the verb is instead followed by an infinitive, this may be either negative or non-negative. Parallel instances of expletive negation are found in modern Slavonic languages, as with the following Russian example:

(14) Ja bojus', kak by on **ne** opozdal.
I fear.PRES.1SG how COND he NEG be.late.PAST.3MSG
'I'm afraid he'll be late.' (Russian) (Brown and Franks 1997: 150)

Constituent negation is with *ne* and its cognates in all the Slavonic languages, and this must have been the case in Common Slavonic too; compare the following example from Old Church Slavonic:

(15) **ne** o sebě bo pridŭ
NEG from REFL PRT come.PAST.1SG
'for it is not from myself that I have come' (OCS) (*Codex Marianus*, John 8:42)

While emphatic negative structures have never given rise to fullscale Jespersen's cycle in the conventional sense in any Slavonic language, use of n-words for emphatic uses is occasionally found. In Old Church Slavonic, *ničĭtože/ničesože* 'nothing (acc./gen.)' (see section 9.5.3 on negative indefinites) can be used as an optional extent argument to emphasize negative polarity with an appropriate verb of success or fearing (Večerka 1993: 136–7, 1995: 515–16):

(16) Viděvŭ že pilatŭ. ěko **ničĭsože** uspěetŭ
see.PP.ACT.NOM.MSG PRT Pilate that nothing.GEN succeed.PRES.3SG
ἰδὼν δὲ ο Πιλᾶτος ὅτι οὐδὲν ὠφελεῖ
'Pilate having seen that it was not working at all...' (OCS) (*Codex Zographensis*, Matt. 27:24)

Bulgarian has loaned Turkish *hiç* 'nothing, anything' as a semi-grammaticalized marker of emphatic negation, *xič* 'not at all' in (17), and as an adnominal quantifier 'no' in (18).

(17) **Xič** ne mi puka.
at.all NEG me care.PRES.3SG
'I couldn't care less.' (Bulgarian)

(18) Ne gledam počti **xič** televizija.
NEG watch.PRES.1SG almost no television
'I watch almost no television at all.' (Bulgarian)

Such phenomena might be thought of as embryonic instances of Jespersen's cycle, but have never developed further.

Fusion of the negative marker and auxiliary 'be' and/or the copula is found in some Slavonic languages: Old Church Slavonic had *něsmĭ* 'I am not' < *ne* + *jesmĭ* 'I am' etc. with parallel formations in all persons of the present tense except the third person plural (Vaillant 1963–4 [1948]: 311–12) (cf. (12) above). The fusion seems to be phonological rather than morphological here, since clitics and other elements may intervene between the two parts (Vaillant 1963–4 [1948]: 356):

(19) **ne** **bo** **jesi** prizŭvanŭ
NEG PRT be.PRES.2SG called
'for you are not called' (OCS) (*Codex Suprasliensis* 21.23)

Fused forms of negation plus 'be' are still present in the modern South Slavonic languages, Serbian, Croatian, and Slovene (Serbian and Croatian *ne* + *sam* = *nisam* 'I am not', *ne* + *je* = *nije* 'it is not'; Slovene *ne* + *sem* = *nisem*, *ne* + *je* > *ni* 'it is not'; etc.). Elsewhere, various irregularities point to similar fusion: in Czech, *ne* + *je* fuses as *není* 'is not'; in Slovak *ne-* is prefixed to lexical verbs in the negative, while *nie* is used with 'be'; in Upper Sorbian, *nje-* is used with lexical verbs, while *njej-* is used with 'be'. Some of these developments are reminiscent of Croft's cycle, where a special negative form of an existential verb comes to be treated as a negative marker (Croft 1991). For instance, Slovak *nie* is historically a fused form of the ordinary negator *ne* + 'be', but has been reanalysed as a special negative marker that co-occurs with 'be' (*ne je* > *nie* > *nie je* 'isn't') and is beginning to spread as an emphatic negator to non-verbal contexts even beyond this environment (Veselinova 2010: 202–3).

Special negative forms of 'have' have also developed across South Slavonic (Bulgarian *ne* + *imam* > *njamam* 'I don't have', Macedonian *ne* + *imam* > *nemam* 'I don't have', Slovene *ne* + *imam* > *nimam* 'I don't have'). These forms are also used as the negative of the existential constructions involving 'have' (Croatian, Macedonian, Serbian *ne* + *ima* 'has, there is' > *nema* 'there isn't'; Bulgarian *ne* + *ima* > *njama* 'there isn't'). Some North and East Slavonic languages have an asymmetric system where existential 'be' in the present tense is replaced by a special negative form based on 'have' (Ukrainian *ne* + *je* 'there is' gives *nemaje* 'there is not', Polish *nie* + *jest* 'there is' gives *nie ma* 'there is not'). Colloquially, Ukrainian has also developed a new negative existential verb *katma*, syntactically equivalent to *nemaje* 'there isn't (any)', from *kat maje* 'the hangman has' (Mel'nyčuk 1982, s.v. *katma*). There are also special suppletive forms for the negative existential verb in the present tense in Russian (*ne* + *est'* 'there is' = *net* 'there is not'), formerly also in the rest of East Slavonic, and in Slovak. In Russian, these have spread beyond their original environments to various elliptical contexts, and to that extent exemplify the emergence of a new negator:

(20) Vse ustali, a ja **net.**
all.PL be.tired.PAST.3PL but I not
'Everyone's tired, but I['m] not.' (Russian) (Švedova 1998, s.v. *net*)

(21) Ideš' ili **net?**
go.PRES.2SG or not
'Are you going or not?' (Russian) (Švedova 1998, s.v. *net*)

In Russian, *net* has also become the anaphoric negator 'no' (normally *ne* or a reflex thereof elsewhere).

Various other verbs also have special negative forms scattered across the language group (Serbian and Croatian *neće* 'doesn't want' < *ne* + *hoće* 'wants', also used for the negative future; Slovene *noče* 'doesn't want'; etc.). For further details of all special negators in Slavonic, see Veselinova (2010).

Bulgarian and Macedonian have developed a distinct way of marking sentential negation in the future. Instead of using a symmetrical pattern with *ne* plus the future particle *šte*, they generally use a construction using *njama*, the negative of 'have', plus the modal complementizer/infinitive marker *da*:

(22) **Šte** xodja na kino.
FUT go.PRES.1SG to cinema
'I'll go to the cinema.' (Bulgarian) (Veselinova 2010: 204)

(23) **Njama** **da** xodja na kino.
NEG.have.PRES.3SG PRT go.PRES.1SG to cinema
'I won't go to the cinema.' (Bulgarian) (Veselinova 2010: 204)

This distinction arose through specialization of two different strategies for marking futurity, both present in Old Church Slavonic. In the affirmative, a strategy based on the verb 'want' (*xoštetŭ* 'he, she wants' > Bulg. *šte*, Mac. *ḱe* 'future particle') largely won out, while in the negative a strategy based on 'have' came to dominate (Ivanova-Mirčeva and Xaralampiev 1999: 141–5). This has been generalized to the conditional perfect, hence Bulgarian has *njamaše da* in place of the expected **ne šteše da*:

(24) Ništo ot tova **njamaše** **da** băde văzmožno bez nego.
nothing of that NEG.have.IMPF.3SG PRT be.FUT.3SG possible without him
'None of that would have been possible without him.' (Bulgarian)

In Macedonian, both sets of options (negative future *nema da* and *ne ḱe*) coexist and are synonymous (Kramer 1997: 411).

9.4 Genitive of negation

One of the most salient features of Slavonic is the phenomenon known as genitive of negation. Direct objects of negative verbs are assigned genitive case if they would be accusative in the equivalent affirmative clauses across most of the Slavonic languages at some stage in their history, with the extent to which assignment is optional and the factors affecting whether it is chosen varying considerably in detail from variety to variety. Genitive is also assigned to some subjects of negative clauses in existential, locative, or unaccusative environments and to some temporal and locative adverbial phrases in the scope of negation. Again, the details vary diachronically and cross-linguistically. This section deals with various historical aspects of the phenomenon. We begin by introducing the phenomenon using straightforward cases of genitive direct objects in negative clauses in the early languages (section 9.4.1). Section 9.4.2 looks at the subsequent development of this phenomenon in Russian, a language where the general decline of the genitive of negation rule has been thoroughly investigated. Sections 9.4.3, 9.4.4, and 9.4.5 introduce more complex environments for genitive of negation, specifically objects in clauses embedded under negative clauses (long-distance negation, section 9.4.3), genitive for subjects in certain existential or unaccusative clauses (section 9.4.4), and genitive for noun phrases used as adjuncts of time (section 9.4.5). Once all of the relevant phenomena have been introduced, we go around the language family in section 9.4.6 to see how genitive of negation has contracted or expanded in each of the subgroups. Section 9.4.7 looks at formal approaches to the genitive of negation rule, while section 9.4.8 considers accounts of its historical origin.

9.4.1 Genitive of negation on direct objects

The genitive of negation with a negative object is illustrated for Old Church Slavonic in (25) and Old East Slavonic in (26).

(25) blǫdite ne vědǫšte **kŭnigŭ ni sily bžiję**
be.mistaken.PRES.2PL NEG knowing [books nor power divine].GEN
'You are mistaken, not knowing the scriptures, nor the power of God.' (OCS) (*Codex Marianus*, Matt. 22:29)

(26) ...on" že ne xotjaše **carstva**
he PRT NEG want.IMPF.3SG empire.GEN
'...but he did not want the empire.' (OES) (*Novgorodskaja pervaja letopis'* p. 47, ll. 12–13)

As in the modern Slavonic varieties with genitive of negation, a noun phrase which would ordinarily be marked by an oblique case must remain in that case under negation; hence, in (27), from Old East Slavonic, *im"* 'them' remains dative even though it is in the scope of negation.

(27) ...i ne da im" knjaz' M'stislav".
and NEG give.PAST.3SG them.DAT [prince Mstislav].NOM
'...and Prince Mstislav did not allow them.' (OES) (*Novgorodskaja pervaja letopis'*, p. 56, l. 29)

Finally, constituent negation never triggers genitive of negation in any Slavonic variety, medieval or modern.

Generally speaking, the genitive of negation on direct objects was more widespread in the earlier Slavonic languages than it is today. It is the norm in the earliest textual records throughout Slavonic and must be reconstructed as an essentially compulsory, syntactically based rule for Common Slavonic. In Old Church Slavonic around 90%

of relevant direct objects are marked genitive: in the data provided by Večerka (1958: 194), only 9.6% of objects are unambiguously accusative. Exceptions to the rule may be due to the influence of the original Greek texts, or may be systematic; for instance, negative rhetorical questions are not treated as negative for assignment of genitive of negation, and fronted objects also escaped its scope (Večerka 1995: 521).

9.4.2 The development of the genitive of negation in Russian

In Russian, consistent genitive marking of direct objects in negative clauses was maintained up to the seventeenth century (Borkovskij 1978: 347), but the accusative has gained ground since then, with the accusative–genitive contrast coming to mark semantic and pragmatic categories such as definiteness and information structure (Mustajoki and Heino 1991, Padučeva 2006, Timberlake 1975). Timberlake (1975, 1977: 157–68) considers the decline of the genitive of negation, in Russian at least (see also below for similar views expressed about Czech), to be part of a wider trend towards replacing genitive objects with accusative ones in all contexts (e.g. partitive objects and intensional verbs such as *izbežat'* 'avoid' or *ždat'* 'wait for', whose objects participate only partially in the event). He shows that the change has been most advanced with individuated (animate, concrete, singular, definite) objects. The underlying unity of the intensional genitive and the genitive of negation has also been advocated by Neidle (1988) and Kagan (2010), the latter noting the parallel recent decline of both constructions and the general, though not exceptionless tendency for individual Slavonic languages to exhibit either both or neither today.

In present-day Russian, genitive vs accusative case selection has come to be associated with interpretative differences. Definiteness and other aspects of information structure (topichood and old–new information) strongly influence case choice, with the genitive case serving to mark an indefinite, non-referential noun phrase (Babby 1980), as shown by the difference in interpretation signalled by case in (28) and (29):

(28) Koška ne est vetčiny.
cat NEG eat.PRES.3SG ham.GEN
'The cat doesn't eat ham.' (Russian) (Pigin 1962: 17)

(29) Koška ne est vetčinu.
cat NEG eat.PRES.3SG ham.ACC
'The cat isn't eating the ham.' (Russian) (Pigin 1962: 17)

The result is that the factors determining whether genitive of negation is preferred have come to be largely the same for both direct objects of negated transitive verbs and subjects of negated existentials and unaccusatives (see below).

The detailed diachronic investigation by Krasovitsky et al. (2011), covering the period from 1801 to 2000, tracks the rise of the accusative from its very limited use

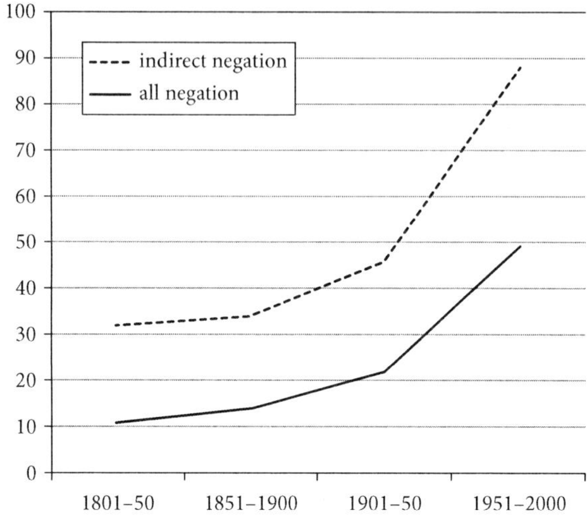

FIGURE 9.2 The rise of accusative direct objects (%) in negative clauses in Russian (adapted from Krasovitsky et al. 2011: 575, 588)

TABLE 9.1 The rise of accusative direct objects (%) in negative clauses in Russian (data adapted from Krasovitsky et al. 2011: 575, 588)

time period	1801–50	1851–1900	1901–50	1951–2000
all negation				
% accusative	11	14	22	49
total tokens	599	758	698	903
indirect negation (object of non-directly negated infinitive)				
% accusative	32	34	46	88
total tokens	71	106	65	105

at the start of the nineteenth century to its very widespread use today, midway along a classic S-curve trajectory (the solid line in Figure 9.2, representing the data in Table 9.1). The spread of the accusative is conditioned by semantic factors, with those factors identified above coming to be significant over the course of the period. For instance, while today concrete objects strongly favour accusative when compared to abstract objects, this effect was much weaker in the nineteenth century and is observed only when the verb is perfective; with imperfective verbs, the effect emerges only in the second half of the twentieth century. Thus in the nineteenth century, we find examples such as (30), where an animate, definite, concrete object appears in the genitive in a negative clause where accusative would be favoured today.

(30) ...ja li ne ljubil moej Duni?
I QU NEG love.PAST.IMPF.3MSG [my Dunja].GEN
'...didn't I really love my Dunja?' (Russian) (Puškin, *Povesti Belkina*, 1831) (Krasovitsky et al. 2011: 585)

Krasovitsky et al. argue that the emergence of a meaningful opposition between accusative and genitive is an epiphenomenon, produced by the spread of accusative within the context of the overall change in which accusative will probably ultimately oust the genitive completely. During the shift, case choice appears to be meaningful for those factors which influence the course of the shift. However, once the shift is complete for that factor, meaningfulness disappears: now that concrete objects almost categorically require the accusative, case choice has limited ability to express other semantic categories in that environment.

Change appears to be ongoing at the present day, perhaps with a continuing narrowing of the domain for genitive of negation to specific (rather than definite) nominals (Babyonyshev and Brun 2002). For some younger speakers, the indefinite, non-referential reading is now available for accusative objects in sentences similar to (29), this reading previously having been the domain of genitive objects (Padučeva 2006: 31–2).

9.4.3 *Long-distance genitive of negation*

In Old Church Slavonic, the genitive of negation could spread into a non-finite complement or adjunct clause. Hence, in (31), *ženy tvoeję* 'your wife' is the object of the infinitival complement of *ne uboi sę* 'do not fear', yet is nevertheless marked genitive; and, in (32), *světilĭnika* 'lamp' is inside an adjunct clause, yet marked genitive because the main clause contains a negative subject *nikto že* 'no one' (Vaillant 1963–4 [1948]: 185, Večerka 1958: 188, 1993: 250–1):

(31) Ne uboi sę prijęti **ženy tvoeję Mariję**.
NEG fear.IMP.2SG REFL take.INF [wife your Maria].GEN
'Do not be afraid to take your wife Mary.' (OCS) (*Evangeliarium Assemani*, Matt. 1:20, ff. 131a.27–131b.1)

(32) Nikto že **světilĭnika** vŭžegŭ pokryvaetŭ i ssǫdomŭ.
no.one PRT lamp.GEN light.PP.NOM.MSG hide.PRES.3SG it vessel.INST
'No one, having lit a lamp, hides it under a vessel.' (OCS) (*Evangeliarium Assemani*, Luke 8:16, f. 55d.14–19)

Genitive marking in these contexts is, however, not compulsory in Old Church Slavonic (Večerka 1993: 253). Elsewhere in Slavonic, some languages have retained this possibility while others have lost it (see further section 9.4.6 below on individual languages). In Russian, genitive of negation could spread into a non-finite complement clause at least up until the nineteenth century (Bulaxovskij 1954: 351), something regarded as archaic at best today. Medieval examples are given in (33), with a subject-control clause; and in (34), with an object-control clause.

(33) ... ne xotja bog"... dati nam" na utexu
NEG want.PAST.3SG God give.INF us.DAT for consolation
groba ego.
[grave his].GEN
'...God did not want...to give us his grave as consolation.' (OES)
(*Novgorodskaja pervaja letopis'*, p. 29, ll. 35–6)

(34) I vsi ljudie ne daša emu složiti **věn'cja**
and all people NEG allow.PAST.3PL him.DAT renounce.INF crown.GEN
'And all the people would not allow him to renounce the crown.' (OES)
(*Novgorodskaja pervaja letopis'*, p. 47, ll. 26–7)

Nineteenth-century Russian examples are given in (35), with a subject-control clause; and in (36), with an object-control clause.

(35) Esli vy ne soglasites' otdat' **ruki** Mar'i Antonovny...
if you NEG agree.PRES.2PL give.away.INF hand Marija Antonova.GEN
'If you do not agree to give Marija Antonova away in marriage...' (Russian)
(Gogol', *Revizor*, 1836) (Bulaxovskij 1954: 351)

(36) Ja vam daže ne sovetuju **dorogi** znat' k ètoj sobake...
I you.DAT even NEG advise.PRES.1SG way.GEN know.INF to that dog
'I advise you not even to know the way to [the house of] that dog...' (Russian)
(Gogol', *Mertvye duši*, ch. 5, 1842) (Bulaxovskij 1954: 351)

Today, licensing into a non-finite clause is more limited, with subject-control clauses allowing it to a greater extent than object-control clauses, where it is generally judged ungrammatical. Judgements are somewhat unclear: Franks cites the subject-control example in (37) as grammatical, while Bailyn (2004: 10) rates a similar example in (38) as double question mark. In any case, this pattern is clearly declining. The object-control pattern in (39) is ungrammatical.

(37) Nataša ne xotela čitat' **knig.**
Natasha NEG want.PAST.3FSG read.INF books.GEN
'Natasha didn't want to read any books.' (Russian) (Franks 1995: 199)

(38) Poljak ne objazan znat' **francuzkogo jazyka.**
Pole NEG obliged know.INF [French language].GEN
'A Pole is not obliged to know French.' (Russian) (Bailyn 2004: 10)

(39)

*Ja	ne	ugovorila	Natašu	čitat'	**knig.**
I.NOM	NEG	persuade.PAST.3FSG	Natasha	read.INF	books.GEN

'I didn't persuade Natasha to read any books.' (Russian) (Franks 1995: 199)

Padučeva (2006: 25) and Krasovitsky et al. (2011: 588–90) note that indirect negation of precisely this type was a factor disfavouring genitive of negation from the start of the nineteenth century (dotted line in Figure 9.2). Its ungrammaticality today can therefore be regarded, in diachronic terms, as a consequence of this effect. Being in the vanguard of change, it is one of the first environments in which the loss of genitive of negation can be regarded as having run to near completion.

9.4.4 Genitive of negation on subjects of negated existential, locative, and unaccusative clauses

In the early Slavonic languages, there is variation as to whether the subject of a negated existential or locative with 'be' appears in the genitive or in the nominative. In Old Church Slavonic, both options are found about equally (Večerka 1993: 75, 1995: 521, 1996: 144–5). In (40), we find the nominative subject *ogńĭ* 'fire', while, in (41), we find a genitive subject *korablě inogo* 'other boat' (against a nominative subject in the Greek original).

(40)

otŭkǫdu	sŭgrěvajetŭ	sę	ašte	něstŭ	**ogńĭ**	podŭ	zemĺejǫ
whence	warm.PRES.3SG	REFL	if	NEG.be.PRES.3SG	fire.NOM	under	ground

'How does it become warm if there is no fire under the ground?'
(translates Greek *πόθεν πυροῦται, εἰ μὴ ἐκβαῖνον ἐν ὑπογαίῳ πυρί*) (OCS) (*Codex Suprasliensis* 129.10–11) (Večerka 1995: 521)

(41)

korablě	**inogo**	ne	bě	tu
[boat	other].GEN	NEG	was	there
Πλοιάριον	*ἄλλο*	*οὐκ*	*ἦν*	*ἐκεῖ*

'... there was no other boat there.' (OCS) (*Codex Marianus*, John 6:22)

Medieval East Slavonic texts also show nominative subjects of negative existential *byti* 'be' alongside genitive subjects (Borkovskij and Kuznecov 2007 [1963]: 401). Compare the following pairs of very similar sentences, in which the first in each case, (42) and (44), has a nominative subject, and the second in each case, (43) and (45), has a genitive one:

(42)

...i	bě	mjatež'	Nověgorodě,	a	**Svjatoslav**	d"lgo	ne	bjaše.
and	was	revolt	Novgorod.LOC	and	Svjatoslav.NOM	long	NEG	was

'... and there was a revolt in Novgorod, and Sjatoslav for a long time was not [there].' (OES) (*Novgorodskaja pervaja letopis'*, p. 25, ll. 32–3)

(43) V Nově že gorodě byst' mjatež' velik": ne bjaše bo **knjazja Jaroslava**...
in Nov- PRT -gorod was revolt great NEG was.IMPF PRT [prince Jaroslav].GEN
'In Novgorod there was a great revolt, for Prince Jaroslav was not [there]...' (OES) (*Novgorodskaja pervaja letopis'*, p. 71, ll. 34–5)

(44) ...i **braci** ne byvaxu v" nix
and marriages.NOM NEG be.IMPF.3PL with them
'...and they did not have marriages.' (OES) (*Povest' vremennyx let*, p. 11, l. 4–5)

(45) ...i **braka** u nix" ne byvaše
and marriage.GEN with them NEG be.IMPF.3SG
'...and they did not have marriage.' (OES) (*Povest' vremennyx let*, p. 11, l. 1)

Parallel examples are found in West Slavonic, suggesting that this variability is an inheritance from Common Slavonic, witness the following example from Old Polish:

(46) Nie jest **zbawienie** w ciele mojem od oblicza gniewu twego.
NEG is salvation.NOM in body my from face.GEN [anger your].GEN
'There is no salvation in my body because of the face of your anger.' (Old Polish) (*Psałterz Floriański*, Psalms 37:3 [=38:3]) (Večerka 1958: 196)

Some other semantically very closely related verbs, such as *ostati sja* 'remain', also allow genitive subjects from an early date in East Slavonic:

(47) ...ne ostasja ni **xoroma**.
NEG remain.PAST.3SG not.even house.GEN
'...no house remained.' (OES) (*Novgorodskaja pervaja letopis'*, p. 57, l. 25)

(48) ...a posle evo **ženy** i **detei** ne ostalos
and after him wife.GEN and children.GEN NEG remain.PP.NSG
'...and after him, no wife or children remained.' (OES) (*Pamjatniki južnovelikorusskogo narečija: Otkaznye knigi* 230) (Kotkov and Popova 1986: 92)

As in modern Slavonic languages, when the subject is marked genitive, the verb is in a default neuter singular form irrespective of the case-number features of the subject, hence *ostalos'*, not plural *ostalis'* in (48). However, with a nominative subject in (44), there is subject-verb agreement (*byvaxu*). For further extensive lists of examples from Old East Slavonic and Old Church Slavonic, see Potebnja (1968 [1899]: 377–9).

Literary Russian has experienced a significant reformulation and expansion of the contexts in which genitive subjects are used in negative existential clauses. In Old East Slavonic, the phenomenon is limited to subjects of *byti* 'be' and some near-synonyms. Vinogradov and Švedova (1964: 304–12) note that, by the end of the eighteenth century, subjects of negative existential clauses involving the verbs *byt'* 'be', *stat'* 'turn out to be', *byvat'* 'be (frequently)', *slučit'sja* 'happen (to be)', and *okazat'sja* 'turn out to be' were obligatorily marked genitive, all contexts where it is strongly favoured or compulsory today. In contrast to present-day Russian, however, in the eighteenth and early nineteenth centuries, subjects in negative existential clauses could be marked genitive irrespective of whether they were definite or indefinite. Consider the following example with a highly definite subject in the genitive:

(49) Togda ešče **pod"ezžačego pod-moskovnogo Petrovskogo dvorca**
then yet [approaching near-Moscow Petrine palace].GEN
postroeno ne bylo.
built.NSG NEG was.NSG
'At that time, the Petrovskij Pod"ezdnoj Palace had not yet been built.' (Russian) (Gavrila Deržavin, *Zapiski* ii, 1808–12) (Vinogradov and Švedova 1964: 307–8)

During the course of the nineteenth century, the genitive gave way to the nominative with definite subjects, disappearing in literary Russian by the end of the century (Vinogradov and Švedova 1964: 308). Conversely, while the genitive contracted with definite subjects, it expanded with indefinite subjects. This is illustrated by the fact that indefinite subjects with *nikakoj* 'no', *ni odin* 'not a single', and *takoj* 'such' allowed a nominative in the early nineteenth century, whereas they strongly favour a genitive today:

(50) V prodolženie četyrex stoletij ne voznikla v Evrope
in course [four centuries].GEN NEG develop.PAST.FSG in Europe
ni odna otrasl' nauk ili xudožestv.
[not.even one branch].NOM [sciences or arts].GEN
'During the course of four centuries not a single branch of the sciences or arts developed in Europe.' (Russian) (*Žurnal Departamenta Narodnogo Prosveščenija*, May 1823) (Vinogradov and Švedova 1964: 309)

A second aspect of change in the nineteenth century is that the range of predicates with which genitive subjects could be used widened to include any unaccusative (including passive) contexts in which the non-existence of the subject was asserted. While at the beginning of the century, genitive was lexically restricted to verbs of motion, appearance/coming into being (*sostojat'sja* 'take place'), and verbs of finding (*naxodit'sja* 'be, be found'), by the end of the century it was found also with predicates of speaking and perception (*videt'sja* 'be seen', *čuvstvovat'sja* 'be felt', *sprašivat'sja* 'be asked') (Vinogradov and Švedova 1964: 309–10). Babby defines the current set of predicates as any that 'denote the subject noun's most *typical action* from the point of view of the human participants in the speech event' (Babby 1980: 134, italics in original). Unaccusativity also seems to be a necessary (but perhaps not sufficient) condition (Pesetsky 1982: 42–69). This is illustrated nicely by (51), which is grammatical only with the unaccusative interpretation of *plavaet* to mean 'float', not with its agentive meaning 'swim':

(51) V bassejne nikakogo rebenka ne plavaet.
in swimming.pool [no child].GEN NEG float.PRES.3SG
'There is no child floating (*swimming) in the pool.' (Russian) (Pesetsky 1982: 45)

The result of this is that the same interpretative effects found with genitive vs accusative case selection on direct objects are now felt with subjects. In the pair in (52) and (53), example (52) is taken to indicate that no frost was felt because there was no frost, while (53) leaves open the possibility that frost was present but could not be detected.

(52) Moroza ne čuvstvovalos'.
frost.GEN NEG be.felt.PAST.NSG
'No frost was felt.' (Russian) (Babby 1980: 59)

(53) Moroz ne čuvstvovalsja.
frost.NOM NEG be.felt.PAST.MSG
'The frost was not felt.' (Russian) (Babby 1980: 59)

Dialectal use is more conservative: nominative survives dialectally in both Russian and Ukrainian in contexts that are prototypical genitive contexts in literary Russian (Borkovskij and Kuznecov 2007 [1963]: 401). In traditional northern Russian dialects and elsewhere, the nominative construction is usual (Šapiro 1953: 140–1, 96–8), hence nominative *vojna* 'war' in the following example:

(54) Dédjuška dváccjat' pet' let služil v saldatax —
grandad twenty five years serve.PAST.3SG in soldiers
né byla vojna.
NEG was.PAST.NSG war.NOM
'Grandad served twenty-five years in the army (but) there was no war.' (Arkhangel'sk Russian)

9.4.5 *Genitive of negation on temporal adjuncts*

Genitive of negation on temporal and some locative adjuncts is possible in Russian, but is less productive than on arguments (Babby 1980: 85):

(55) Togda ja ne probyl doma daže pjati dnej.
then I NEG stay.PAST.MSG at.home even [five days].GEN
'I didn't stay at home then even for five days.' (Russian) (Babby 2001: 40)

(56) Ja ni odnu minutu / odnoj minuty ne spal.
I not.even [one minute].ACC [one minute].GEN NEG sleep.PAST.MSG
'I didn't sleep even for a minute.' (Russian) (Pesetsky 1982: 64)

This use goes back to the early varieties, as witnessed by the Old Church Slavonic example in (57), corresponding to a Greek accusative and an expected accusative in corresponding affirmative sentences (Večerka 1993: 251).

(57) Tako li ne vŭzmože edinogo časa pobŭděti sŭ mŭnojǫ
thus QU NEG can.PAST.2SG [one hour].GEN keep.watch.INF with me
'Thus could you not keep watch with me for one hour?' (OCS) (*Codex Marianus*, Matt. 26:40)

Different views have been expressed as to whether this should be regarded as a subcase of the genitive of negation or as a distinct phenomenon. Borovikoff (1997) argues for the former position. Franks and Dziwirek (1993) and Franks (1995: 205–9) argue for the latter, treating it as a special type of partitive licensed only in negative environments. They note that the adjunct genitive of negation is found in Serbian and Croatian, which no longer have genitive of negation generally on direct objects; and, in Polish, while direct objects in the scope of negation are obligatorily genitive, temporal adjuncts in the scope of negation are only optionally genitive. Similarly, Slovene has obligatory genitive of negation on direct objects, but does not allow it on temporal adverbs in the scope of negation. These pieces of evidence lead to the conclusion that it is not possible to assimilate adjuncts to direct objects with respect to genitive of negation.

9.4.6 The development of the genitive of negation across Slavonic

Polish has generally been conservative with respect to the genitive of negation, retaining it as an obligatory rule with direct objects in the scope of negation, (58). It is also found on temporal adverbs in the scope of negation, negated subjects of *być* 'be', (59), and *zostać* 'remain', but not of unaccusative verbs in general, (60) (Błaszczak 2001: 60–65, Guiraud-Weber 2003: 365–6):

(58) Ewa nie karmi ptaków / *ptaki.
Ewa NEG feed.PRES.3SG birds.GEN birds.ACC
'Ewa doesn't feed the birds.' (Polish) (Błaszczak 2001: 61)

(59) Na stole nie było gazet / *gazety.
on table NEG be.PP.NSG newspapers.GEN newspapers.NOM
'There were no newspapers on the table.' (Polish) (Błaszczak 2001: 63)

(60) Studenci / *Studentów nie przyszli na wykład.
students.NOM students.GEN NEG come.PP.PL to lecture
'The students didn't come to the lecture.' (Polish) (Błaszczak 2001: 63)

Polish also retains widespread genitive of negation within non-finite complements (Witkoś 2008: 249–50):

(61) Maria nie pozwoliła Janowi pić kawy / *kawę.
Maria NEG let.PAST.3FSG Jan.DAT drink.INF coffee.GEN coffee.ACC
'Maria did not let Jan drink coffee.' (Polish) (Witkoś 2008: 249)

This system has been largely historically stable, as shown by the data in Figure 9.3 and Table 9.2. Genitive of negation remains a purely syntactic phenomenon in Polish and has not gained the ability to reflect semantic or information-structural distinctions as it has in Russian. While there has been a slight shift towards use of the

TABLE 9.2 **The development of accusative direct objects (%) in negative clauses in Polish (data calculated from Harrer-Pisarkowa 1959: 12)**

time period	Old Polish	16th c.	17th c.	18th c.	19th c.
direct negation					
% accusative	7	1	1	3	4
total tokens	244	379	520	450	1069
indirect negation (object of non-directly negated infinitive)					
% accusative	11	1	2	13	11
total tokens	53	121	130	174	378
***ani* 'nor'-clauses (not on graph)**					
% accusative	50	0	6	20	71
total tokens	22	24	16	5	7

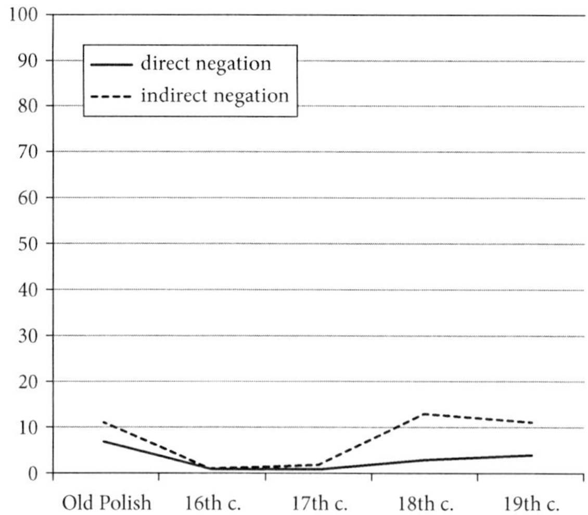

FIGURE 9.3 The development of accusative direct objects (%) in negative clauses in Polish (constructed using data from Harrer-Pisarkowa 1959: 12)

accusative of direct objects in the modern period, it is not clear that this is due to systematic language change, rather than fluctuating foreign influence on the literary language or fluctuating presence of certain sentence types (e.g. rhetorical questions) favourable to the accusative. The contrast with the recent history of Russian in Figure 9.2 above is striking. Accusative case is becoming increasingly acceptable in the context for long-distance genitive of negation (sentences parallel to (61) above), and this may reflect the start of real change beyond the nineteenth-century sample in Figure 9.3 (Przepiórkowski 1999: 143–50, see Rybicka-Nowacka 1990). With subjects of existential and unaccusative predicates, the system has also largely been stable historically, and Polish has not experienced the expansion in the use of the genitive of negation that Russian has in this environment. The restriction to *być* 'be' and a few other verbs seems to reflect the Common Slavonic situation fairly accurately.

Genitive of negation in Ukrainian broadly follows the patterns found in Russian, including a wide range of subjects of unaccusatives, except that genitive direct objects are acceptable in more contexts than in Russian (Kryshevich 2010). It thus occupies a position intermediate between Russian and Polish.

The genitive of negation has almost completely disappeared in Czech. Its decline has been charted in detail in an extensive study by Hausenblas (1958). In Old Czech, genitive of negation was ubiquitous with both main-clause and embedded direct objects:

(62) Proto **kněžě** **Otty** nerodichu
thus [prince Ota].GEN NEG.heed.PAST.3PL
'Thus they did not heed prince Ota.' (Old Czech) (*Dalimilova kronika*, ch. 57, l. 45) (Hausenblas 1958: 49)

(63) nekazuj nám přejíti **Jordana**
NEG.order.IMP.2SG us.DAT cross.INF Jordan.GEN
'Do not make us cross the Jordan.' (Old Czech) (*Bible Olomoucká*, Numbers 32:5, 1417) (Gebauer 1929: 333)

With subjects of existential clauses using *býti* 'be', both nominative and genitive are possible and well attested. For instance, variation is found in the early Bible translations, contrast nominative *jalová* 'barren' in (64) with genitive *jalové* in (65).

(64) ...a jalová nenie mezi nimi. (*Sborník Mikulovský*, 1406)

(65) ...a jalové nenie mezi nimi. (*Bible Pražská*, 1488)
and barren.GEN/NOM.FSG NEG.is among them
'...and there is no barren one among them.' (Old Czech) (Song of Songs 4:2) (Hausenblas 1958: 55)

Some examples of the genitive marking are also found on subjects with other existential verbs such as *nezůstalo* 'did not remain' (Gebauer 1929: 333–4). In the sixteenth century, use of the genitive strengthens in existential constructions, while exceptions to genitive on transitive objects remain limited, common only in translated texts. This suggests that the two constructions had been structurally distinct up to this point, but were reanalysed as structurally parallel around this time. It seems that, in the initial system, in terms of information structure, nominative subjects of 'be' coincided with a topic interpretation of the subject, while genitive subjects coincided with an interpretation of the clause as an undivided comment. Thenceforth their development is parallel: from around 1620, the genitive declines rapidly in both contexts in vernacular sources from Prague, disappearing almost completely by 1730. An early example of accusative (*pozor* 'attention'), where genitive would be expected historically, is given in (66), from vernacular court testimony. The loss of genitive subjects in negative existential clauses runs parallel to that of genitive objects in negative transitive clauses.

(66) ... na to sem **pozor** nedal, co více mluvil.
to that be.PRES.1SG attention.ACC NEG.give.PP.3MSG what more say.PP.3MSG
'... I didn't pay attention to what else he said.' (Czech) (*Knihy Svědomí Nového Města Pražského*, MS 1123, f. 62a, 1625) (Hausenblas 1958: 66)

Literary Czech retains the genitive much longer, but nevertheless shows gradual decline in its use from the seventeenth century onward as it came closer into line with spoken usage. Today, the remaining contexts for it in literary Czech are emphatic, signalling complete absence, especially in fixed expressions. Among vernacular dialects, it survives well in eastern Moravia, although never as consistently as in Polish (Hausenblas 1958).

This contraction is generally attributed to two factors: the general decline of the genitive as a case for marking direct objects in Czech (cf. the parallel, but independent, general trend discussed for Russian above); and contact with German, which both lacks any genitive of negation rule and itself witnessed a decline in the use of the genitive as a case to mark direct objects.

In Slovak, negative direct objects in transitive clauses have also largely switched to the accusative, while negative existential subjects retain the genitive, more robustly in the present with *niet*(*u*) 'there is not' than in other tenses (Hausenblas 1958: 116). There has been extensive variation in case usage since the fifteenth century (Stanislav 1967–73: iv.160–78).

The genitive of negation remains operative in Sorbian, but is in decline. It was fully present in the seventeenth century (Schuster-Šewc 2001: 293), but today is optional in all varieties. It is better preserved in Lower Sorbian than in Upper Sorbian: Faßke and Michalk (1981: 459) provide figures of 28.5% for use of the genitive with relevant direct objects in Lower Sorbian dialects, but only 4.9% in Upper Sorbian dialects. Within Upper Sorbian, the literary language preserves it better, above all with the object of certain verbs, such as *měć* 'have', *dać* 'give', *dostać* 'get', or *wědźeć* 'know', where it correlates with the use of the genitive as the case for partitive objects. The nominative may also be replaced by the genitive in negative existential constructions, predominantly with *być* 'be', but also with synonyms and verbs of related meaning, such as *zwostać* 'remain', *wobstać* 'exist', and *falować* 'be lacking' (Faßke and Michalk 1981). As with Czech, contact with German may have played a role in the decline of the genitive of negation in Sorbian.

In South Slavonic, Slovene retains obligatory genitive of negation with direct objects:

(67) Ne vidim **hiše** / ***hišo**.
NEG see.PRES.1SG house.GEN / house.ACC
'I don't see the house.' (Slovene)

This is also the case with the subject of *biti* 'be':

(68) **Jajc** ni bilo v hladilniku.
eggs.GEN NEG.be.3SG be.PP.3NSG in fridge
'There were no eggs in the fridge. (Slovene) (Herrity 2000: 63)

However, Slovene has not innovated it with other unaccusatives:

(69) a. **Nihče** ni prišel.
no.one.NOM NEG.be.3SG come.PP.3MSG
b. ***Nikogar** ni prišlo.
no.one.GEN NEG.be.3SG come.PP.3NSG
'No one came.' (Slovene)

Genitive of negation spreads obligatorily into non-finite complements of modal and subject-control verbs (*Ne morem zatisniti* ***očesa*** 'I can't close my eye (gen.)', *Jan ni pozabil omeniti* ***svojih otrok*** 'Jan didn't forget to mention his children (gen.)'), and optionally into non-finite complements of object-control verbs (*Marija ni prisilila Jana piti* ***kave/kavo*** 'Maria didn't force Jan to drink coffee (gen./acc.)'. It thus essentially retains the inherited Common Slavonic situation unchanged, in parallel with Polish.

Serbian and Croatian manifest genitive case in all existential constructions with both affirmative and negative forms of existential clauses (involving *ima* 'there is/are' and negative *nema* 'there isn't/aren't' in the present tense, and forms of *biti* 'be' in other tenses):

(70) Da, knjige ima, ali markice nema.
yes book.GEN have.PRES.3SG but stamp.GEN NEG.have.PRES.3SG
'Yes, there is a book, but there isn't a stamp.' (Serbian) (Hartmann and Milićević 2008: 173)

Even here though, negation plays a role, as there are some environments where genitive is optional or disallowed in the affirmative version of an existential clause, but compulsory in the negative one. Specifically a singular noun phrase headed by *jedan* 'one' may occur in the nominative in an affirmative existential construction, but must be genitive in its negative counterpart:

(71) Ima jedna zanimljiva knjiga.
have.PRES.3SG [one interesting book].NOM
'There is one interesting book.' (Serbian)

(72) Nema nijedne knjige / *nijedna knjiga.
NEG.have.PRES.3SG [not.one book].GEN [not.one book].NOM
'There isn't a single book.' (Serbian)

Thus, there is a mechanism for assigning genitive of negation to existential subjects; however, genitive of negation on direct objects of the kind found elsewhere in Slavonic has been lost. Likewise, Bulgarian has no remnants of genitive of negation, having given up case marking on lexical noun phrases in Middle Bulgarian (Ivanova-Mirčeva and Xaralampiev 1999: 185–7).

9.4.7 Theoretical approaches to genitive of negation

The dominant synchronic approaches to genitive of negation in Russian have tended to focus on the semantic conditioning of case choice. This makes them rather difficult to reconcile with the constantly shifting diachronic situation and with the comparative perspective that shows semantic conditioning to be only a fleeting property of genitive-of-negation systems under conditions of change.

For Babby (1980), genitive of negation in Russian is assigned to the sister of V when both the verb and the noun phrase are in the scope of negation, $[_{ne}$ V $NP_{gen}]$. This configuration sums up the following conditions that he proposes:

(a) the NP must be in the scope of negation;
(b) the V must also be in the scope of negation;
(c) the NP must be non-oblique (the direct-case condition);
(d) the NP must be indefinite/non-referential;
(e) the verb must be semantically empty/light by denoting the subject's typical action e.g. 'thunder is heard', 'cold is felt', 'rain falls', 'trees grow', etc.

Pesetsky (1982) attributed genitive of negation to a null quantifier, arguing that genitive-marked phrases were in fact QPs. When they appear to be subjects of existential/unaccusative predicates, they do not trigger subject–verb agreement because the phrase as a whole lacks Case and is not a true subject, failing to act in other ways as a subject (e.g. in terms of binding). This is developed further by Bailyn (2004), who proposes that the negative head (Neg) selects a complement marked with a quantified feature [+q]. This feature is selected in all of the subsequent phrases, and a verb with this feature selects a QP, whose head Q assigns genitive to its NP complement. He addresses the difference between Russian and Polish, the latter allowing long-distance genitive of negation, by suggesting that Polish allows V to select for complement clauses marked with [+q], while Russian does not. Thus the selection chain is broken by a CP in Russian but not in Polish. Historically, since the Polish pattern is clearly the historically prior one, this would mean that Russian has innovated an exception to the chain selection patterns. While descriptively accurate, it seems unlikely that acquirers of Russian would innovate such an exception, and the account fails to integrate the change into the more general decline in the genitive of negation in Russian, of which it surely forms an integral part.

Another approach is to associate genitive and accusative arguments (or nominative in the case of negated existentials) with different structural positions, with genitive being checked inside the scope of existential closure and accusative being checked outside it. Bailyn (1995: 331–4, 1997) argues that the genitive of negation is associated with a unique structural position, [Spec, VP]. The basic architecture of the system is given in (73): a VP-shell structure is posited where (agentive) subjects are merged into the specifier of a predication phrase, [Spec, PrP]; direct objects and unaccusative subjects are merged into [Spec, VP] and oblique complements are merged as complements to V.

(73)

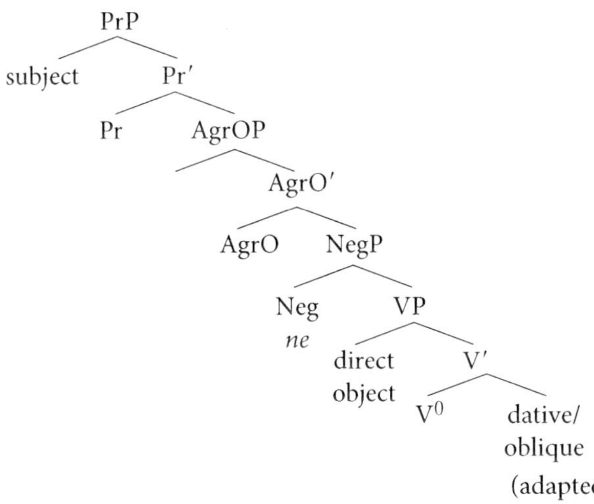

(adapted from Bailyn 1997: 88, 95, 105)

Bailyn argues that the negative head assigns genitive case (under government) to [Spec, VP]. This allows genitive to be assigned to direct objects and unaccusative subjects. Existential closure is assumed to take place at NegP, hence any element that remains within VP receives an existential reading. A direct object may escape genitive of negation by checking Case with AgrOP, an accusative-Case assigner. In doing so it moves to [Spec, AgrOP], thereby escaping existential closure and receiving a presupposed rather than an existential reading. Agentive (transitive and unergative) subjects in [Spec, PrP] and oblique objects in [Comp, V] escape this entire system and so can only be marked in their usual way as nominative or an oblique case respectively. Bailyn suggests that, in Polish, the Neg head raises to AgrO, where it must be checked overtly, forcing genitive on all objects in the scope of negation; however, in Serbian and Croatian, Neg does not check genitive. Historically, this would mean that Russian has lost raising of Neg to AgrO, making genitive of negation optional. While this seems to model the development of direct-object genitive of negation, it is difficult to see how it can deal with the differential development of existential subjects in the scope of negation: here the tendency in Russian has been to extend the use of the genitive (but see section 9.4.8 below, where it is argued that the two are, in any case, historically distinct). Another approach broadly along these lines is that of Brown (1999), who argues that genitive of negation is checked in SpecNegP by a combination of the features [+Neg], found on Neg, and [+Vmax] found on predicates that take an internal argument (that is, transitives and unaccusatives). Harves (2002) analyses genitive of negation as involving projection of a defective V that fails to value accusative case on its complement, allowing a higher Neg-head to value genitive on the complement of the verb.

9.4.8 Origin of the genitive of negation

The genitive of negation on direct objects clearly goes back to Common Balto-Slavonic, hence its emergence cannot be observed directly. It has generally been suggested that it emerged from partitive uses of the genitive (Kuryłowicz 1971, Meillet and Vaillant 1965). The idea goes back to Minkov (1911), who suggested that the emergence of the partitive of negation was the Slavonic counterpart to the emergence of strengthening particles such as French *pas*, English *not*, or German *nicht* in Western European languages (see section 1.4); the Slavonic genitive expressed the same partitive meaning as minimizers in other languages. A partitive in the scope of negation amounts to a minimizer, making a stronger and hence more emphatic claim than an ordinary accusative direct object by suggesting that the negation holds even of a part of the direct object's referent. Use of the genitive was initially therefore a way of conveying emphatic negation with mass nouns and plural count nouns. According to this hypothesis for the reconstructed development, use of the genitive was subsequently grammaticalized as a means to convey emphatic negation with any noun phrase. However, this effect was reduced through overuse to such an extent that genitive case of all direct objects in the scope of negation, including for instance objects headed by singular count nouns, became grammaticalized as a syntactic rule (Levinson 2006: 4–6, Minkov 1911: 389–94). That is, the source of the genitive case is reanalysed as being syntactic (assigned by negation) rather than semantic (an expression of partitivity). Kuryłowicz already noted the parallel here with partitive objects developing in Old French, suggesting the same pathway of historical development for them too:

(74) de s' espee ne volt mie guerpir
of his sword NEG want.PRES.3SG NEG abandon.INF
'He does not want to abandon his sword.' (Old French) (*Chanson de Roland* 465) (Kuryłowicz 1971: 13)

While a partitive source is likely, the link with partitivity has been broken. Where genitive of negation remains compulsory in most contexts, as in Polish, it has become a syntactic rule unconnected with partitive; even in Russian, where genitive of negation is optional and linked to quantitative interpretations, genitive of negation remains morphologically distinct from partitive for the few nouns that have innovated a distinction between genitive and partitive (Franks 1995: 198–9, Klenin 1978).

This account deals only with the genitive of negation on direct objects, and the relationship between this and other kinds of genitive of negation is by no means historically straightforward. Comparison of the histories of the different Slavonic languages shows that genitive of negated direct objects and genitive of negated existential subjects have rather different histories and did not arise in the same way. The genitive of negated existential subjects seems to derive historically from the use of the genitive in existential sentences generally. While this use is most clearly attested today in Bosnian, Croatian, and Serbian, it is found sporadically across the early Slavonic languages, as in the examples in (75) from Old Church Slavonic and (76) from (northern) Old East Slavonic, and in northwestern Russian dialects today, where it may be an innovation (Meščerskij 1972, Trubinskij 1984: 199). It is also ultimately related to partitive uses of the genitive in other contexts. The availability of genitive marking on the subjects of existentials may therefore be the correct reconstruction for Common Slavonic (Vondrák 1928 [1906–8]: 243–4).

(75) ...da ljubŭve vŭ nixŭ bǫdetŭ.
that love.GEN in them be.FUT.3SG
'...that love may be in them.' (OCS) (*Savvina kniga*, John 17:26)

(76) ino u tebe solodu bylo.
then with you.GEN malt.GEN be.PP.NSG
'...then you had malt.' (OES) (Novgorod birchbark 363, second half of the 14th c.) (Zaliznjak 2004: 159, 606–7)

Hartmann and Milićević (2008), following Błaszczak's (2008) analysis of Polish existentials, propose a distinct configuration for the assignment of genitive to the subject/internal argument of existential predicates in present-day Serbian. They treat existentials as involving a special existential predicate (Pred$_{ex}$), which selects a genitive-assigning quantificational head F, as in (77).

(77)

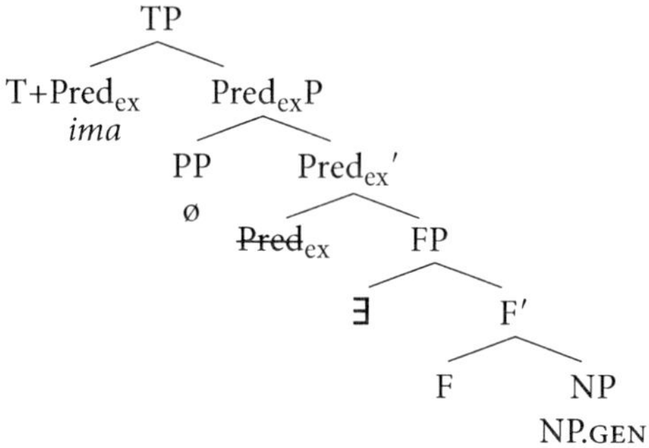

While this has similarities to some proposals for genitive of negation, notably Pesetsky's assignment by a quantificational Q-head, it potentially provides for a very different configuration for the two phenomena which allows us to account for the historical developments. A separate configuration can be proposed for assignment of genitive to objects in the scope of negation, for instance, using Bailyn's proposals discussed in section 9.4.7 above. In Common Slavonic, assignment of genitive to existential subjects and assignment of genitive to direct objects in the scope of negation were structurally different phenomena to be accounted for by distinct syntactic mechanisms (perhaps ultimately related, but only historically, via an earlier link to the partitive genitive). In some languages, they have remained distinct and have experienced different fates: in Czech, they merged and then declined in parallel, while, in Russian, they have largely merged and taken on the same semantic conditioning as they have declined. The development in Serbian and Croatian is less clear: while the modern languages have genitive existential subjects, there is little evidence of the historical continuity of this pattern.

9.5 Negative concord and negative indefinites

We now turn to indefinites, focusing on negative clauses and related contexts, and beginning with the main indefinite series found in negative clauses, the *ni*-series. Negative concord is a characteristic of this series throughout Slavonic. We look at the phenomenon of negative concord, historical exceptions to it, and current analyses of it in section 9.5.1. Later sections focus on the indefinites themselves, beginning with an outline of the system found in Old Church Slavonic (section 9.5.2), which must be close to that of the ancestral language. The development of the two series of indefinites found in all Slavonic languages, the negative *ni*-series and the non-negative *ně*-series, forms the focus of sections 9.5.3 and 9.5.4, after which we turn to more localized developments which have often seen extensive renewal via creation of new series of indefinites. Such developments in East Slavonic are discussed in section 9.5.5 and 9.5.6, while sections 9.5.7 and 9.5.8 concentrate on West and South Slavonic respectively.

9.5.1 *Negative concord*

All modern Slavonic languages have a dedicated set of indefinites which are found (essentially) only in negative contexts and which manifest strict negative concord with the verb. This *ni*-series (Russian *nikto*, Polish *nikt*, Bulgarian *nikoj*, etc. 'no one') obligatorily co-occurs with the marker of sentential negation (negative doubling), as in (78), and a number of *ni*-items may co-occur in the same clause yielding a single-negation interpretation (negative spread), as in (79):

(78)	Nikt	nie	przyszedł. (Polish)
	Nikto	ne	prišel. (Russian)
	Nikoj	ne	dojde. (Bulgarian)
	no.one	NEG	came
	'No one came.'		

(79)	Nikt	nigdy	nie		widział	jednorożca. (Polish)
	Nikto	nikogda	ne		videl	edinoroga. (Russian)
	Nikoj	nikoga	ne	e	viždal	ednorog. (Bulgarian)
	no.one	never	NEG	(is)	saw/seen	unicorn
	'No one has ever seen a unicorn.'					

These patterns clearly go back to Common Slavonic and are present in the earliest varieties. The following Old Church Slavonic example illustrates the earlier syntax of multiple indefinites:

(80)	vŭ	nemĭže	ne	bě	nikogdaže	nikŭtože	položenŭ
	in	which	NEG	was	never	no.one	put.PP.PASS.MSG
	'in which no one had ever been put' (OCS) (*Codex Marianus*, Luke 23:53)						

In general—exceptions are discussed under individual languages—the negative marker must be in the same clause as the *ni*-item (clausemate negation) (including infinitival complements of modal and control verbs). The following Polish example is ungrammatical because the negative marker is in an indirect negative context, that is, one clause higher than the *ni*-item *nikomu*:

(81)	*Ewa	nie	powiedziała,	że	ona	**nikomu**	pokazała	ten	artykuł.
	Ewa	NEG	say.PAST.3FSG	that	she	no.one.DAT	show.PAST.3FSG	this	article
	'Ewa didn't say that she had shown this article to anyone.' (Polish) (Błaszczak 2001: 147)								

In most languages (except Serbian and Croatian), in addition to clausemate negation, *ni*-series items are also licensed inside the complement of *bez* 'without'.

Ni-items are not licensed outside of direct negative contexts and 'without'-phrases or clauses, even in negative polarity environments such as yes–no questions or conditionals. In these environments a range of other indefinites is used, many of which are restricted to negative polarity environments (that is, they are not freely available in assertive environments). While the *ni*-series goes back to Common Slavonic and is essentially uniform across the languages, most of the other series have been innovated within the historical period and vary significantly from language to language.

Present-day Slavonic languages all have strict negative concord; that is, a marker of sentential negation is required with a *ni*-item, irrespective of its syntactic position. However, earlier stages of a number of them show evidence for non-strict negative concord of a type reminiscent of certain Romance languages (Italian, Catalan, and Spanish; see sections 1.10 and 3.8). In this non-strict system, where a *ni*-item follows the verb, the negative marker *ne* is compulsory; however, when the *ni*-item precedes, whatever its grammatical role in the clause, *ne* is optional. Such a system is well documented for Old Church Slavonic (Večerka 1995) and Old East Slavonic (Borkovskij and Kuznecov 2007 [1963]: 403–5). Křížková (1968: 24–5) considers the evidence from both, as well as from Old Czech. Absence of the negative particle with a preposed *ni*-item, whether a subject or an object, is illustrated for Old Church Slavonic in (82) and (83).

(82) nĭ niktože vŭzloži na nĭ rǫku
but no.one lay.PAST.3SG on him hand
'but no one laid a hand on him.' (OCS) (*Evangeliarium Assemani*, John 7:44, f. 32a.28–30) (Křížková 1968)

(83) ničĭsože otŭvěštavaaše
nothing answer.IMPF.3SG
'he answered nothing' (OCS) (*Codex Marianus*, Matt. 27:12) (Křížková 1968)

In (82), the optionality is shown by the fact that another Old Church Slavonic gospel manuscript, the Codex Marianus, has the equivalent phrase with *ne* (*nŭ nikŭtože ne vŭzloži na nĭ rǫku*). Contrast this with (84), where *ne* is present.

(84) nikoliže zapovědi tvoeję **ne** prěstǫpixŭ. i mĭně
never command.GEN your NEG transgress.PAST.1SG and me.DAT
οὐδέποτε ἐντολήν σου παρῆλθον, καὶ ἐμοὶ
nikoliže **ne** dalŭ esi kozĭlęte
never NEG give.PP be.PRES.2SG kid.GEN
οὐδέποτε ἔδωχας ἔριφον
'I never broke your command but you have never given me a kid.' (OCS) (*Codex Marianus*, Luke 15:29) (Večerka 1995: 516)

With a postverbal *ni*-item, negative concord is obligatory:

(85) ... ne jęsomŭ ničesože
NEG take.PAST.1PL nothing.GEN
'... we did not take anything.' (OCS) (*Codex Marianus*, Luke 5:5)

Křížková (1968: 24–5) shows that, in Old Church Slavonic, the pattern with *ne* predominates even when the *ni*-item precedes, but that the pattern without *ne* is nevertheless a significant minority pattern, occurring in around a third of instances in the texts she examined. Večerka (1995: 516) gives a frequency of 1 : 2.6 for the ratio of absence of concord to presence of concord in all negative environments (all orders) in the canonical Old Church Slavonic gospel texts. For East Slavonic examples, see the discussion below.

There has been some debate as to whether non-strict negative concord in Slavonic is an inherited indigenous feature or a literary phenomenon due to contact with Greek, as originally suggested by Buslaev (1959 [1858]: 523). Buslaev suggested that absence of negative concord arose in Old Church Slavonic translations of Greek texts; from there it was imported into Russian, where it was maintained up until the sixteenth and seventeenth centuries in the most elevated literary style. Sprinčak (1960: 225–6) subscribes to this view, as does Bulaxovskij (1958: 406), who nevertheless accepts that its occasional use in vernacular texts (chancery documents) may reflect genuine northern dialect usage. Others have been more sceptical: Jordal (1973: 155–6), in discussing Greek influence on early Russian more generally, leaves the question open, while other more recent views (Borkovskij 1963: 249–51, 1978: 319–36, Borkovksij and Kuznecov 2007 [1963]: 403–5, Meščerskij 1972: 253, Stecenko 1972: 17–19, Zenčuk 1972: 58–61) have tended to accept it as an early Slavonic vernacular feature which has receded across Slavonic during the historical period. Most likely to be correct is the intermediate position adopted by Vaillant (1963–4 [1948]: 357), who suggests that while the non-strict negative-concord pattern is inherited (in OCS) from Common Slavonic, it was in decline from the earliest period and maintained in Old Church Slavonic texts only under Greek influence.

There was register variation from the outset in East Slavonic, with medieval religious texts manifesting absence of negative concord with a preverbal *ni*-item to a far higher degree than vernacular texts: Borkovskij (1978: 324) suggests impressionistically that, in Old East Slavonic religious texts, absence of negative concord where a *ni*-item precedes the verb is about as frequent as its presence, while in secular texts examples of absence of negative concord are rare. From the sixteenth century until its disappearance in literary texts in the eighteenth century it was restricted to bookish, classicizing, or Church Slavonic-influenced contexts in Russian (Křížková 1968: 25). Nevertheless, there are arguments to suggest that the pattern itself may once have been a vernacular one inherited from Common Slavonic. First, it is found, albeit infrequently, in Old East Slavonic in relatively vernacular texts, such as example (86), where the adjunct *ni*-item *nikako* 'in no way' precedes the verb, from a chancery document, or as in (87), from a military account.

(86) Ašče rus'skyi gost' v" rizě... izvinit sja, nikako že
if Russian merchant in Riga commit.offence.3SG REFL no.way PRT
ego v"saditi v dybu.
him put.INF in stocks
'If a Russian merchant in Riga... commits an offence, he is in no way to be put in stocks.' (OES) (*Smolenskie gramoty*, D 36, 1229)

(87) Ni edin ot nix vozvratisja vspjat'...
NEG one of them return.PAST.3SG back
'None of them returned...' (OES) (*Povest' o razorenii Rjazani Batyem v 1237 g., Voinskie povesti drevnej Rusi*, p. 12, ll. 1–2)

It is also found in Ludolf's Russian grammar of 1696 (Borkovskij and Kuznecov 2007 [1963]: 404–5), and is furthermore attested in a number of modern Russian dialects (Borkovskij 1963: 249–51, Bulaxovskij 1958: 366, Meščerskij 1972: 253), as shown in the following example from Šapiro (1953: 198):

(88) Nikto i prišol.
no.one PRT come.PAST.3SG
'No one came.' (Vologda Russian)

Finally, examples of absence of negative concord with a preverbal *ni*-word are found across all of the early Slavonic languages, including Old Czech (Křížková 1968: 24) in (89) and Old Polish (Greszczuk 1993: 44–5) in (90):

(89) Nic mi ... jest v nich známo
nothing me.DAT is in them known.NSG
'I know nothing of them.' (Old Czech) (*Alexandreida* 1163)

(90) Nikt może jego jimienią brać.
no.one can.PRES.3SG his property take.INF
'No one may take his property.' (Old Polish) (Urbańczyk 1953–2002: vol. 270, s.v. *nikt*)

The shift between the two systems has been variously interpreted. Haspelmath (1997: 210–13) argues that the introduction of a marker of sentential negation with postverbal negative indefinites is motivated by Jespersen's Negative First Principle, the tendency, to aid effective spoken communication, for languages to inform the hearer that the sentence is negative as early as possible, and, above all, before the verb. This imperative does not apply with a preverbal negative indefinite, hence the emergence of an asymmetric system. The re-establishment of symmetry, as has happened in the attested history of Slavonic, may then be the result of the analogical spread of the sentential negation marker to all contexts with an indefinite.

Several formal analyses have been proposed. Brown (2003) argues that the shift represents a change in the feature specification of the *ni*-items from the older to the modern Slavonic languages. She posits two possible systems: in one, n-words are marked for polarity as [uPol], while, in the other, they are marked for negation [uNeg]. She further suggests that negation is represented by a null negative operator in Neg, but that this operator bears a [uPol] feature to ensure some overt representation of negation in the clause. Markers of sentential negation, such as *ne*, also bear a [uPol] feature. In languages where they are [uPol], such as Italian, n-words may move to SpecNegP to check the [uPol] feature on Neg (in her system two uninterpretable features may apparently check each other successfully and trigger obligatory movement to a Spec-head configuration in doing so). The result is that a preverbal n-word alone suffices to check negation and a marker of sentential negation such as *ne* is not required (hence is disallowed). In languages where n-words are [uNeg], such as Russian, they cannot check the [uPol] feature on Neg, so, irrespective of where they move, a marker of sentential negation must be inserted. Historically, she argues that early Slavonic languages were in a state of transition between the two systems, allowing both in competition, with the latter system winning out in all the modern languages. This could be thought of as an instance of the general trend for n-words to become more negative cross-linguistically over time.

Zeijlstra (2004, 2008) implies an analysis of the historical development by positing a system to capture the cross-linguistic variation between strict and non-strict negative-concord languages, applied subsequently to Slavonic by Tsurska (2009). For a strict negative-concord language, such as Czech, Zeijlstra adopts the following structures:

(91) Nevolá nikdo.
NEG.call.PRES.3SG anyone
'No one is calling.' (Czech) (Zeijlstra 2004: 251)

(92)

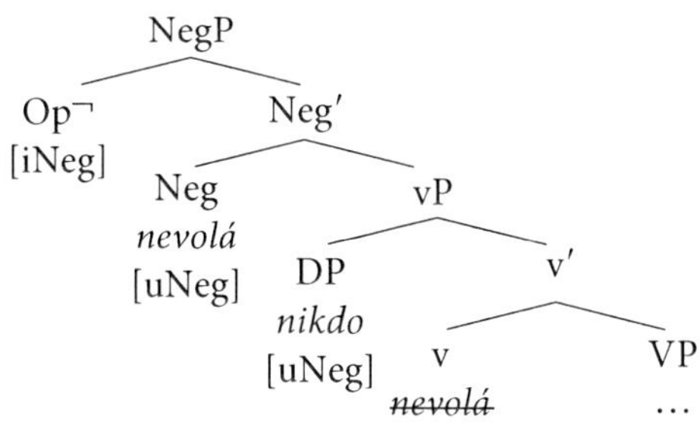

(adapted from Zeijlstra 2004: 251)

Here, the negative marker *ne-* is a prefix associated with an uninterpretable [uNeg] feature that must move to NegP to form a local checking relationship with an abstract negative operator with an interpretable [iNeg] feature. N-words also bear a [uNeg] feature, which is checked by multiple Agree at the same time.

In a non-strict negative-concord language, the negative marker itself is a negative operator in this analysis. Adapting Zeijlstra's Italian structures to the constructed Old Church Slavonic example in (93) gives us the following:

(93) Ne zovetŭ nikŭtože.
NEG call.PRES.3SG anyone
'No one is calling.' (OCS)

(94)

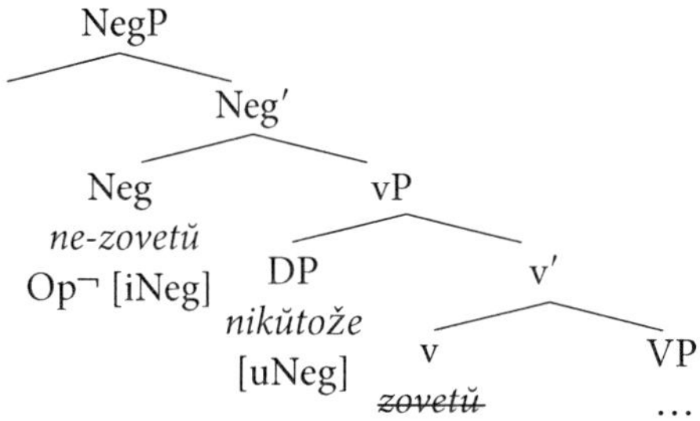

(adapted from Zeijlstra 2004: 251)

If the subject raises beyond this operator to give SV order, it will be outside the scope of negation and not bound by it. The only way for it to do this is to insert an abstract negative operator with which it can form a compound. However, this means that *ne* cannot also be inserted, otherwise we would have two interpretable negations. The relevant structure is:

(95) Nikŭtože zovetŭ.
anyone call.PRES.3SG
'No one is calling.' (OCS)

(96)

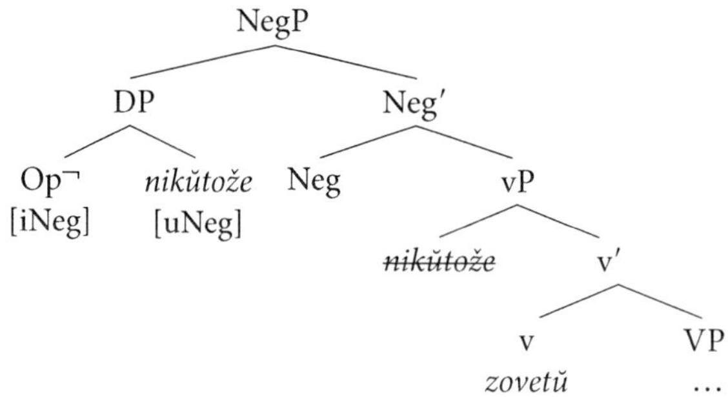

These two systems would have to have co-existed in the early Slavonic languages, with the [uNeg] feature on *ne* eventually winning out over the version of *ne* with [iNeg]. Tsurska speculates that this represents a weakening of the negative marker, which would link this process to Jespersen's cycle.

9.5.2 *Negative indefinites: Old Church Slavonic and the ancestral system*

Having looked at negative concord, we now turn to consider the individual systems of negative indefinites across the Slavonic languages, with a focus on those aspects most relevant for negative and negative-polarity contexts. We begin by considering the Old Church Slavonic system, probably close to the ancestral one, before looking at developments within each language.

Old Church Slavonic exhibits a system of indefinites that must be more or less identical to that of Common Slavonic. The major series of indefinites are based on interrogative *wh*-elements. All indefinites are morphologically either identical to interrogative elements (*kŭto* 'who', *čĭto* 'what', etc.) or derived from them. In negative contexts, the *ni*-series is used:

(97) nikŭto(že) 'no one'
nіčĭto(že) 'nothing'
nikyjĭ(že) 'no'
nikoteryjĭ(že), nikotoryjĭ(že) 'no'
nikŭde(že) 'nowhere'
nikamože 'to nowhere'
nikoli(že), nikogda(že) 'never'
nikakože 'in no way'

(98)

Bljuděte	sę	niktože	vasŭ	da	ne	prělĭstitŭ.
be.careful.IMP.2PL	REFL	no.one	you.PL.ACC	PRT	NEG	decieve.PRES.3SG

'Be careful that no one deceives you / lest anyone deceive you.' (OCS) (*Codex Marianus*, Matt. 24;4)

In assertive, affirmative contexts, another series, the *ně*-series, is used, formed from the prefix *ně*- plus an interrogative (see section 9.5.4 below for detailed discussion of the etymology):

(99) někŭto 'someone'
něčĭto 'something'
někyjĭ 'some'
někotoryjĭ 'some'
někŭde 'somewhere'
někamo 'to somewhere'
několi, někogda 'at some time'
někako 'in some way'

(100) Simone imamŭ ti něčŭto rešti…
Simon.voc have.pres.1sg you.dat something say.inf
'Simon, I have something to say to you…' (OCS) (*Codex Marianus*, Luke 7:40)

In conditionals with *ašte* 'if' or *eda* 'lest', in interrogatives and in some similar environments (*egda* 'when, if', *jako* 'for, as'), ordinary *wh*-pronouns are generally used as indefinites:

(101) kŭto 'who, someone'
čĭto 'what, something'
kyjĭ, kotoryjĭ 'which, some'
kŭde 'where, somewhere'
kamo '(to) where, (to) somewhere'
kogda, koli 'when, at some time'
kako 'how, somehow'

(102) Nŭ ašte kto tę udaritŭ vŭ desnǫjǫ lanitǫ, obrati
but if someone you strike.pres.3sg in right cheek turn.imp.2sg
emu i drugǫjǫ.
him.dat and other
'But if someone strikes you on the right cheek, turn to him also the other.' (OCS) (*Codex Marianus*, Matt. 5.39) (Ivanova-Mirčeva 1999: i.776)

(103) Li kako možetŭ kŭto vĭniti vŭ domŭ krěpŭkaago
qu how can.3sg anyone enter.inf in house strong.gen
i sŭsǫdy ego rasxytiti…
and possessions his plunder.inf
'How can anyone enter into a strong man's house and plunder his possessions…?' (OCS) (*Codex Marianus*, Matt. 12.29) (Ivanova-Mirčeva 1999: i.776)

Plain interrogatives are also used as indefinites in ordinary affirmative contexts, as illustrated for *kŭto* 'who, someone' in (104), but less frequently, and other items seem generally to have been preferred.

(104) Prikosnǫ sę mně kŭto
touch.past.3sg refl me.dat someone
'Someone touched me.' (OCS) (*Codex Zographensis*, Luke 8.46) (In the manuscript, *prikosnǫ vy* has been emended by the scribe to *prikosnǫ*.)

The use of plain interrogative pronouns as indefinites has declined historically throughout Slavonic and is fairly restricted in most of the modern languages.

In comparatives, it seems likely that universal quantifiers were used in place of indefinite pronouns:

(105) skaži namĭ… věru, jaže jestĭ lučĭ vŭsěxŭ
tell.IMP.2SG us.DAT religion.ACC REL.NOM.FSG is better all.GEN.PL
'Tell us… the religion that is better than any.' (OCS) (*Constantinus et Methodius thessalonicenses* 11.1)

A number of other less fixed combinations also occur, for instance, free-choice indefinites with *ljubo* 'pleasing' (< the verb *ljubiti* 'love'), *kŭto ljubo* 'anyone, whoever', *kyjĭ ljubo* 'any, whichever', etc.:

(106) ašte i k'to načĭnetŭ otŭ zagradĭnyixŭ…
if PRT anyone begin.PRES.3SG from gardens
ne t'čiję otŭ ploda nŭ i otŭ listvija i **otŭ kojego**
NEG only from fruit but PRT from foliage and from [which
ljubo sěmene kǫjǫ ljubo vinǫ da ukradetŭ…
ever seed].GEN [which ever reason].ACC PRT steal.PRES.3SG
'if anyone begins to steal from gardens… not only from the fruit but also from foliage or from any seed [anything sown] for any reason…' (OCS) (*Codex Suprasliensis* 42.16–20)

The adjective *eterŭ* 'a certain' is also used very frequently (Rusek 1978), to the extent that it could be considered a semi-grammaticalized member of the *ně*-series in place of *někyjĭ* 'some' (cf. also Večerka 1993: 60–1).

This distribution is summarized in Figure 9.4 using Haspelmath's (1997: 63–4) system of implicational maps for indefinite pronoun functions (for full details, see section 1.9). Dotted lines indicate less common usage. It is unclear what forms were used with indirect negation.

9.5.3 *The* ni-*series*

The *ni*-series goes back to Common Slavonic, derived from the negative particle *ni* 'not even' and an interrogative pronoun. The origin of the series as two distinct words that could at one time be separated by other syntactic material is reflected in the fact that, in the older Slavonic languages and some of the modern ones, *ni* can be split from the interrogative pronoun by a preposition, as in Old Church Slavonic:

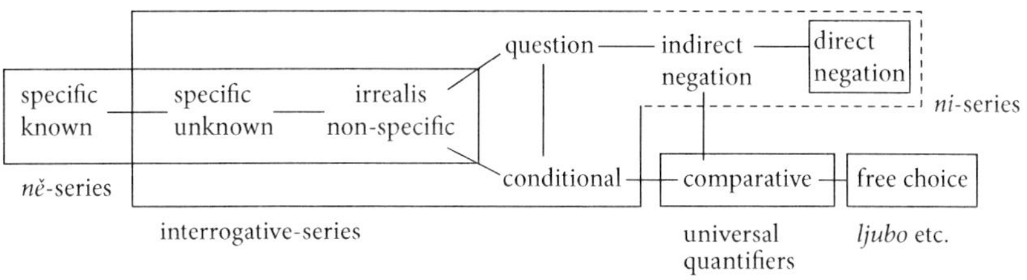

FIGURE 9.4 Expression of indefinites in Old Church Slavonic

(107) i **ni** o **komĭže** ne rodiši
and NEG about who.LOC NEG care.PRES.2SG
'and you do not care about anyone' (OCS) (*Codex Marianus*, Matt. 22:16)

There are also contexts in Old Church Slavonic where an interrogative pronoun alone is used as an indefinite in a negative clause; for instance, use of a plain interrogative in negative-spread contexts, as with *kŭde*, rather than *nikŭdeže* for '(n)ever' in the following example (Večerka 1993: 135–6):

(108) ...žrěbę... na ńeže nikŭtože **kŭde** otŭ č[love]kŭ ne vŭsěde
foal on which no.one ever of people NEG sit.PAST.3SG
'...a foal...on which no one has ever sat...' (OCS) (*Codex Marianus*, Luke 19:30)

These facts suggest that this series arose from merger of the negative particle with interrogatives used as indefinites, the latter thereby disappearing from negative environments to survive above all in non-assertive negative polarity contexts.

The situation with prepositions governing indefinites develops differently across the language group. In some of the modern languages, *ni* now prefixes to the interrogative directly, for instance, throughout West Slavonic (Czech and Slovak *s nikým*, Polish *z nikim*, Upper Sorbian *z nikim* 'with no one'), and much of South Slavonic (Bulgarian *s nikoj, s nikoga*, Slovene *z nikómer* 'with no one'). The remainder of South Slavonic is more complicated: while standard Serbian requires the preposition to split *ni* from the pronoun, hence *ni za šta* 'not for anything', regularized forms such as *za ništa* 'for nothing' are now also possible (Belić 1962: ii.1, 117). East Slavonic languages retain the ancestral system unchanged, for instance, Russian *ni s kem*, Ukrainian *ni z kym* 'with no one'). Earlier forms of all the languages showed the conservative pattern, for instance, Old Czech *ni o čemž jiném* 'about nothing else' (*Alexandreida* l. 2453, Křížková 1968: 29) (cf. also Stanislav 1967–73: ii.656 for earlier stages of Slovak). Even in those languages with splitting constructions, such as Russian, narrow-scope and lexicalized interpretations require *ni-* to be affixed to the pronoun and escape negative concord:

(109) Iz **nikogo** ona prevratilas' v važnuju figuru.
from no.one she turn.PAST.3FSG to important figure
'From a mere nobody she turned into someone important.' (Russian) (Billings 1997: 125)

The same is true after *bez* 'without', which generally does not split a *ni*-item in any of the languages, for instance, Russian *bez nikogo*, Ukrainian *bez nikoho* 'without anyone' (see Hill 1977: 227–9 for confirmation of this on the basis of a historical corpus of Russian and discussion of other cases where splitting fails). These seem to escape negative concord (lacking also the otherwise compulsory concord expressed by *ne* before the verb) because the scope of negation does not reach up to the verb and hence does not reach NegP (or TP), where negative concord is expressed. For other analyses and discussion of the peculiarities of the syntax of indefinites with narrow-scope negation in prepositional phrases, adjuncts, and small clauses, see Progovac (2000) and Fitzgibbons (2010).

Historically then, splitting of *ni*-items has been lost in the history of various Slavonic languages, (presumably) in three independent parallel innovations, one in West Slavonic, one in Slovene, and one in Bulgarian–Macedonian. The simplest interpretation of this is that these languages have fused *ni* + the *wh*-item as a single lexical item (Billings 1997), whereas it was formerly composed of two syntactically active elements, a negative marker *ni* followed by a *wh*-item used as an indefinite. This reanalysis will also have been favoured by the decline in the use of bare *wh*-items as indefinites in the history of Slavonic generally.

Harves (1998) treats the synchronic difference between Russian (with splitting) and Czech (without) as a parametric difference reflecting the presence or absence of movement of *ni* from a position within the prepositional phrase to a polarity phrase generated outside it. This is illustrated in (110), for Russian *ni s kem* and Czech *z nikým* 'with no one'.

(110)

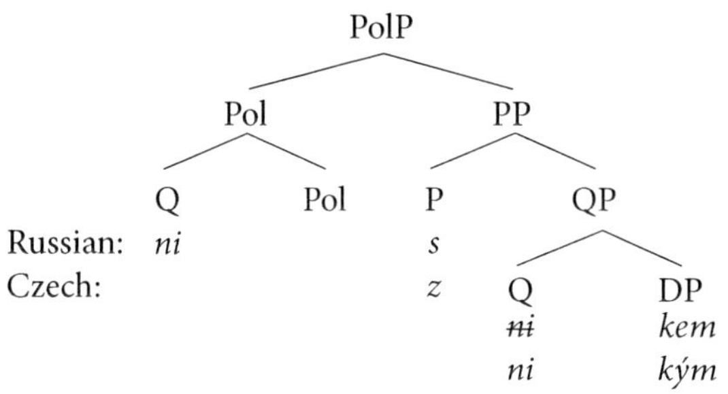

(adapted from Harves 1998: 179–80)

For Harves, Russian *ni* raises from Q to Pol because it bears a strong interpretable [neg] feature that needs to be checked against an interpretable [neg] feature on Pol. The equivalent feature in Czech is weak and does not trigger overt movement of *ni*. Using more recent theoretical assumptions, we can invert the relationship and propose that Pol is generated in quantified prepositional phrases and bears an unvalued polarity feature valued by the quantifier within the prepositional phrase. In Russian, it triggers movement of the Q-head (being marked with an EPP movement-triggering diacritic), while in Czech no movement is triggered. This also allows us to account for the unique behaviour of *bez* 'without', which fails to trigger movement in any language, even Russian: *bez*, being pseudo-negative, can itself value the polarity feature of Pol and move to Pol, freeing *ni* from the need to do this. The natural historical interpretation of such an analysis is that, in the history of Czech and all of the other languages that have lost *ni*-splitting, movement of *ni* to the Pol-head has been eliminated, a common correlate of ongoing grammaticalization.

On this latter account, both *ni* and the *wh*-item remain syntactically active in all the languages. Some evidence against the view that the *ni-* in Polish *nikt* or Czech *nikdo* 'no one' is syntactically active comes from the fact that *ni* in these languages may not occur within a prepositional phrase except as part of an indefinite; that is, Polish displays a contrast between *z nikim* 'with no one' and:

(111) *z ni jednym lingwistą
with not.even [one linguist].INST
'with not even a single linguist' (Polish)

With the reverse order of negative and preposition, judgements are reversed (**ni z kim*, but *ni z jednym lingwistą* is grammatical). This is unexplained if *ni* has the same syntactic status in both cases, suggesting that *nikt* 'no one' is morphologically generated, while *ni jeden* 'not even one' is syntactically generated. There is no such contrast in the 'splitting' languages; for instance, in Russian, both *ni s kem* 'with no one' and *ni s odnim lingvistom* 'not even with a single linguist' remain grammatical.

9.5.4 The ně*-series*

The origin of the *ně*-series is disputed. The second element is clearly a *wh*-item, but the nature of the first element, *ně-*, is less clear. It is not the same as the marker of sentential negation, *ne*, having a historically different vowel, /ě/ (*jat'*), derived ultimately from a long /ē/. This difference is demonstrated by distinct reflexes of the two items in some of the modern languages, for instance, *ně- gives Bulgarian *nja-* (*njakoj* 'someone') while *ne gives *ne*.

One hypothesis does, however, link it to negation, suggesting it derives from a lengthened grade of the negative particle (Trubačev 1974: xxiv.92–3, Vasmer 1953, s.v. *nekij, nekto*). The implication of this hypothesis, developed most recently by Jäger (2010: 813–16), is that the *ně*-series was originally a series of negative indefinites, but had already reversed its distribution in Common Slavonic to become NPI-indefinite series. She claims it remains an NPI today in Slovene, but has developed further in Serbian and Croatian to become a 'normal' indefinite (i.e. one no longer occurring in negative polarity environments at all). The original mirror-image shift (negative indefinite > NPI indefinite) seems unlikely, and the developments proposed for Slovene, Serbian, and Croatian are not attested (*ně*-items are ordinary indefinites, not NPIs, in Slovene, see section 9.5.8 below). Furthermore, since both *ni*-series and *ně*-series must be reconstructed for Common Balto-Slavonic, this raises the question of how the parent language operated with two series of pronouns with identical distributions. Vasmer originally also cited parallels in Celtic and Baltic indefinites, namely Welsh *neb* and Irish *nech* (see section 7.9.2) and Lithuanian *kas ne kas* 'anyone'. However, these merely demonstrate that indefinites may arise from combinations of negative markers plus interrogatives and do not demonstrate that these indefinites will be negative in interpretation or restricted to negative environments.

A second approach derives the *ně*-series from a Common Slavonic (or perhaps Balto-Slavonic) sequence parallel to Old Church Slavonic *ne vě kŭto*... 'I don't know who...' (Haspelmath 1991: 107, 1997: 131–2, Miklosich 1886: 214). The evolution of new indefinites via a grammaticalization pathway from 'I don't know'-sequences is well-attested cross-linguistically, as Haspelmath demonstrates extensively. The appearance of the vowel /ě/ is explained; it derives from the /ě/ in the verb *vě* 'know', while the /e/ in *ne* is dropped. The only problem is the high degree of loss of phonological material. However, this is often found with grammaticalization of new indefinite pronouns. In fact, it seems to be the norm, cf. English *whatchamacallit* < *what you may call it*. If this second hypothesis is correct, *ně*-series items have never occurred in direct-negation contexts, and have always been ordinary affirmative indefinites with a distribution approximating to English *someone*. For further discussion, see Willis (2012). Note also that Russian develops a completely new construction, the *negde spat'* 'there's nowhere to sleep' construction, which appears to use these indefinites as negatives, but which is, in fact, historically unrelated to them (see section 9.5.5 below).

No modern Slavonic language splits *ně*-series items around a preposition. There are some medieval examples in Russian (Bulaxovskij 1958: 184–5), but these could have developed analogically on the basis of splitting with *ni*-items. Stanislav (1967–73: ii.656) reports such cases for historical stages of Slovak, but provides no examples.

9.5.5 *Overview of the development of indefinites in East Slavonic*

East Slavonic retains the *ni*-series essentially intact. The forms in Old East Slavonic are *nik"to* 'no one', *nič'to* 'nothing', *nikakov"*, *nikakyj*, *nikotoryj*, *nikyj* 'no', *nigdě* 'nowhere', *nikako* 'in no way', *nikuda* '(to) nowhere', *ni ot" kudu* 'from nowhere', *nikoli* and *nik"gda* 'never' (Borkovskij and Kuznecov 2007 [1963]: 401). With minor developments, these basically survive into the modern languages. In Ukrainian, *nikyj* is replaced by *žodnyj* (earlier *žadnyj*) 'no' as in Polish (see section 9.5.7 below) from the end of the fourteenth century (Hryščenko et al. 1983: 18). The *ně*-series remains, but is marginal (see full discussion in section 9.5.6).

A number of new indefinite series arise: in Russian, the *to*-series (*kto-to* 'someone', *čto-to* 'something', etc.), the *libo*-series (*kto-libo* 'someone', *čto-libo* 'something', etc.), the *nibud'*-series (*kto-nibud'* 'anyone', *čto-nibud'* 'anything', etc.), the *by (to) ni byl*-series (*kto by (to) ni byl* 'whoever', *čto by (to) ni bylo* 'whatever', etc.). Ukrainian shares innovation of the *nibud'*-series with its own *nebud'*- and *bud'*-series (*xto-nebud'* 'anyone' etc.) and the *by (to) ni byl*-series with its own equivalent (the *b ce ne buv*-series, e.g. *xto b ce ne buv* 'whoever') while adding various others: the free-choice *aby*-series (*aby xto* 'anyone' < *aby* 'in order that, lest', also loaned into Yiddish, van der Auwera and Gybels forthcoming), the pejorative free-choice *kazna*-series *(kazna-xto* 'anybody' etc. < *kat znaje* 'the hangman knows') and the free-choice *xtozna*-series (*xtozna xto* < *xto zna* 'who knows'). While none of these series impinge on direct negation, the *libo*- and *nibud'* series have spread to negative polarity environments and to embedded contexts with superordinate negation, replacing earlier plain interrogatives. In both cases, they derive from original free-choice items whose distribution has spread. In fact, they have been replaced in their original free-choice function by new items (the adjective *ljuboj* 'any' and the particle *ugodno* 'whichever (is pleasing)' < the verb *goditi* 'please', see also section 9.5.8 below).

The *libo*-series derives from *ljubo* 'pleasing' (from the verb *ljubiti* 'love') (cf. its use as a free-choice particle in Old Church Slavonic above); however, it has also been connected with particles *li* 'interrogative particle' and *bo* 'therefore' (Bulaxovskij 1958: 185), which, while phonologically more straightforward, makes little sense semantically. As it spread out from free-choice meanings into the negative polarity domain, it abandoned its original free-choice meaning.

The *nibud'*-series derives from the Old East Slavonic sequence negative particle *ni* plus *budi*, imperative of *byti* 'be', the free-choice interpretation deriving from the meaning 'whoever it may be'. While *ni budi* could originally be separated from the *wh*-item, the two are now inseparable. The transition is shown in the sixteenth-century *Domostroj*, where both patterns are found:

(112) ili kakoe platno ni budi
or which.N cloth NEG be.IMP.2SG
'or whichever cloth it may be, or any (kind of) cloth' (Russian) (*Domostroj* §31) (16th c.) (Bulaxovskij 1958: 185)

(113) ili kakoi ni budi ryby
or which.PL NEG be.IMP.2SG fish.PL
'or whichever fish it may be, or any (kind of) fish)' (Russian) (*Domostroj* §43) (16th c.)

These examples also illustrate the semantic shift by which a free-choice item could come to be an NPI/ordinary indefinite. The *nibud'*-series (with support from Polish free-choice *kto(kolwiek) bądź* 'anyone' and similar items elsewhere) has been calqued into eastern Yiddish as *ver(-'s)-nit-iz* 'anyone, whoever it might be' (van der Auwera and Gybels forthcoming).

As a result of these changes, the modern Russian indefinite system looks like the one represented in Figure 9.5 (Haspelmath 1997: 273, modified in the light of Pereltsvaig 2006: 156–7).

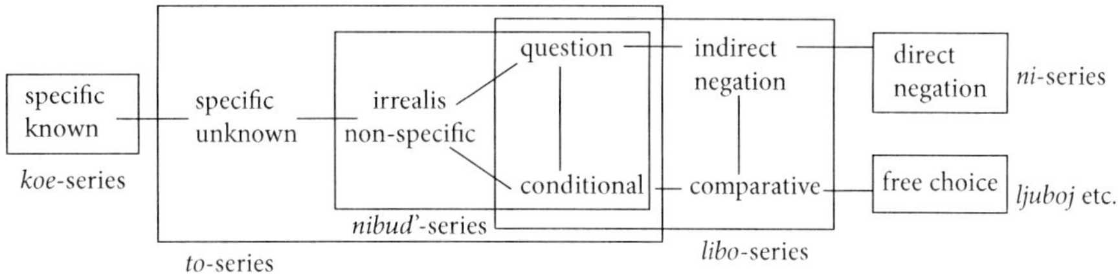

FIGURE 9.5 Expression of indefinites in Modern Russian

9.5.6 *The Russian* negde spat' *construction and new negative* nĕ*-indefinites in Russian*

Special mention should be made of some developments specific to Russian that lead to a confusing synchronic picture with *nĕ*-items. The inherited *nĕ*-items have a limited distribution in Russian today, being available only in a small range of cases: *nekto* 'someone', for instance, is available in the nominative, as in (114), but in the accusative some other item, such as *kogo-to* (accusative of *kto-to* 'someone') must be used, as in (115).

(114) Nekto prišel ko mne.
someone come.PAST.MSG to me
'Someone came up to me.' (Russian)

(115) Ja uvidel kogo-to / *nekogo u dveri.
I see.PAST.MSG someone.ACC / *someone.ACC at door
'I saw someone at the door.' (Russian)

Similarly, *nečto* 'something' is available only in the nominative and accusative, and has no forms for oblique cases. Some items have adverbial uses (e.g. *nekogda* 'at some time'), while others do not (e.g. *negde* in the meaning 'somewhere' is obsolete, *BASRJa* s.v. *negde*).

A completely different construction has arisen, creating forms of an identical morphological structure, such as *nekogo* 'no one (gen.)' or *nečego* 'nothing (gen.)', but available in all cases except the nominative (examples from Rappaport 1986: 1):

(116) Nam **nečego** delat'.
us.DAT nothing.GEN do.INF
'There is nothing for us to do.' (Russian)

(117) Nam **ne** o **čem** budet dumat'.
us.DAT NEG about what.INST be.FUT.3SG think.INF
'There will be nothing for us to think about.' (Russian)

(118) **Negde** bylo spat'.
nowhere be.PAST.NSG sleep.INF
'There was nowhere to sleep.' (Russian)

The interpretation of these sentences is as negative existentials, 'for us there is not anything to do', which is directly reflected in the equivalent sentences in other Slavonic languages, for instance, Czech or Belarusian:

(119) Ale není kam jít.
but NEG.be.PRES.3SG to.where go.INF
'But there isn't anywhere to go.' (Czech) (Růžička 1994: 59)

(120) Njama, kudy exac'.
NEG.have.PRES.3SG to.where go.INF
'There isn't anywhere to go.' (Belarusian) (Lomtev 1956: 78)

The Russian *nĕ*-items have lexicalized as a new series of indefinites. They cannot simply be derived via a phonological merging of *ne-*, analysed as a negative existential verb, and an interrogative pronoun. However, this is the historical origin of the current situation (Lomtev 1956: 78–81, Peškovskij 1956 [1927]: 361, 363–4). In Old East Slavonic, we find regular negation of the existential verb followed by a *wh*-item introducing an infinitival free relative clause:

(121) …i [ne byst' im" [kuda pereĕxati]].
and NEG be.PAST.3SG them.DAT to.where cross.INF
'…and there wasn't anywhere for them to cross.' (OES) (*Ipat'evskaja letopis'*, s.a. 1183)

In the present tense, the negative marker *ne* merged with the form of 'be', *e*(*st'*), to give *nĕ* (found independently of this construction). Where this appeared immediately before the *wh*-item, as in (122), it provided the context for a reanalysis of the *nĕ* + *wh*-item sequence as a single item, a new indefinite.

(122) Uže nam" nĕ [kamo sja dĕti]…
already us.DAT NEG.be.PRES.3SG to.where REFL put.INF
'There is nowhere for us to go…' (OES) (*Povest' vremennyx let*, p. 33, ll. 33–4)

In the new analysis, a null form of existential 'be' (normal in the present tense in Russian) was posited before the indefinite, now interpreted as fronted (cf. modal predicates such as *Prijatno bylo uznat'* 'It was pleasant to learn…', with fronting of *prijatno* 'pleasant'). This reanalysis, with much detail omitted, such as the derivation of the *wh*-dependency, is given in (123).

(123) [$_{TP}$ [$_{Neg+T}$ nĕ] [$_{VP}$ [$_{CP}$ kamo [$_{TP}$ sja dĕti…]]]] =>
[$_{TP}$ nĕkamo [$_{T}$ ø] [$_{VP}$ [$_{CP}$ ~~nĕkamo~~ [$_{TP}$ sja dĕti…]]]]

In tenses other than the present, forms of 'be' must be overt. Given the reanalysis, these could no longer intervene between the negative marker and the *wh*-item, but instead came to occupy a position after the new indefinite, the position where they are found today. Consequently, while Old East Slavonic patterned in every respect with Czech and Belarusian above and manifested complete parallellism between affirmative and negative versions of the construction, Modern Russian shows significant asymmetries of unmarked word order between affirmative and negative, and differs from the other modern languages in apparently having a set of truly negative indefinites in *nĕ*-:

(124) Vozduxu bylo kuda det'sja.
air.DAT be.PAST.NSG to.where go.INF
'There was somewhere for the air to go. / The air had somewhere to go.' (Russian)

(125) Vozduxu nekuda bylo det'sja.
air.DAT to.nowhere be.PAST.NSG go.INF
'There was nowhere for the air to go. / The air didn't have anywhere to go.' (Russian)

Thus, most instances of *nĕ*-indefinites in Modern Russian (all except the type in (114) and (115)) are not cognate with *nĕ*-series items in the other Slavonic languages.

9.5.7 *West Slavonic*

The *ni*-series is retained throughout West Slavonic. It is licensed only by clausemate negation, and generally not by negation in a higher clause except in very limited circumstances. For instance, *ni*-items are licensed in embedded clauses when they are a complement to a noun phrase. In the Czech example in (126), there is optionality between a *ni*-item and other indefinites (*ně*-series *něco* and *koli*-series *cokoli*).

(126) Nemá právo [nic říkat].
have.3SG right nothing say.INF
'He/she doesn't have the right to say anything.' (Czech) (Křížková 1968: 38)

Non-finite complements to modal and volitional verbs also generally require *ni*-items:

(127) Nemohu nic / *něco dělat.
NEG.can.1SG nothing do.INF
'I can't do anything.' (Czech) (Křížková 1968: 38)

While Polish retains inherited forms for most items in the *ni*-series (*nikt* 'no one', *nic* 'nothing', *niczyj* 'no one's', *nigdy* 'never', *nigdzie* '(to) nowhere', *znikąd* 'from nowhere', *nijak* 'in no way'), it innovates one completely new item, namely *żaden* 'no' replacing the expected inherited form *nijaki* 'no' (now obsolete). The same innovation is found in the equivalent series in Czech (*žádný*), Upper Sorbian (*žadyn*), and Ukrainian (*žoden/žodnyj*). The source of this innovation is disputed, with two hypotheses under consideration. One derives it from *ni že jedĭnŭ 'not even one' (*ni* 'scalar negative particle, not even' + *že* 'contrastive particle' + *jedĭnŭ* 'one') (Brückner 1957 [1927], Otrębski 1966). A second hypothesis links it to the verb *žędati 'demand, desire' (cf. Cz. *žádati*, Po. *żądać* 'desire, want') (Miklosich 1886: 409, Vondrák 1928 [1906–8]: 343). Although different versions of this hypothesis exist, the most plausible story is that an adjective, Old Polish *żądny*, developed with a free-choice meaning 'desired, whichever you like' (Greszczuk 1993: 48). Błaszczak (2001: 135) envisages the semantic development to be 'desirable' > 'valuable' > 'rare' > 'no', which seems unlikely, especially in the face of a plausible free-choice pathway. In Old Polish, forms both with and without a nasal vowel are found, witness the nasal vowel in the Old Polish example in (128), and there are dialect forms with a nasal vowel today (Siatkowski 1978).

(128) A nie poznał **żądni** jego.
but NEG know.PAST.3MSG any.MSG him
'But no one recognized him.' (Old Polish) (*Legenda o świętym Aleksym*, l. 139, mid 15th c.) (Greszczuk 1993: 48)

There is also a negative form *niżadny* in Old Polish (Greszczuk 1993: 50), which could represent 'not' plus free-choice 'any'. Although the loss of the nasal vowel in standard Polish is phonologically irregular, these facts seem to suggest that this second hypothesis is correct, encompassing both a plausible phonological and a plausible semantic account. This would mean its development amounted to the extension of an item from free-choice into the general NPI indefinite domain (cf. Russian *kto-libo* 'anyone' and other examples throughout Slavonic).

Polish additionally develops a new series of negative polarity indefinites, the *kolwiek*-series (*ktokolwiek* 'anyone', *cokolwiek* 'anything', *jakikolwiek* 'any', *gdziekolwiek* 'anywhere', *skądkolwiek* 'from anywhere', *kiedykolwiek* 'ever', and *czyjkolwiek* 'anyone's'). Today, these are found in all negative polarity environments except for direct negation (the 'bagel' distribution):

(129) *Ewa nie widziała czegokolwiek.
Ewa NEG see.PAST.3FSG anything.GEN
'Ewa didn't see anything.' (Polish) (Błaszczak 2002a: 381)

(130) Widziałeś tam kogokolwiek?
see.PAST.2SG there anyone
'Did you see anyone there?' (Polish)

(131) Jeżeli ktokolwiek przyjdzie, daj mi znać.
if anyone come.PRES.3SG let.IMP me know.INF
'If anyone comes, let me know.' (Polish) (Błaszczak 2002a: 382)

(132) Ewa napisała test lepiej niż ktokolwiek by się spodziewał.
Ewa write.PAST.3FSG test better than anyone COND REFL expect.PAST.3MSG
'Ewa did better in the test that anyone had expected.' (Polish) (Błaszczak 2002a: 382)

Researchers have generally treated them as ungrammatical with direct (clausemate) negation, although in emphatic contexts they are possible:

(133) Nie wykazała jakiegokolwiek zainteresowania.
NEG show.PAST.3FSG [any interest].GEN
'She didn't show any interest (at all)/even the slightest interest.' (Polish) (Błaszczak 2002a: 387)

Kolwiek-series items develop as free-relative markers in Old Polish and can still be used in this environment today. Etymologically the form is composed of *koli* 'ever' (cf. Old East Slavonic *nikoli* 'never'), later reinforced by *wiek* 'ever' (< *věk). As with other new indefinites, the two elements were formerly independent of one another and could be separated by intervening material (Cieślikowa 1965: 47). There seem to have been two cycles of innovation: in Old Polish, the earliest form was with *koli* or *kole* alone, yielding a *koli*-series of indefinites (*ktokoli* 'anyone' etc.):

(134) Paknyąly kthokole bącz kmyecza zabye...
if anyone ever serf kill.PRES.3SG
'If anyone kills a serf...' (Old Polish) (*Tłumaczenia polskie statutów ziemskich*, p. 67) (Urbańczyk 1953–2002)

Initially, *kolwie(k)*-items seem only to have been used in free relatives:

(135) Ktokolvye chcze pyrvy bycz, then bądzye naposlyednyeyschy
whoever want.PRES.3SG first be.INF he be.FUT.3SG last
'Whoever wants to be first will be last.' (cf. Mark 9:34) (Old Polish) (*Rozmyślanie o żywocie Pana Jezusa*, 370) (Urbańczyk 1953–2002)

However, they spread to use as general NPI-indefinites, ousting *koli*-items in NPI-contexts, especially conditionals, in the sixteenth century (Cieślikowa 1965: 679):

(136) Ieslizebi kthokolwie drugiego w sądzie ranyl...
if anyone.NOM another.ACC in court injure.PAST.3MSG
'If anyone has injured someone else in court...' (Old Polish) (*Historia prawodawstw słowiańskich* VI 273, 1498) (Urbańczyk 1953–2002)

This second innovation extended also into Slovak, where we find a *koľvek*-series, and partially into Czech, where both the standard *koliv*-series and archaic forms in *-kolivěk* are found, as well as the original *koli*-series (Cieślikowa 1965: 77). The modern Polish system after these innovations is given in Figure 9.6. Additionally, in

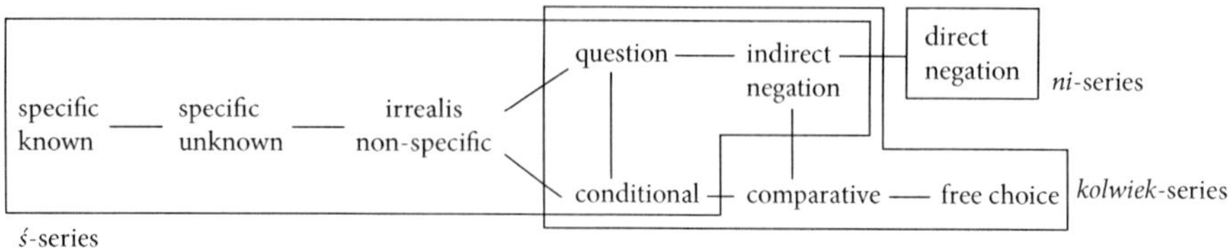

FIGURE 9.6 Expression of indefinites in Modern Polish (Haspelmath 1997: 271)

colloquial Polish, the Common Slavonic use of ordinary *wh*-items as indefinites is retained (Błaszczak 2001: 14–15, 313). The Common Slavonic *ně*-series is partially inherited (e.g. as *nieco* 'somewhat', *niejaki* 'a certain'), but is now marginal and is omitted. Losing out to the *kolwiek*-series in non-assertive contexts, it has been replaced by a new *ś*-series in affirmative declarative ones.

The 'bagel' distribution, typical of Slavonic although not exclusively confined to it, has been seen as problematic, because the most straightforward analysis of *kolwiek*-items would treat them as weak NPIs, yet it is odd that a weak NPI should be inadmissible with direct negation, the most prototypical NPI context. To explain why the Polish *kolwiek*-series, along with the Russian *libo*-series (see section 9.5.5 above), Serbian *i*-series (see section 9.5.8 below), and other related series in Slavonic should be absent from direct negative clauses, two factors seem to be relevant. The first is their origin as free-choice items (or free relatives). Błaszczak (2002b: 10–11) suggests essentially that Polish *kolwiek*-items retain their free-choice meaning today, and that they are hence felicitous in contexts where 'a given proposition holds even for an arbitrarily chosen element (from the range denoted by the common noun)'. She argues that Polish free-choice items cannot appear under direct negation because the arbitrary element of free choice is rendered vacuous in many cases (not (x or y) > not (x) and not(y)). A second factor is the existence of the *ni*-series of strong NPIs/ negative quantifiers in all Slavonic languages, inducing a blocking effect. Items like the Polish *kolwiek*-series seem to be blocked from direct negation precisely because there exists another series specialized for this function. Pereltsvaig (2006) formalizes this intuition for the Russian *libo*-series within a Distributed Morphology approach, where insertion of Russian *libo*- or Polish *kolwiek*-series items would be blocked in direct negative clauses because their licensing requirements are a subset of those of the *ni*-series. Specifically, where two items may be inserted, the more specific one wins.

Czech has two series cognate with and distributionally similar to the two main Polish series in the negative polarity domain, namely the inherited *ni*-series (*nikdo* 'no one', *nic* 'nothing', *nikde* 'nowhere', *žádný* 'no') and the innovated *koli*(*v*)-series (*kdokoli*(*v*) 'anyone', *cokoli*(*v*) 'anything', *kdekoli*(*v*) 'anywhere') found in free-choice and negative polarity contexts other than direct negation. In addition, there are various other minor *wh*-based indefinites (Šimík 2008). In ordinary affirmative environments, Czech has also retained the *ně*-series (*někdo* 'someone', *něco* 'something', *někde* 'somewhere', etc.), alongside a new *si*-series (*kdosi* 'someone', *cosi* 'something', *kdesi* 'somewhere', etc.), parallel to the Polish *ś*-series. As in Polish, the new *koli(v)*-series spreads out historically from free-choice meaning to its current wider distribution.

Upper Sorbian largely retains the inherited system, with two main series, a negative-only *ni*-series (*nichtó* 'no one', *ničo* 'nothing', *žadyn* 'no', *nihdźe* '(at/to) nowhere', etc.) and a non-negative *ně*-series (*něchtó* 'anyone', *něšto* 'anything', *někajki* 'any', *něhdźe* '(at/to) anywhere', etc.). It innovates a new free-choice *žkuli*-series (*něchtóžkuli* 'anyone', *něštožkuli* 'anything', *někajkižkuli* 'any', *něhdźežkuli* '(at/to) anywhere'). Etymologically this series derives from the Common Slavonic contrastive particle *že* plus a generalizing particle *kuli*, cognate with *koli*, found in the Polish *kolwiek*-series, the Czech *koli(v)*-series and sporadically across Slavonic (Schuster-Šewc 1978: 721, s.v. *-kuli*). While the *žkuli*-series must have begun life as a free-choice series, its current use includes some ordinary (non-NPI) indefinite contexts, where it seems to have been influenced by German *manch* 'some, many a':

(137) Něchtóžkuli to nochcyše wěrić.
some that NEG.want.IMPF.3PL believe.INF
'Some didn't want to believe that.' (Upper Sorbian) (Schuster-Šewc 1978: ii.46, s.v. *manch*).

9.5.8 South Slavonic

In South Slavonic, the inherited *ni*-series survives in all languages (Bulgarian *nikoj* 'no one', *ništo* 'nothing', *nikakăv* 'no (kind)', *nikăde* 'nowhere, to nowhere', *nikoga* 'never'; Macedonian *nikoj* 'no one', *ništo* 'nothing', *nikoj/nikakov* 'no', *nikade* '(at/to) nowhere', etc.; Slovene *nihče* 'no one', *nič* 'nothing', *nobeden, noben* 'no', *nikakršen* 'no kind of, no', *nikamor* 'nowhere', *nikdar, nikoli* 'never'; Serbian and Croatian *ni(t)ko* 'no one', *ništa* 'nothing', *nikakav* 'no', *nikud* 'nowhere', *nikad* 'never'). The inherited adnominal quantifier (OCS *nikyjĭ* 'no') replaces the inherited form for 'no one' (OCS *nikŭto* 'no one') in Bulgarian, yielding *nikoj*. Slovene *nihče* 'no one', attested already in the fifteenth century, derives from a variant form of the pronoun in Common Slavonic, *ni-kŭtŭ-že; the base *ni-kŭtŭ also serves as the ancestor of Polish *nikt* (Snoj 2003). Slovene *noben* 'no' is, however, an innovation, whose origin is disputed. The earliest forms (15th and 16th centuries) are *obeden* and *oben*, showing that *n-* is an addition, analogical on the rest of the paradigm. One suggestion derives *obeden* from *ob 'around, without' + *(j)edĭnŭ 'one' (Slovene *en, eden*) (Bezlaj 1976: ii.225–6, Music 1921, Snoj 2003: 449). This would imply that it was negative from the start. Phonologically this hypothesis is unproblematic, but it requires a rather strange syntactic reanalysis of a prepositional phrase as a noun phrase (e.g. 'I did not see without one house' > 'I did not see any house'). A second suggestion, due originally to Stanislav Škrabec and developed by Mirčev (1932), is that it derives from a free-choice item *ljubo edĭnŭ 'any one (you like)'. According to this hypothesis, it began life as a free-choice item, spreading to all NPI-contexts including direct negation, ultimately joining the existing *ni*-series. The historical input to this hypothesis is more secure, since *ljubo* is well attested as a free-choice marker in Old Church Slavonic (see section 9.5.2 above), and it is syntactically straightforward, there being no category reanalysis; phonologically, however, it requires some irregular loss of material. This second pathway would make an interesting parallel for one of the hypothesized developments for Polish *żaden* 'no' (see section 9.5.7 above) and dialectal Bulgarian *boedin* 'some, any, no' (see examples (167) and (168) and the discussion thereof), and seems more in keeping with general pathways of development across Slavonic indefinites.

The *ni*-series retains its inherited distribution across South Slavonic, being restricted to clausemate negation and a few related contexts, as shown in Serbian (138) and Bulgarian (139).

(138) Milan nikoga / *ikoga ne voli.
Milan no.one / anyone NEG love.PRES.3SG
'Milan loves no one.' (Serbian) (Progovac 2005: 180–1)

(139) Nikoj nikoj / *njakoj ne običa.
no.one no.one someone NEG love.PRES.3SG
'No one loves anyone.' (Bulgarian)

They are not found in clauses in the scope of superordinate negation in a higher clause, where an *i*-series indefinite, *ikoga*, is required in Serbian, (140), and a *nĕ*-series indefinite, *njakoj*, is required in Bulgarian, (141).

(140) *Ne kažem da vidim nikoga.
NEG say.PRES.1SG that see.PRES.1SG no.one
'I'm not saying that I can see anyone.' (Serbian) (Milićević 2008: 110)

(141) *Ne mislja, če nikoj e povlijal na izborite.
NEG think.PRES.1SG that anyone be.PRES.3SG influence.PP on elections.the
'I don't think that anyone has influenced the elections.' (Bulgarian)

However, a subjunctive-like complement clause extends the domain of licensing, allowing a *ni*-series item in both Serbian (142) and Bulgarian (143):

(142) Ne želim da vidim nikoga/?ikoga.
NEG want.PRES.1SG that see.PRES.1SG no.one/anyone.
'I don't want to see anyone.' (Serbian) (Milićević 2008)

(143) Ne iskam da obiždam nikoj/?njakoj.
NEG want.PRES.1SG PRT insult.PRES.1SG no.one
'I don't want to insult anyone.' (Bulgarian)

Their availability in 'without'-clauses varies from language to language, being acceptable in Bulgarian, (144), and in some contexts in Slovene, but not in Serbian or Croatian, (145). Again, *i*-series items are used instead in Serbian and Croatian.

(144) Bez nikoj ništo da znae
Without no.one nothing PRT know.PRES.3SG
'without anyone knowing anything' (Bulgarian)

(145) bez da iko / *niko išta / *ništa zna
with PRT anyone / no.one anything / nothing know.PRES.3SG
'without anyone knowing anything' (Serbian)

The inherited *ně*-series survives in all languages (Bulgarian *njakoj* 'someone', *nešto* 'something', *njakoj* 'some (of some type)', *njakakăv* 'some kind of', *njakăde* 'somewhere, to somewhere', *njakoga* 'at some time, formerly'; Macedonian *nekoj* 'someone', *nešto* 'something', *nekoj/nekakov* 'some', *nekade/negde* 'somewhere', etc.), being impossible with clausemate negation (except where it takes wide scope with respect to negation), but possible elsewhere.

In Serbian and Croatian, however, a new series has arisen, competing with the *ně*-series items in non-negative NPI contexts, namely the *i*-series (*iko* 'anyone', *išta* 'anything', *ikakav* 'any', *ijedan* 'any (one)', *igde* 'anywhere', *ikad(a)* 'ever', *ikoliko* 'ever', *ičiji* 'anyone's', *ikako* 'any way' etc.). Like the *ně*-series, these items are found in all other negative polarity contexts, including interrogatives in (146) and conditionals in (147), but not ordinary affirmative contexts.

(146) Da li Milan ikoga / *nikoga voli?
COMP QU Milan anyone / no.one love.PRES.3SG
'Does Milan love anyone?' (Progovac 2005: 180–1)

(147) Ako Milan ikoga / *nikoga voli, neka nam oprosti.
if Milan anyone no.one love.PRES.3SG OPT us.DAT forgive.PRES.3SG
'If Milan loves anyone, let him forgive us.' (Progovac 2005: 180–1)

I-items thus have the familiar 'bagel' distribution, occurring in all NPI environments except the central one, direct negation. While *ně*-series items occur in these contexts, they often have an affirmative, referential interpretation with interpretation of 'some' outside of the scope of negation, as in (148), although non-referential, narrow-scope interpretations are also possible, as in (149).

(148) Da li su uopšte nekoga videli?
QU be.PRES.3PL at.all someone see.PP
'Did they (perhaps) see someone?'

(149) Da li neko zna nešto o tome?
QU someone know.PRES.3SG something about that
'Does anyone know anything about that?'

With a direct negative, *ně*-items must be interpreted outside of the scope of negation, and are thus fundamentally incompatible with negation (Progovac 2005: 183).

Progovac (1994) offers a binding-based analysis of the distribution of Serbian indefinites, suggesting that *ni*-items must be locally bound by negation within their governing category (typically IP), while *i*-items must be locally free, but bound within the sentence. This means that *ni*-items require direct clausemate negation, while negation in a higher clause is consistent only with an *i*-item in a lower clause (indirect negation). For questions and conditionals, she assumes a null polarity operator in C, which is outside of the governing category for *ni*-items, hence fails to license them, licensing only *i*-items. *Ně*-items must be free, hence outside of the scope of both negation and the negative polarity operator. While this gives a straightforward account of Serbian, it requires unmotivated LF-raising operations to derive the distribution of English *any*-items. In later work, Progovac (2005: 167–96) converts this into a feature-based system, eliminating binding. She analyses the *i*-series items as being neither negative nor positive, hence bearing the features [–pos, –neg]. Negative declarative clauses have a polarity phrase specified as [–pos, +neg] located high in the clausal structure. Indefinites need to check their features against those of the polarity phrase, and, in doing so, raise to that phrase. However, *i*-series items cannot check their [–pos, –neg] features against the [–pos, +neg] features of a negative clause; only those of a weak NPI-clause, specifed as [–pos, –neg] will suffice, hence limiting the distribution of *i*-series items to this context. Correspondingly, Progovac interprets the *ně*-series as being positive polarity items marked as [–neg], hence clashing with a c-commanding negative feature on sentential negation, unless they raise in Logical Form outside of the scope of negation. This allows them in non-assertive but non-negative contexts such as interrogatives and conditionals.

Figure 9.7, showing the distribution of indefinites in Serbian is adapted from Haspelmath (1997: 270), but, following Progovac (2005), differs in the distribution of the *ně*-series.

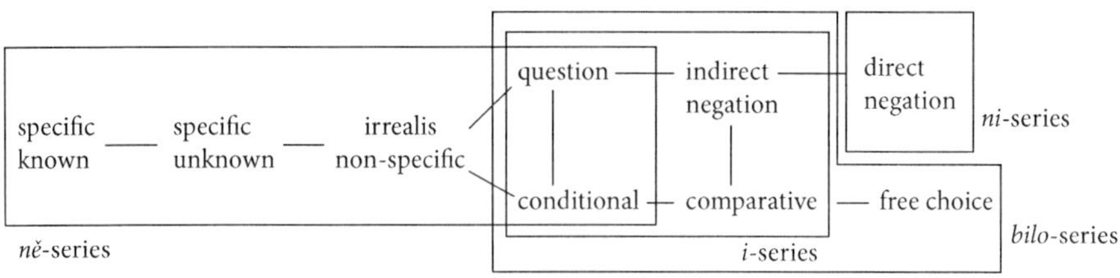

FIGURE 9.7 Expression of indefinites in Modern Serbian

While the *ni*-series and the *ne*-series go back to Common Slavonic, the *i*-series is a South Slavonic (Serbian/Croatian) innovation, deriving from the particle *i* 'and, also, even' plus the relevant *wh*-element, for instance, *i* + *tko* 'who' > *itko*, later (eighteenth century) *iko* 'anyone'. These items are found in their modern usage, written as one word, from the fifteenth or sixteenth centuries (Daničić 1880–1975, s.v. *itko, išto*; Skok 1971: 702, s.v. *i*). By this time, they seem to have reached their modern distribution, being found extensively in yes–no questions, comparatives, conditionals, with superordinate negation, and with downward-entailing verbs (e.g. *prestati* 'cease'). This raises the question of why they should have grammaticalized immediately to the 'bagel' distribution. Presumably the role of the pre-existing particle *ni* 'not even' and the pre-existing *ni*-series (*nitko* 'no one') was crucial. The current distribution of *i*- and *ni*-series items in Serbian correlates with that of the particles *i* 'even' and *ni* 'not even'. This is clearest with 'without'-clauses. In Serbian and Croatian, we find *i*-series items in 'without'-clauses, just as we find *i* rather than *ni*, while in languages that allow *ni*-series items in 'without'-clauses, such as Russian and Bulgarian, we find also *ni* being permitted in these clauses:

(150)	bez	*ni / i	najmanje	sumnje	(Serbian)
(151)	bez	ni / *i	malejšego	somnenija	(Russian)
(152)	bez	ni / i	naj-malko	sămnenie	(Bulgarian)
	without	not.even / even	slightest	doubt	
	'without even the slightest doubt'				

As with the *ni*-series in some languages, the origin of *i*-series indefinites is reflected in their behaviour with a preposition, where *i* precedes the preposition, hence *i za šta* 'for anything' (although regularized forms such as *za išta* are now also possible) (Belić 1962: ii.1, 117).

Various new free-choice series occur throughout South Slavonic, varying in their details from one language to another. Series built on *bilo* are found as the Serbian and Croatian *bilo*-series (*bilo ko* 'anyone', *bilo što/kaj* 'anything'; and *ko bilo* 'anyone', *što/kaj bilo* 'anything' < *bilo*, the neuter singular perfect of *biti* 'be'), the Macedonian (*i da*) *bilo*-series (*koj* (*i da*) *bilo* < *koj* 'who' *i* 'even' *da* 'modal particle' *bilo* 'perfect of "be"', hence 'whoever it may be') (see Progovac 1994: 122–31). Slovene has a free-choice and free-relative series built on *koli* 'ever' (cf. West Slavonic above) (*kdorkoli* 'whoever, anyone', *karkoli* 'whatever, anything'). Bulgarian and Macedonian have free-choice series built on (*to*) (*i*) *da e* (*bilo*) 'it may be': Bulgarian *kojto i da e* (*bilo*) 'whoever it may be, anyone', *kakvoto i da e* (*bilo*) 'whatever it may be, anything'; Macedonian *koj* (*i*) *da e* 'anyone', *što* (*i*) *da e* 'anything'. These may all in principle be used to express free-choice meaning in negative and non-assertive contexts (the meaning being close to that of an ordinary indefinite), although they have not spread to general use in the negative.

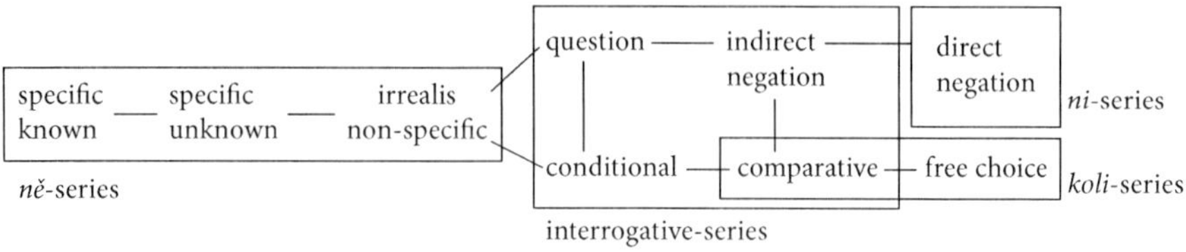

FIGURE 9.8 Expression of indefinites in Modern Slovene

Slovene is the most conservative of the South Slavonic languages in the domain of indefinites, retaining a system whose overall structure is quite similar to that of Old Church Slavonic. As well as retaining the *ni*- and *ně*-series with more or less their reconstructed distributions, it also retains robust use of interrogative pronouns as indefinites in non-assertive environments, a feature found in Old Church Slavonic, but which has tended to decline elsewhere. The overall distribution is shown in Figure 9.8, with typical patterns of data given in (153)–(157). *Ně*-series items are also often possible in questions and conditionals, but with referential interpretations.

(153) **Nihče** **nikogar** ne ljubi.
no.one no.one NEG loves
'No one loves anyone.' (Slovene)

(154) Je **kdo** tam?
be.PRES.3SG anyone there
'Is there anyone there?' (Slovene)

(155) Če je **kdo** tam, naj pride.
if be.PRES.3SG anyone there OPT come.PRES.3SG
'If there's anyone there, let them come in.' (Slovene)

(156) Ne pričakujem, da **kdo** pride.
NEG expect.PRES.1SG that anyone come.PRES.3SG
'I don't expect anyone to come.' (Slovene)

(157) To veš bolje kot **kdo / kdorkoli** drug.
that know.PRES.2SG better than anyone else
'You know that better than anyone else.' (Slovene) (FidaPLUS corpus)

Some marginal series present interesting historical developments across South Slavonic. First is the *god*-series, present in Serbian and Croatian and to some extent also in Bulgarian and Macedonian. Synchronically, there are two different structures with *god*, one a free-relative structure ((*t*)*ko god* 'whoever', *što god* 'whatever'), the other an indefinite pronoun ((*t*)*kogod* 'someone', *štogod*/*štagod* 'something'). The free-relative structure, illustrated in (158), has independent stress on each element, with phrase-level stress on *god*, and allows clitics to intervene between the *wh*-element and *god*, while the indefinite pronoun, illustrated in (159) and (160), has word-initial stress on the *wh*-element and does not allow clitics to intervene:

(158) Što me god upitaju…
what me ever ask.PRES.3PL
'Whatever they ask me…' (Cindrić 1955: 32)

(159) Je li me tkogod tražio? (*tko me god)
be.PRES.3SG QU me someone look.for.PP.MSG
'Did anyone look for me?' (Cindrić 1955: 32)

(160) Ako tkogod ima pitanja, rado ću odgovoriti.
if anyone have.PRES.3SG questions happily FUT.1SG reply.INF
'If anyone has questions, I'll happily respond.'

The free-relative structure is historically primary, derived ultimately from a free-choice item based on *godě*, in medieval Serbian both an impersonal verb 'it pleases' and an adjective 'pleasing' (Daničić 1880–1975: xvii.835):

(161) …da ju prodastъ, kamo jemu godje.
PRT her.ACC sell.3SG where him.DAT please.PRES.3SG
'…let him sell her wherever pleases him / wherever he likes.' (*Monumenta serbica*, p. 14, 1222–8) (Daničić 1962 [1863–4]: i.214, s.v. *godě*)

(162) ili ino, što imъ je godě.
or else what them.DAT is pleasing
'or else, whatever they like.' (*Monumenta serbica*, p. 37, 1253) (Daničić 1962 [1863–4]: i.214, s.v. *godě*)

Examples such as (163), where the *wh*-element and *godě* are adjacent, provided the context for a reanalysis of these items as a single unit:

(163) ili u Brъskovo ili inъdě gdě godě
or in Brskovo or elsewhere where ever
'either in Brskovo or anywhere else' (*Monumenta serbica*, p. 16, 1222–8)

Already in Old Serbian, this reanalysis seems already to have taken place, since we find *godě* being attached to quantifiers in prenominal position:

(164) i sъ srъblinomъ u koemъ godě sudě nigdě inъdě da
and with Serbs in which ever court nowhere elsewhere PRT
se ne prju…
REFL NEG dispute.3PL
'…and not to dispute with Serbs in any court anywhere else.' (*Monumenta serbica*, p. 16, 1222–8)

Somewhat later it spreads to act as a free-relative marker:

(165) ... i vsaka ina zla čineće nimь, kako godi mogaše.
and all other evils doing them.DAT how ever could.3SG
'... and doing all other sorts of wickedness, however he could.' (*Monumenta serbica*, p. 444, 1451) (Daničić 1962 [1863–4]: i.214, s.v. *godě*)

Once this reanalysis as a single unit had taken place, the distribution of these new items widened to all non-negative contexts by the seventeenth and eighteenth centuries, therefore with the same distribution and semantics as the *ně*-series:

(166) Podobno mi se vidi reći štogod od darova ńegovi(h).
Suitable me.DAT REFL see.3SG say.INF something of gifts his
'It seems appropriate to me to say something of his gifts.' (J. Filipović 1, 172a) (Daničić 1880–1975: xvii.835, s.v. *štogod*)

The cognate *gode*-series in Macedonian (*koj gode* 'anyone, whoever', *što gode* 'anything, whatever') is marginal and retains the more conservative use as a free-choice item and a free-relative marker, but has not developed further along the path found in Serbian and Croatian. The series occurs marginally in Bulgarian too.

All Slavonic languages maintain a major division between negative indefinites (*ni*-series items) and other indefinites (*ně*-series, *i*-series, *kolwiek*-series, etc.), with the former never occurring outside negative contexts and the latter generally not occurring within negative contexts. This division seems to be a stable, recurrent feature of all historical stages. However, a development in some southern Bulgarian dialects has led to the breakdown of this division in some varieties. Dialectally, there is an indefinite *boedin* 'any, anyone' based on the free-choice item *ljubo edin* (cf. OCS above). This has spread to negative clauses in many dialects that have it, giving it the distribution of a weak negative polarity item, spanning interrogatives, conditionals and negative clauses.

(167) kăkto e vozmožno boedin taka da mislet
how is possible anyone thus PRT think.PRES.3PL
'How is it possible for anyone to think that?' (Gotse Delchev (Nevrokop) dialect, Mirčev 1932: 10)

In some other dialects it has gone further and simply joined the negative system, forming part of the *ni*-series:

(168) Boadin ni dòjde tam.
no.one NEG go.PRES.3SG there
'No one goes there.' (Western Rup dialect, Vitanova 2002)

Since, in general, Slavonic free-choice items spread only to non-negative environments, the appearance of this item in negative contexts may well be due to language contact with Greek, acquiring the distribution of *kanenas* 'any, no' (Giannakidou 2000) or Turkish *hiç bir* 'any one, no'. For details, see Mirčev (1932) and Vitanova (2002, 2005).

9.6 Conclusion

The main features of the negative system have remained remarkably stable in the historical period across the Slavonic languages: the inherited negative marker *ne* has not been replaced or renewed, and a distinction between a *ni*-series of negative indefinites and other negative polarity indefinites has also remained constant across the group. No new series of negative indefinites have been created, although individual items have sometimes emerged to supplement the existing *ni*-series. However, this does not mean that there has been no change. Case has been one area of constant change: starting from a parent language with categorical marking of genitive case on direct objects in the scope of negation, we have observed differential developments in the daughter languages, genitive marking of direct objects disappearing in some but remaining steady in others, and even spreading to or interacting with existential clauses. In the indefinite system, the stability of direct negative indefinites has contrasted with instability of indefinites in other negative polarity contexts. Constant innovation of new indefinites in these contexts, typically emerging from the extension of former free-choice items to negative polarity environments, has been characteristic of almost all members of the group. The result is a lexically diverse set of new indefinites that nevertheless show a good deal of structural similarity.

Acknowledgements

My thanks to the following people for grammaticality judgements and discussion of their native languages: Luna Filipović, Pavel Iosad, Anna Kibort, Olesya Kryshevich, Iliya Manolov, Moreno Mitrović, Nik Nikolić, Martina Ožbot Currie, and Natalia Petrovskaia; and to Björn Hansen, Chris Lucas, Mary MacRobert, and Johan van der Auwera for useful comments on the chapter as a whole. Naturally none of them is responsible for the uses to which I have put their discussion. The FidaPLUS corpus (<http://www.fidaplus.net>) and the Russian National Corpus (<http://www.ruscorpora.ru>) were also used to confirm grammaticality patterns for Slovene and Russian respectively.

Primary sources

Alexandreida, Václav Vážný and František Svejkovský (eds.). Prague: Československá Akademie Věd, 1963.

Constantinus et Methodius thessalonicenses: Fontes, Franc Grivec and Franc Tomšič (eds.). Zagreb: Radovi Staroslavenskog Instituta, 1960.
Domostroj, V. V. Kolesov, M. V. Pimenova, and V. V. Roždestvenskaja (eds.). Moscow: Sovetskaja Rossija, 1990.
Evangeliarium Assemani. Evangeliář Assemanův: Kodex Vatikánský 3. slovanský (vol. 2), Josef Vajs and Josef Kurz (eds.). Prague: Nakladatelství Československé Akademie Věd, 1955.
Ipat'evskaja letopis', A. A. Šaxmatov (ed.). Moscow: Jazyki Russkoj Kul'tury, 1998 [1908].
Monumenta serbica spectantia historiam Serbiae, Bosnae, Ragusii, Franz Miklosich (ed.). Vienna: Braunmüller, 1858.
Novgorodskaja Pervaja Letopis' staršego i mladšego izvodov, A. N. Nasonov (ed.). Moscow: Izdatel'stvo Akademii Nauk SSSR, 1950.
Quattuor evangeliorum codex glagoliticus olim Zographensis nunc Petropolitanus, Vatroslav I. Jagić (ed.). Berlin: Weidmann, 1879. [*Codex Zographensis*]
Quattuor evangeliorum versionis palaeoslovenicae Codex Marianus glagoliticus, Vatroslav I. Jagić (ed.). Berlin: Weidmann, 1883.
Savvina kniga, Vjačeslav Ščepkin (ed.). St Petersburg: Izdatel'stvo Otdelenija Russkogo Jazyka i Slovesnosti Imperatorskoj Akademii Nauk, 1903.
Smolenskie gramoty XIII–XIV vekov, T. A. Sumnikova, Vladimir Vladimirovič Lopatin, and Ruben Ivanovič Avanesov (eds.). Moscow: Izdatel'stvo Akademii Nauk SSSR, 1963.
Suprasălski ili Retkov sbornik, Jordan Zaimov and Mario Kapaldo (eds.). Sofia: Izdatelstvo na Bălgarskata Akademija na Naukite, 1982. [*Codex Suprasliensis*]
Voinskie povesti drevnej Rusi, Varvara Pavlovna Adrianova-Peretc (ed.). Moscow: Izdatel'stvo Akademii Nauk SSSR, 1949.

10

Negation in the history of Arabic and Afro-Asiatic

CHRISTOPHER LUCAS

10.1 Introduction

10.1.1 Scope

Afro-Asiatic is among the largest and most heterogeneous of the widely accepted language families. It is generally sub-divided into six genera: Semitic, Berber, Egyptian–Coptic, Cushitic, Chadic, and Omotic,[1] comprising in excess of 300 languages between them. Naturally, only a small subset of these can be discussed in detail here, and the primary focus of this chapter will be Arabic, together with languages whose developments in the expression of negation are arguably linked to those of Arabic through contact (including Coptic, a range of Berber and Modern South Arabian languages, and the Indo-Aryan language Domari). Other Semitic, and certain Cushitic languages will be addressed to the extent that written records or comparative reconstruction allow us to observe, and offer explanations for, significant changes in the syntax of negation.

There are several practical reasons motivating a focus on Arabic in particular. First, a comparison of the present-day spoken varieties with Classical Arabic shows that many of the former have undergone significant developments in the expression of sentential negation (Jespersen's cycle);[2] and developments in the polarity of

[1] There is no consensus as to the correct sub-grouping of these genera, and the status of Omotic within Afro-Asiatic is the subject of particular controversy. Once thought to be a sub-group within Cushitic (Greenberg 1963), there is now serious doubt as to whether it really belongs to Afro-Asiatic at all (Newman 1980, Diakonoff 1996, Theil 2006). A compromise view is that Omotic was the first genus to split from Afro-Asiatic, such that the remaining genera together form a sub-group that has been labelled 'Erythraean' (Ehret 1995).

[2] The same is not true of some of the other Semitic languages with long and extensive written histories, such as Hebrew, Aramaic, or Akkadian. Discussion of these is mostly limited to section 10.3 on the development of indefinites and negative concord.

spoken Arabic indefinites, while less clear-cut, raise some interesting theoretical questions. Second, the large number of relatively well-described modern varieties with minor but significant variations in the expression of negation allows for detailed and relatively secure comparative reconstruction of various areas of morphology and syntax (see, e.g., Owens 2006) compared with many other Afro-Asiatic languages. This is necessary because of the near total lack of earlier written records of vernacular Arabic: (an approximation to) the language of the Qur'an became the written standard of the Arabic-speaking world from the earliest days of Islam and has remained so ever since,[3] such that textual evidence of the development of the spoken varieties is restricted to occasional minor deviations (chiefly phonological and lexical) from this standard. Finally, although there has been very little expressly historical work on negation in Arabic, it has attracted the interest of a number of synchronic linguists, both descriptive and theoretical, especially as regards its intersection with indefinites (e.g. Woidich 1968, Haspelmath and Caruana 1996, Benmamoun 1997, 2000, Ouhalla 2002, Hoyt 2005, 2006). Thus we have a more solid basis for a detailed understanding of the mechanisms of language change that brought about the development of negation in Arabic than we do for any other Afro-Asiatic language.

The remainder of this chapter is structured as follows. The rest of section 10.1 provides a critical survey of the small amount of previous literature on the diachrony of negation in Afro-Asiatic. Section 10.2 deals with Jespersen's cycle in Arabic, Coptic, Modern South Arabian, Berber, Jerusalem Domari, Ethiopian Semitic, and Lowland East Cushitic. Section 10.3 deals with the development of indefinites in the scope of negation in Arabic and Maltese primarily, as well as in Hebrew, Aramaic, and Ethiopian Semitic.

10.1.2 Previous literature

The existing literature on the diachrony of negation across Afro-Asiatic is very slight. Two articles will be discussed here.

The first is by Faber (1991), who presents a reconstruction of the negative and interrogative markers of Proto-Afro-Asiatic, using a methodology akin to the 'mass comparison' school of reconstruction (e.g. Greenberg 1987, Ruhlen 1994). She states (1991: 41) that her article 'has its origin in a trivial observation: the interrogative pronouns and adverbs of the Semitic languages... bear a striking resemblance to one of the negative markers. The interrogative words begin in *m*- and the NEG is *mā*.'

[3] The language of literature from the seventh to ninth centuries (including the Qur'an) is generally referred to as Classical Arabic. The form of this language was rigidly maintained in almost all writing up to the late nineteenth or early twentieth centuries, from which time written Arabic tends to be referred to as Modern Standard Arabic. The differences between Classical and Modern Standard Arabic are chiefly lexical and stylistic, rather than morphological or syntactic.

While interrogatives in *m-* are indeed common in Semitic and wider Afro-Asiatic, it is far from clear that the statement concerning negation is valid as it stands. Within Semitic it is only in Arabic that there is uncontroversially a negator (as well as an interrogative pronoun) *mā*.

Faber's claim that a similar negator *ma* is found in relic expressions in Biblical Hebrew is problematic. The standard grammars and dictionaries do not recognize a negative function for this item. It seems likely that this claim is based on an idiosyncratic analysis of certain obscure passages containing the Hebrew *wh-* pronoun *ma* (cf. Lipínski 1997: §47.15, Rubin 2005: 50). Faber also points to the enclitic negator *-m* in certain Ethiopian Semitic languages. We will see in section 10.2, however, that this represents a (relatively speaking) recent innovation within Ethiopian Semitic that appears to have its origin in the reanalysis of an enclitic conjunction. It is unlikely to be derived from a Common Semitic negator.

Within Semitic, the only other evidence that Faber adduces in support of her 'observation' is that Harari (Ethiopian Semitic) has a word *mēʔ* 'no' and a negative copula *elum* (this is in fact the negation of the existential verb *ḥāl* featuring the same enclitic negator *-m* just mentioned, see section 10.2.2.1; Wagner 1997: 507), and that Akkadian (East Semitic), Hebrew, Phoenician, and Xanthos Aramaic (Northwest Semitic) all have negative polarity items (NPIs) featuring one or more bilabial nasals: *mimma* 'anything', *məʔuma* 'anything', *mnm* 'anything', and *mtwm* 'ever', respectively. The oddity of listing these NPIs under the heading of negative elements is partially explained by reference to a (1988) paper, in which Faber shows that these items are indeed NPIs, and that they all probably involve an original generalizing *-ma* suffix which can be reconstructed to Proto-Semitic. However, based on no more evidence than that just discussed, Faber derives this suffix from a hypothesized Proto-Semitic negator **ma*. As Haspelmath (1997: 231) points out, this scenario runs counter to a sensible 'general rule of diachronic typology that reconstructed changes can never disprove a proposed universal': overwhelmingly, empirically attested developments in indefinite pronoun systems are from less to more negative, and not vice versa. Given that the only certain case of negative *mā* is in Arabic, it is far from clear that one can reasonably talk of a Common Semitic negator *mā*, and thus especially unlikely that this should be the source of the generalizing suffix *-ma*. Despite these problems within Semitic, Faber (1991) also points to the presence of a prohibitive *m-* proclitic in Egyptian and a negative suffix *-ma* in Hamer (which is apparently the only Omotic language to have this item), and lists a number of Lowland East Cushitic languages which all have a proclitic *ma-* negator (presumably present in Common Lowland East Cushitic). On this basis she states, without obvious justification, that there is 'ample evidence that the Sem[itic] NEG *mā* was inherited from A[fro]-A[siatic]' (Faber 1991: 412). Given that *wh*-pronouns with initial *m-* are widespread in Semitic, Berber, Chadic, and Cushitic, and are also found in Egyptian, it does seem plausible to reconstruct a *wh*-formative in *m-* to Proto-Afro-Asiatic as

Faber (1991: 412) suggests. In view of the above, however, Faber's main proposal to explain the resemblance she perceives between Afro-Asiatic negative and interrogative markers is hardly warranted. She suggests, albeit 'tentatively' (1991: 420), that 'in all of A[fro]-A[siatic] but Om[otic], a word meaning "what?" developed from prefixation of a [question particle] derived from the NEG **(m)ba* to a word meaning "thing"'. She does not explain how or why a question particle should have developed from a negator, nor what the status of the parentheses in her reconstructed negator **(m)ba* is, nor what the form of the word meaning 'thing' might have been. Given all this, it seems unwise to treat her reconstruction as definitive. Rather, it seems more likely, a) that the Semitic generalizing suffix is straightforwardly derived from an *m-* interrogative, a change also attested by, e.g., Russian *koe-* (Haspelmath 1997: 232), and b) that the two certain instances of homophony between interrogative and negative markers in *m-*, in Arabic and Lowland East Cushitic, represent separate instances of the change *wh*-pronoun > negator (attested separately also by Spanish Arabic *iš* < *aš* 'what' < *ʔayy šayʔ* 'which thing', according to Corriente 1977: 145), where the resemblance of the Arabic and Cushitic *wh*-pronouns (but not the negators) is probably due to common inheritance from Proto-Afro-Asiatic.

The only other work I am aware of that is of direct relevance to the overall scope of the present chapter is van Gelderen (2008), which is concerned with negative cycles cross-linguistically. It features a section on Afro-Asiatic languages with discontinuous negative constructions, giving data on Central Atlas Tamazight, Kabyle, Tarifit (Berber), Standard and Moroccan Arabic, Zway, Amharic (Semitic), Koorete (Omotic), Somali, Beja (Cushitic), and Hausa (Chadic).

Among these data, van Gelderen identifies three main diachronic developments. In the first, exemplified by the above-mentioned Berber languages as well as Moroccan Arabic, an older negator that is a prefix or proclitic on the verb (*mā* in Arabic, *ur* or a cognate form in Berber) is joined by a second, postverbal negator that is clearly a more recent innovation (*-š(i)* in Arabic, and variations on *ara/kra/ša* in Berber). Van Gelderen analyses the negative prefix in each case as occupying the head of NegP, while the postverbal element occupies the specifier position. In all of these languages this postverbal negator is derived from a word for 'thing'. As we will see in section 10.2, there is good reason to think that this similarity is not coincidental, but rather that the Berber construction developed under the influence of Arabic.

The second development that van Gelderen mentions is from an interrogative marker to a preverbal negator occupying the head of PolP: the last-mentioned change in the discussion of Faber (1991) above. She notes this development for Arabic (following Rubin 2005: 50) and also for Somali (Lowland East Cushitic), though, as we saw above, this development probably took place already at the stage of Common Lowland East Cushitic. In addition to the preverbal element *ma*, van Gelderen notes that Somali negative sentences also involve a special negative form of the verb. She hypothesizes that this special negative form is the remnant of an earlier negator that predates the introduction of *ma*. We will see in section 10.2 that this is unlikely to be correct: rather *ma* is the original negator in Lowland East Cushitic and the special negative form of the verb is derived from a former perfect tense negative auxiliary that is still identifiable as such in the related language Afar. This is in fact, then, more closely related to the third development that van Gelderen mentions: that of a negative auxiliary from a lexical verb, as in Koorete (Omotic) *ba* < 'to disappear' and Beja (North Cushitic) *rib* < 'to refuse'. This particular development is relatively common cross-linguistically (van Gelderen identifies it as a central feature of the history of negation in a number of Uralic languages for example, cf. chapter 11), but to the extent that it occurs in Afro-Asiatic languages beyond Lowland East Cushitic, there is insufficient diachronic data to comment further here.

Van Gelderen also gives data on negation in a couple of Afro-Asiatic languages without making suggestions as to how the structures in question developed. Concerning Amharic and Zway (a related Ethiopian Semitic language not usually thought of as a variety of Amharic as van Gelderen labels it), she notes that these languages feature a bipartite negative construction that appears structurally similar to that of Arabic: there is a proclitic on the verb *al-* (in both languages) and an enclitic *-u* in Zway, *-m* in Amharic. As noted above, we will see in greater detail in the following section that this *-m* enclitic (as well as its vocalized Zway cognate, and cognates in other Ethiopian Semitic languages) has its origin in the contact-induced reanalysis of an 'and' conjunction.

Additionally, van Gelderen gives data on negation in Hausa (Chadic), which again features a bipartite construction as shown in (1) (= van Gelderen's (76)), though here the second element occupies a clause-final or clause-late position and, except for the tonal contrast, is identical to the preverbal negator.

(1) **bà** kà kāwō àbinci **ba**
NEG you bring food NEG
'You didn't bring food.' (Hausa; Kraft and Kirk-Greene 1973: 38)

Van Gelderen makes no suggestion as to the origin of this construction, but these data look similar to the 'resumptive' negative construction (see section 1.4) found in Afrikaans (den Besten 1986: 51–81, Bernini and Ramat 1996, Biberauer 2008), Brazilian Portuguese (Schwenter 2006), certain northern Italian dialects (see sections 3.2.2.1 and 3.3.2), and arguably Modern South Arabian (see section 10.2). As such we might speculate that the path of development for the Hausa construction was similar to that suggested for Brazilian Portuguese by Schwegler (1988), whereby a second negator routinely appended to a negative sentence for emphasis (as in *I don't like that, not* (*at all*)) is reanalysed as functioning as part of the negation of the main clause.

10.2 Developments in the expression of negation

The principal theme of this section is that of Jespersen's cycle as an areal phenomenon in Afro-Asiatic (cf. the discussion of Jespersen's cycle as a European areal phenomenon in Bernini and Ramat 1996; Breitbarth, Lucas, and Willis forthcoming). We will consider two areas in which Jespersen's cycle looms large in the histories of a number of languages: coastal North Africa from northern Morocco to Palestine; and the southwestern part of the Arabian peninsula (parts of present-day Yemen and Oman). The final part of the section deals with some unusual variations on the theme of Jespersen's cycle in certain Semitic and Cushitic languages of Ethiopia.

10.2.1 Jespersen's cycle in Arabic and neighbouring languages

10.2.1.1 Jespersen's cycle in Arabic Classical Arabic had a number of different negators specialized for different syntactic environments. All but one of these have been lost, or have become highly restricted in their distribution, in the majority of modern Arabic dialects. The one that has been retained and generalized is *mā*.[4] As noted in section 10.1, this item appears to have its origin in the reanalysis of a homophonous interrogative pronoun which is also found in Classical Arabic, but which has been lost in the modern dialects. Lipínski (1997: §47.15) and, following him, Rubin (2005: 50) plausibly suggest that the bridging context for this reanalysis would have been rhetorical questions such as 'what do I know?' > 'I know nothing'. By asking what a given predicate holds of, when it is mutually manifest to the speaker and hearer that the speaker believes that the predicate holds of nothing, a speaker is able in cases such as these to communicate his belief as an implicature, rather than as part of the literal content of his utterance. It is not hard to imagine, however, that frequent use of such a communicative strategy could result in the semanticization of the implicature and hence the reanalysis of a former interrogative pronoun as a negator.[5] Precisely what factors provoked this reanalysis in Arabic in particular, when presumably such implicatures are available in all languages, remains unclear, however.

[4] Some dialects have retained the long *ā* in this item, others have shortened it. The representation *mā* is used here to refer to the item cross-dialectally.

[5] Given an example along the lines of 'what do I know?' > 'I know nothing', one would expect an intermediate negative quantifier stage: interrogative > negative quantifier > negator. Since we have no textual evidence of the prehistory of this change, we have no way of knowing whether this was the case or not. Note, however, that another possibility is that the reanalysis took place in contexts where *mā* was the pseudo-argument of a verb such as 'to profit', 'to succeed', etc., which often feature an expression of the extent of profit/success that is potentially ambiguous between an adjunct and an argument, e.g. 'What does it profit a man, if...?'. An interrogative pronoun could in such contexts plausibly be reanalysed directly as a non-argument (i.e. a negator) rather than an argument (i.e. a negative quantifier) (cf. Breitbarth, Lucas, and Willis 2013).

In any case, while *mā* is now the principal sentential negator in all modern Arabic dialects, only a subset have undergone Jespersen's cycle, such that *mā* has been joined (and, in a couple of cases, superseded) by a second element. These are the dialects spoken in the two areas mentioned above: coastal North Africa from Morocco to Palestine and the southwest of the Arabian peninsula. Those spoken outside these areas, such as the dialect of Damascus (2), negate with *mā* alone.

(2) ʔal-lon **mā**-ḥabbēt-ha
say.PRF.3MSG-to.them NEG-love.PRF.1SG-her
'He told them, "I didn't fall in love with her."' (Damascene Arabic; Brustad 2000: 284)

The postverbal element which has been grammaticalized as part of the bipartite negative construction in the dialects spoken within these areas is the word for '(any) thing': *šayʔ* in Classical Arabic, generally *ši* in those modern dialects where this form retains its original meaning. As a negator, this form has generally become an enclitic *-š*, as in (3) from Cairene.[6]

(3) **ma**-baḥibb-ⁱ**š** migiyy-u hina ktīr
NEG-like.IMPF.1SG-NEG coming-his here much
'I don't like his coming here a lot.' (Cairene Arabic; Woidich 1968: 33)

This is the standard sentential negative construction in the non-Bedouin[7] dialects of Morocco, Algeria, Tunisia, Malta, Libya, Egypt, Israel/Palestine (as well as adjacent areas of countries neighbouring Israel/Palestine), plus much of Yemen (4) and quite likely the interior and south of Oman.[8]

(4) bih nās **mā** yiʕjib-hum-**š** aš-šāy
there.is people NEG please.IMPF.3MSG-them-NEG the-tea
'There are people who don't like tea.' (Ṣanʿāni (Yemen); Watson 1993: 261)

[6] Heath (2002: 212) notes that an unreduced form *-ši* is common in northern Moroccan dialects, while Khalafallah (1969: 100–2) gives *-šey* as the ordinary form for the variety of Ṣaʕīdi (southern) Egyptian that he describes.

[7] A reasonable generalization in Arabic dialectology is that a Bedouin dialect will tend to have more in common with another spoken in a different region than it does with the sedentary dialects of the same region. One typical feature of Bedouin dialects is conservatism in the expression of negation (as well as quite generally). There are of course individual exceptions to both of these generalizations, particularly in cases where Bedouin have become sedentarized and (partially) integrated into sedentary communities, when dialect hybridization tends to result.

[8] The most recent grammar that describes a bipartite construction as in (3) and (4) for Omani is Reinhardt (1894). While it seems reasonable to take this information at face value concerning *some* Omani variety, it is not clear which one is described in this grammar. More recent grammars, such as Brockett (1985), which describes a dialect of the northeast coast of Oman, make no mention of a bipartite construction, and it seems likely that this feature (among others) is in retreat throughout Oman under the influence of Gulf Arabic.

A related development is that of a constituent negator and negator of non-verbal (including participial) sentences *muš~miš* (5) from a reduced form of the third person masculine singular negative copula *mā-hū-š*.

(5) šuft[i] ḥāga **miš** maʕʔūla
see.PRF.1SG thing NEG thinkable
'I saw something unbelievable.' (Cairene Arabic; Woidich 2006: 338)

The negative copula itself is composed of a pronoun and negation marking, and presumably arose through reanalysis of structures containing a (negated) resumptive pronoun following a left-dislocated subject:

(6) aḥmad **mā**-hū-**š** ɣabi
Ahmad NEG-he-NEG stupid
Conservative interpretation: 'Ahmad, he's not stupid.'
Innovative interpretation: 'Ahmad isn't stupid.'

Dialects which have not undergone Jespersen's cycle tend to have a negative copula composed of *mā* plus a pronoun without enclitic *-š*, and a corresponding constituent negator *mū* (not *muš*).

In Palestinian Arabic and, to a certain extent, Cairene Arabic (as well as the variety of Omani Arabic that Reinhardt 1984 describes and the variety of southern Egyptian Arabic that Khalafallah 1969 describes), Jespersen's cycle has proceeded to the third stage, whereby the original preverbal negator *mā* is now optional:

(7) ana (mā) bašrab-š il-ʔahwa
I (NEG) drink.IMPF.1SG-NEG the-coffee
'I don't drink coffee.'

(8) (mā) txaf-[i]š
(NEG) fear.IMPF.IRR.2MSG-NEG
'Don't be afraid.' (Palestinian Arabic)

A stage III construction is only possible, however, in certain well-defined contexts in both Palestinian and Cairene Arabic, as detailed in section 10.2.1.4.

In Lucas (2009) and Lucas and Lash (2010), I argued for two separate innovations of bipartite negation in dialectal Arabic, once in North Africa and once in the southern Arabian Peninsula, both of which were triggered by contact with languages that were themselves undergoing Jespersen's cycle: Coptic and Modern South Arabian languages, respectively.

10.2.1.2 Jespersen's cycle in Coptic and Modern South Arabian Coptic is the latest stage of the Egyptian language, written in a Greek-derived script and attested from the first century onwards, although it is thought to have died out as a spoken language by the sixteenth century at the latest (Watterson 1988). As noted in section 1.2, it has been known since Gardiner (1904) that Coptic underwent developments in the expression of negation comparable to those found in French and Arabic—what we would now call Jespersen's cycle. The negation system of Egyptian–Coptic is complex and heterogeneous and cannot be treated in detail here (see Reintges 2004 for an overview of Coptic). Suffice it to say that one of the commonest negative constructions in all Coptic texts, used for the negation of present, imperfect, and future tenses of main verbs, as well as in certain other contexts, is bipartite, with a preverbal element *ən* and a postverbal element *an*:[9]

(9) a. **ən** ti-ouōš dōron ənto'ot tēutən **an**
neg 1SG-desire gift from you NEG
'I do not desire a gift from you.' (Coptic; Seven Coptic Homilies 1.8.5)

b. **ən** ti-na-tsabo-ou **an** e-amənte
NEG 1SG-FUT-teach-them NEG on-hell
'I will not teach them about hell.' (Coptic; Seven Coptic Homilies 4.66.3)

As can be seen in (9), the position of the second element is variable: while it sometimes appears immediately following the verb and any clitic object pronouns as in (9b) it more usually occupies a clause-late position as in (9a).

Coptic also shows evidence of moving to the third stage of Jespersen's cycle, whereby the original element *ən* can be optionally omitted, as in (10) (where the parentheses indicate the position one would expect *ən* to occupy had it been present).

(10) a. (ən) ne-u-na-paht-əf **an** ejō pe
(NEG) IMPF-3PL-FUT-pour-it NEG over.2FSG PRT
'It would not be poured over you.' (Coptic; Four Martyrdoms 2.70Vi)

b. (ən) ənta-i a'a-u **an** hən oumənt-magos
(NEG) PRF-1SG do-them NEG by magic
'It is not by magic that I did them.' (Coptic; Four Martyrdoms 3.119Rii)

However, as demonstrated in Lucas and Lash (2010), *ən* is dropped almost exclusively in the environment of a following nasal (or occasionally a following non-nasal apical), suggesting that this is primarily a phonological rather than a syntactic phenomenon, at least in the ninth-century texts examined.

The question, then, is whether a claim of Coptic influence in the development of Arabic negation makes sense from a linguistic, historical, and present-day distributional point of view, and, if so, what the mechanics of that influence might have been. There might initially be some doubt on all three counts. First, the negative

[9] There is no consensus as to the correct etymology of *an*. It is thought to derive immediately from the particle *iwnȝ*, but this is first attested in Late Egyptian texts where it already has the function of a strong NPI adverb '(not) at all'. There is no textual evidence of a possible earlier history as a less grammaticalized item.

constructions in the two languages are not entirely congruent in terms of positioning of the second element (this is variable in Coptic, as in (9), whereas in most Arabic varieties *-š* always immediately follows the verb and any clitic object pronouns), and the Arabic construction cannot be seen as a straightforward calque on that of Coptic (Coptic *an* has no non-functional meaning to be calqued). Second, owing to the lack of written records of earlier varieties of spoken Arabic we cannot know for certain when the Arabic bipartite construction first came into being. Third, we also cannot know where it came into being, and, in any case, it is found in Arabic dialects far beyond Egypt, including in the Arabian peninsula. Nevertheless, what we know about the Coptic–Arabic situation, particularly when seen in the context of other languages in the same area which have also undergone Jespersen's cycle, means that contact-induced change must be viewed as a strong possibility in this case.

First of all, sociohistorical conditions appear conducive to the possibility of Coptic syntactic influence on Arabic: between the Islamic invasion of Egypt in 639 and the eventual death of Coptic as a spoken language in the sixteenth century at the latest, many tens of thousands of native speakers of Coptic will have learnt Arabic as a second language, and we would expect many of them to have acquired it imperfectly, including imposing aspects of the syntax of their native Coptic on their second-language Arabic. Thus, we should not be surprised to find evidence of this influence in present-day Egyptian Arabic at least (see Bishai 1962 for some examples).

Concerning the linguistic evidence for Coptic influence in the development of (Egyptian) Arabic negation, the lack of total congruence in the positioning of the innovative elements in Coptic and Arabic need not rule out the possibility of transfer here. In cases such as this, where the agents of contact-induced change are assumed to be second language learners of the recipient language,[10] there is no good reason to suppose that their imperfect acquisition of the second language should always, or even often, result in the perfectly faithful replication of (elements of) the syntax of their native language. Trivially, second language acquisition is guided not only by the native grammar of the acquirer, but also by the data from the second language that the acquirer is exposed to. On the assumption that, before reanalysis, *ši* 'thing' could only function as an argument and not an adverb, as appears to have been the case in Classical Arabic (though see Lucas and Lash 2010 for some discussion), it could only have occupied the same slot in the clause as present-day negative *-š*: immediately following the verb, with only direct and indirect pronominal object

[10] Though theoretically possible, it is unlikely that the agents of contact-induced change in this case were native speakers of Arabic from the Arab community: the Arabs appear to have remained a small and, socially-speaking, relatively isolated military–political elite in Egypt for several centuries after their arrival. They are highly unlikely to have learnt Coptic in any great numbers and thus to have been in a position to borrow from it. However, it cannot be ruled out, of course, that, once the Copts had begun to abandon Arabic and shift to Coptic, native speakers of Arabic from the Coptic community could have played a role in borrowing this construction from their second-language Coptic.

clitics intervening, given that Arabic appears always to have maintained a fairly rigid Verb–Object–Adjunct word order. This would therefore be in contrast to the frequently clause-late position of Coptic *an*. Perhaps, then, *ši* would not have been a sufficiently good candidate for the role of '*an*-equivalent' in the acquisition of Arabic as a second language by native speakers of Coptic. Note, however, that excessive scepticism here forces us into the paradoxical position of denying that Coptic learners of Arabic could have reanalysed *ši* as a negator, while claiming that first language acquirers, by contrast, were capable of this, since this reanalysis must have taken place at some stage. This is despite the fact that first language acquirers of Arabic would have lacked the specific motivation for positing a bipartite construction that native speakers of Coptic would have had, and that children evidently tend to converge on a grammar of their first language that is a great deal closer to that of other native speakers of that language than do adult second language learners. Moreover, we face the difficulty of explaining why first language acquirers of Arabic outside the areas in which we find bipartite negation today were not similarly capable of making this reanalysis.

Hence, given the present state of our knowledge of the syntax of early spoken Arabic, the hypothesis of a Coptic origin for Arabic bipartite negation (as spoken in Egypt at least) seems more economical than one based on a purely internal innovation.

Finally, the present-day distribution of bipartite negation across all of North Africa, but only the small part of the Levant closest to Egypt, is also consistent with an innovation in Egypt and the subsequent diffusion to a succession of neighbouring dialects from there. A steady spread westward is what we would expect as this follows the prevailing flow of migration in this region; similarly, the very limited spread eastward against this flow.

What the Coptic-origin hypothesis leaves unexplained, however, is the presence of bipartite negation in the Arabic dialects of Yemen and Oman. Given that diffusion from North Africa just to this region is implausible, and that, as in North Africa, there is no obvious reason why first language acquirers of Arabic here should have reanalysed *ši* as a negator, while others further north did not, a separate explanation is required. This can be found by considering the Modern South Arabian languages, which are a sub-group of Semitic most closely related to the Ethiopian Semitic languages and which are spoken in Yemen and Oman, but in no other Arabic-speaking countries. Significantly, the Modern South Arabian languages have themselves also undergone Jespersen's cycle, and they have clearly been in contact with Arabic for a considerable period, all speakers of mainland varieties of these languages being bilingual in the local variety of Arabic.[11]

[11] The Ethnologue entry on Soqoṭri states that speakers are mostly monolingual (<http://www.ethnologue.com/show_language.asp?code=sqt>).

All stages of Jespersen's cycle are observable in different Modern South Arabian languages. The most conservative is Soqoṭri, which is spoken on the island of Soqoṭra, 200 miles southeast of mainland Yemen. Here negation is with a preverbal item *al* alone:

(11) **al** fśek
NEG lunch.PRF.1SG
'I didn't eat lunch.' (Soqoṭri; Simeone-Senelle 1997: 414)

Other varieties, such as Jibbāli, spoken in the mountains of Dhofar in southwestern Oman, feature the same preverbal negator *al* in a bipartite construction with a clause-final item *laʔ*:

(12) ãxṭɛr **al** kse mih her yafhəs ti-hum **laʔ**
caravan NEG find.PRF.3MPL water COMP boil.IMPF.3MPL meat-their NEG
'The caravan didn't find water to boil their meat.' (Jibbāli; Simeone-Senelle 1997: 413)

This item *laʔ*, which is the new negator in all the mainland Modern South Arabian varieties, is homophonous with the anaphoric negator 'no'. As such, this appears to be an instance of the same type of 'resumptive' negative construction as suggested for Hausa in section 10.1.2.

Still other varieties have lost the original preverbal negator in all but a few restricted contexts, for example Ḥarsūsi:

(13) əkhōl əġəter **laʔ**
can.IMPF.1SG speak.IMPF.1SG NEG
'I cannot speak.' (Ḥarsūsi; Simeone-Senelle 1997: 414)

To summarize the situation, then, two geographically separate groups of Arabic dialects have undergone Jespersen's cycle. In both regions there has been close contact with other languages which have themselves also undergone Jespersen's cycle. Arabic varieties spoken outside these regions are, and have been, in contact with a great many different languages, but as far as we know none of these have undergone Jespersen's cycle during the period of contact. Thus, it seems to be an accurate descriptive generalization that only those Arabic varieties that have been in contact with a language with bipartite negation (or those neighbouring such a variety) have themselves gone on to develop bipartite negation, which suggests that contact is the most likely cause of Jespersen's cycle in Arabic.

10.2.1.3 Jespersen's cycle in Berber and Jerusalem Domari The spread of Jespersen's Cycle in this region does not end with Arabic. There are two other (groups of) languages that have themselves been in intensive contact with Arabic over many centuries and have also undergone Jespersen's cycle. These are: a subset of the Berber languages spoken in the more northerly parts of Morocco and Algeria, as well as in Tunisia and a few pockets in Libya; and also the Jerusalem variety of the Indo-Aryan language Domari.

Berber Berber languages which are conservative with respect to the expression of negation feature a single preverbal marker *ur* (or a cognate thereof):

(14) **ur** igle
NEG leave.PRF.3MSG
'He didn't leave.' (Tuareg; Chaker 1996 16)

The most prominent representatives of this conservative negative construction are Tuareg (spoken in southern Algeria and Libya, Niger, Mali, and Burkina Faso) and Tashelhiyt (spoken in southern Morocco). In other varieties, most prominently represented by Central Atlas Tamazight (central Morocco) and Tarifit (northern Morocco), there is a bipartite construction, of which the second element is *ša* in Central Atlas Tamazight and central Tarifit varieties; other Tarifit varieties have *-š/ši/šay* (Lafkioui 2007: 235). The form *ša* can be traced back via the regular sound change /k/ > /š/ to an item *kra* 'thing' (Brugnatelli 1987), which remains the form of the innovative negator in certain Kabyle varieties of Algeria (see (17) below). The phonological resemblance of Central Atlas Tamazight *ša* to Arabic *-š* is thus coincidental, but *-š/ši/šay* in other Tarifit varieties is clearly a direct loan from Arabic. Although *ša* in Central Atlas Tamazight can still function as an argument 'thing', at least in the scope of negation (Boumalk 1996: 41), it has undoubtedly developed a non-argumental function as a pure, non-emphatic negator (15). The same is true of Tarifit *-š/ši/šay* (16).

(15) **ur** iffiɣ **ša**
NEG exit.PRF.3MSG NEG
'He didn't go out.' (Central Atlas Tamazight; Boumalk 1996: 36)

(16) **ur** iẓri **ši** imma-s
NEG see.PRF.3MSG NEG mother-his
'He hasn't seen his mother.' (Tarifit; Boumalk 1996: 36)

Nevertheless, the second element is always optional in these varieties. This contrasts with the situation in the two major varieties of northern Algeria, Kabyle and Shawia, where the second element (*š(a)* in Shawia and a range of forms in different dialects of Kabyle, including *kra*, *ara* and *ani*) is obligatory, except in certain well-defined contexts, to be discussed below.

(17) **ul** ittaggad **kra**
NEG fear.IMPF.3MSG NEG
'He is not afraid.' (Kabyle; Rabhi 1996: 25)

Finally, there are at least three Berber languages—Sened (Tunisia; Provotelle 1911), Aujila (Libya; Paradisi 1961), and Ghadames (the border of Libya, Tunisia, and Algeria; Motylinski 1904, Mettouchi 1996)—all of which are endangered or extinct, in which the postverbal element is an obligatory verbal enclitic *-ka/-č* and the preverbal element is optional or entirely absent.

(18) akellim iššen-**ka** amakan w-iššen-ka tebārut
servant know.PRF.3MSG-NEG place and-know.PRF.3MSG-NEG road
'The servant didn't know either the area or the road.' (Aujila; Paradisi 1961: 82)

A number of factors make it highly likely that Jespersen's cycle in Berber is the result of contact with Arabic. First, just as the Arabic postverbal negator is homophonous with a word meaning 'thing' (or would have been before it underwent phonological erosion, see n. 6), so too are the various postverbal negators in Berber. The immediate postverbal positioning of the innovative Berber negators, preceding nominal objects, is also directly parallel to that of Arabic. Furthermore, all of the Berber languages which have undergone Jespersen's cycle are spoken in areas where the local Arabic variety has too. The two major Berber languages that lack a bipartite construction, Tashelhiyt and Tuareg, have had relatively little contact until recently with the sedentary Maghrebi Arabic dialects that do have a bipartite construction. They have been in contact with a number of languages, but the principal exposure to Arabic has been through the bedouin Hassānīyya dialect, which is not (typically) spoken further north than southern Morocco and southern Algeria, and which also lacks the bipartite construction (Iaaich 1996). In addition, we may note that there is a great deal of overlap concerning the contexts in both Berber and Arabic where the postverbal element is typically absent. Many of these (such as the presence of indefinites, negative coordination, or the negation of a verb taking a sentential complement) are familiar from the history of Jespersen's cycle in European languages, and might be considered 'natural'. However, there are some that are shared by Berber and Arabic that are rather more idiosyncratic: for example, when the negation is in the context of a statement whose truth is strongly guaranteed by the speaker, usually by invoking God as a witness:

(19) w-aḷḷāh **mā** hdar maʕā-ya
by-God NEG speak.PRF.3MSG with-me
'By God, (I swear) he didn't speak with me.' (Mazouna Algerian Arabic; Elhalimi 1996: 146)

(20) w-əlləh **ur** t-swiɣ
by-God NEG it-drink.PRF.3MSG
'By God, (I swear) I didn't drink it.' (Kabyle; Mettouchi 1996: 192)

Taken together, the above considerations strongly suggest that Jespersen's cycle in Arabic and Berber were not independent parallel developments. Moreover, the general trend for Jespersen's cycle to be less advanced in Berber varieties the further west and south one travels from Libya and Tunisia is consistent with a gradual spread westwards and southwards of the cycle in the local contact varieties of Arabic, as suggested above. Regarding the mechanics of this transfer, it seems likely that speakers of the relevant Berber varieties, when they became bilingual in the local Arabic variety, recreated in their native Berber the bipartite construction that they had become familiar with in their second language Arabic. Noticing (presumably unconsciously) that the Arabic postverbal negator was homophonous with the Arabic word for 'thing',[12] the speakers in question would have recruited the word for 'thing' in their native Berber variety to fulfil the same double function (what Heine and Kuteva 2005 call 'polysemy copying', or 'contact-induced grammaticalization', cf. Lucas 2007). The motivation for this (again, probably unconscious) was perhaps the expressive force of a new emphatic negative construction, or perhaps simply the strong routinization of expressing negation in this way.

Jerusalem Domari The same kinds of processes appear to have been at work in the spread of bipartite negation to Jerusalem Domari. Domari is an Indo-Aryan language, closely related to Romani, spoken by the Dom people, a marginalized ethnic group of the Middle East and South Asia. So far, the only variety of the language to have been described in any detail is the highly endangered Jerusalem variety, sketched in Matras (1999, 2007). Here too we find a bipartite negative construction, consisting of a proclitic *n*-, inherited from Proto-Indo-Aryan and ultimately Proto-Indo-European, and a stressed enclitic *-eʔ*, whose etymology is unknown, though Yaron Matras (personal communication) points out that the similarity with the numeral 'one' *ek* is suggestive of a derivation from some former indefinite item, conceivably an indefinite pronoun.

(21) **n**-mangam-**eʔ**
NEG-want.1SG-NEG
'I don't want.' (Jerusalem Domari; Matras 2007: 152)

It is tempting to see this bipartite construction as a borrowing from Palestinian Arabic. Unfortunately, however, the current state of our knowledge of other varieties of Domari makes it impossible to rule out an earlier innovation unconnected to that of Arabic. That said, there is sufficient record of at least the Turkish and Armenian varieties of the language to suggest that a bipartite construction is absent from these. The information we have on these varieties comes in the form of 46 sentences in the Bosha dialect of Armenia, published by Patkanoff (Patkanoff 1907/8), and a glossary of the Zapari dialect of Turkey, published in Paspati (1870). In the first of these works

[12] Here we must make the reasonable assumption that unreduced reflexes of the postverbal negator were more widespread formerly than they are today, such that identification of the negator with the word for 'thing' was possible during the initial period of transfer of the bipartite construction.

there are three negative sentences, all of which contain a preverbal negator *na* and no other negative element. For example:

(22) charav **na** tharem
money NEG have.1SG
'I have don't have any money.' (Armenian Domari; Patkanoff 1907/8: 247)

In the second source, there is a five-page entry on negation, with numerous examples listed of a number of forms: *na*, *nana*, *in*, *nanai*, *nasti*, *nastik*, *nanasti*, *nanastik*, *ne ne*, and *ma* (Paspati 1870: 384–9). All of these are preverbal and no mention is made of any postverbal item (though, incidentally, it is interesting to note the extensive strengthening and reduplication of the preverbal negator itself here).

This is clearly insufficient evidence to make strong claims, but it does speak more in favour of bipartite negation being confined to Jerusalem Domari (and thus possibly the result of contact with Palestinian Arabic) than against it, thereby strengthening the impression of Jespersen's cycle as an areal feature of North Africa and the southwestern Levant.

10.2.1.4 Stage III of Jespersen's cycle in Palestinian and Cairene Arabic As noted above, a number of dialects of Arabic, Berber, Coptic, and Modern South Arabian have progressed to the third stage of Jespersen's cycle, in which the original negator has become optional or obsolete. For most of these languages this development has not been investigated in detail, an exception being Palestinian and Cairene Arabic (Lucas 2010a), which we now consider.

There are contexts in both Palestinian and Cairene in which a stage III negative construction, featuring only the innovative element *-š*, is possible, but these contexts are quite different in the two dialects and the two will be dealt with separately here.

In Palestinian there is optionality between a stage II (bipartite) and a stage III negative construction in the context of verbs with the imperfect indicative prefix *b-* (23), imperatives with the second person prefix *t-* and no *b-*, hence irrealis-marked (24), and most members of the class of 'pseudoverbs' (25), a small, closed-class set of highly irregular verbs derived mainly from prepositional phrases.[13]

(23) **(mā)** b-aḥibb-^{i}š il-fūl
(NEG) IND-like.IMPF.1SG-NEG the-beans
'I don't like beans.' (Palestinian Arabic)

[13] The two tense forms found in many Semitic and Afro-Asiatic languages are traditionally labelled 'perfect' and 'imperfect', a convention which is maintained here. Semantically, perfect and imperfect tense forms align reasonably closely with past and non-past time reference, respectively. Although the two tense forms differ in their aspectual properties, they do not, in general, align straightforwardly with perfective and imperfective aspect.

(24) **(mā)** txaf-^{i}š
(NEG) fear.IMPF.IRR.2MSG-NEG
'Don't be afraid.' (Palestinian Arabic)

(25) **(mā)** biddī-š maṣāri
(NEG) want.1SG-NEG money
'I don't want money.' (Palestinian Arabic)

A stage III construction is not possible (i.e. *mā* is obligatory) with the perfect of any regular verb (26), including those beginning with (bi)labials (27), nor with the pseudoverb *ʕind-* 'to have' (28). Note that all the Palestinian pseudoverbs other than *ʕind-* 'to have'—i.e. *fī* 'there is', *maʕ-* 'to have (on one's person)', and *bidd-* 'to want'—begin with labial consonants and allow a stage III construction (i.e. do not require *mā*), and that none of the pseudoverbs have a non-periphrastic past/perfect tense form.

(26) *akalt-^{i}š il-fūl
eat.PRF.1SG-NEG the-beans
'I didn't eat the beans.' (Palestinian Arabic)

(27) *mesaḥnā-š
wipe.PRF.1PL-NEG
'We didn't wipe.' (Palestinian Arabic)

(28) *ʕindī-š maṣāri
have.1SG-NEG money
'I don't have any money.' (Palestinian Arabic)

It is clear from these data that no simple synchronic rule, whether phonological, morphosyntactic, or semantic, can specify all and only the contexts in which the stage III construction is possible in Palestinian Arabic, *pace* earlier suggestions (e.g. by Hoyt 2006) that omission of *mā* is a phonological deletion phenomenon, occurring before all and only those verbs beginning with labial consonants. Instead, I argued in Lucas (2010a) that these synchronic facts must be accounted for in terms of a series of natural diachronic developments, as follows.

In the first phase, *mā* does indeed undergo phonological deletion immediately before a labial consonant, but only in the context of the highest-frequency verbs, high token frequency being a well-known promoter of phonetic reduction (e.g. Bybee 2003). There can be no doubt that all the Palestinian pseudoverbs—*fī* 'there is', *ʕind-* 'to have', *maʕ-* 'to have (on one's person)', and *bidd-* 'to want'—are among the highest-frequency verbs there are in the language (just like their English translation equivalents), in addition to their beginning with labials (except pharyngeal-initial *ʕind-* 'to have'). So deletion of *mā* at this initial stage will have been particularly associated with the pseudoverbs (but not *ʕind-*), as well, perhaps, as the highest-frequency regular verbs taking the present indicative *b-* prefix.

In the second phase, surface strings where *mā* is lacking, as in (25), are reanalysed such that they no longer represent the output of a phonological deletion operation on *mā*, but instead are simply taken at face value, with enclitic *-š* being the sole marker of negation. At this stage, the applicability of this new stage III construction is not generalized beyond the context in which it first arises: that of the highest-frequency labial-initial verbs. Notably, pharyngeal-initial *ʕind-* 'to have' will not be among the verbs that permit this new stage III construction.

The final phase concerns the extension of the new stage III construction beyond the contexts in which it originally arose. What we observe in (23)–(28) is that this construction has spread so that it can now occur with all non-perfect verb forms, except for *ʕind-*, which must presumably be learnt as an individual lexical exception to this rule. This extension makes sense, in that a key property of the pseudoverbs, which, by hypothesis, were the core context for the initial development of the stage III construction, is that they lack a non-periphrastic past/perfect tense form and are in this sense inherently non-past. The present system, then, should be seen as the outcome of a fairly conservative analogical extension of the stage III construction from the context in which it originally appeared, to the set of verbal predicates sharing a salient property of the original context—the property of being non-past.

Turning to stage III negation in Cairene, the contexts in which this is possible are entirely orthogonal to what we find in Palestinian, having to do not with tense-marking and other properties of individual verbs, but with clause type. It also appears to be much less common than in Palestinian, where stage III negation is at least as common in the contexts that permit it as stage II negation.

As in Palestinian, negation with a bipartite (stage II) construction is possible in all contexts in Cairene. Negation with *-š* alone (a stage III construction) is not possible in any declarative clause (29), but may optionally occur in (embedded) interrogative (30) and conditional clauses (31), regardless of the phonological or tense features of the predicate.

(29) *baḥibb-^{i}š il-fūl
like.IMPF.1SG-NEG the-beans
'I don't like beans.' (Cairene Arabic)

(30) ma aʕraf-š^{i} kān mawgūd wala kan-š
neg know.IMPF.IRR.1SG-NEG be.PRF.3SG present or be.PRF.3SG-NEG
'I don't know if he was present or not.'
Lit. 'I don't know: was he present or not?' (Cairene Arabic; Willmore 1901: 298)

(31) law kunt-^{i}š šuft-ak...
if AUX.PRF.1SG-NEG see.PRF.1SG-you
'If I hadn't seen you...' (Cairene Arabic; Woidich 2006: 336)

It is important to note here that enclitic -*š* can also optionally appear in interrogative clauses without any negative meaning, instead giving the clause merely 'a doubtful sense' (Woidich 2006: 306):

(32) bēt abū-ya huwwa fēn walla akun-š[i] ġliṭt[i]
house father-my it where or be.IMPF.IRR.1SG-QU err.PRF.1SG
fi š-šāriʕ
in the-street
'Where's my father's house? Or have I got the wrong street?' (Cairene Arabic; Woidich 2006: 358)

The data illustrated in (29)–(32) are summarized in Tables 10.1 and 10.2.

How stage III negation in Cairene came to have this distribution can best be understood in the wider context of the functions of enclitic -*š* (or a less reduced, non-cliticized form *ši*) in other Arabic dialects, including those which have not undergone Jespersen's cycle.

The form *ši* (< *šayʔ*) 'thing' has grammaticalized in numerous different ways in the various Arabic dialects, but the grammaticalization that concerns us here is as a negative polarity adverb (NPA) roughly equivalent to English *at all*. The use of *ši* as an NPA is widespread in the dialects and seems to occur particularly frequently in interrogatives. Significantly, *ši*/-*š* as an NPA is found both in dialects which have not undergone Jespersen's cycle, such as Damascene (33), and those which have, but

TABLE 10.1 Grammaticality of negative constructions by context in Cairene Arabic

context	stage II construction	stage III construction
declarative clauses	✓	*
interrogative clauses	✓	✓
conditional clauses	✓	✓

TABLE 10.2 Distribution of -*š* without *ma* in Cairene Arabic

context	grammaticality of -*š* without *ma* with a negative interpretation	grammaticality of -*š* without *ma* with an affirmative interpretation
declarative clauses	*	*
interrogative clauses	✓	✓
conditional clauses	✓	*

which have not progressed to stage III, such as Eastern Libyan (34), as well as in dialects such as Cairene which do feature stage III constructions.

(33) ʕam-təʔṣod šì ʔənn-i kazzāb
PROG-intend.IMPF.2MSG NPA COMP-I liar
'Are you implying that I'm a liar?' (Damascene Arabic; Cowell 1964: 378)

(34) šiftū-š muḥammad
see.PRF.IRR.2MPL-NPA Muhammad
'Have you seen Muhammad?' (Eastern Libyan Arabic; Owens 1984: 102)

As such it is clear that the grammaticalization of *šī* as an NPA is separate from, and presumably prior to, its grammaticalization as the second element in a bipartite negative construction.

This gives us a basis for understanding the Cairene data in Tables 10.1 and 10.2. We can envisage the following stages in the development of the present system from the ancestor variety of modern Cairene. 1) The only negative marker is preverbal *ma*. Separately from its nominal use meaning 'thing', adverbial *šī* has an NPI distribution and no negative meaning. 2) Contact with Coptic causes the form *šī* to grammaticalize as the second element in the bipartite negative construction *ma... š(i)*.[14] NPA *š(i)* remains and retains its non-negative meaning. 3) As a result of the development of the bipartite negative construction, the form *š(i)* becomes strongly associated with negation, with the result that NPA *šī*, even in the absence of *ma*, comes to be optionally interpreted as negative wherever it occurs (i.e. in interrogatives and conditionals as well as negative declaratives). 4) Possibly because *š(i)* is less common in conditionals than in interrogatives, or because conditionals are less frequent than interrogatives overall, *š(i)* without *ma* comes to function only as a negator in conditionals, whereas *š(i)* without *ma* in interrogatives continues to be able to function either as a negator or a non-negative NPA. This is the present-day situation.

Note that an exactly parallel development can be observed in the history of French. As Price (1993) demonstrates, the commonly held belief that the original French negator *ne* came to be dropped first in interrogative clauses (e.g. Ashby 1991) is incorrect. Rather, French had a range of NPAs such as *pas*, *point*, and *mie*, and when these began to grammaticalize as part of a bipartite negative construction with the original French negator *ne*, their growing association with negation led to them being able to express negation also in contexts where they had not previously co-occurred with *ne*, that is, interrogatives and conditionals.

There is even a French precedent for *šī* becoming obligatorily negative in conditionals while remaining optionally non-negative in interrogatives. Hansen (this

[14] It is difficult to say whether it was *šī* in its nominal or its adverbial guise that was reanalysed as a negator. Conceivably the relevant Coptic-speaking second language learners of Arabic could have seen evidence for a bipartite negative construction in either nominal or adverbial *šī* or both.

volume, section 2.3.1) shows that *jamais* 'ever' in present-day Standard French has become obligatorily negative in conditional clauses (apart from in the fossilized expression *si jamais* 'if by any chance') while in interrogative clauses it continues to be interpretable as either negative or non-negative, depending on the context.

Stage III of Jespersen's cycle in Palestinian and Cairene Arabic is thus a good example of how a set of data which seem highly irregular and idiosyncratic from a synchronic point of view can be shown to be a natural outcome of regular and cross-linguistically common diachronic developments.

10.2.2 Jespersen-type developments in Afro-Asiatic languages of the Horn of Africa

As noted in section 10.1, a bipartite negative construction is also found in several Ethiopian Semitic and Lowland East Cushitic languages. Family trees of the relevant genera and sub-groups are given in Figures 10.1 and 10.2 (based on Lamberti 1991, Faber 1997), where names of languages (as opposed to sub-groups) are indicated in boldface.

10.2.2.1 Ethiopian Semitic In many (but not all) of the languages of the Ethiopian Semitic group, we find a bipartite construction whose precise form varies slightly

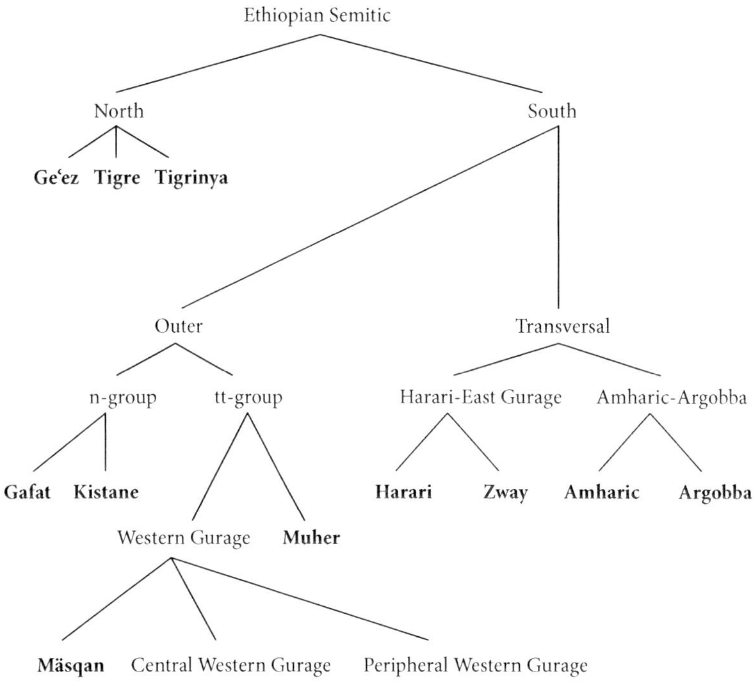

FIGURE 10.1 Ethiopian Semitic family tree

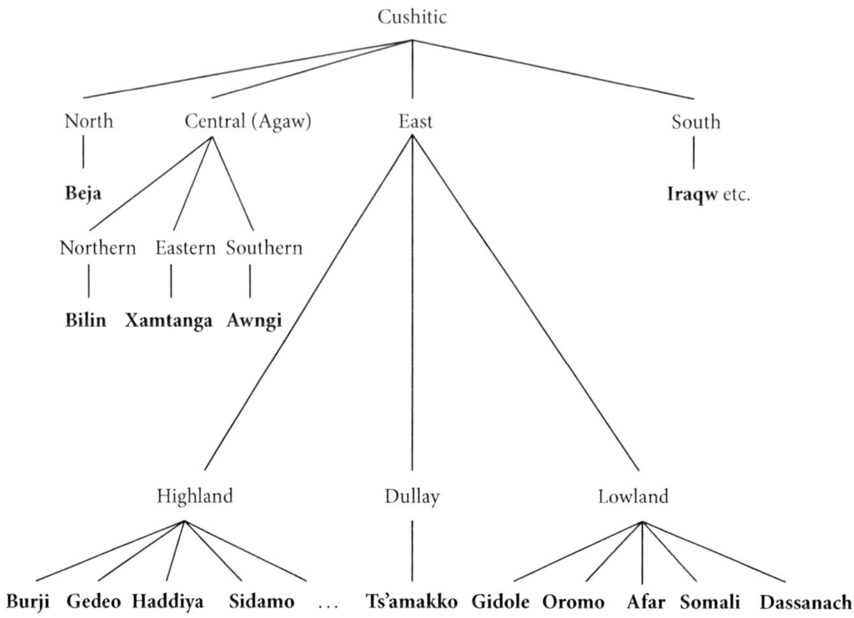

FIGURE 10.2 Cushitic family tree

from language to language, but whose syntax is essentially identical in all of them. Thus in Tigrinya (North Ethiopian Semitic) we have a proclitic *ʔay* and an enclitic *-n*:

(35) nəḥna mənəm ḥadä nägär **ʔay**-gäb̠ärna-**n**
we any one thing NEG-do.PRF.1PL-NEG
'We didn't do anything.' (Tigrinya; Kogan 1997: 442)

In Geʿez and Tigre, the other two North Ethiopian Semitic languages, we have just a preverbal construction with *ʔi* (a reduced form of *ʔay*). In Amharic and Harari (Transversal South Ethiopian Semitic) negation is expressed with a proclitic *al-* and an enclitic *-m*:

(36) **al**-näggärku-**m**
NEG-tell.PRF.1SG-NEG
'I didn't tell.' (Amharic; Hudson 1997: 471)

In Zway (also Transversal South Ethiopian Semitic) we have proclitic *al-* (often reduced to *a-*) and enclitic *-u* (reduced to zero in forms with an original final vowel):

(37) **a**-yfärək-**u**
NEG-be.patient.IMPF.3MSG-NEG
'He is not patient.' (Zway; Leslau 1999: 75)

That the Zway enclitic *-u* is the result of vocalization of original /m/ is suggested by the intermediate position of Argobba (also Transversal South Ethiopian Semitic), in which the enclitic negative has the original form /m/ following /u/, but /w/ following other vowels, /aw/ following nasal consonants, and /u/ following other consonants (Hudson 1997: 471–2). Bipartite negation seems in general to be absent from the Outer South Ethiopian Semitic subgroup.[15]

Thus, the Ethiopian Semitic languages can be divided into three types with respect to the expression of negation: those that do not have bipartite negation, those that have it with an enclitic *-m* or a derivative thereof, and Tigrinya, which has bipartite negation with an enclitic *-n*. Representatives of the first type (no bipartite negation) are found in both North and South Ethiopian Semitic, suggesting that the bipartite construction cannot be explained by means of a single innovation in a common ancestor.

Historical records of Ethiopian Semitic languages can shed a limited amount of light on the development of the bipartite construction(s). The Ethiopian Semitic language with by far the oldest and most extensive written attestation is Geʿez, but this has been conservative with respect to Jespersen's cycle. The only other languages for which we have records going back more than about a century are Amharic and Harari. The oldest extensive and easily accessible record of Amharic is Ludolf's (1698) grammar. Here we find precisely the same situation as today: a bipartite construction *al-…-m* found with all main clause verbs (Ludolf 1698: 52). Harari is attested in documents from approximately the mid-eighteenth century or possibly earlier (Wagner 1997: 486). Significantly, the situation in the oldest Harari documents is quite different to what obtains in Modern Harari: in the texts collected in Wagner (1983), out of the 69 instances of negation that I counted in contexts where a bipartite construction with *-m* would be obligatory in Modern Harari, all but four lacked a negative *-m*. An example of each variant is given in (38).

(38) a. **al**-xätäre-w zi-jalīl bärʔi
NEG-prevent.PRF.3MSG-him of-exalted gate
'The gate of the Exalted One did not block his way.' (Old Harari; Wagner 1983: 59)

b. **al**-qäbäṭa-**m** mauʕidata
NEG-fail.PRF.3MSG-NEG promise
'He did not break his promise.' (Old Harari; Wagner 1983: 79)

Additionally, there are a handful of examples, such as (39), where there is an *-m* enclitic present, but it is difficult to say whether this is the negative *-m* or a different,

[15] An exception to this generalization is the Peripheral Western Gurage dialect cluster which features a bipartite construction *a-…-ka/-ta/-da* (Hetzron 1997: 545). It is unclear what the origin of this second element is.

homophonous item that functions as a conjunction. In considering (39), note that asyndetic coordination is common in Ethiopian Semitic languages, also that in Modern Harari the position of negative *-m* is variable and need not cliticize to the verb, and that the negative of the existential verb *ḥal* is an irregular form *ēl(um)* (Wagner 1997: 502).

(39) ḥoji bi-dinät ge-m **el**-bä-na way, geš
today in-property world-and/NEG there.is.NEG-in-us woe tomorrow
bi-āxirat-**um** el-bä-na way
in-end-and/NEG there.is.NEG-in-us woe
'We do not suffer (have) misery on Earth today, neither will we suffer (have) misery in the hereafter tomorrow.' (Old Harari; Wagner 1983: 214)

Examples such as (39) look like a plausible bridging context for the hypothetical reanalysis of the *-m* conjunction as part of a bipartite negative construction, and this hypothesis is given added weight by the fact that Amharic also has an identical enclitic conjunction *-m* (Leslau 1995: 882).

At first glance, the Tigrinya negative enclitic *-n* appears to be a problem for this hypothesis, given that it cannot be cognate with *-m* in other Ethiopian Semitic languages (Proto-Semitic **m* and **n* are preserved unchanged in all Ethiopian Semitic languages). However, in Tigrinya we find that the commonest 'and'-conjunction is not the same as in Harari and Amharic, but is in fact a different item that is homophonous with the Tigrinya negative enclitic *-n* (Kogan 1997: 442).

Hetzron (1972: 94–8) was the first to suggest a link between negative and conjunctive clitics in Ethiopian Semitic. However, he went further and hypothesized that this development was not internal to Semitic, but was in fact the result of contact with Agaw languages (Central Cushitic). Hetzron (1972: 98) points out that in Awngi (a Southern Central Cushitic language spoken by several hundred thousand people southwest of Lake Tana, and undoubtedly an important contact language for both Amharic and Tigrinya in particular) there is a bipartite negative construction in which the second element is an enclitic *-kí*. The crucial point is that Awngi also features an identical enclitic conjunction. This appears, therefore, to be another case of polysemy copying: the fact that the sound string serving as an 'and'-conjunction in Awngi is also used as part of a negative construction has been transferred to a number of different contact languages, each of which has reanalysed the form ordinarily used for coordination such that it can also function as a negator. Significantly, negative *-kí* in Awngi is possible only in matrix clauses, and precisely the same is true of its counterpart *-m* in Amharic and Harari (Hudson 1997: 471, Wagner 1997: 503).

Too little is known about the sociolinguistic situation in Ethiopia in the period when this transfer would have taken place to speculate about whether the agents of change in this case would have been native speakers of Awngi or of the relevant Ethiopian Semitic languages, or perhaps a combination of the two. However, a clear advantage of Hetzron's contact explanation for Jespersen's cycle in Ethiopian Semitic is that it can explain its sporadic occurrence in both North and South Ethiopian Semitic, as well as the formal difference of the bipartite construction in Tigrinya versus the other languages discussed above.

The situation here is similar to what we saw for Jespersen's cycle in Arabic above: an internal explanation based on reanalysis in contexts such as that given in (39) may be plausible (though it is worth pointing out that conjunction > negator is hardly a commonly observed grammaticalization path) but it leaves us unable to explain the uneven distribution of this construction across a range of closely related languages. By contrast, the contact-based explanation makes a clear and potentially falsifiable prediction: Jespersen's cycle will only have occurred in those Semitic languages which were in intensive contact with Southern Central Cushitic languages, or in those languages to which bipartite negation could have diffused more recently through contact with Amharic. Bipartite negation in Harari, a language which has historically been surrounded by Lowland East Cushitic, rather than Central Cushitic languages, and which seems to have developed the bipartite construction only relatively recently, would appear to be a case of the latter.

10.2.2.2 Bipartite negation in Lowland East Cushitic A number of Lowland East Cushitic languages also show a Jespersen-type development, though of a rather different type to any we have seen thus far.

Starting with Oromo, the situation here is as follows (cf. Owens 1985, Bader 2006). Affirmative verbs are fully inflected for person and number in both past and non-past tenses (40a,b), and the paradigms for both tenses are very similar (though there appears to be more variation in the realization of the vowels of the inflections in the non-past than in the past). Negative non-past verbs are inflected like their affirmative counterparts (albeit apparently without the variability noted for the affirmative non-past) and are marked negative by prefixing *hin* plus high tone on the root syllable of the verb. Negative past verbs, by contrast, take the same prefix *hin* as well as high tone on the root, but do not inflect for person and number, adding instead an invariable *-ne* suffix to the root.

(40) a. dēmte ~ dēmti ~ dēmtu
go.NONPAST.3FSG
'She is going.'

b. dēmte
go.PAST.3FSG
'She went.'

c. hin-dḗmtu
NEG-go.NONPAST.3FSG
'She isn't going.'

d. hin-dḗmne
NEG-go.PAST
'I/you/he/she/we etc. didn't go.' (Oromo; Owens 1985: 66)

The situation in Somali is similar. Affirmative past and non-past verbs are fully (and similarly) inflected, as is the negative non-past verb together with the preverbal negator *má*, while the negative past is formed with the same negator and by the addition to the bare infinitive form (*súgi* in the case of the verb in (41)) of an invariable suffix *-n* (where the infinitive ends in /n/ the past negative form of the verb is identical to the infinitive form).

(41) a. sugtaan
wait.for.PRES.2PL
'You wait for (it).'

b. má-sugtàan
NEG-wait.for.PRES.2PL
'You don't wait for (it).'

c. sugteen
wait.for.PAST.2PL
'You waited for (it).'

d. má-sugín
NEG-wait.for.PAST
'I/you/he/she/we etc. didn't wait for (it).' (Somali; Saeed 1999: 86–8)

This surprising lack of inflection in the negative past of these languages becomes clearer if we consider a third Lowland East Cushitic language, Afar, which appears to have been more conservative in its expression of negation than Somali and Oromo.

In Afar, precisely the same situation obtains with respect to the non-past and to the past affirmative. The negative past, however, is clearly composed of the negator *ma*, plus the infinitive form of the verb, plus a clitic auxiliary *-inna* which is fully (but irregularly) inflected for number and person (Bliese 1981: 85). Note that Cushitic languages are largely head-final, with Verb–Aux order as standard. Moreover, the negative copula in Afar is expressed with a form *hinna*, which is inflected identically to the negative past auxiliary. This is shown in Table 10.3 (after Bliese 1981: 85, 111–12).

If it is right to assume that Afar more closely resembles the situation in Common Lowland East Cushitic (CLEC) than do Oromo and Somali, then the picture is clear. Pre-CLEC negated both past and non-past matrix clause verbs simply by prefixing the negator **ma* (itself derived from an earlier interrogative pronoun; see

TABLE 10.3 **Inflection of negative past tense and negative copula in Afar**

	m-aggaf-inna 'didn't kill'	hinna 'is not'
1SG	m-aggaf-inniyo	hinniyo
2SG	m-aggaf-innito	hinnito
3SG	m-aggaf-inna	hinna
1PL	m-aggaf-innino	hinnino
2PL	m-aggaf-innitōnu	hinnitōnu
3PL	m-aggaf-innōnu	hinnōnu

section 10.1). In CLEC itself the negative copula *hinna* was grammaticalized as a negative perfect tense auxiliary, following a common grammaticalization path (copula > 'be'-perfect auxiliary), albeit one that is more usually neutral with regard to polarity. As expected, the inflected auxiliary selects the bare infinitive form of the main verb. In Afar, the CLEC situation is preserved, except that the auxiliary becomes a clitic on the verb and the initial /h/ is lost. In Oromo and Somali, this clitic presumably first became an inflection and then underwent further phonological erosion, such that it lost its inflection and became a frozen form *-ne* and *-n* respectively. It appears that Pre-Oromo underwent an additional related development such that a bare form of the copula/auxiliary was then regrammaticalized as a new preverbal negator *hin*, in place of the original CLEC negator *ma*.

To summarize, then, the quasi-bipartite past-tense negative construction of Oromo and Somali appears to be the result of the grammaticalization (and concomitant heavy phonological reduction) of a negative copula, first as a negative past-tense auxiliary, and then as an uninflected past-tense negative morpheme. A slightly less reduced form of this morpheme then appears to have ousted *ma* as the preverbal negator in all tenses in Oromo.

This concludes our discussion of developments in sentential negation in Afro-Asiatic. We now turn to consider the development of indefinites in the scope of negation.

10.3 Developments in indefinite systems

10.3.1 Arabic

This section describes the system of NPI and n-word indefinites of Classical Arabic and shows how this system has evolved differently in the present-day dialects of Morocco, Malta, and Egypt/Palestine (the dialects of Egypt and Palestine are very similar as far as indefinites are concerned). We will see that, with the exception of Maltese, the indefinite systems of each of these Arabic varieties are internally rather heterogeneous in comparison with those of well-studied European languages, such that Arabic dialects resist straightforward classification as negative-concord or non-negative-concord languages (cf. Hoyt 2010).

There is a small amount of previous literature dealing with the issue of indefinites and negative concord in Arabic dialects, much of it focusing on Morrocan Arabic. The lack of agreement in this literature on the nature of the elements under discussion illustrates the analytical difficulties that the aforementioned heterogeneity of the Arabic indefinite system presents. For example, Benmanoun (1997, 2000, 2006) discusses the Moroccan Arabic determiner *ḥətta* 'even/any/no', which he analyses as a non-negative item with an NPI distribution. By contrast, Ouali (2008) analyses *ḥətta* as an inherently negative quantifier and, following Watanabe (2004), appeals to the syntactic mechanism of Agree to explain the lack of a double negation reading when it co-occurs with sentential negation. Finally, Hoyt (2005) follows Benmamoun's analysis of *ḥətta* as a non-negative NPI, while for the Palestinian Arabic equivalent to *ḥətta* (namely, *wala*) he combines the two analyses and posits two distinct, homophonous determiners with very similar semantics apart from their polarity, such that one is inherently negative and the other is a non-negative NPI.[16]

10.3.1.1 Issues in the terminology and analysis of indefinites A further difficulty encountered in discussions of negative concord in Arabic or any other language is widespread disagreement and confusion concerning the definitions of some of the relevant technical terms. The terminology used in this chapter is all consistent with the definitions provided in chapter 1, except that I prefer to slightly rephrase Giannakidou's (2006: 328) definition of n-word, as in (42).[17]

(42) N-word:
An expression α is an n-word iff:
(a) α can be used in structures containing sentential negation or another α-expression yielding a reading equivalent to one logical negation; and
(b) α can be interpreted negatively in non-sentential utterances lacking a negator.

In line with the argumentation in Lucas (2010b, 2011), in which I addressed the problem of negative concord from the perspective of Dynamic Syntax (Kempson,

[16] Hoyt (2010) offers an in-depth analysis of negative concord in Levantine Arabic, which maintains his multiple-item analysis for *wala*, but drops the claim that one of these items is a non-negative NPI.

[17] I have rephrased the second clause of Giannakidou's (2006: 328) original definition 'α can provide a negative fragment answer' in terms that are both more general (n-words can appear in non-sentential utterances which are not answers to *wh*-questions) and more theory-neutral (we need not necessarily analyse non-sentential utterances as having undergone ellipsis).

Meyer-Viol, and Gabbay 2001, Cann, Kempson, and Marten 2005), the position adopted here is that sentential negation should not be seen as a semantic property of individual lexical items, but as a property of the propositions associated with them. This view is based on the intuition (which Dynamic Syntax takes as axiomatic) that semantic interpretation is an incremental, real-time process in which individual lexical items make contributions to the emerging proposition that are inherently context-dependent, varying according to what contributions, if any, have already been made by other lexical items earlier on in the parse.

If we take this view, then indefinites can be categorized as follows: (a) those which always rigidly contribute one logical negation to the proposition expressed, regardless of whether any other item has previously done the same; (b) those which will contribute a logical negation only if another item has *not* already done the same; and (c) those which never contribute a logical negation themselves, which can then be subdivided into those which do or do not require such a contribution from some other item in the parse string. From this perspective, the items usually labelled 'negative quantifiers' or 'inherently/semantically negative' will fall into category (a), but should not themselves be seen as having negation as part of their own meaning. Henceforth items of this type will be referred to as 'rigid negative indefinites'. N-words on the definition in (42) will fall into category (b), on this view. Like the rigid negative indefinites of category (a), these also should not be seen as having negation as part of their meaning, though they will contribute a logical negation to interpretation in at least some contexts. Unlike the rigid negative indefinites, they will not blindly make this contribution in all possible contexts. All other indefinites, including weak and strong NPIs, fall into category (c).

While this kind of context-sensitive lexical approach to indefinites can straightforwardly accommodate systems with no n-words (such as Standard English), or systems with only n-words and no rigid negative indefinites (as in Slavonic, chapter 9), we will see in what follows that such an approach is particularly well suited to systems that contain indefinites from each of the three above-mentioned categories, and which therefore resist neat classification as 'negative-concord languages' or 'non-negative-concord languages'.

10.3.1.2 N-words and NPIs in Classical and Modern Standard Arabic (MSA)[18] Classical/MSA has a rather impoverished indefinite system. It has one clear example of an n-word according to the definition in (42), namely *ʔabadan* '(n)ever', 'by no/any means'. This is used principally in the context of negation and with reference to the future:

[18] There is no appreciable difference in the indefinite systems of Classical Arabic and its present-day descendant Modern Standard Arabic.

(43) a. wa-lan yatamannaw-hu **ʔabad-an**
and-NEG wish.IMPF.3MPL-it eternity-ADV
'And they will never long for it.' (Qur'an 2: 95)

b. A: hal satatruku-nī
QU leave.FUT.2MSG-me
'Will you leave me?'

B: **ʔabad-an**
eternity-ADV
'Never!' (Classical/MSA)

ʔabadan is derived from *ʔabad* 'eternity' plus the adverbializing suffix *-an*. As such, it is not surprising that it is also occasionally found in affirmative sentences meaning 'forever, for all eternity':

(44) wa-jannātin la-hum fī-hā naʕīmun muqīmun
and-gardens.ACC for-them in-them comfort.NOM permanent.NOM
xālidīna fī-hā **ʔabad-an**
enduring.ACC.PL in-them eternity-ADV
'... and gardens for them in which they will have permanent comfort, enduring in them forever.' (Classical/MSA; Qur'an 9: 21–22)

Note that in both negative and affirmative sentences *ʔabadan* takes scope over the whole proposition: 'it is forever the case that (it is not the case that) *P*'. Clearly *ʔabadan* in (44) is not negative, and yet its semantic contribution is apparently identical to that of *ʔabadan* in (43). Rather than positing two distinct homophonous items with identical meanings apart from their polarity, we can analyse *ʔabadan* along the lines outlined above: as a single item which contributes a logical negation to the proposition associated with it, if and only if it occurs as the first word in a non-sentential utterance, as in (43b).

Turning to NPIs, Classical/MSA has just two clear examples of these: *ʔaḥad* 'anyone', which is weak (45), and *qaṭṭu* 'ever' (46). *Qaṭṭu* differs from *ʔabadan* in that it is used mainly with reference to the past, it never seems to occur in upward entailing contexts like (44), it is almost entirely restricted to negative contexts,[19] and it appears not to be possible in non-sentential utterances. It is therefore (virtually) a strong NPI.

[19] Lane (1863: s.v. *qaṭṭu*) gives a rare example of *qaṭṭu* from *Ṣaḥīḥ al-Buxārī* in the non-negative downward entailing context of a superlative:

ʔaṭwalu ṣalātin ṣallaytu-ha **qaṭṭu**
longest prayer.OBL pray.PRF.1SG-it ever
'the longest prayer I ever prayed'

(45) a. lam ʔara **ʔaḥadan**
NEG see.JUSS.1SG anyone.ACC
'I didn't see anyone.'

b. hal raʔayta **ʔaḥadan**
QU see.PRF.2MSG anyone.ACC
'Did you see anyone?'

(46) lam yaʕud **qaṭṭu**
NEG return.JUSS.3MSG ever
'He never returned.' (Classical/MSA)

We may also note here the existence of an adverb *baʕdu* 'yet, still', which appears predominately in the scope of negation (47a), but not exclusively (47b), thus making it a semi-NPI.

(47) a. lam yaʔti **baʕdu**
NEG come.JUSS.3MSG yet
'He hasn't come yet.'

b. huwa **baʕdu** ṣaɣīr
he yet small
'He's still young.' (Classical/MSA; Wehr 1979: s.v. *baʕdu*)

There is also a marker of exclusive disjunction *ʔam* (48) that is restricted to questions (hence a restricted kind of NPI) and a negative conjunction *walā* (49).

(48) hal turīdu qahwatan **ʔam** šāyan
QU want.IMPF.2MSG coffee.ACC or tea.ACC
'Do you want coffee or tea?'

(49) a. lan tuɣniya ʕan-hum ʔamwālu-hum **wa-lā**
NEG avail.IRR.3FSG from-them riches.NOM-their and-NEG
ʔawlādu-hum šayʔan
children.NOM-their thing.ACC
'Neither their riches nor their children will avail them anything.' (Classical/MSA; Qur'an 3: 10)

b. ʔinna-hā baqaratun lā fāriḍun **wa-lā** bikrun
COMP-it cow.NOM NEG old.NOM and-NEG virgin.NOM
'It is a cow that is neither old nor immature.' (Classical/MSA; Qur'an 2: 68)

As we will see, *walā* undergoes some important developments in the transition to present-day spoken Arabic.

Classical/MSA can form rigid negative indefinites (of category (a)) by means of the negative determiner *lā* and an appropriate (pro)noun in the accusative case. These indefinite phrases are not n-words, since, like Standard English *nothing/nobody*, they cannot be used in structures containing sentential negation to yield a reading equivalent to one logical negation (clause (a) of (42)). If they are used in such structures the result is ungrammaticality, or perhaps marginally double negation—(50a–b). These negative quantifier phrases are also only possible in preverbal subject position—(50c).

(50) a.

lā	**ʔaḥada**	(*mā)	jāʔa
NEG	one.ACC	NEG	come.PRF.3MSG

'No one came.'

b.

lā	**šayʔa**	(*mā)	ḥadatha
NEG	thing.ACC	NEG	happen.PRF.3MSG

'Nothing happened.'

c.

(mā)	ḥadatha	šayʔun	/	***lā**	**šayʔa**
NEG	happen.PRF.3MSG	thing.NOM		NEG	thing.ACC

'Something (/nothing) happened.' (Classical/MSA)

Finally, Classical/MSA also has a free choice and weak NPI determiner *ʔayy* 'any'. The same item also functions as a *wh*-interrogative 'which?'. The path of evolution is therefore presumably: *wh*-interrogative > free-choice item > NPI. The use of *ʔayy* as an NPI differs from that of English *any* in that it is always optional and emphatic, and has no effect on the aspectual interpretation of the predicate:

(51) a.

hal	ištarayta	kutuban
QU	buy.PRF.2MSG	books.ACC

'Have you bought (any) books?' [Note aspectual effect of (absence of) English *any*.]

b.

hal	ištarayta	**ʔayya**	kutubin
QU	buy.PRF.2MSG	any.ACC	books.OBL

'Have you bought any books (at all)?' (Classical/MSA)

The Classical/MSA data presented in this section are summarized in Table 10.4. Blanks there represent absent forms. Parentheses represent phrases which are not clearly lexicalized.

10.3.1.3 N-words and NPIs in Palestinian/Cairene Arabic There are two lexicalized rigid negative indefinites in Cairene and one in Palestinian. These are *maḥaddiš* 'no one' (Cairene), *maḥadāš* 'no one' (Palestinian), and *mafīš* 'nothing' (Cairene). These items are clearly morphologically negative. In the case of *maḥaddiš/maḥadāš*, this is transparently derived from the NPI indefinite *ʔaḥad* 'anyone', described for Classical/MSA above, plus bipartite *mā ... -š* negation, while *mafīš* is derived from the existential expression *fī* plus the same negative construction ('there is not' > 'nothing'). Both *maḥaddiš/maḥadāš* and *mafīš* are rather restricted in their distribution: *maḥaddiš/maḥadāš* can only occupy the subject position and must stand before

TABLE 10.4 N-words and NPIs in Classical Arabic / Modern Standard Arabic

	rigid neg. indef.	n-word		non-n-word		
		free distribution	NPI distribution	semi-NPI	strong NPI	weak NPI
Determiner	lā	—	—	—	—	ʔayy
Thing	(lā šayʔ)	—	—	—	—	(ʔayy šayʔ)
Person	(lā ʔaḥad)	—	—	—	—	ʔaḥad
Extent adverb	—	ʔabadan	—	—	—	—
Future time	—	ʔabadan	—	—	—	—
Past time	—	—	—	—	qaṭṭu	—
'Still'/'yet'	—	—	—	baʕdu	—	—
Conjunction	walā	—	—	—	—	ʔam

the verb (52)–(53), while *mafīš* appears not to be able to occupy argument positions at all (54)–(55). Nevertheless, both rigidly contribute one logical negation to the proposition with which they are associated, so whenever they co-occur with sentential negation double negation results.

(52) **maḥadāš** ʔaja
no.one come.PRF.3MSG
'No one came.' (Palestinian)

(53) **maḥaddiš** min al-bašar **ma-lū-š** maḥāsin
no.one from the-mankind NEG-have.3MSG-NEG good.qualities
'No one in existence doesn't have some good qualities.' (Cairene; Woidich 1968: 73)

(54) wa-māδa kull hāδihi š-šawšara ʕala **mafīš**
and-what all this the-noise about nothing
'What's all this fuss about nothing?' (Educated Cairene)

(55) in-naḍḍarāt dōl **miš** ʔaḥsan min **mafīš**
the-glasses this.PL NEG better than nothing
'These glasses are not better than nothing.' (Cairene)

On the basis of (52)–(55) one might be tempted to classify Palestinian and Cairene as non-negative-concord languages. This seems inappropriate, however, given that these dialects also have n-words in the categories of determiner, extent adverb, and 'still/yet' words: *wala* '(not) even a' (both Cairene and Palestinian), *ʔabadan* 'in no/any way' (both), *b-il-marra* 'in no/any way' (lit. 'in-the-time'; Palestinian only), *lissa* 'still/yet' (both) and *baʕd* 'still/yet' (Palestinian only).

Starting with the last category, the Classical/MSA semi-NPI adverb *baʕdu* 'still/yet' has become the Palestinian n-word *baʕd* 'still, (not) yet'. It has also undergone other syntactic developments, such that it now predominately appears sentence-initially and has been partially reanalysed as an impersonal verb taking the logical subject of the sentence it appears in as a pronominal object. *Lissa* (< *l-is-sāʕa* 'to the (current) time') is very similar to *baʕd* in its meaning, function and distribution (56) and (57). Neither of these words shows any polarity sensitivity, and neither is associated with a negative interpretation in sentences without predicate negation:

(56) a. hiyya **lissa** (/**baʕ^{i}d**-ha) txīna
she still (still-her) fat.F
'She's still fat.'

b. ṭayyib bass ana **lissa** (/**baʕ^{i}d**-ni) **ma**-šuft-^{i}**š** il-ʕarūsa
ok but I still (still-me) NEG-see.PRF.1SG-NEG the-bride
'OK, but I haven't seen the bride yet.' (Cairene (/Palestinian); Woidich 2006: 167, 349)

However, both items satisfy clause (b) of (42) and are therefore n-words:

(57) A: huwwa mayyit
he dead
'Is he dead?'

B: **lissa** (/**baʕd**-u)
still (/still-him)
'Not yet.' (Cairene (/Palestinian))

As such, an analysis along the same lines as given for Classical/MSA *ʔabadan* '(n)ever' seems most appropriate. *Lissa/baʕd* are single items whose negative contribution to the proposition they are associated with is context-dependent: iff they feature in a non-sentential response to a yes–no question then they will contribute one logical negation; in all other contexts they will not.

Turning to Palestinian/Cairene *ʔabadan*, this remains an n-word as it was in Classical Arabic (58a–b), and the analysis of its negative contribution will be the same. However, in Palestinian/Cairene it can no longer appear in affirmative declarative sentences, and it is used principally as an extent adverb (58b), rather than a temporal adverb (but see (58c); the meaning of 'ever' has in general been taken over by a construction with *ʕumr* < 'age, life', which is a weak NPI, not an n-word). *ʔabadan* can also occur in interrogative clauses such as (58c), where it is clearly non-negative. The Palestinian extent adverb *b-il-marra* is an extent adverb used interchangeably with *ʔabadan*, but it has a strictly strong NPI distribution.

(58) a.

inni	saḫbit-na	tiskut,	**ʔabadan**	(/**b-il-marra**)
COMP	friend-our	be.silent.IMPF.IRR.3FSG	ever	(/ever)

'That our friend would keep quiet? Never!'

b.

di	masʔala	miš	sahla	**ʔabadan**	(/**b-il-marra**)
this.F	issue	NEG	easy.F	ever	(/ever)

'This is an issue which is not at all easy.'

c.

huwwa-nta	maʕā-na	**ʔabadan**
QU-you.MSG	with-us	ever

'Do you ever agree with us?!' (Cairene (/Palestinian); Woidich 2006: 162, 349)

The final n-word to be discussed here is the scalar focus particle *wala*, which typically has the syntax of a determiner (the Classical/MSA rigid negative determiner *lā* has been lost in the present-day dialects). *Wala* contributes a logical negation to the proposition it is associated with in a wider range of contexts than do the n-words in (56)–(58). As well as always being associated with a negative interpretation in non-sentential utterances without a predicate negator (59), it can also occur in preverbal subject position, also in the absence of a predicate negator, again always being interpreted as negative (60).

(59)

wala	kilma
not.even	word

'Don't say a word!'

(60)

wala	taksi	wiʔif
not.even	taxi	stop.PRF.3MSG

'Not a single taxi stopped.' (Cairene/Palestinian; Woidich 2006: 342)

As can be seen from these examples, determiner *wala* always modifies a singular count noun, but its scalar properties vary according to those of the noun it modifies. Where the reference of the noun itself constitutes the minimal endpoint of a relevant scale, as in (59), *wala* serves to emphasize this minimality. Otherwise, as (60), *wala* emphasizes the zero cardinality of the noun it determines.

The reason *wala* is an n-word, rather than a rigid negative indefinite, is that, when it appears as a postverbal subject or (indirect) object, it must co-occur with predicate negation and the sentence is then interpreted as containing just one logical negation.

(61)

miš	sāmiʕ	**wala**	kilma
NEG	hear.ACTP.MSG	not.even	word

'I can't hear a single word.' (Cairene/Palestinian; Woidich 2006: 342)

The appropriate characterization of *wala* thus seems to be the inverse of the n-words we have examined thus far: rather than only contributing a logical negation to interpretation in certain specific circumstances, *wala* will always make a negative contribution, except when it follows some item which has already done so.

This behaviour makes sense from a diachronic perspective, since *wala* is clearly etymologically derived from the Classical Arabic negative conjunction *walā* (cf. (49)), in a common development from additive focus particle to scalar focus particle (cf. Haspelmath 1997: 157). In fact the same item also maintains the negative conjunctive function of its Classical ancestor in the present-day dialects:

(62) ma-ʕandu-hum-š^{i} zōʔ **wala** tarbiyya
NEG-have-they-NEG taste and.not upbringing
'They have neither taste nor manners.' (Cairene; Woidich 2006: 344)

Utterances such as (61) would therefore presumably first have been interpreted as a conjunction of two negated propositions, with an unexpressed (or ellipsed) indefinite pronoun argument in the first conjunct, and with the material in the second that is identical to that in the first also unexpressed (or ellipsed):

(61)′ I can't hear ~~anything~~, not even a word.

(61)′′ $\neg\exists x(\text{hear}(Me, x)) \wedge \neg\exists y(\text{word}(y) \wedge \text{hear}(Me, y))$

At this stage *wala* would still have been rigidly negative, becoming an n-word once such structures were reanalysed as monoclausal and *wala* was reanalysed as a scalar focus particle rather than a conjunction.

The rest of the Palestinian/Cairene indefinite system is very similar to that of Classical/MSA. *ʔayy* remains a free-choice and emphatic weak NPI determiner, and the Classical/MSA weak NPI pronoun *ʔaḥad* 'anyone' remains a weak NPI in Palestinian *ḥada* and Cairene *ḥadd*. The Classical/MSA marker of exclusive disjunction *ʔam* has been lost and replaced with a functionally equivalent item *willa~walla* (< *wa-ʔin lā* 'and if not'), which, like *ʔam*, is restricted to questions. Note the geminate /l/ – this is not the same item as the negative conjunction and negative scalar focus particle *wala*.

The Palestinian (P) and Cairene (C) data are summarized in Table 10.5. Blanks represent absent forms. Parentheses represent phrases which are not clearly lexicalized.

10.3.1.4 N-words and NPIs in Moroccan Arabic The indefinite system of Moroccan is rather different to that of Palestinian/Cairene. It has no rigid negative indefinites and has innovated two new n-words, a 'thing'-pronoun *wālu* (whose etymology is not certain; but see below) and a determiner *ḥətta* (< Classical Arabic *ḥattā* 'even, until'):

(63) a. ʕazīz ma taygūl li-ya **wālu**
Aziz NEG say.IMPF.3MSG to-me n.thing
'Aziz doesn't tell me anything.' (Moroccan; Adila 1996: 110)

TABLE 10.5 N-words and NPIs in Palestinian/Cairene Arabic

	rigid neg. indef.	n-word		non-n-word	
		free distribution	weak NPI	strong NPI	weak NPI
Determiner	—	—	—	wala	ʔayy
Thing	mafīš (C)	—	—	(wala ḥāga - C)	(ʔayy[i] ḥāga - C)
				(wala ʔiši - P)	(ʔayy ʔiši - P)
Person	maḥaddiš (C)	—	—	(wala wāḥid)	(ʔayy[i]) ḥadd (C)
	maḥadāš (P)			(wala ḥada - P)	(ʔayy) ḥada (P)
Extent adverb	—	—	ʔabadan	b-il-marra (P)	—
Time	—	—	—	—	ʕumr-
'Still'/'yet'	—	baʕd (P)	—	—	—
		lissa			
Conjunction	wala	—	—	—	walla

b. A: mā-l-ək
what-to-you
'What's the matter?'

B: **wālu wālu**
'Nothing, nothing.' (Moroccan; Durand 2004: 111)

(64) a. ana ma klīt **ḥətta** ḥāja mən əl-bārəḥ
I NEG eat.PRF.1SG even thing since yesterday
'I haven't eaten anything/a thing since yesterday.'

b. ma kayswa **ḥətta** bəṣla
NEG be.worth.IMPF.3MSG even onion
'It's not worth a penny.' (Moroccan; Adila 1996: 111–12)

c. A: škun kayskən mʕ-ak
who live.IMPF.3MSG with-you
'Who lives with you?'

B: **ḥətta** wāḥəd
even one
'No one.' (Moroccan; Ouali 2008: 9)

Examples (63) and (64a) illustrate the two Morrocan n-word expressions translatable as 'nothing': *wālu* and *ḥətta ḥāja*.[20] Ouali (2008) views both of these as inherently negative. Despite the considerable overlap in their meaning and behaviour, however, there are also clear differences in their respective contributions to the negativity of the propositions they are associated with. Most strikingly, *wālu* always signals the presence of negation, whereas *ḥətta*, which, as illustrated in (64), can co-occur with a range of nominal expressions, is common in affirmative sentences, where it broadly retains the original meaning of the Classical Arabic scalar focus particle *ḥattā* 'even':

(65) **ḥətta** š-šībāni kayḥəbb lə-bnāt
even the-old.man like.IMPF.3MSG the-girls
'Even an old man (still) likes girls.' (Moroccan; Harrel 1966: 249)

Moreover, its interpretation as negative or affirmative in certain non-sentential utterances is dependent on the context:

(66) a. A: ana ɣādi nṣāfər š-šhər l-māji
I FUT travel.IMPF.1SG the-month the-coming
'I'm taking a trip next month.'

B: **ḥətt**-ana
even-I
'So am I.'

b. A: ḥna ma xārjīn-š
we NEG leaving.MPL-NEG
'We're not leaving.'

B: **ḥətt**-ana
even-I
'Neither am I.' (Moroccan; Harrel 1966: 249)

It is clear from these data that *ḥətta* is an n-word, but Ouali's negative-quantifier analysis of *ḥətta ḥāja* seems wide of the mark, unless we posit an unparsimonious analysis of *ḥətta* in affirmative declarative contexts, as in (65), as a separate but homophonous lexical item to the *ḥətta* in *ḥətta ḥāja*. At the same time, Benmamoun's (1997, 2000, 2006) characterization of *ḥətta* simply as an NPI is also mistaken: this characterization offers no explanation of this item's negative contribution in (66b), as well as also forcing an unparsinonious homophony analysis for the data in (65).

In fact, *ḥətta* clearly has similar negative properties to Palestinian/Cairene *baʕd* and *lissa* 'still, yet'. Like them, it is an n-word which only contributes a logical

[20] Note that the presence of an indefinite obligatorily triggers the absence of negative *-š* in Moroccan. The same is true of Maltese, below. Compare French (section 2.3.1), where the postverbal negator *pas* is in complementary distribution with n-word indefinites.

negation to the proposition it is associated with when it occurs utterance-initially in a non-sentential utterance lacking predicate negation. Unlike *baʕd* and *lissa*, however, *ḥətta* only makes a negative contribution in a subset of dialogue contexts: namely those in which the proposition associated with the utterance just made either contains a *wh*-variable, as in (64c), or is itself negative, as in (66b). Where this is not the case, as in (66a), there is no negative contribution.

A further noteworthy property of Moroccan *ḥətta* is that it is an NPI licenser, to the extent that it licenses the only two NPI pronouns that occur in Moroccan: the strong NPI *šay* 'anything' and the weak NPI *ḥədd* 'anyone' (cf. Classical/MSA *ʔaḥad*, Egyptian *ḥadd*). Compare (67b), which is ungrammatical because *ḥədd* is unlicensed in initial position, with (67a), where the addition of *ḥətta* provides the necessary licensing. The same is illustrated for *šay* in (68).

(67) a. **ḥətta** **ḥədd** ma gāl-li-k ʔaji
even anyone NEG say.PRF.3MSG-to-you come.IMP.2SG

b. *__ḥədd__ ma gāl-li-k ʔaji
anyone NEG say.PRF.3MSG-to-you come.IMP.2SG
'No one told you to come.'

(68) a. **ḥətta** **šay** ma yfərrəq-na
even anything NEG separate.IMPF.3MSG-us

b. *__šay__ ma yfərrəq-na
anything NEG separate.IMPF.3MSG-us
'Nothing will separate us.' (Moroccan; Adila 1996: 111)

Ouali (2008) cites this property of *ḥətta* as further evidence of its status as an inherently negative quantifier. But the lack of any negative interpretation associated with *ḥətta* in (65) and (66a) remains problematic for such an analysis. Instead, it seems preferable to analyse (67) and (68) as an instance of 'parasitic' NPI licensing (Hoeksema 2007b), especially since, outside of non-sentential utterances such as (66b), *ḥətta* cannot license NPIs in the absence of the predicate negator *ma*:

(69) *__ḥətta__ **ḥədd** gāl-li-k ʔaji
even anyone say.PRF.3MSG-to-you come.IMP.2SG
'No one told you to come.'

(70) *__ḥətta__ **šay** yfərrəq-na
even anything separate.IMPF.3MSG-us
'Nothing will separate us.' (Moroccan)

Concerning the other innovative Moroccan n-word *wālu* 'nothing', the negative properties of this item closely resemble those of Palestinian/Cairene *wala*, in that it is always associated with a negative interpretation and it contributes a logical negation to the proposition it is associated with in all contexts, except where it is preceded by some item which has already made this contribution, as in (63a). This similarity with *wala* is not surprising, inasmuch as *wālu* appears to be derived from *wala* plus some other element such as *hu(wa)* 'he, it' (although the long /ā/ preceding the /l/ is unexpected on this scenario).

The fact that Moroccan lacks rigid negative indefinites, together with the fact that *wālu*, an item which is always associated with a negative interpretation, regularly co-occurs with predicate negation giving only a single logical negation in interpretation, as in (63a), means that a characterization as a negative-concord language seems reasonable. On the other hand, it is unclear whether, given the above discussion, it is accurate (or helpful) to describe the interaction of *ḥətta* 'even' with predicate negation as an instance of negative concord.

The interaction of the weak NPI *ʕammər̥-* 'ever' with predicate negation provides perhaps the clearest example of negative concord in Moroccan; albeit the negative-concord construction is only one of three possible synonymous configurations, all apparently in free variation:[21]

(71)	a. ma	ʕammər̥-ni		šəft-u
	b.	ʕammər̥-ni	ma	šəft-u
	c. ma	ʕammər̥-ni	ma	šəft-u
	(NEG)	ever-me	(NEG)	see.PRF.1SG-him

'I've never seen him.' (cf. Caubet 1996: 91, Adila 1996: 105, Durand 2004: 198)

Here we see that predicate negation can occur just on *ʕammər̥-* (71a), just on the main verb (71b), or on both at once (71c), with precisely the same interpretation (one logical negation) in each case.

The rest of the Moroccan indefinite system is more straightforward. Like Cairene/Palestinian, Moroccan retains the Classical n-word *ʔabadan* 'ever' as an extent adverb restricted to negative contexts. It has also innovated a non-n-word strong NPI extent adverb, *gāʕ* (< *qāʕ* 'bottom'):

(72)	gāʕ	ma	kanəbɣī-h
	at.all	NEG	like.IMPF.1SG-him

'I don't like him at all.' (Moroccan; Harrel 1966: 42)

There is also an n-word 'still/yet' adverb *mazāl* (< Classical Arabic *mā zāla* 'has not stopped') with the same free distribution as Palestinian/Cairene *lissa* and Palestinian *baʕd*.

Finally, there is a very frequent indefinite determiner *ši* 'some, any' (< *šayʔ* 'thing') which seems to be totally unrestricted by clause type:

[21] Like Palestinian *baʕd* 'still, yet', *ʕammər̥-* seems to have developed into an impersonal verb taking the logical subject of the clause it appears in as an accusative-marked object suffix. It clearly derives from the form *ʕumr-* 'age, life', described above for Cairene/Palestinian.

(73) a. səlləf-ni ši flūs
lend.IMP.2SG-me some money
'Lend me some money.'

b. wāš ʕəndu ši wlād
QU have.3MSG any sons
'Does he have any sons?'

c. ʕammṛ-u ma ləbbā-l-hom ši ṭalab
ever-him NEG answer.PRF.3MSG-to-them any request
'He never consented to any request of theirs.' (Moroccan; Harrel 1966: 69, 153, 185)

The Moroccan data are summarized in Table 10.6. Blanks represent absent forms. Parentheses represent phrases which are not clearly lexicalized.

10.3.1.5 N-words and NPIs in Maltese The final Arabic variety to be considered here is Maltese, a variety which has undergone numerous changes in all domains relative to the North African dialects it is most closely related to, largely as a result of its isolation from these varieties and heavy contact with Sicilian since about the thirteenth century.

Maltese seems to be the only Arabic variety that may accurately and straightforwardly be described as a negative-concord language. The Maltese indefinite system has been described by Haspelmath and Caruana (1996). They contrast two series of indefinites, shown in Table 10.7, which they call the *xi*-series[22] and the *ebda*-series (Haspelmath and Caruana 1996: 215).

TABLE 10.6 **Indefinites in Moroccan Arabic**

	n-word		non-n-word		
	free distribution	strong NPI distribution	strong NPI	weak NPI	free distribution
Determiner	ḥətta	—	—	—	ši
Thing	—	wālu (ḥətta ḥāja)	šay	—	—
Person	—	(ḥətta ḥədd)	—	ḥədd	—
Extent adverb	—	ʔabadan	gāʕ	—	—
Time	—	—	—	ʕamməṛ-	—
Still/yet	mazāl	—	—	—	—

[22] All reference to Maltese here is made using Maltese orthography, the most important detail of which for present purposes is that the symbol <x> represents the sound /ʃ/.

TABLE 10.7 Haspelmath and Caruana's (1996) Maltese *xi*- and *ebda*-series

	ebda-series	*xi*-series
Determiner	ebda	xi
Thing	xejn	xi ħaġa
Person	ħadd	xi ħadd
Time	qatt	xi darba
Place	imkien	xi mkien

As can be seen from Table 10.7, the members of the *xi*-series are all composed of the determiner *xi* plus some other element. This determiner is the Maltese equivalent of the Moroccan indefinite determiner *ši*. The distribution of Maltese *xi* is almost as wide as that of Moroccan *ši*, except that the former cannot occur in the scope of negation. This means that they are, in general, in complementary distribution with the members of the *ebda*-series, which are all n-words: when these function as arguments (or predicate-level adjuncts in the case of *qatt* '(n)ever' and *imkien* 'nowhere') in a full sentence they must co-occur with the sentential negator *ma* and these sentences are interpreted as having a single semantic negation. However, members of the *ebda*-series can appear without *ma* in non-sentential utterances which are nevertheless still interpreted as negative, as can be seen in (74) (= Haspelmath and Caruana's (16a) and (17)). There are no rigid negative indefinites in Maltese.

(74) a. It-tifla ma rat **xejn.**
the-girl NEG see.PRF.3FSG n.thing
'The girl didn't see anything.'

b. A: X'rat?
what-see.PRF.3FSG
'What did she see?'

B: **Xejn!**
n.thing
'Nothing!' (Maltese)

Haspelmath and Caruana (1996: 217) consider the *ebda*-series to be 'inherently negative'. However, in line with the discussion above, it seems preferable to say that, as n-words which are always associated with a negative interpretation, the *ebda*-series will always contribute a logical negation to the proposition they are associated with, unless some other item has already made such a contribution earlier on in the parse.

Moreover, it is not in fact the case that all members of the *ebda*-series are always associated with a negative interpretation (though there is no doubt that they are all n-words). In particular, *qatt* '(n)ever', unlike the other members of the *ebda*-series, is regularly found with non-negative meaning in non-veridical contexts such as questions and conditionals, as in (75) (= Haspelmath and Caruana's (24)).[23]

(75) a. Jekk **qatt** tiġi Londra, ejja ara-ni.
if ever come.IMPF.2SG London come.IMP.2SG see.IMP.2SG-me
'If you ever come to London, come and see me.'

b. **Qatt** mort Londra?
ever go.PRF.2SG London
'Have you ever been to London?' (Maltese)

Similarly, *imkien* 'nowhere', while much more common in negative sentences, can also be used in affirmative declarative contexts, simply meaning 'place' (although the plural *imkeyyen* 'places' is far more common than the singular, which must be regarded as an archaism in such a context; Michael Spagnol, p.c.). *Imkien* could thus be said to have a 'virtual-NPI' distribution (i.e. closer to a true NPI than a semi-NPI).

The same is not true, however, of the determiner *ebda*, the 'person'-pronoun *ħadd*, or the 'thing'-pronoun *xejn*, which are invariably associated with a negative interpretation. The extent to which Maltese speakers have come to associate these items with negation can be illustrated by the denominal verb *xejjen* 'to make nothing, destroy':

(76) Iżda **xejjen** lilu n-nifs-u.
but nullify.PRF.3MSG to.him the-self-his
'But he made himself nothing.' (Maltese; Philippians 2:7)

Comparison of the members of the *ebda*-series with etymological equivalents in other Arabic varieties clearly shows that their association with negation is innovative, as illustrated below. But even within Maltese, a number of fixed relic expressions containing these items also show that this association with negation is comparatively recent. For example: *qabelxejn* 'first of all' (*qabel* 'before'), *kull xejn* 'everything' (*kull* 'all'), *kulħadd* 'everyone', and *ħadd ieħor* 'someone else' (*ieħor* 'other'; cf. *ħadd iżjed* 'no one else' (*iżjed* 'more')).

[23] *Qatt* 'ever' and *xi darba* 'ever, sometime' are thus interchangeable and synonymous in non-veridical contexts other than negation, except that members of the *xi*-series can always optionally outscope co-occurring non-veridical operators, whereas members of the *ebda*-series can never do so. When members of the *xi*-series occur as arguments of negated verbs, they obligatorily outscope negation in the same way as the *some*-series in English:

(i) jekk ma fhimt-x xi.ħaġa…
if NEG understand.PRF.2SG-NEG something
'If you didn't understand something…' (Maltese; Michael Spagnol, p.c.)

Note also that Maltese shares with Moroccan strict complementarity between the negative suffix *-x/-š* and indefinites (cf. (75a)), but that this only applies to indefinites in the scope of negation, unlike in (i).

This raises the question of how the present Maltese indefinite system evolved from one which we must assume originally more closely resembled that of Classical Arabic and the North African dialects. Starting with the etymology of the individual items, as already noted, the determiner *xi* is identical to Moroccan *ši* in all respects other than its interaction with negation (see Table 10.6). This item is evidently derived from Classical Arabic *šayʔ* 'thing', presumably through reanalysis of a genitive construct with partitive meaning: 'a thing of *X*' > 'some/any *X*'.

Ħadd is also quite clearly the same item etymologically as Moroccan *ḥədd* and Classical *ʔaḥad*. As we have seen, however, *ħadd* has undergone a marked contraction in the environments in which it can occur: from all weak NPI contexts as expected by comparison with *ḥədd* and *ʔaḥad*, to essentially only negative contexts, culminating in its becoming an n-word, except in certain fossilized phrases including *xi ħadd* 'someone/anyone'.

The etymology of the remaining items in the *xi*-series which collocate with *xi*—*ħaġa*, *darba*, and *(i)mkien*—is trivial, since these are all still lexical items (meaning 'thing', 'time', and 'place', respectively) in their own right.

Qatt 'ever' is clearly derived from Classical Arabic *qaṭṭu* with the same meaning (see Table 10.4), which is interesting in itself, given that this item is not found in other North African Arabic dialects. Also, the fact that *qatt* is an ordinary weak NPI, frequently found in non-negative non-veridical contexts, is interesting given that its ancestor *qaṭṭu* appears to have been strong. Either this is a rare instance of a polarity item becoming weaker over time, or, perhaps more likely, the restriction of *qaṭṭu* virtually only to negative contexts could have been a development of immediately post-Islamic written Arabic (i.e. what became Classical Arabic), not shared by the contemporaneous spoken varieties.

The remaining n-words *ebda* and *xejn* have undergone a somewhat greater shift. *Xejn* is clearly also derived from *šayʔ* 'thing', apparently with a vestige of the indefinite accusative suffix *-an*. A number of offshoots of *šayʔ* have developed into a range of weak and strong NPIs in a range of Arabic dialects, as we have seen, but it is only in Maltese that one of these has become an n-word. *Ebda* appears to be derived from *ʔabadan* '(n)ever, (not) at all', and, if this is correct, this would make Maltese also the only Arabic variety in which this item has become a determiner. This development from 'time'-adverb to determiner does not seem to be common in other languages, but it is attested in English, for example, in the colloquial or archaic forms *nary (a)/ne'er (a)* 'no X' < *never*.

Concerning the development of negative concord, we have seen that while there are clear elements of negative concord in Moroccan, Maltese appears to be the only Arabic variety with clearly lexicalized n-words in the categories determiner, 'thing'-pronoun, and 'person'-pronoun that are always associated with a negative interpretation. It also the only variety in which no non-n-word indefinites are grammatical in the scope of negation. In accounting for this unique position, it seems reasonable to point to the centuries of intensive contact between Maltese and the Romance varieties of Sicily, which have always exhibited negative concord since their earliest attestations (Adam Ledgeway, p.c.), as in the following example from the 14th century:

(77) **nullu** homu **non** la vitti
no man NEG it see.PAST.3SG
'no man had seen it' (14th-century Sicilian; ex. (92) section 3.8)

In an indefinite system in which no indefinites are marked negative morphologically, negative sentences containing indefinite pronouns will always be analysable either as featuring negative concord between sentential negation and an n-word, or as featuring a non-n-word indefinite and therefore no negative concord. Maltese speakers, like speakers of other Arabic varieties, will originally have made the latter analysis. However, intensive second-language exposure to Sicilian could have prompted a reanalysis of the polarity of the indefinites of their native Maltese on the basis of the morphologically transparent negative-concord system of Sicilian. This would amount to borrowing of the rule that the presence of sentential negation requires indefinite pronouns in the scope of negation to be n-words. Alternatively, and perhaps more plausibly, this change could have occurred under the agentivity of L1 Sicilian speakers (for example in mixed Sicilian–Maltese marriages) who identified the non-n-word indefinites of their L2 Maltese with the n-word indefinites of their L1 Sicilian. On the other hand, it cannot be ruled out that the development of negative concord in Maltese was a purely internal process that was unaffected (or perhaps merely accelerated) by contact with Sicilian.[24]

TABLE 10.8 **Indefinites in Maltese**

	n-word			non-n-word
	strong NPI distribution	virtual NPI distribution	weak NPI distribution	free distribution except negation
Determiner	ebda	—	—	xi
Thing	xejn	—	—	xi ħaġa
Person	ħadd	—	—	xi ħadd
Time	—	—	qatt	xi darba
Place	—	imkien	—	xi mkien

[24] Data on the indefinite systems of Tunisian and western Libyan Arabic, the dialects most closely resembling Maltese, are scarce. It does seem, however, that Tunisian has developed an n-word *šay* 'n.thing' (cf. Maltese *xejn*), while in western Libyan both *šay* and *ḥadd* 'n.body' are n-words (Borsley and Krer 2011, Miriam Bouzouita, p.c.). This suggests that the shift of certain indefinites to n-word status was perhaps

Table 10.9 The evolution of Maltese indefinites

		n-word		non-n-word			
		strong NPI distribution	weak NPI distribution	strong NPI	weak NPI	free except neg	free distribution
Determiner	inherited						ši
	current	ebda				xi	
Thing	inherited						šayʔan, ḥāja
	current	xejn				xi ħaġa	
Person	inherited				ḥadd		
	current	ħadd				xi ħadd	
Time	inherited			ʔabadan qaṭṭu			ḍarba
	current		qatt			xi darba	
Place	inherited						makān
	current	imkien				xi mkien	

The synchrony of the Maltese indefinite system is summarized in Table 10.8 and its evolution in Table 10.9. Blanks represent absent forms. Arrows indicate diachronic developments from the presumed inherited to the present-day state of affairs.

10.3.2 Other Afro-Asiatic languages

For Afro-Asiatic languages other than Arabic, scarcity of data, especially of former states of languages, is a significant barrier to detailed investigation of developments in indefinite pronoun systems. Among the languages for which we do have extensive attestation, however, one striking feature is that neither Akkadian nor Egyptian–Coptic ever develop anything resembling negative concord throughout the entire (very lengthy) period of their attestation.

already underway in the mainland central North African dialect that is the common ancestor of Maltese, Tunisian, and western Libyan, and from which Maltese split, starting in about the 13th century.

Ancient Egyptian lacks dedicated indefinite pronouns distinct from the words for 'man' and 'thing' (Haspelmath 1997: 324, citing Gardiner 1957), while Coptic developed a 'person'-pronoun *laau* 'someone, anyone' with free distribution but no NPI pronouns or n-words (Reintges 2004). Akkadian *mimma* 'something, anything, everything' and *mamman* 'someone, anyone' are attested, with minor variations, from Old Akkadian (third millennium BCE) to Neo-Babylonian/Neo-Assyrian (first millennium BCE), and at no stage is there evidence to suggest a development into n-words, despite frequent co-occurrence with negation (Black, George, and Postgate 2000).

There is a widespread perception that negative concord is in some sense a default option and that there is strong functional pressure for languages which lack negative concord to acquire it over time. Typological investigations of this issue (Kahrel 1996, van der Auwera and van Alsenoy 2011) reveal that it is far less common worldwide than a Eurocentric perspective would suggest, however;[25] and the long-term evidence of Egyptian–Coptic and Akkadian shows that whatever functional pressure there might be to develop negative concord, this is clearly far from irresistible.

The remainder of this section focuses on Hebrew and Ethiopian Semitic.

10.3.2.1 Developments in Hebrew indefinites In contrast to Egyptian–Coptic and Akkadian, Hebrew shows significant developments in its indefinite system, with various originally non-polarity-sensitive items becoming n-words over time. Biblical Hebrew did not have negative concord, and appears only to have had one item that was restricted to weak NPI contexts, the 'thing'-pronoun *məʔuma* mentioned in section 10.1.2. Brown, Driver, and Briggs (1999: s.v. *məʔuma*) suggest two possible etymologies for this item. The first is *ma u-ma* 'what and what' (cf. Latin *quidquid* 'anything'), which is favoured by Faber (1988). The second is a feminine form of *məʔum* 'blemish, speck, particle' used as a minimizer, which Faber (1988) dismisses, probably correctly given that minimizers tend to be restricted just to negative contexts, while *məʔuma* is found both in the context of negation (78a) as well as other non-veridical contexts such as questions (78b). It is not found in affirmative declarative sentences.

(78) a. wa-ha-mmeθim ʔen-om yodʕim **məʔuma**
and-the-dead.PL NEG-they knowing.PL anything
'But the dead do not know anything.' (Eccl. 9: 5)

b. ha-yaxol ʔuxal dabber **məʔuma**
COMP-be.able.INF be.able.IMPF.1SG say.INF anything
'Can I really say anything?' (Num. 22: 38) (Biblical Hebrew; Faber 1988: 225–6)

[25] Just five out of the forty languages in Kahrel's (1996) stratified sample have negative concord, while, in a genetically and areally balanced sample of eighty-one languages of Africa and the Americas, van der Auwera and van Alsenoy (2011) found only seven clear cases.

In Mishnaic and Medieval Hebrew we witness the development of two new indefinites, a 'thing'-pronoun *klum* (79) and a determiner *šum* (80), while *məʔuma* becomes less common.

(79) ʔen l-o **klum**
there.is.not to-him anything
'He doesn't have anything.'

(80) li-vn-o ʔexad lo natan **šum** matana
to-son-his one NEG give.PRF.3MSG any present
'To one of his sons he didn't give any present at all.' (Mishnaic Hebrew; Fernández 1999: 175)

The etymology of *klum* is not known. Segal (2001: 210) suggests that it arises from annexation of *kol* 'all, any' to *məʔuma* with subsequent phonological erosion of the latter, while Faber (1988: 222) sees it as derived from a reconstructed (pre-Hebrew?) form **kullu* 'all' + the generalizing suffix *-ma* (see section 10.1.2). Neither of these proposals can be seen as definitive.

The etymology of *šum* is clearer: it is a borrowing of the Aramaic word *šema* 'name', frozen in its construct-state form. Before reanalysis as a determiner it must have functioned as a minimizer: the name of a given entity viewed as the minimal quantity of that entity (cf. Akkadian *mimma šumšu* 'anything at all', lit. 'anything, its name'). Both *klum* and *šum* are found (almost)[26] exclusively in the context of negation (and are therefore strong NPIs), as expected in the case of *šum*, given its derivation from a minimizer. There is, however, no evidence to suggest that these items were n-words in the pre-revival period.

Modern Israeli Hebrew has a number of n-words, however. These are: the determiners *af*[27] (81) and *šum* (82), the 'thing'-pronouns *klum* and *šum davar* (83), the 'person'-pronoun *af exad* (84), the 'time'-adverb *af paʔam* (85), and the 'place'-adverb *be-šum makom* (86).

(81) a. **lo** raʔiti **af** xatul
NEG see.PRF.1SG *af* cat
'I didn't see a single cat.'

b. hayu harbe morim aval **af** mora
be.PRF.3PL many teacher.MPL but *af* teacher.FSG
'There were many male teachers, but no female teacher(s).'

[26] Rubin (2005: 51) gives an example of *šum* in a conditional sentence from a 12th-century translation of an earlier Arabic work.

[27] *Af* is derived from a homophonous Biblical Hebrew scalar focus particle meaning 'also, even' (cf. Moroccan Arabic *ḥətta* 'even, a single, no').

(82) a. **lo** raʔiti **šum** xatul
NEG see.PRF.1SG *šum* cat
'I didn't see any cat(s).'

b. šamanu neʔumim aval **šum** xidušim
hear.PRF.1PL speeches but *šum* new.ideas
'We heard speeches but no new ideas.'

(83) a. **al** tikne **klum** / **šum davar**
NEG buy.FUT.2MSG n.thing *šum* thing
'Don't buy anything.'

b. A: ma kara
what happen.PRF.3MSG
'What happened?'

B: **klum** / **šum davar**
n.thing *šum* thing
'Nothing.'

(84) a. **lo** raʔiti **af exad**
NEG see.PRF.1SG *af* one
'I didn't see anyone.'

b. A: mi ba
who come.PRF.3MSG
'Who came?'

B: **af exad**
af one
'No one.'

(85) a. ani **af paʔam lo** hayiti šam
I *af* time NEG be.PRF.1SG there
'I've never been there.'

b. A: hayita šam
be.PRF.2MSG there
'Have you been there?'

B: **af paʔam**
af time
'Never.'

(86) a. **lo** hayiti **be-šum makom**
NEG be.PRF.1SG in-*šum* place
'I haven't been anywhere.'

b. A: eyfo hayita
where be.PRF.2MSG
'Where have you been?'

B: **be-šum makom**
in-*šum* place
'Nowhere.' (Modern Hebrew; Glinert 1982: 434, 450, 454)

Unlike some of the dialectal Arabic n-words considered above, all of these Modern Hebrew n-words are restricted to negative contexts. As with the Arabic items with similar properties, we can analyse their behaviour as a context-dependent contribution to the negativity of the proposition with which they are associated: they will always make such a contribution, except if some other item has already done so earlier on in the parse.

Since Modern Hebrew has no rigid negative indefinites, its widespread classification as a negative-concord language (e.g. Tonciulescu 2007) seems reasonable. Tracking how this system developed from the non-negative-concord system of earlier Hebrew is in fact made rather easy by the fact that Hebrew did not exist as a spoken language from the early part of the first millennium until its revival in the late nineteenth century. Those who learnt to speak it in the latter period were necessarily therefore native speakers of other languages, predominantly Yiddish (Zuckerman 2009: 43). On this basis, given that a) negative concord appears not to have been a feature of pre-revival literary Hebrew, b) it does appear, by contrast, to have been a feature of Modern Hebrew since the earliest days of its revival as a native language (cf. Chanoch 1930: 71–2), and c) Yiddish has long been a negative-concord language (see (87) for some examples), this looks like a clear case of syntactic transfer.

(87) a. **keyner** hot **nit** gezen ire trern
nobody AUX.PAST.3SG NEG see.PP her tears
'No one saw her tears.'

b. du vest dayn tsil **keyn** mol **nit** dergreykhn
you AUX.FUT.2SG your goal no time NEG reach.INF
'You will never reach your goal.'

c. zi hot geboyrn a kind **on keynems** hilf
she AUX.PAST.3SG bear.PP a child without nobody.GEN help
'She bore a child without anyone's help.' (Lockwood 1995: 130–1)

Native Yiddish speakers must have found evidence in the strict co-occurrence with negation of Hebrew indefinites such as *klum* and *šum* for the negative-concord system familiar to them from their native language and interpreted these indefinites as n-words rather than strong NPIs.

10.3.2.2 Developments in Ethiopian Semitic indefinites There is no language in the Ethiopian subgroup of Semitic for which we can track significant developments in its indefinite system in written texts. However, in Geʿez we have an early snapshot of the grammar of one Ethiopian Semitic language before it died out as a spoken variety some time before the end of the first millennium (Gragg 1997: 243). Here we have a clear negative-concord system, where n-words that are morphologically marked as negative with the general-purpose negator *ʔi—ʔimənt(-hi/-ni)* 'nothing' and *ʔimännu(-hi/-ni)* 'no one'—co-occur with sentential negation:

(88) wä-**ʔi**-täʔämməxu wä-**ʔi-männa**-hi ba-fənot
and-NEG-greet.IMPF.2MPL and-NEG-what.ACC-PRT in-way
'and don't greet anyone on the way.' (Geʿez, Luke 10:4; Tropper 2002: 148)

Tigre and Tigrinya, the modern Ethiopian Semitic languages most closely related to Geʿez, do not appear to be able to form morphologically n-marked indefinites in this way. Leslau (1945) provides some data on the indefinite pronoun system of Tigre. It does not appear to have any rigid negative indefinites. Leslau does describe one 'thing'-pronoun indefinite *sema*, which appears to be an n-word. He glosses it as '(with negation) nothing' and gives the following examples:

(89) a. **sema**-ma ʔi-räkäbko
sema-EMPH NEG-find.PRF.1SG
'I haven't found anything.'

b. **sema** əmbäl däḥan
sema except good
'nothing but good.' (Tigre; Leslau 1945: 192)

Tigrinya does not in general appear to have n-word indefinites (cf. example (35)), with the exception of the negative scalar focus particle *wala*, borrowed from Arabic (see section 10.3.1):

(90) a. A: ḥadä säb rəʔaka-do
one man see.PRF.2MSG-INTERROG
'Did you see anyone?'

B: **wala** ḥadä
not.even one
'No one.'

b. **wala** ḥadä melsi zə-fälliṭ **yällo-n**
not.even one answer REL-know.IMPF.3MSG NEG.be.3MSG-NEG
'There's nobody who knows the answer.' (Tigrinya; Bernini 2003: 94–5)

Amharic, whose indefinite system is better described than that of any other Ethiopian Semitic language thanks to Leslau's (1995) comprehensive grammar, also lacks morphologically negative indefinites of the type found in Geʿez. It does, however, have one n-word, the 'person'-pronoun *mannəmm*:

(91) a.

balläfäw	sammənt	bet-ä	krəstiyan	**mannəmm**
in.last	week	house-CONSTR	Christian	n.one

al-hedä-**mm**
NEG-go.PRF.3MSG-NEG

'No one went to church last week.'

b. A:

yəhən	man	näggärä-h
this	who	say.PRF.3MSG-you

'Who told you this?'

B:

mannəmm
n.one

'No one.' (Amharic; Leslau 1995: 122–3)

This item *mannəmm* occurs without a negative interpretation in questions and conditionals, and as a free-choice item, as well as in the standard of comparison:

(92)

amarəñña	kä-mannəmm	yəbälṭ	ənnaggärallä^{w}h
Amharic	than-*mannəmm*	more	speak.COIMPF.1SG

'I speak Amharic better than anyone.' (Amharic; Leslau 1995: 121)

To summarize, while Geʿez is a clear case of a straightforward negative-concord language, the vernacular Ethiopian Semitic languages surveyed are more reminiscent of Cairene or Palestinian Arabic in the behaviour of their indefinites: each of these langauges seems to have one or possibly more n-words with varying properties, but none has a clear paradigm of n-words that are predominantly associated with a negative interpretation. As such, they are not straightforwardly characterizable as negative-concord languages in the same way as Geʿez.

Thus we have an interesting situation among the Ethiopian Semitic languages that is the inverse of what has been the case in Europe. Like Latin, Geʿez is a high language, is no one's mother tongue, and was for a long time the only or principal language of literature and liturgy. Unlike Latin, however, it has a clear negative-concord system. The fact that a number of languages in Europe lack negative concord, particularly in the case of standard languages, despite an assumed functional pressure to acquire it, is often attributed to the influence of Latin (e.g. Haspelmath 1997: 205, 220). But the vernacular Ethiopian Semitic languages we have looked at have not developed the straightforward kind of negative-concord system familiar from European languages despite both the hypothetical functional pressure to do so and potential influence from the high variety Geʿez.

Together with the failure of both Akkadian and Egyptian–Coptic to develop negative concord over the course of several millennia, the evidence presented from Afro-Asiatic thus far must therefore cast some doubt on the strength of the functional pressures that have been proposed to account for the prevalence of negative concord cross-linguistically.[28] For example, Haspelmath (1997: 193–234) invokes two relevant universal preferences: i) for negation to be marked on the verb in ordinary sentence negation (assuming Jespersen's view of negation as predicate denial and Haiman's 1980 principle of form–meaning isomorphism), and ii) Jespersen's (1917) and Horn's (1989) Neg-First Principle, which states that the semantic importance of negation is sufficiently great that it needs to be expressed as early as possible in a sentence. The first of these preferences is upheld in all the languages investigated in this section. The same cannot be said for the second. In fact, we have four languages with SOV order—Akkadian, Tigre, Tigrinya, and Amharic—all of which lack morphologically negative indefinites (with the exception of Tigrinya's borrowed negative determiner *wala*), despite the fact that sentential negation in these languages routinely follows subjects and objects. Geʿez, on the other hand, has developed negative-marked indefinites despite their not being necessary to satisfy the Neg-First Principle, which is automatically satisfied by its VSO basic word order and negation being a proclitic on the verb.

Of course, proposed functional-typological universals are not invalidated by individual exceptions, but the evidence presented must cast some doubt on the Neg-First Principle. First, verb-final Akkadian did not develop a means of adhering to the Neg-First Principle at any stage during the more than two millennia of its recorded history. Second, other than Geʿez, for which the Neg-First Principle appears to be irrelevant in its development of negative concord, both the clear cases of Semitic languages that have developed negative concord (Maltese and Hebrew) have only done so after intensive contact with other negative-concord languages. Finally, the modern Ethiopian Semitic languages reviewed here, like Akkadian, all violate the Neg-First Principle by having no morphologically negative-marked indefinites while predicate negation is a clitic on the verb in final position; and this despite the influence one might expect from the high variety, Geʿez.

10.4 Conclusion

In the sample of Afro-Asiatic languages surveyed in this chapter, three trends are apparent in the diachronic syntax of sentential negation and indefinite systems. First, we have seen that a number of languages undergo little or no change in one or both of these domains throughout the lengthy period of their attestation. Hebrew, Aramaic,

[28] But see n. 25: negative concord seems to be far less common worldwide than extrapolating from European languages would suggest.

and Akkadian, for example, always maintain a single preverbal marker of negation, while Akkadian and Egyptian-Coptic never develop anything resembling n-words. Second, there are several instances of languages undergoing significant developments in these areas apparently as a result of purely internal change. Both Egyptian-Coptic and Modern South Arabian languages undergo Jespersen's cycle, while Palestinian and Cairene Arabic, as well as several Berber varieties, progress to stage III of the cycle, all seemingly without external influence. Similarly, all the Arabic dialects discussed witness internally motivated changes of various sorts in the structure and membership of their indefinite systems. Most strikingly, however, the third trend observed is the crucial role played by contact in triggering many of the specific developments investigated here. We have seen that Jespersen's cycle is a feature of the histories of Egyptian-Coptic, Modern South Arabian, certain Arabic varieties, certain Berber varieties, Jerusalem Domari, and a number of Semitic and Cushitic languages of the Horn of Africa. I argued that Jespersen's cycle was triggered through contact in all of these cases except Egyptian-Coptic, Modern South Arabian, and Cushitic. We also saw that, despite various changes in the indefinite systems of all the languages we looked at in detail, the two languages to develop the kind of negative-concord systems familiar from European languages—Maltese and Hebrew—did so only as a result of contact with European languages.

Primary sources

Al-Qur'ān, ed. Ali Ahmed. Princeton, NJ: Princeton University Press, 1988.

Biblia Hebraica Stuttgartensia, ed. K. Elliger, R. Kittlel, and W. Rudolph. Stuttgart: Deutsche Bibelgesellschaft, 1991.

Four Martyrdoms from the Pierpont Morgan Coptic Codices, ed. E. Reymond and J. Barns. Oxford: Clarendon Press, 1973.

Homiletica from the Pierpont Morgan Library: Seven Coptic Homilies Attributed to Basil the Great, John Chrysostom, and Euodius of Rome, ed. L. Depuydt. Leuven: Peeters, 1991.

Il-Bibbja, ed. Carmel Sant. Floriana: The Malta Bible Society, 1984.

11

Negation in the history of the Mordvin languages

ARJA HAMARI

11.1 Introduction

The Mordvin languages, Erzya and Moksha, belong to the Uralic language family. According to the 2010 Russian census, the number of ethnic Mordvins was 744,237 (Federal'naja Služba Gosudarstvennoj Statistiki 2010), but there is no reliable data as to the number of speakers of Erzya and Moksha. The last relatively trustworthy approximation is from the census in 1989, when the number of ethnic Mordvins was 1,154,000 and the number of speakers about 750,000—two-thirds of them speakers of Erzya and the rest speakers of Moksha (Bartens 1999: 9–11). It can be assumed that the proportions of the two groups have remained the same, although the numbers have reduced significantly. About 27% of Mordvins live in the autonomous Republic of Mordovia (Russia) situated in the Middle-Volga region, while the rest of them are scattered in the surrounding provinces of the Russian Federation and the neighbouring republics of Tatarstan, Chuvashia, and Bashkortostan (Salo 1991: 156, Bartens 1999: 10, 13).

The Mordvin languages are in many ways typical Uralic languages: they are agglutinating with suffixation being more characteristic for them than prefixation, postpositions are used rather than prepositions, the modifier precedes the head, and there is an elaborate system of case-marking of nouns and conjugation of verbs. On the other hand, Erzya and Moksha also have certain features that are more or less unusual for the languages of the Uralic family. These include, for example, the subject (or indefinite) vs object (or definite) conjugation of verbs; the subject conjugation is used in intransitive verbs as well as in transitive verbs with indefinite objects, while the object conjugation is applied only to transitive verbs with definite objects. In the subject conjugation verbs agree with the subject person and number, whereas in the object conjugation both the subject and the object person and number are indicated in the personal endings of verbs (see Keresztes 1999 for details). Other features rather uncommon for Uralic languages include the indefinite vs definite declension of nouns, as well as the so-called non-verbal (or nominal) conjugation, which means that a noun, adjective, or adverb in predicate position can be conjugated according to the verbal paradigm, to a certain extent (Alhoniemi 1982, Turunen 2011). The basic word order of the Mordvin languages is SVO, while the original word order of the Uralic languages was SOV. However, the word order is fairly flexible in Erzya and Moksha (e.g. Vilkuna 1998: 178).

As argued by Grünthal (2007), the Mordvin languages show a relatively large number of especially grammatical similarities with the Finnic and Saami languages and some of these similarities can be considered innovations that have taken place after the disintegration of the Uralic protolanguage. Consequently, a common subbranch of these language groups can be assumed. On the other hand, earlier assumptions of a Volgaic subbranch (Mordvin and Mari languages) as well as a Finno-Permic subbranch (Finnic, Saami, Mordvin, Mari, and Permic languages) have been abandoned by many scholars.

It is assumed that the Proto-Mordvin period came to an end approximately 1000 years ago in the separation of Erzya and Moksha (Bartens 1999: 13, 15–17). The two sister languages are still mutually intelligible to a certain extent, but there are several phonological, morphological, lexical, as well as syntactic, differences between them, and they both have their own literary variety.

11.2 Negation in Proto-Uralic vs negation in modern Mordvin languages

The most prominent feature of negation in the Uralic languages is the use of negative auxiliaries that can be found in the majority of languages of the family (see, e.g., Comrie 1981). Therefore, there is no reason to doubt that a negative auxiliary was present already in the Uralic protolanguage, although not all daughter languages have preserved it (for an overview, see section 1.7). In fact, as assumed by for example Janhunen (1982: 37), there were probably two stems of a negative auxiliary in Proto-Uralic, the seemingly basic negative auxiliary *e-, along with its variant *elV- used in the imperative; the latter may have been some type of extension or a supplementary form of *e-. Janhunen further states that the negative auxiliaries probably had extensive conjugational properties, as they carried the marking of the subject person, tense, and mood, while the lexical verb had a fixed connegative form marked with an ending in *-k. The ending was originally a nominalizer of verbs and, moreover, it was also used as a marker of the imperative of the second person singular in the affirmative.

Considering that the finite element of the Proto-Uralic construction of standard negation[1] was the negative auxiliary, it is possible to state that this construction was a representative of a pattern that Miestamo (2005: 51, 73, 81–86) refers to as asymmetric. In this pattern the affirmative expression and its corresponding negative expression

[1] For the definition of standard negation, see section 1.1, and, for further discussion, Payne (1985: 198, 206–7) and especially Miestamo (2005: 42).

have more differences than merely the presence of a negative marker under negation. Furthermore, according to Miestamo's classification, the Proto-Uralic negative system could be categorized as A/Fin/NegVerb, that is, an asymmetric negative construction in which the marking of finiteness is associated with a negative verb. However, as the suffix of the connegative form was most likely an original nominalizer of verbs, Honti (1997: 241–2) has suggested that the negative constructions could have originated as copular clauses with a negative copula verb, for example *e-m mene-k 'I am not a goer' = 'I don't go' (*-m '1SG', *mene- 'to go').

Compared to the rather simple pattern reconstructed for the Uralic protolanguages, the negation system of the modern Mordvin languages is strikingly complex when it comes to the number of negative markers and their morphosyntactic characteristics. In both languages, several markers are found. These include negative auxiliaries, negative particles, and negative markers with nominal features. Moreover, there is a negative suffix in Moksha, although it is not frequently used in the present-day language. The choice of a negative marker and the formation of the negative construction depends on factors such as clause type, predicate type or the tense, and the mood of the predicate (see Hamari 2007 for a detailed description).

The Mordvin negative markers are introduced in Table 11.1. The type of the marker is indicated in the first column, while the second and third columns contain the actual forms of the markers in Erzya and Moksha, respectively. (A tilde is used to denote dialectal or stylistic variants and a slash to denote paradigmatic alternation.) The Erzya and Moksha markers are juxtaposed in those cases where they are generally considered to have a common origin. In the fourth column the possible developmental paths of the markers are described briefly, and the last column lists the contexts of the uses of the markers in the present-day Mordvin languages.

Many negative markers are regarded as cognates in the two sister languages and, as the functions of the markers are also very similar, one can assume that a large part of the system is of common origin in the Mordvin languages. However, since many of the negative markers do not have clear cognates in other Uralic languages, much of the complexity of the system must have taken shape only during the Proto-Mordvin period. Moreover, there are also differences in the system that indicate that some further development has taken place since the separation of Erzya and Moksha.

In this chapter, I will treat the development of the Mordvin negation system from the viewpoint of the morphosyntactic characteristics of the negative markers and their development:[2] the best-preserved old negative auxiliaries are considered in

[2] In this chapter, the Finno-Ugric transcription system is used in the transcription of Mordvin. For example, the palatalization of alveolar consonants is marked with an apostrophe or an acute accent. In both Mordvin languages, all consonants are palatalized when followed by a front vowel and, moreover, all word-final consonants are palatalized when preceded by a front vowel. However, in the case of alveolar consonants, palatalization is a phonological feature and not totally dependent on the surrounding sounds. Therefore, in this chapter, only the palatalization of alveolars is indicated.

TABLE 11.1. **The Mordvin negative markers**

Type of negative marker	Erzya	Moksha	Reconstructed development	Use of the negative marker in modern languages
Auxiliary	*eź*	*əź-* ~ *iź-*	< Proto-Uralic auxiliary *e- + past-tense markers *ś and *j	1st past tense of verbs
Auxiliary	*il'-*	*ťa-* ~ *ďa-*	< Proto-Uralic auxiliary *elV-	Imperative and optative moods of verbs
Particle	*a* ~ *at*	*af* ~ *ajaf*	< ? nominal form of a back vowel variant of Proto-Uralic auxiliary *e-	Present tense and 2nd past tense of verbs and non-verbal predicates, conditional and conditional–conjunctive moods of verbs, in Erzya: conjunctive and desiderative moods of verbs
Particle	*apak*	*apak*	< ? nom. form of a back vowel variant of Proto-Uralic auxiliary *e-	Participles and gerunds
Auxiliary	*avol'-*	*afəl'-*	< neg. particle E *a*, M *af* + verb *uľə- 'be'	Conjunctive and conditional–conjunctive moods of verbs
Auxiliary	–	*afəľəksəl'-*	< neg. auxiliary *afəl'* + desiderative suffix *-ksəl'-*	Desiderative mood of verbs
Particle	*avol'*	–	< neg. particle E *a* + verb *uľə- 'be'	Present tense and 2nd past tense of non-verbal predicates
Noun	*araś*	–	1) < ? neg. particle *a* + nom. form of *eŕams* 'live, be' 2) < ? nom. form of *arams* 'disappear, get lost; be destroyed; die' 3) < ? noun *ara 'non-existent, nonsense'	Existential, 'have'-possessive, and some locative clauses
Noun	–	*aš* ~ *ajaš*	1) < ? neg. particle *a + *ašt- 'be, exist, stay' 2) < ? noun *aš* 'non-existence; poverty'	Existential, 'have'-possessive, and some locative clauses
Auxiliary	–	*aš-*	< neg. noun *aš* + neg. auxiliary *əź-* ~ *iź-*	1st past tense of verbs
Suffix	–	*-fťäŕa-/ -fťäŕä-*	< neg. particle *af* + *ťäŕa- 'try'	Conditional mood of verbs
Suffix	–	*-fťäŕaľə-/ -fťäŕäľə-*	< neg. particle *af* + verb *ťäŕa- 'try' + verb *uľə- 'be'	Conditional–conjunctive mood of verbs

section 11.3, the development of negative particles in section 11.4, the constructions originally based on a fusion of a negative particle and the verb meaning 'be' in section 11.5, the negative nouns in section 11.6, and the Moksha negative suffix in section 11.7. In section 11.8, the formation of negative indefinites will be dealt with and, finally, conclusions will be presented in section 11.9.[3]

11.3 The preserved negative auxiliaries

Negative auxiliaries play an important role in the negation system of the present-day Mordvin languages, although they are not the only way of expressing negation. Both Erzya and Moksha have preserved the reflexes of the Proto-Uralic negative auxiliaries fairly well in certain contexts, but it seems that in others, the auxiliaries have fossilized into particles (see section 11.4). Moreover, there has been a tendency in the Mordvin languages to develop new negative auxiliaries from other negative markers (see sections 11.5.1 and 11.6.2).

In the Mordvin languages, the best-preserved reflexes of the Uralic negative auxiliaries are found in the negative constructions of the first past tense of the indicative mood and in the imperative and optative moods. Regardless of certain changes, the basic asymmetric negative structure A/Fin/NegVerb of Proto-Uralic has been maintained in both languages; the auxiliaries are conjugated, while lexical verbs appear in fixed connegative forms.

11.3.1 The auxiliaries of the first past tense

In both Mordvin languages, the negation of the unmarked past tense, the so-called first past tense, is formed by using a negative auxiliary: Erzya *eź*-, Moksha *əź*- ~ *iź*- (dialect variants).[4] The auxiliaries can be conjugated in both the subject and object conjugation. The personal endings of both conjugations are attached to the auxiliary, while the connegative form of the lexical verb is invariant, that is, it is the mere stem of the verb. However, the Proto-Uralic *k-ending of connegatives has been preserved dialectally in both languages (Paasonen 1953: 011, Bartens 1999: 140). Table 11.2 (adapted from Bartens 1999: 140, Paasonen 1953: 07, 011, and Grammatika 1980: 291, 319) gives the subject conjugation of the verb *morams* 'sing' in Erzya and Moksha; both the affirmative and negative paradigms are introduced to enable comparison.

Etymologically, the first and second person forms of the negative auxiliaries of Erzya and Moksha consist of four elements—the reflexes of the Proto-Uralic *e-stem (Erzya *e*-, Moksha *ə*- ~ *i*-), two past-tense markers, which go back to Proto-Uralic

[3] I would like to thank Niina Aasmäe, a linguist and a native speaker of Erzya, for her comments on an earlier draft of this article. Any remaining flaws are, of course, my own.

[4] In Moksha there is also another negative auxiliary *aš*- that is used in the negation of the first past tense (see section 11.6.2).

TABLE 11.2. **The first past tense subject conjugation of Erzya and Moksha *morams* 'sing' in the affirmative and negative**

	Erzya				
	Affirmative		Negative		
1SG	*moriń*	'I sang'	*eźiń*	*mora*	'I didn't sing'
2SG	*morit́*	'you sang'	*eźit́*	*mora*	'you didn't sing'
3SG	*moraś*	'(s)he sang'	*eź*	*mora*	'(s)he didn't sing'
1PL	*moryńek* ~ *morime*	'we sang'	*eźińek* ~ *eźime*	*mora*	'we didn't sing'
2PL	*moriďe*	'you sang'	*eźiďe*	*mora*	'you didn't sing'
3PL	*moraśt́*	'they sang'	*eźt́*	*mora*	'they didn't sing'
	Moksha				
	Affirmative		Negative		
1SG	*morań*	'I sang'	*əźəń*	*mora*	'I didn't sing'
2SG	*morat́*	'you sang'	*əźət́*	*mora*	'you didn't sing'
3SG	*moraś*	'(s)he sang'	*əź*	*mora*	'(s)he didn't sing'
1PL	*moramä*	'we sang'	*əźəmä*	*mora*	'we didn't sing'
2PL	*moraďä*	'you sang'	*əźəďä*	*mora*	'you didn't sing'
3PL	*moraśt́*	'they sang'	*əśt́*	*mora*	'they didn't sing'

(*-ś and *-j), and a personal suffix—while in the third person singular forms, only the appropriate reflex of the suffix *-ś is attached to the auxiliary stem (e.g. Bartens 1999: 140–1). In the third person plural forms, the sibilant is followed by a plural marker in *-t́*. In Erzya the present-day stem is *eź-*, while the Moksha equivalent is *əź-* ~ *iź-*in all persons except for the third person plural form, which has a stem-final *ś* due to the devoicing effect of the plural marker *-t́*.

As pointed out by for example Korhonen (1967: 211–15) and Bartens (1999: 140–1), the presence of two original past-tense markers *-ś and *-j in the past-tense negative auxiliary is typical not only for the Mordvin languages, but can be found in some related languages as well, such as the Estonian Kodavere dialect (see also Savijärvi 1977: 17–18) and Southern Saami (see also Korhonen 1981: 266–7, 309). Some other Uralic languages seem to have based their past-tense marking of the negative auxiliary on only one of these: for example Mari on *-ś (Bereczki 1988: 346) and the Permic languages on *-j (Korenchy 1972: 159–160, Rédei 1988a: 391).[5] These past-tense

[5] Most Finnic and Saami languages have lost the past-tense marking of the negative auxiliary altogether (Korhonen 1973), while in the Northern Samoyedic languages Nenets and Enets, the use of the past-tense marker *-ś in the negative auxiliary is regarded as a relatively recent innovation (Hajdú 1988: 13–14, Künnap 1993: 208).

markers are also found in the affirmative verbal paradigms of the Uralic languages. In the affirmative, the languages usually rely on only one of the suffixes, but the Mordvin languages make use of both: in the first and second person forms, the *i*-element of Erzya (e.g. *moriń*, *morit́*) is a reflex of *-j, while in Moksha, only the palatalization of the personal suffixes (e.g. *morań*, *morat́*) indicates that a front vowel originally preceded these endings (Bartens 1999: 123, 128–9). In the third person forms (Erzya and Moksha *moraś*, *moraśt́*), on the other hand, a reflex of *-ś appears (Bartens 1999: 129).

According to Janhunen (1982: 36), both *-ś (or *-śå/-śä) and *-j go back to original Proto-Uralic nominalizers of verbs and their past-tense meanings developed only later. Building on this assumption, Künnap (1993: 203–9, 1994: 83–6) has argued that the negative marker *eś was probably originally an uninflected verbal noun form. Considering that in many Uralic languages third person forms of verbal paradigms especially are based on original verbal nouns, I assume that this nominal form of the negative auxiliary could have been subsumed into the verbal personal paradigm. As argued in Hamari (2007: 115–17), considering the distribution of the past tense markers *-ś and *-j and their intertwined relationship in negative auxiliaries in the Uralic languages, it seems possible that, when the suffixes were still in the process of becoming past tense markers, the negative marker *eś was reanalysed as a stem form for a past tense negative auxiliary and acquired a (redundant) past tense marking in *-j.

Finally, it should be noted that, in some Erzya dialects, the conjugation of the negative auxiliary follows the same distribution of the past-tense markers that can be seen in the affirmative paradigm: *eiń*, *eit́*, *eś*, *eińek*, *eid́e*, *eśt́* (Bartens 1999: 140). In this suppletive paradigm, only the first and second persons carry the reflex of the original *-j and the third person forms that of *-ś.

11.3.2 The auxiliaries of the imperative and the optative moods

In present-day Erzya and Moksha, the negation of the imperative and optative moods is also produced by using negative auxiliaries. In Erzya, the auxiliary of these moods is *il'a-* and in Moksha *t'a-* (dialectally *d'a-*). The negative imperative forms are only possible with a second person subject, but within these limits, both the subject and object conjugations are possible. The formation of the imperative subject conjugation is illustrated in Table 11.3 (Grammatika 1980: 295, 319).

Erzya *il'a-* and Moksha *t'a-* are regarded as the reflexes of the Proto-Uralic negative marker *elV- of the imperative (e.g. Bartens 1999: 143) (see section 1.12 for background). A regular Proto-Mordvin sound change *e- > *i-* (when originally followed by *ä*) has produced the word-initial *i* of the Erzya stem (cf. Bartens 1999: 58–9), but the Moksha form seems to have gone through some additional changes. As suggested by Bereczki (1988: 328–9) and Alhoniemi (2001: 361), the word-initial *i-* may have

TABLE 11.3. **The imperative subject conjugation of Erzya and Moksha *morams* 'sing' in the affirmative and negative**

		Erzya			
	Affirmative		Negative		
2SG	*morak*	'sing!'	*iľa*	*mora*	'don't sing!'
2PL	*morado*	'sing!'	*iľado*	*mora*	'don't sing!'
		Moksha			
	Affirmative		Negative		
2SG	*morak*	'sing!'	*ťat*	*mora*	'don't sing!'
2PL	*morada*	'sing!'	*ťada*	*mora*	'don't sing!'

disappeared from the Moksha form,[6] but since *l'* could not appear in a word-initial position, an irregular change *ľ > *ď* > *ť* occurred (i.e. Proto-Uralic *elV- > Proto-Mordvin *ilV- > Moksha *ľa- > *ďa-* > *ťa-*).

The lexical verb of the negative imperative always appears in the invariant connegative form in both Erzya and Moksha. Dialectally, however, the ending in *-k* can be attached to the connegative form in both languages. Moreover, in some Erzya dialects the *k*-suffix appears in the negative auxiliary instead of the lexical verb (e.g. *iľak pala* 'don't kiss') (Paasonen 1953: 013). This is not surprising, since this ending is also the original marker of the imperative and is also still preserved in the affirmative paradigm. In the literary language, however, the second person singular form *iľa* does not have a person marking in Erzya subject conjugation (Bartens 1999: 143).

Unlike in the imperative, all six personal categories are possible in the optative mood, although, as stated in for example Mokšəń käľ (2000: 135), the first and second person plural forms are not used in present-day literary Moksha. Furthermore, the use of the object conjugation in this mood is rare in both modern languages (Bartens 1999: 139). The optative paradigms of the subject conjugation are given in Table 11.4 (adapted from Paasonen 1953: 010, 013 and Grammatika 1980: 296, 319).

According to Bartens (1999: 138), the formation of the optative mood was originally built on the possessive suffix of the third person (*-sA); in Moksha, the *s* and *z* vary depending on the voicing of the following sound.[7] The same suffixes that are used in the lexical verb in the affirmative appear on the negative auxiliary in the negative constructions.

[6] An auxiliary in *ľa-* is also attested dialectally in Erzya (Bereczki 1988: 328).

[7] In Erzya dialects, the *z*-form of the optative marker has had a voicing effect on the personal endings (1PL) *morazdano* and (2PL) *morazdado* in the affirmative, while in the negative paradigm, the initial consonants of the plural personal endings seem to have had a devoicing effect on the optative marker, which appears as *-s* (*il'astano, il'astado, il'ast*).

TABLE 11.4. **The optative subject conjugation of Erzya and Moksha *morams* 'sing' in the affirmative and negative**

	Erzya				
	Affirmative		Negative		
1SG	*morazan*	'let me sing'	*il'azan*	*mora*	'let me not sing'
2SG	*morazat*	'let you sing'	*il'azat*	*mora*	'let you not sing'
3SG	*morazo*	'let her/him sing'	*il'azo*	*mora*	'let him/her not sing'
1PL	*moraztano*	'let us sing'	*il'aztano*	*mora*	'let us not sing'
2PL	*moraztado*	'let you sing'	*il'aztado*	*mora*	'let you not sing'
3PL	*morazt*	'let them sing'	*il'azt*	*mora*	'let them not sing'
	Moksha				
	Affirmative		Negative		
1SG	*morazan*	'let me sing'	*t'azan*	*mora*	'let me not sing'
2SG	*morazat*	'let you sing'	*t'azat*	*mora*	'let you not sing'
3SG	*moraza*	'let her/him sing'	*t'aza*	*mora*	'let him/her not sing'
1PL	*morazama*	'let us sing'	*t'astama*	*mora*	'let us not sing'
2PL	*morazada*	'let you sing'	*t'astada*	*mora*	'let you not sing'
3PL	*morast*	'let them sing'	*t'ast*	*mora*	'let them not sing'

As mentioned above, the imperative mood was already present in the Uralic protolanguage, but the optative mood is younger. The present-day Mordvin personal paradigms must have taken their shape during the Proto-Mordvin period, but there are signs that at least the third person forms probably developed as a part of the imperative paradigm even earlier; they have cognates in the third person imperative forms of, for example, Mari, Finnic, and some Saami languages (Korhonen 1981: 273). The use of the negative auxiliary of the imperative in the optative is explicable in terms of the common origin of these two moods.

11.4 From negative auxiliaries to negative particles

Despite their ancient history, the negative markers of the first past tense and the imperative and optative moods are quite well preserved in the Mordvin languages and, therefore, relatively easily comparable to the cognates of more distantly related languages. Strangely, however, the more recently developed negative markers of Erzya and Moksha are much more opaque and have resisted all fully satisfactory etymological explanations in spite of recurrent effort. The problematic cases include a series of markers that all have a word-initial *a-* in both languages; these are the multi-functional particles Erzya *a* (dialectally *at*), Moksha *af* (emphatically *ajaf*), the particle *apak* of Erzya and Moksha gerunds and participles, as well as the negative existential nouns, Erzya *araś* and Moksha *aš* (or *ajaš*). In this section, the evolution of the negative particles is treated, while the assumptions concerning the development of the negative nouns will be considered in section 11.6.1.

11.4.1 The development of Erzya a *and Moksha* af ~ ajaf

The particles Erzya *a* and Moksha *af* (emphatically *ajaf*) are employed in the negation of the most unmarked verbal category, the present tense indicative mood. The negative constructions formed by using these particles deviate radically from the Proto-Uralic pattern of negation; as the particles are invariant, the lexical verb is conjugated according to either the subject or object conjugation in exactly the same way as in the corresponding affirmative expression. Therefore, in Miestamo's (2005: 51) terminology, the negative patterns can be characterized as symmetric in that the only difference between the affirmative and negative constructions is the presence of the negative marker under negation, as can be seen in examples (1) and (2).

(1) a. affirmative
mora-n
sing-PRES.1SG
'I sing / am singing.' (Erzya) (Grammatika 1980: 290)

b. negative
a mora-n
NEG sing-PRES.1SG
'I don't sing / am not singing.' (Erzya) (Grammatika 1980: 318)

(2) a. affirmative
mora-n
sing-PRES.1SG
'I sing / am singing.' (Moksha) (Grammatika 1980: 290)

b. negative
af mora-n
NEG sing-PRES.1SG
'I don't sing / am not singing.' (Moksha) (Grammatika 1980: 318)

The Mordvin negative particles are generally regarded as fossilized forms of an original negative auxiliary. As implied by for example Bartens (1999: 141–2), the development from an auxiliary into a particle is likely to have taken place during the Proto-Mordvin period. In the negation system of the Mari language, there is a negative auxiliary with a back vowel stem in the present tense and another stem that goes back to an original front vowel stem in the past tense, which has been seen as an indication that the two stems for two different tense forms may have developed even before the Proto-Mordvin period (e.g. Bartens 1999: 141–2).[8] In Proto-Mordvin, the auxiliary with the front vowel was preserved, but the variant with a back vowel developed into a particle.

However, the etymological relation of Erzya *a* and Moksha *af* is not clear (see Hamari 2007: 98–101 for a detailed discussion). As argued by Bartens (1999: 142), the most likely form to fossilize would be the third person singular—a development that has taken place for example in Estonian, where the former third person singular form *ei* is now used with all subject persons (see section 1.7). The Mordvin negative particles do not carry any sign that would suggest that they are related to the third person singular forms attested in the modern languages, but since the suffixes of this particular person are often based on verbal nouns in the Uralic languages, it has been assumed that the predecessors of the negative particles could have carried some nominalizers other than those found in the personal paradigms of lexical verbs (see e.g. Bartens 1999: 142). The *-f* of the Moksha particle is probably connected to the *f*-ending of the past tense participles (e.g. Klemm 1934: 386–7, Bereczki 1988: 326–7). Heikkilä (1968: 185–6) has suggested that this ending could have developed in verbs with a derivational suffix in *-v* followed by the aforementioned nominalizer *-k. The Moksha *f*-ending has a cognate *-vt* of past-tense participles in Erzya;[9] according to Heikkilä, a change *-vk > *-vχ > *-vf > *-f* could have produced the Moksha variant and a change *-vk > *-vt* the Erzya form.[10] In this way, the Erzya dialectal particle *at* (< *avt < *avk) could be cognate with the Moksha particle *af*. However, it has also been suggested that the *-t* in Erzya *at* could go straight back to the Proto-Uralic nominalizer *-k (Ščemerova 1972: 174, Honti 1997: 90, Bartens 1999: 142). On this assumption, the reconstructed *ak could be connected to the third person singular forms of present-tense negative auxiliaries in Mari (Eastern Mari *ok* ~ Western Mari *ak*) and the Permic languages (e.g. Udmurt *ug*). The suffix in *-k could have been dropped in some parts of the Erzya linguistic area, while in others an irregular change *ak > *at* occurred.

However, in addition to the form *a*, the Erzya particle can also appear in the form *aj*, when preceding a word with an initial vowel, for example *aj eŕavi* 'is not needed' instead of *a eŕavi* or *aj aščan* 'I am not staying' instead of *a aščan*; the *j* has probably been introduced in this position to prevent hiatus (Klemm 1934: 388, Ščemerova 1972: 174). In some dialects *aj* even appears when the following word begins with a consonant (e.g. Meľsetevo dialect: *aj moľan* 'I don't go; I am not going'),

[8] There are two different negative auxiliary stems also in the Permic languages, a back vowel stem for the present and future tenses and a front vowel stem for the past tense (Bartens 2000: 184–6). However, the back vowel variants of the Permic languages can be explained through regular sound changes (Itkonen 1954: 272, Bartens 2000: 191).

[9] In modern Erzya, the past-tense participial ending in *-vt* is rather marginal; usually, a *ź*-suffix identical to that of gerunds is used (Bartens 1999: 152–3; see also section 11.4.2).

[10] According to Bereczki (1988: 236), the Moksha *-f* is a reflex of the Uralic present-tense participial suffix in *-pA. However, the Erzya equivalent *-vt* cannot be derived from this ending. On the other hand the change *k > *t* after a consonant is regular (Bartens 1999: 53).

which has led for example Ščemerova (1972: 175) to consider the possibility that there was a Proto-Mordvin negative auxiliary form in *aj, which could have been the origin of both the Erzya particle *a* (and its variant *aj*) and the Moksha prefixal *aj-* in *ajaš* (see section 11.6.1). This *aj could then be regarded as a descendant of an earlier form *ejä. This reconstruction has been built on evidence provided by those modern Finno-Saamic languages that possess a third person singular form of the negative auxiliary formed by the ancient nominalizer *-jA (< Proto-Uralic *j) (Bergsland 1960: 328–30, Savijärvi 1977: 11–12). However, in the case of Moksha, *aj* can be prefixed to either *af* (> *ajaf*) or *aš* (> *ajaš*), which has led Bartens (1996: 79) to suggest that the Moksha element could rather have originated as an emphatic marker.

Finally, it has also been suggested that Erzya *a* and Moksha *af* might be original loans from other languages. In Rédei (1988b: 69) the possibility is mentioned that at least some of the Uralic negative markers with an initial back vowel could have been borrowed from an Iranian language; in this language family a negative prefix *a-*, *an-* is attested. According to Künnap (2002: 36), on the other hand, the Moksha particle *af* could be parallel to Middle Turkic negative markers such as *ab*...*ab* 'neither...nor' and *aw* 'not' (cf. Räsänen 1969: 1). However, these suggestions also remain an open question, since they cannot be verified by attested data. Although both Mordvin languages do have old Iranian loans, it is unclear when and from which language or branch of the Iranian language family the words were borrowed (Bartens 1999: 14–15). In the case of the Turkic languages, on the other hand, none of the modern Turkic languages spoken in the vicinity of Mordvin or elsewhere has an attested reflex of the Middle Turkic negative markers (Clauson 1972: 3).

As was seen in Table 11.1, the negative particles of the modern Mordvin languages are multifunctional, in that their uses extend far beyond the present-tense indicative mood; first of all, they are also employed in the negation of the indicative second past tense (examples (3) and (4)) and the conditional mood (examples (5) and (6)). In the conditional–conjunctive mood (examples (7) and (8)), these symmetric constructions can be used, but also asymmetric constructions based on the negative auxiliaries Erzya *avol'-*, Moksha *afəl'-* are possible. Moreover, in Erzya, the particle *a* is also employed in the desiderative mood (example (9)). The auxiliaries Erzya *avol'-* and Moksha *afəl'-* (as well as Moksha *afəl'əksəl'-*) will be considered in section 11.5.1.

(3) a. affirmative
mori-l'i-ń
sing-PAST2-1SG
'I (usually) sang.' (Erzya) (Grammatika 1980: 292)

b. negative
a mori-l'i-ń
NEG sing-PAST2-1SG
'I didn't (usually) sing.' (Erzya) (Grammatika 1980: 318)

(4) a. affirmative
mora-ľə-ń
sing-PAST2-1SG
'I (usually) sang.' (Moksha) (Grammatika 1980: 292)

b. negative
af mora-ľə-ń
NEG sing-PAST2-1SG
'I didn't (usually) sing.' (Moksha) (Grammatika 1980: 318)

(5) a. affirmative
mori-ńďeŕa-n
sing-COND-1SG
'if I sing' (Erzya) (Grammatika 1980: 298)

b. negative
a mori-ńďeŕa-n
NEG sing-COND-1SG
'if I don't sing' (Erzya) (Grammatika 1980: 318)

(6) a. affirmative
mora-ńďäŕa-n
sing-COND-1SG
'if I sing' (Moksha) (Grammatika 1980: 298)

b. negative
af mora-ńďäŕa-n
NEG sing-COND-1SG
'if I don't sing' (Moksha) (Grammatika 1980: 318)

(7) a. affirmative
mori-ńďeŕavľi-ń
sing-CONDCONJ-1SG
'if I had sung / were to have sung' (Erzya) (Grammatika 1980: 299)

b. negative
a mori-ńďeŕavľi-ń
NEG sing-CONDCONJ-1SG
'if I hadn't sung / weren't to have sung' (Erzya) (Grammatika 1980: 318)

(8) a. affirmative
mora-ńďäŕäľə-ń
sing-CONDCONJ-1SG
'if I had sung / were to have sung' (Moksha) (Grammatika 1980: 299)

b. negative
af mora-ńďäŕäľə-ń
NEG sing-CONDCONJ-1SG
'if I hadn't sung / weren't to have sung' (Moksha) (Grammatika 1980: 318)

(9) a. affirmative
mori-kseľi-ń
sing-DES-1SG
'I wanted to sing.' (Erzya) (Grammatika 1980: 300)

b. negative
a mori-kseľi-ń
NEG sing-DES-1SG
'I didn't want to sing.' (Erzya) (Grammatika 1980: 318)

Moreover, in both languages the particles also appear in the negation of non-verbal predicates. In the Mordvin languages, nominal predicates can be conjugated according to the verbal paradigm in the present tense as well as in the second past tense (see e.g. Alhoniemi 1982, Turunen 2006, 2011) and, since these categories are negated using the particles Erzya *a*, Moksha *af* in the verbal paradigm, their use in the non-verbal conjugation is understandable. Examples (10) and (11) represent cases of non-verbal conjugation in Erzya, but the same relationship holds in Moksha.

(10) a. affirmative
viška-n
small-PRES.1SG
'I am small.'

b. negative
a viška-n
NEG small-PRES.1SG
'I am not small.' (Erzya) (Bubrix 1953: 97)

(11) a. affirmative
viška-ľi-ń
small-PAST2-1SG
'I was small.'

b. negative
a viška-ľi-ń
NEG small-PAST2-1SG
'I was not small.' (Erzya) (Bubrix 1953: 97)

In addition to nominative forms of adjectives and nouns, nominal predicates with, for example, inessive or elative case endings can also acquire non-verbal conjugation as well as certain adverbs in predicate position. In all these instances, the negation is formed by the particles Erzya *a*, Moksha *af* (see Hamari 2007: 125–47, 236–43 for a closer analysis).[11]

What is common to the predicational categories dealt with in this section is that they all developed during the Proto-Mordvin period. The second past tense developed when an earlier auxiliary verb *ul'ə-* (< *vole-) 'be' was fused into a nominalized form of the lexical verb and the desiderative mood was built on a fusion of a translative form (*-ks*) of the nominalized lexical verb and the verb *ul'ə-* 'be'. The conditional mood was likewise built on a fusion of a lexical verb and an auxiliary, only this time the auxiliary was *ťäŕa- 'try'.[12] The conditional–conjunctive mood, on the other hand, developed through the fusion of both *ťäŕa- 'try' and *ul'ə-* 'be' with the lexical verb (see Bartens 1999: 129–37 for a detailed description of the development of these tenses and moods). Moreover, the non-verbal conjugation also took shape during the Proto-Mordvin period, and it seems that, in its development, reciprocal influence can be seen in its relation to verbal conjugation both in the present tense and in the second past tense (Ravila 1929: 104–5, Turunen 2006: 177–8, Hamari 2007: 65–73).

It seems that, as the new predicational categories brought increased complexity to the Proto-Mordvin conjugation, the most simple negative pattern available was chosen, that is, a symmetric pattern in which the only difference between the affirmative and negative forms is the presence of the negative marker under negation. The complexity of the new system could even be regarded as at least one of the factors leading to the fossilization of an earlier auxiliary into a multifunctional particle. However, as will be seen in sections 11.5.1 and 11.7, there has also been a tendency for the earlier auxiliaries of the Proto-Mordvin moods to be fused to the negative particles to give rise to new negative markers.

11.4.2 *The development of Erzya and Moksha* apak

Another possible fossilization of an earlier negative auxiliary can be seen in the development of the negative particle of participles and gerunds, which has an identical shape in both Erzya and Moksha: *apak*. Within the system of participles, the corresponding affirmative form is the past-tense participle, which is formed by the suffix *-ź* in Erzya and *-f* in Moksha. The corresponding affirmative gerund, on the other hand, is formed by *-ź* in both languages. The following examples illustrate the use of *apak* as a negative marker of past participles, (12) and (13), and that of gerunds, (14) and (15):

(12) a. soka-ź paksa
plough-PAST.PTCP field
'a field that has been ploughed' (Erzya) (Grammatika 1962: 323)

[11] In Erzya, the particle *avol'* is also possible (see section 11.5.2).

[12] The verb has been preserved dialectally in Erzya (*ďeŕams* 'try, strive') (Bartens 1999: 134).

b. apak soka pakśa
NEG plough.CNG field
'a field that hasn't been ploughed' (lit. 'a non-ploughed field') (Erzya) (Grammatika 1962: 322)

(13) a. soka-f pakśa
plough-PAST.PTCP field
'a field that has been ploughed' (Moksha) (Grammatika 1962: 323)

b. apak soka-k pakśa
NEG plough-CNG field
'a field that hasn't been ploughed' (lit. 'a non-ploughed field') (Moksha) (Grammatika 1962: 322)

(14) Ejkakšo-ś avarďe-ź / apak avarďe sa-ś kudo-v.
child-SG.DEF.NOM cry-GER NEG cry.CNG arrive-PAST1.3SG home-LAT
'The child came home crying / without crying.' (Eryza) (Grammatika 1962: 325)

(15) Šaba-ś avarďə-ź / apak avarť-ť sa-ś kud-u.
child-SG.DEF.NOM cry-GER NEG cry-CNG arrive-PAST1.3SG home-LAT
'The child came home crying / without crying.' (Moksha) (Grammatika 1962: 325)

As can be seen in the examples, the lexical verb is in the connegative form in these constructions. The connegative ending in *-k* is preserved on the lexical verb in Moksha, while in Erzya it is often missing, but can optionally be employed, as in *apak palak* 'non-kissed; without kissing' (Bartens 1999: 144).[13] In these cases, however, the connegative forms are exceptional compared to the ones seen thus far; when the negative participial form appears in predicate position, it is possible to attach the endings of the non-verbal conjugation to the lexical verb in connegative form, a pattern which is not attested in the connegatives of other negative markers (Hamari 2007: 153–4) and which should therefore be regarded as a relatively recent innovation:

(16) a. apak orča-k-an
NEG dress-CNG-PRES.1SG
'I am not dressed.' (Erzya) (Evsev'ev 1931: 190)

b. apak ščа-k-an
NEG dress-CNG-PRES.1SG
'I am not dressed.' (Moksha) (Hamari 2007: 154, from Valentina Ščankina, p.c.)

Moreover, as shown in Hamari (2007: 153–7), when the participial form appears in predicate position, there is also a possible conjugational paradigm in which only the third person singular form is formed with the construction in which a connegative

[13] In example 15, *-k* > *-t* after a stem-final consonant.

form of the lexical verb appears. With other personal forms, the lexical verb is in the corresponding affirmative form and the endings of the non-verbal conjugation are attached to the lexical verb. In other words, although the third person singular follows an asymmetric pattern in these cases, the other persons display a symmetric pattern. Moreover, in modern Mordvin languages, the negative particles Erzya *a*, *avol'* and Moksha *af* are also possible with a participial verb form in predicate position. In these cases, the third person singular is likewise formed using the symmetric pattern with an affirmative form of the lexical verb. In the following examples, the possible variants for the first person singular in the present tense are given:

(17) a. apak čavo-ź-an
NEG hit-PAST.PTCP-PRES.1SG
'I am not hit.'

b. a čavo-ź-an
NEG hit-PAST.PTCP-PRES.1SG
'I am not hit.'

c. avol' čavo-ź-an
NEG hit-PAST.PTCP-PRES.1SG
'I am not hit.' (Erzya) (Hamari 2007: 154, from Svetlana Gausheva, p.c.)

(18) a. apak šav-f-an
NEG hit-PAST.PTCP-PRES.1SG
'I am not hit.'

b. af šav-f-an
NEG hit-PAST.PTCP-PRES.1SG
'I am not hit.' (Moksha) (Hamari 2007: 155, from Valentina Ščankina, p.c.)

In examples (12)–(18) above, the negative marker *apak* appears as an invariant negative copula, but it is sometimes treated as a nominal element. In Erzya, it can be conjugated according to the non-verbal conjugation when negating a participial in predicate position, for example, *apak-an čav-t* (NEG hit-CNG)[14] (Gheno 1994–5: 135), and in both languages, non-verbal conjugation can be employed when *apak* is used as a one-word answer (Sivonen 1977: 108, Hamari 2007: 160–1).

The etymology of the particle *apak* is as poorly understood as that of the other negative particles; there are no cognates for this particle in other related languages. Considering that the lexical verb often carries the connegative suffix, the most likely source of the negative particle is a negative auxiliary, as originally suggested by Klemm (1934: 396) and later by, for example, Bartens (1999: 144). According to Klemm, the particle could be an original participial form of the ancient

[14] In this example, *-k* > *-t* after a stem-final consonant.

negative auxiliary *e-, or rather, the hypothetical variant with a back vowel; the participial negative marker could have been formed using a participial ending in *-pA. However, as discussed in Hamari (2007: 120–1), this etymology is problematic in that, firstly, the Uralic deverbal ending in *-pA is probably not otherwise preserved in the Mordvin languages and, secondly, the regular sound change *p > *v* between vowels does not allow the continuation of *p in modern Erzya and Moksha (see Bartens 1999: 38). Likewise, the word-final *-ak* of the particle eludes etymological explanation, although it could be connected to the *k*-ending of the connegative forms or the enclitic particle of the indefinites (see section 11.8). Since the form and use of this negative marker is so similar in both Mordvin languages, it must be regarded as one of the innovations of the Proto-Mordvin period.

11.5 The fusion of a negative particle and a verb meaning 'be'

The negative particles Erzya *a* and Moksha *af* have had a role also in the development of new negative markers in the Mordvin languages. What is common to both Erzya and Moksha is that the particle has fused together with the verb *uľə-* 'be' to produce a negative auxiliary of certain moods (e.g. Bartens 1999: 142). However, in Erzya, a negative marker going back to the same fusion is also employed in the negation of non-verbal predicates, although in this context, the marker is synchronically a negative particle rather than a negative verb. The negative auxiliaries will be dealt with in section 11.5.1 and the Erzya negative particle in section 11.5.2.

11.5.1 New negative auxiliaries

In the previous section it was stated that during the Proto-Mordvin period, new predicational categories such as the non-verbal conjugation and the second past tense developed. Moreover, the modal categories of conditional, conditional–conjunctive, and desiderative arose through the agglutination of auxiliaries onto lexical verb stems or their nominalizations. In the negation of all these new categories, negative particles began to be used, in which case the negative pattern became symmetric.

During the Proto-Mordvin period, there was also a tendency for new negative auxiliaries to arise for the negation of some of the new moods. These, too, were built on the fusion of an auxiliary verb *uľə-* 'be'—only in these cases it was fused with the negative particles Erzya *a* and Moksha *af* or their predecessor. The modern Mordvin auxiliaries Erzya *avoľ-*, Moksha *afəľ-* of the conjunctive mood are the reflexes of this Proto-Mordvin negative marker. They carry the personal marking of the subject person and number and, in the object conjugation, also the marking of the object person and number. Given the past-tense forms of the personal suffixes of the first and second persons, the form of the auxiliary was most likely originally in the past tense (Bartens 1999: 133). The lexical verb, again, appears in the fixed connegative form, making the pattern a case of asymmetric negation, as can be seen from the full paradigms given in Table 11.5 (adapted from Bartens 1999: 142 and Grammatika 1980: 297, 319–320).

TABLE 11.5. **The conjunctive mood subject conjugation of Erzya and Moksha *morams* 'sing' in the affirmative and the negative**

	Erzya				
	Affirmative		Negative		
1SG	*moravlïń*	'I would sing/ would have sung'	*avolïń*	*mora*	'I wouldn't sing/ wouldn't have sung'
2SG	*moravlïť*	'you would sing/ would have sung'	*avolïť*	*mora*	'you wouldn't sing / wouldn't have sung'
3SG	*moravol'*	'(s)he would sing/ would have sung'	*avol'*	*mora*	'(s)he wouldn't sing / wouldn't have sung'
1PL	*moravlïńek*	'we would sing/ would have sung'	*avolïńek*	*mora*	'we wouldn't sing / wouldn't have sung'
2PL	*moravlïďe*	'you would sing/ would have sung'	*avolïďe*	*mora*	'you wouldn't sing / wouldn't have sung'
3PL	*moravolť*	'they would sing/ would have sung'	*avolť*	*mora*	'they wouldn't sing / wouldn't have sung'
	Moksha				
	Affirmative		Negative		
1SG	*moral'əń*	'I would sing/ would have sung'	*afəl'əń*	*mora*	'I wouldn't sing/ wouldn't have sung'
2SG	*moral'əť*	'you would sing/ would have sung'	*afəl'əť*	*mora*	'you wouldn't sing / wouldn't have sung'
3SG	*moral'*	'(s)he would sing/ would have sung'	*afəl'*	*mora*	'(s)he wouldn't sing / wouldn't have sung'
1PL	*moral'əmä*	'we would sing/ would have sung'	*afəl'əmä*	*mora*	'we wouldn't sing / wouldn't have sung'
2PL	*moral'əďä*	'you would sing/ would have sung'	*afəl'əďä*	*mora*	'you wouldn't sing / wouldn't have sung'
3PL	*moraʟť*	'they would sing/ would have sung'	*afəʟť*	*mora*	'they wouldn't sing / wouldn't have sung'

The affirmative paradigm of the conjunctive mood is also built on the basis of the agglutination of *ul'ə-* 'be'. In the case of Moksha, this mood is identical to the paradigm of the indicative second past tense (cf. example (4)), but in the negative, a distinction is made between the two verbal categories: in the second past tense, the particle *af* is used, while in the conjunctive mood the auxiliary *afəl'-* appears. As pointed out by Givón (1978: 97), this kind of a distribution is typologically rare. According to him, languages tend to be more innovative in their affirmative forms, and only slowly do these innovations spread to the negative forms. According to Miestamo (2005: 161), negative expressions do not usually have a greater number of distinctions of grammatical categories than the affirmative. Consequently, in the case of the Mordvin languages, I consider the Erzya pattern, where the distinction between the verbal categories is made both in negation and in affirmation, to be the original. For some reason, Moksha has lost this distinction in the affirmative pattern but maintained it under negation (Hamari 2007: 105–6).

The Erzya second past tense and the conjunctive mood are also clearly based on the verb *uľə-* 'be', but in the conjunctive, there is an element *-v-* preceding the original auxiliary *uľə-*. This element also appears in the negative marker, where it seems to correspond to the *-f-* of the Moksha equivalent. Nevertheless, the Erzya *-v-* is etymologically confusing, as it does not appear in the negative particle *a* (~ Moksha *af*). It has been considered a derivational suffix (Bereczki 1988: 327–8), a secondary consonant, which has developed in a position preceding a labial vowel (Bartens 1999: 104), or a former participial suffix (Hamari 2007: 106), but none of these explanations is fully satisfactory (see Hamari 2007: 104–6 for a discussion). Therefore, the actual correspondence between the Erzya and Moksha forms remains an open question.

In Moksha, *afəľ-* is the only option to express negation in the conjunctive mood. In Erzya, on the other hand, *avoľ-* is in free variation with the particle *a* in the conjunctive and the conditional-conjunctive. The negative constructions of the conditional-conjunctive formed by the particles were already presented in examples (7) and (8), while Table 11.6 (adapted from Bartens 1999: 142, Grammatika 1980: 299, 319–20, Eŕźań keľ 2000: 190) illustrates the alternative constructions formed by the auxiliary in Erzya. Moreover, according to Mokšəń käľ (2000: 148), Moksha *afəľ-* is likewise used in the conditional-conjunctive (see also section 11.7 for the Moksha negative suffix of the conditional and conditional-conjunctive).

TABLE 11.6. **The conditional-conjunctive mood subject conjugation of Erzya *morams* 'sing' in the affirmative and the negative**

	Affirmative		Negative		
1SG	*morińďeŕavľiń*	'if I had sung'	*avoľiń*	*morińďeŕa*	'if I hadn't sung'
2SG	*morińďeŕavľiť*	'if you had sung'	*avoľiť*	*morińďeŕa*	'if you hadn't sung'
3SG	*morińďeŕavoľ*	'if (s)he had sung'	*avoľ*	*morińďeŕa*	'if (s)he hadn't sung'
1PL	*morińďeŕavľińek*	'if we had sung'	*avoľińek*	*morińďeŕa*	'if we hadn't sung'
2PL	*morińďeŕavľiďe*	'if you had sung'	*avoľiďe*	*morińďeŕa*	'if you hadn't sung'
3PL	*morińďeŕavoľť*	'if they had sung'	*avoľť*	*morińďeŕa*	'if they hadn't sung'

TABLE 11.7. The desiderative mood subject conjugation of Mordvin *morams* 'sing' in the affirmative and the negative

	Affirmative		Negative		
1SG	*moraľəksəľəń*	'I wanted to sing'	*afəľəksəľəń*	*mora*	'I didn't want to sing'
2SG	*moraľəksəľəť*	'you wanted to sing'	*afəľəksəľəť*	*mora*	'you didn't want to sing'
3SG	*moraľəksəľ*	'(s)he wanted to sing'	*afəľəksəľ*	*mora*	'(s)he didn't want to sing'
1PL	*moraľəksəľəmä*	'we wanted to sing'	*afəľəksəľəmä*	*mora*	'we didn't want to sing'
2PL	*moraľəksəľəďä*	'you wanted to sing'	*afəľəksəľəďä*	*mora*	'you didn't want to sing'
3PL	*moraľəksəʟť*	'they wanted to sing'	*afəľəksəʟť*	*mora*	'they didn't want to sing'

As was described in section 11.4.1, the conditional–conjunctive mood developed during the Proto-Mordvin period, when two auxiliary verbs were agglutinated to a presumably nominal form of the lexical verb. The agglutinated auxiliaries were *ťäŕa- 'try' and *uľə-* 'be' (Bartens 1999: 129–37). The variation between the two possible negative constructions has obviously resulted from the potential for the verb *uľə-* 'be' to be attached either to the negative marker or the lexical verb. The personal conjugation that was originally a feature of the auxiliary *uľə-* 'be' still follows the element Erzya *-(v)oľ-*, Moksha *-əľ-* developed from this auxiliary. In the negative forms, when the element is attached to the lexical verb, the negative marker remains invariant and the result is a symmetric negative construction. On the other hand, when it is attached to the negative marker, the resulting construction is an asymmetric negation. The new negative constructions followed the old pattern of negative auxiliaries: the negative auxiliaries carried the markers of finiteness, while the lexical verb was in the fixed connegative form. As pointed out by Bartens (1999: 142), in the new constructions the former *uľə-* 'be' has been reanalysed as a modal marker, which is why it does not require a nominal form of the lexical verb, as the original auxiliary did.

The moveable nature of the modal markers that developed in Proto-Mordvin is evident also in the case of the desiderative mood. The affirmative and negative paradigms of this mood are given in Table 11.7 (adapted from Bartens 1999: 142 and Grammatika 1980: 300, 320). The modal markers, Erzya *-kseľ-* and Moksha *-ksəľ-*, go back to a combination of the translative case suffix *-ks* and the auxiliary *uľə-* 'be', which were attached to the lexical verb (Bartens 1999: 129–37).[15] In Moksha, the modal marker is also attached to the negative marker, in which case the personal endings are likewise shifted to the negative marker, which, consequently, behaves as a negative

[15] The Moksha affirmative desiderative is exceptional in that it seems that the verb *uľə-* 'be' has been attached to the form twice: *-ľə-ks-əľ-*. According to Bartens (1999: 142), this could be due to the influence of the negative auxiliary.

auxiliary. However, in this case the negative marker that the modal ending is agglutinated to is the auxiliary *afəl'-* and the resulting auxiliary form for the desiderative is *afəl'əksəl'-*. Paasonen (1953: 012) and Bartens (1999: 142) also give an Erzya equivalent *avol'ksel'-* for the desiderative, but this is not attested in the present-day language. As was seen in section 11.4.1, the desiderative of Erzya is negated with the particle *a*, while the lexical verb carries the modal marker *-ksel'-* and the personal endings.

11.5.2 The Erzya particle avol'

Unlike the Moksha negative marker *afəl'-*, the Erzya form *avol'* has another function outside the sphere of verbal paradigms; it functions as a negative marker of non-verbal predicates. As noted in Hamari (2007: 125–30), it is in these cases in free variation with the particle *a*. Examples (19) and (20) illustrate the use of *avol'* with a non-verbal predicate:

(19) son ńej uš avol' viški-ńe
3SG now anymore NEG small-DIM
'Now he is not small anymore.' (Erzya) (UPTMN 1967: 198)

(20) Mon avol' koldun-an.
1SG NEG warlock-PRES.1SG
'I am not a warlock.' (Erzya) (UPTMN 1967: 338)

However, this use of Erzya *avol'* poses certain problems for the etymology of both this marker and the negative auxiliaries Erzya *avol'-* and Moksha *afəl'-*. First, as can be seen in example (20), unlike in cases in which the Erzya auxiliary *avol'-* is used in the negation of certain moods, in clauses with a non-verbal predicate the negative marker does not agree with the subject person. Therefore, it can be argued that, synchronically, *avol'* is a particle in this clause type and not an auxiliary. Secondly, as was mentioned above, the negative auxiliaries Erzya *avol'-* and Moksha *afəl'-* probably go back to the combination of the negative particles with the first past tense forms of the verb 'be'. However, as can be seen in examples (19) and (20), the Erzya particle *avol'* is employed in the present tense. If the past tense is needed, the construction follows the same pattern that was seen in the case of Erzya *a* (and Moksha *af*) —that is, the non-verbal predicate acquires the suffix of the second past tense and the negative marker stays unchanged.[16] This can be seen in example (21).

(21) son avol' bojece-l'
3SG NEG fighter-PAST2.3SG
'He was not a fighter.' (Erzya) (Śatko 2000: 96)

[16] Another possibility to form a past-tense negative construction in both languages is to use the verb 'be' accompanied by the auxiliary of the first past tense, but these constructions are quite uncommon (Hamari 2007: 130–1).

As suggested in Hamari (2011: 254–6), the development of this Erzya particle may be parallel to the developmental pattern that is often seen in the evolution of another type of negative marker, that is, negative existentials. According to Croft (1991), it is cross-linguistically common for negative existential markers to arise from the fusion of an ordinary negative marker and an affirmative existential marker (see section 1.6 for background). In the case of Erzya, no existential expressions are involved, but the pattern of development is similar to that described by Croft: a negative marker is attached to an affirmative copula and, consequently, a new negative marker arises. A further parallel could be drawn to Croft's theory, according to which the negative existentials can sometimes acquire functions of negating verbal predicates; this intrusion may take place by a gradual development in which the element is initially employed only in a restricted area of verbal predication, but can finally become the sole marker of negation in the particular language. If we assume that the Proto-Mordvin negative marker that developed through a fusion of a negative particle and the verb 'be' was first used in the negation of non-verbal predicates and acquired its functions as a negative auxiliary of certain moods only later, the development could be seen as similar to that described by Croft. What is confusing, however, is the fact that the Moksha cognate form is only used with verbal predicates and cannot confirm this direction of development.

11.6 Negative nouns

11.6.1 The negative existential-possessives

In both Erzya and Moksha there is a special negative marker for existential clauses. In Erzya this marker is *araś*, and in Moksha *aš* or its extended form *ajaš*. In addition to existential expressions, the markers are also employed in the negation of possessive clauses of the 'have'-type and in the negation of clauses expressing location, as described in detail in Hamari (2007: 163–228). Both markers have been characterized as negative nouns (or adjectives), as they behave like non-verbal predicates in existential clauses with only the non-verbal conjugation being possible in these contexts (Bartens 1996: 77). For example, the Erzya marker *araś* agrees with the number of the subject in existential expressions (see example (22)) and in the past tense, only the marking of the second past tense is possible (as in (23)).

(22) a. mastor lank-so **araś** iśťamo źveŕ
earth top-INE NEG such animal
'There is no such animal in the world.'

b. mastor lank-so **araś-ť** iśťamo źveŕ-ť
earth top-INE NEG-PL such animal-PL
'There are no such animals in the world.' (Erzya) (Bartens 1999: 162)

(23) Iśťa-t soldat-t moda-ńť lang-so źars **araś-el'-ť.**
such-PL soldier-PL earth-SG.DEF.GEN top-INE so.far NEG-PAST2-3PL
'So far, there haven't been such soldiers in the world.' (Erzya) (Šatko 1978: 22)

In Moksha, on the other hand, the negative existential has two variants, *aš* and *ajaš*, although the latter can be regarded as rare in the modern language (Hamari 2007: 190–1). The following examples illustrate the fact that these two variants can be used in exactly the same kind of contexts in the present tense:

(24) a. mol'-ś uzfə-zə-nza, **aš** kelaś, **aš** kal
go-PAST1.3SG load-ILL-POSS.3SG NEG fox NEG fish
'He went to his load, there is no fox, there is no fish.' (Moksha) (Paasonen and Ravila 1947: 843)

b. mol'-ś vozə-zə-nzə, **ajaš** kelaś, **ajaš** kal
go-PAST1.3SG load-ILL-POSS.3SG NEG fox NEG fish
'He went to his load, there is no fox, there is no fish.' (Moksha) (Paasonen and Ravila 1947: 837)

The Erzya and Moksha negative existentials are not etymologically connected to each other, but they are usually described as functional equivalents (e.g. Bartens 1996: 77–9, Bartens 1999: 162–3). Their usage is unquestionably similar to a large extent, but there are also certain aspects of their characteristics that make a considerable difference between them, as stated in Hamari (2007: 189–94). Firstly, both the Moksha negative marker *aš* and its longer form *ajaš* are invariable in present-tense existential constructions, that is, they do not agree with the subject in number:

(25) Śťir-ńä-t mala-sə-nk **aš,...**
girl-DIM-PL near-INE-POSS.1PL NEG
'There are no little girls near us, [–]' (Moksha) (Mokša 2000: 60)

(26) I pop-t tos(a) **ajaš.**
and priest-PL there NEG
'And there are no priests there.' (Moksha) (Paasonen and Ravila 1947: 809)

This inability to inflect for number could be regarded as a feature that makes the Moksha markers less noun-like and more particle-like than the Erzya marker *araś*. However, this is an accurate view only in the case of the Moksha extended form *ajaš*, which rejects all conjugational endings and can therefore be considered a particle that appears in present-tense expressions only. On the other hand, in the case of Moksha *aš*, marking of the second past tense *is* possible, which means that the marker is not totally invariant:

(27) Śä pinkńəń **aš-əʟ'ť** ťelevizoʀ-t-viďik-t, **aš-əʟ'ť**
that at.time NEG-PAST2.3PL television-PL-video-PL NEG-PAST2.3PL
mekśikanskaj śeriaʟ-t, . . .
Mexican serial-PL
'At that time there were no televisions or videos, there were no Mexican serials, [–]' (Moksha) (Mokša 2000: 4)

In addition to prototypical existential clauses with an indefinite third person subject, both Erzya *araś* and Moksha *aš* can be employed in certain locative clauses with a definite subject. In these cases, the subject can also be one of the first or second persons. Therefore, in locative constructions, the negative markers are conjugated in the present and second past tense throughout the personal paradigm (see Table 11.8). Consequently, unlike in existential clauses, the third person plural form is also possible for the Moksha marker *aš* in the present tense when the subject is marked definite.

As mentioned above, Erzya *araś* and Moksha *aš* ~ *ajaš* are not etymologically connected to each other. However, it seems that not much more can be said about their origin. Several suggestions have been made about their development, but all explanations are problematic in one way or another.

First of all, Erzya *araś* has been regarded as an original fusion of the negative particle *a* and a nominal form of the verb *eŕams* 'live, be'—a fusion in which one would have to suppose an assimilation of *a* and *e-* (*a eŕaś* > *araś*) (Klemm 1934: 398, Ščemerova 1972: 178–9, Honti 1997: 162). Cross-linguistically, this explanation is plausible since, as shown by Croft (1991: 6–13), negative existentials frequently arise from fusions of negative markers and affirmative existentials. However, as argued in Hamari (2007: 108), this suggestion is weakened by three facts. Firstly, there is no deverbal derivational suffix in *-ś* in Erzya. A participial ending in *-ź* does appear, but it does not normally become unvoiced. Secondly, the sound change *ŕ* > *r* is not regular even in cases in which the consonant ends up between two back vowels—a phonetic context that has been suggested as a reason for the loss of palatalization of *ŕ* in this

TABLE 11.8. **The conjugation of Erzya *araś* and Moksha *aš***

	Present tense		Second past tense	
	Erzya	Moksha	Erzya	Moksha
1SG	*araśan*	*ašan*	*araśeľiń*	*ašəľəń*
2SG	*araśat*	*ašat*	*araśeľiť*	*ašəľəť*
3SG	*araś*	*aš*	*araśeľ*	*ašəľ*
1PL	*araśťano*	*aštama*	*araśeľińek*	*ašəľəmä*
2PL	*araśťado*	*aštada*	*araśeľiďe*	*ašəľəďä*
3PL	*araśť*	*ašət*	*araśel'ť*	*ašəʟ'ť*

particular word (Ščemerova 1972: 178–9). Thirdly, the verb *eŕams* 'live, be' is not used as an affirmative existential in Erzya, apart from in the Shoksha dialect, where an affirmative existential-possessive predicate *eŕme* has been attested instead of *uľi*, which is used in other parts of the linguistic area (Paasonen 1990–9: 380, Hamari 2007: 50–2). Since the nominalization of *eŕme* is formed by the suffix *-me* instead of *-ś* (or *-ź*), the hypothesis can be considered even weaker.

Another etymological explanation of Erzya *araś* has been proposed by Bartens (1996: 79) and Mosin (2005: 182–3), who considered the possibility of *araś* being a nominalized form of a verb *arams* 'disappear, get lost; be destroyed; die' or, more specifically, its assumed participial form *araź* 'lack', 'non-essential, irrelevant, useless'. According to Bartens (1996: 79) the forms *araś* and *araź* could have specialized in two different functions. Despite being a reasonable assumption, this suggestion is not without problems either. Firstly, the problematic change *-ź* > *-ś* that was discussed above is present in this proposal as well, although Bartens does plausibly suggest that it could have taken place in the plural form in which the sibilant appears before a voiceless consonant, that is, the plural marker in *-ť* (*arasť*). Even more problematic, however, is the semantics of the verb *arams*, as pointed out in Hamari (2007: 109). This verb has the meaning 'disappear, get lost; be destroyed; die' only when it appears in a pair-verb construction with another verb *jomams* 'disappear, get lost; be destroyed; die': *jomams-arams*. Otherwise, *arams* means 'place onself; go; change; become'. Mordvin pair verbs typically consist of two verbs with two different meanings and, when combined, acquire a third meaning which is not necessarily the sum of the two constituents (Bartens 1999: 157). Therefore, there is no need to assume that the verb *arams* ever had any other meaning than 'place onself; go; change; become' and, furthermore, a verb with this type of semantics could hardly be considered a source for a negative existential. The relation of *arams* 'place onself; go; change; become' and *araź* 'lack', 'non-essential, irrelevant, useless' remains unclear.

The third proposal for the origin of Erzya *araś*, tentatively given by Bartens (1996: 79) in a footnote, is, again, semantically reasonable but difficult to prove. In this proposal the existential is connected to a hypothetical noun *ara 'non-existent, nonsense', which could be the source of the inexplicable word form *arań*, which appears in an idiomatic Moksha expression *arań koʀtaj* 'ignorant, silly babbler, slanderer'. In this expression *koʀtaj* is the present-tense participial form of the verb *koʀtams* 'speak, talk, discuss', but *arań* is without explanation, as it only appears in this construction. However, if there once was a noun *ara 'non-existent, nonsense' in Mordvin, *arań* could be its genitive form and the expression could be understood as 'the speaker of nonsense'. Still, *arań* is only attested in Moksha, so its connection to the Erzya negative existential *araś* is questionable (Hamari 2007: 109–10).

The origin of the Moksha negative existential *aš* ~ *ajaš* has likewise been dealt with on several occasions but no definitive explanation is available. Firstly, it was suggested by Klemm (1934: 388) that the marker *aš* has developed from a combination of a negative particle in *a and the verb Moksha *aščəms* 'be, exist, stay' (~ Erzya *ašťems*), which originally must have carried the connegative suffix in *-k (*a-ašt-k > *ašk > *aš*), see also Ščemerova (1972: 177) and Honti (1997: 161–2). The variant in *ajaš* could have developed from a form in which *j* was introduced to prevent hiatus (*a-j-ašk > *ajaš*). As argued in Hamari (2007: 112–13), the problem with this etymology is that it has been built on evidence attested in Erzya only; the negative particle *a* only appears in Erzya—where, in fact, a *j*-sound sometimes does appear, as mentioned above in section 11.4.1. There is no proof of whether there once was a negative particle or an auxiliary form in *aj in Moksha, which means that this explanation is uncertain. However, if it could be proved, the etymology could explain the fact that *ajaš* is still invariant, since in Mordvin negative constructions formed with a negative auxiliary, no endings can be attached to the connegative verb form—in this case, the latter part *-aš* of *ajaš* would represent that connegative form.

The most likely suggestion of the origin of the Moksha *aš* is the one given by Bartens (1996: 79), who assumes that it originated as a noun. According to her, the non-verbal conjugation of the form can be seen as a proof of this origin, as well as the fact that there is a diminutive noun form *ašəńä* 'non-existence, smallness' in Moksha and an adjective *ašu* 'poor', which could have been derived from the same original noun *aš. Moreover, as pointed out in Hamari (2007: 269–70), the negative existential *aš* can still sometimes be used as a noun with meanings such as 'non-existence; poverty' and the like.

11.6.2 The development of Moksha auxiliary aš-

In addition to the existential in *aš*, there is also a negative auxiliary in Moksha that has the same form, that is, the negative marker of the first past tense. This auxiliary can be used interchangeably with the auxiliary *əź- ~ iź-* discussed above, and the negative constructions formed by these two auxiliaries are formally similar; *aš-* is also conjugated both according to the subject and the object conjugation, while the bare stem of the lexical verb appears in the connegative form (or dialectally labelled with *-k*) (e.g. Paasonen 1953: 011), see Table 11.9 (adapted from Bartens 1999: 141, Grammatika 1980: 319).

TABLE 11.9. Negation of Moksha *morams* 'to sing' in the first past tense subject conjugation

1SG	*ašəń*	*mora*	'I didn't sing'
2SG	*ašət́*	*mora*	'you didn't sing'
3SG	*ašəź*	*mora*	'(s)he didn't sing'
1PL	*ašəmä*	*mora*	'we didn't sing'
2PL	*ašəd́ä*	*mora*	'you didn't sing'
3PL	*ašəśt́*	*mora*	'they didn't sing'

Considerations of the origin of the negative auxiliary in *aš-* have always taken into account its relation to the negative existential. Traditionally, the negative auxiliary *aš-* has been explained as a development of verbal conjugation in the negative existential marker (originally proposed by Szinnyei 1884: 148). However, this explanation is problematic in that the development of a nominal negative marker primarily used in present-tense expressions of existence, location, and possession into a verbal marker primarily used in past-tense action clauses is hard to explain. Therefore, another, typologically more plausible, development is described in Hamari (2007: 273–5), where it is suggested that the origin of the negative auxiliary *aš-* could be in a construction of emphasis; the negative existential marker *aš* could have been attached to the older negative auxiliary in *əź- ~ iź-* to emphasize negation (e.g. 1SG *aš-əźəń, 2SG *aš-əźəť, etc.). Later, the construction lost its emphatic force and was analogically adapted to the conjugation of the auxiliary in *əź- ~ iź-* by dropping the element in *əź- ~ iź-* (e.g. 1SG *aš-əźəń > ašəń, 2SG *aš-əźəť > ašəť, etc.). One reason for this formal reduction could have been the fact that the forms of the object conjugation would have been relatively long (e.g. 3SG > 3PL *aš-əźəźəń).

As argued in Hamari (2007: 273–5), there are three facts that support the etymological explanation given to the Moksha auxiliary in *aš-* just described. Firstly, although the first and second person forms as well as the third person plural form of the auxiliary in *aš-* show the regular first past tense forms of lexical verbs (cf. 1SG PAST1 *palań* 'I kissed' ~ *ašəń* 'I did not'), the third person singular form is irregular (cf. 3SG *palaś* '(s)he kissed' ~ *ašəź* '(s)he did not'). In fact, the only verb showing a voiced palatalized sibilant as a third person singular marker is the negative auxiliary in *əź- ~ iź-* (3SG *əź* '(s)he did not'). Consequently, the fusion of the existential in *aš* and the auxiliary in *əź-* is still visible in the third person forms of the subject conjugation of the auxiliary in *aš-* (3SG *ašəź* < *aš-əź). In fact, the same goes for the third person plural form (3PL *ašəśť* < *aš-əśť; cf. *əśť* 'they did not'). Secondly, there is some dialectal evidence to support the fusion; according to Ščemerova (1972: 178), in the district of Kovylkino the paradigm of the auxiliary in *aš-* has an element in *-əź-* also in the first and second person forms of the subject conjugation (Kovylkino: 1SG *ašəźəń* 'I did not', 2SG *ašəźəť* 'you did not'). Therefore, these dialectal forms could be regarded as transparent relics of the fusion (1SG *aš-əźəń, 2SG *aš-əźəť) described above. Finally, according to Croft (1991: 10, 13–14) it is relatively common crosslinguistically for negative existential markers or identical negative interjections to acquire the function of negating verbs by first appearing in negative constructions as emphatic markers and then, as their emphatic force decreases with frequent use, they may even become obligatory negative markers or parts of negative constructions in negation of verbs. In Moksha, this kind of development seems to have led to a situation in which the language has two negative auxiliaries for the negation of past-tense actions: the ancient *əź- ~ iź-* and a more recent *aš-*.

11.7 The negative suffix of Moksha

In this description of the Mordvin negation system, we have so far encountered negative auxiliaries, negative particles, and negative nouns. Finally, there is one more type of negative marker to be treated, namely the negative suffix, which is only found in Moksha. The suffix is *-fťäŕa-*/*-fťäŕä-* in the conditional (28) and *-fťäŕaľə-*/*-fťäŕäľə-* in the conditional–conjunctive mood (29):

(28) pala-fťäŕa-n
kiss-NEG.COND-PRES.1SG
'if I don't kiss' (Moksha) (Bartens 1999: 141)

(29) pala-fťäŕaľə-n
kiss-NEG.CONDCONJ-PRES.1SG
'if I hadn't kissed; if I weren't to kiss' (Moksha) (Bartens 1999: 141)

As was discussed in section 11.4.1 above, the conditional mood was originally built through the fusion of the verb *ťäŕa- 'try' and the conditional–conjunctive mood through the fusion of both *ťäŕa- 'try' and *uľə-* 'be'. The negative suffixes carry these same elements, but it seems that in these negative forms, the particle *af* has ended up between the lexical verb and the fused auxiliaries (Klemm 1934: 392–3, Pall 1957: 221); that is, in the conditional *pala af ťäŕan > *palafťäŕan*. I would see this as another indication of the moveability of the original auxiliaries of these particular moods (see sections 11.4.1 and 11.5.1 for alternative constructions for the conditional and conditional–conjunctive).

11.8 The indefinites

So far, this presentation of the system of negation of the Mordvin languages has involved only the actual negative markers of sentential negation, but to create a coherent picture of the system it is necessary to view other possible negative polarity items. These include, self-evidently, the use of indefinite pronouns in negative contexts.

Actually, there are no inherently negative indefinite pronouns or adverbs in the Mordvin languages, nor are there indefinite pronouns or adverbs that are specifically marked negative. Instead, both languages have a set of allomorphic suffixes which are attached to interrogative pronouns or adverbs to produce indefinites, and which can be used either in negative or affirmative clauses.[17] In Erzya the suffix forms are *-jak* (after vowels), *-gak* (after voiced consonants), and *-kak* (after voiceless consonants)

[17] However, there are also indefinite pronouns that can only appear in affirmative clauses. These are formed either by reduplication (e.g. Erzya *kije-kije*, Moksha *kijä-kijä* 'someone') or by using enclitic particles borrowed from Russian (e.g. Erzya *kije-buťi*, Moksha *kijä-bədi* 'someone'; Erzya *koj-kije*, Moksha *koj-kijä* 'id.') or the Turkic languages spoken in the vicinity of Mordvin (e.g. Erzya *ta-ki*, Moksha *tag-kijä* 'someone') (Bartens 1999: 117). Erzya and Moksha *ki* and Erzya *kije* ~ Moksha *kijä* are variants of the same interrogative pronoun meaning 'who' (Bartens 1999: 115).

and in Moksha *-vək/-gək* (after vowels), *-ga/-gä* (after vowels or voiced consonants), and *-ka/-kä* (after voiceless consonants), for example Erzya *ki* 'who', *ki-jak* 'someone, anyone, no one'; Moksha *ki* 'who', *ki-vək* ~ *ki-gä* 'someone, anyone, no one'. The suffixes are regarded as enclitic particles, and they are always attached after the case ending or plural suffix, for example SG. GEN: Erzya *ki-ń=gak*, Moksha *ki-ń=gä* (who-GEN=CL); PL. NOM. Erzya *ki-ť=kak*; Moksha *ki-t=ka* (who-PL=CL) (e.g. Bartens 1999: 118). As can be seen in the following, the indefinite pronouns can be used in affirmative (30) as well as in negative (31) contexts and, moreover, in non-assertive sentences such as interrogatives (32).

(30) a. Bazar-sto meźe=jak rama-n.
marketplace-ELA what=CL buy-PRES.1SG
'I'm buying something from the marketplace.' (Erzya) (Grammatika 1962: 238)

b. Bazar-sta meźə=vək rama-n.
marketplace-ELA what=CL buy-PRES.1SG
'I'm buying something from the marketplace.' (Moksha) (Grammatika 1962: 238)

(31) a. Meźe=jak ton a soda-t.
what=CL 2SG NEG know-PRES.2SG
'You don't know anything.' (Erzya) (Eŕźań keľ 2000: 142)

b. Meźə=vək iź az.
what=CL NEG.PAST1.3SG say.CNG
'(S)he didn't say anything.' (Moksha) (Bartens 1999: 118)

(32) a. Son tońeť jovta-ś meźe=jak?
3SG 2SG.ALL say-PAST1.3SG what=CL
'Did (s)he say anything/something to you?' (Erzya) (Grammatika 1962: 238)

b. Son ťejť az-ś meźə=vək?
3SG 2SG.ALL say-PAST1.3SG what=CL
'Did (s)he say anything/something to you?' (Moksha) (Grammatika 1962: 238)

The particles are assumed to be connected to, for example, the Finnish enclitic particles *-kin* (~ dial. *-kik*) and *-kAAn* (Paasonen 1953: 66–7, Hakulinen 1979: 237–8, Bartens 1999: 167), although the exact etymological relation has not been established. In the languages of this branch, these particles are not restricted to the formation of indefinite pronouns only; they are also used as scalar as well as additive focus particles. Haspelmath (1997: 224) discusses the use of these particles in Finnish and Estonian, which are closely related to each other. In Finnish, example (33), the distribution of the particles is such that, as scalar focus particles, *-kin* 'even' appears in affirmative contexts and *-kAAn* 'not even' in negative ones. In Estonian, example (34), on the other hand, only one focus particle *-gi/-ki* (cognate with Finnish *-kin*) is employed in both affirmative and negative contexts.

(33) a. Se voi maksaa 100-**kin** markkaa.
it can cost 100-even marks
'That can cost even 100 marks.' (Finnish) (Haspelmath 1997: 224)[18]

b. Hän ei edes tahtonut kuulla-**kaan** mitä minulla olisi
3SG NEG.3SG even wanted listen-even what on.me was
ollut sanomista.
been to.say
'(S)he didn't even want to listen to what I had to say.' (Finnish) (Haspelmath 1997: 224)

(34) a. Lapsed-**ki** teavad seda.
children-even know it
'Even children know it.' (Estonian) (Haspelmath 1997: 224)

b. Ta ei julge piuksatada-**gi**.
3SG NEG dare make.sound-even
'(S)he doesn't dare to even make a sound.' (Estonian) (Haspelmath 1997: 224)

As further pointed out by Haspelmath, the same correlation holds when the particles are used as formatives of indefinite pronouns; Finnish negative indefinites are formed by using *-kAAn*, while *-gi*/*-ki* is used in Estonian:

(35) Hän ei palannut milloin-**kaan.**
3SG NEG.3SG returned when-INDEF
'(S)he never returned.' (Finnish) (Haspelmath 1997: 224)

(36) Ta ei kohanud keda-**gi.**
3SG NEG met whom-INDEF
'(S)he did not meet anybody.' (Estonian) (Haspelmath 1997: 224)

Moreover, in the formation of affirmative pronouns, only *-kin* can be used in Finnish, whereas Estonian uses the same *-gi*/*-ki* that appears in negative forms.

The uses of the enclitic particles of the Mordvin languages display the same pattern as can be seen in Estonian; the same particles can be used in both affirmative and negative contexts, be it the formation of indefinites or as scalar or additive focus particles. In affirmative contexts the meaning of the Mordvin equivalents is 'even, also, too' and in negative 'not even' (Bartens 1999: 167). An affirmative context is illustrated in the Erzya example (37), where the element functions as an additive focus particle, while in the Moksha example (38) a negative context appears and the element functions as a scalar focus particle:

(37) Mon=gak ťe-sa-n.
1SG=CL this-INE-PRES.1SG
'I am here, too.' (Erzya) (Grammatika 1962: 352)

[18] I have slightly revised the glossings and translations of Haspelmath's examples.

(38) T'ev=ga aš, a t́ä-sa-t.
work=CL NEG but this-INE-PRES.2SG
'There is not even work, but here you are.' (Moksha) (Grammatika 1962: 352)

Finally, as mentioned, for example, by Bartens (1999: 167), the Mordvin enclitic particles can also be employed as means of expressing coordination:

(39) oj=gak lovco=jak
butter=CL milk=CL
'butter and milk' (Erzya) (Bartens 1999: 167)

As argued by König (1991: 66), the relatedness of especially additive particles and markers of coordination is not really surprising given that they both 'link separate but parallel information to the preceding discourse'.

11.9 Conclusions

Mordvin negative markers or negative constructions can be divided according to their origin into three groups which represent three different layers in the history of the Mordvin languages. Firstly, there are markers that go back to a pre-Proto-Mordvin period and, consequently, have cognates in distantly related languages. These include the negative auxiliaries Erzya *eź-*, Moksha *əź-* ~ *iź-* of the first past tense (section 11.3.1), and the auxiliaries Erzya *il'a-*, Moksha *t́a-* ~ *d́a-* of the imperative/optative mood (section 11.3.2). Secondly, there are markers that most likely originated in the Proto-Mordvin period; these include the multi-functional particles Erzya *a* ~ *at* and Moksha *af* ~ *ajaf* (section 11.4.1), the particle Erzya, Moksha *apak* of participles and gerunds (section 11.4.2), as well as the auxiliaries Erzya *avol'-*, Moksha *afəl'-*, Erzya *avol'ksel'-*, and Moksha *afəl'əksəl'-*, which were originally combinations of negative particles and a verb 'be' (section 11.5.1). Thirdly, there are markers in both Erzya and Moksha that do not have a cognate in the other Mordvin language and, therefore, must be regarded as relatively recent innovations, that is, they must have originated after the break-up of Proto-Mordvin. These include the negative nouns/particles Erzya *araś* and Moksha *aš* ~ *ajaš* (and the use of Moksha *aš-* as a negative auxiliary) (section 11.6.1) and the Moksha suffix *-ft́äŕa-*/*-ft́äŕä-* and its extended form *-ft́äŕal'ə-*/*-ft́äŕäl'ə* (section 11.7). Furthermore, in Erzya, *avol'* can also be employed as a particle in the negation of non-verbal predicates, a function that does not exist in the case of Moksha *afəl'-* (section 11.5.2).

As has become evident in the course of this chapter, much of the development of the rather complex system of negation in the Mordvin languages is most likely intertwined with the development of several new predicate categories during the Proto-Mordvin period. For the negation of these categories, two different developmental paths were taken; the functions of the fossilized negative auxiliary forms Erzya *a*, Moksha *af* were enlarged to cover some of the categories, but, for others, new negative auxiliaries developed. The predicate categories which stayed relatively unchanged during the Proto-Mordvin period were the first past tense and the imperative/optative mood. Consequently, it should not be surprising that these are the categories that have remained closest to the negative patterns reconstructed for the earliest protolanguage.

Still, what sticks out in the historical reconstructions of the negative markers is the peculiar fact that the younger the negative markers are considered to be, the more obscure their form. The most ancient auxiliaries can relatively easily be recognized as reflexes of the Uralic negative auxiliaries, while the etymologies of the negative particles Erzya *a*, Moksha *af*, and Erzya, Moksha *apak* are much more controversial, and the development of the negative nouns/particles Erzya *araś* and Moksha *aš* ~ *ajaš* can also be regarded as uncertain. Nevertheless, since none of these forms can be positively shown to be original loans from the surrounding languages, they are usually considered to be the results of language-internal development.

Primary sources

Mokša 2000 = *Mokša* 1/2000. Literaturno-xudožestvennyj i obščestvenno-političeskij žurnal. Saransk: Pravitel'stvo Respubliki Mordovija.

Paasonen, Heikki and Paavo Ravila. 1947. *Mordwinische Volksdichtung IV*. Helsinki: Suomalais-Ugrilainen Seura.

Śatko 1978 = *Śatko* 1/1978. Literaturno-xudožestvennyj i obščestvenno-političeskij žurnal. Saransk: Sojuz pisatelej Mordovskoj ASSR.

Śatko 2000 = *Śatko* 1/2000. Literaturno-xudožestvennyj i obščestvenno-političeskij žurnal. Saransk: Pravitel'stvo Respubliki Mordovija.

UPTMN 1967 = Maskaeva, A. I., V. A. Evseev, and L. S. Kavtas'kina (eds.). 1967. *Ustno-poètičeskoe tvorčestvo mordovskogo naroda. Tom tretij, čast' vtoraja. Èrzjanskie skazki.* Saransk: Mordovskoe Knižnoe Izdatel'stvo.

References

Abondolo, Daniel (ed.) (1998). *The Uralic languages*. London: Routledge.

Abraham, Werner (2003). Autonomous and non-autonomous components of 'grammatic(al)-ization': Economy criteria in the emergence of German negation. *Sprachtypologie und Universalienforschung: Language Typology and Universals* 56, 325–65.

Acquaviva, Paolo (1992). The representation of negative 'quantifiers'. *Rivista di Linguistica* 4, 319–81.

Adila, Aziz (1996). La négation en arabe marocain. In Salem Chaker and Dominique Caubet (eds.), *La négation en berbère et en arabe maghrébin*, 99–116. Paris: L'Harmattan.

Admoni, Wladimir (1990). *Historische Syntax des Deutschen*. Tübingen: Niemeyer.

Ageno, Franca Brambilla (1955). L'uso pleonastico della negazione nei primi secoli. *Studi di filologia italiana* 13, 339–61.

Alhoniemi, Alho (1982). Eräitä näkökohtia mordvan nominaalisen predikaatin käytöstä. In Kaisa Häkkinen (ed.), *Lauseenjäsennyksen perusteet. Seminaari Seilissä 9.–10.9.1982*, 47–58. Turku: Suomen kielitieteellinen yhdistys.

Alhoniemi, Alho (2001). Eine Gesamtdarstellung der mordwinischen Sprachen. Review of Bartens, Raija: *Mordvalaiskielten rakenne ja kehitys*. *Finnisch-Ugrische Forschungen* 56, 359–63.

Andrew, Samuel (1940). *Syntax and style in Old English*. Cambridge: Cambridge University Press.

Aranovich, Raúl (2007). Negative polarity and scalar semantics in Spanish. *Lingvisticæ Investigationes* 30, 181–216.

Armstrong, Nigel (2002). Variable deletion of French *ne*: A cross-stylistic perspective. *Language Sciences* 24, 153–73.

Armstrong, Nigel and Alan Smith (2002). The influence of linguistic and social factors on the recent decline of French *ne*. *Journal of French Language Studies* 12, 23–41.

Ashby, William J. (1981). The loss of the negative particle *ne* in French: A syntactic change in progress. *Language* 57, 674–87.

Ashby, William J. (1991). When does variation indicate linguistic change in progress? *Journal of French Language Studies* 1, 1–19.

Ashby, William J. (2001). Un nouveau regard sur la chute du *ne* en français parlé tourangeau: s'agit-il d'un changement en cours? *Journal of French Language Studies* 11, 1–22.

Avery, J. B. (1978). A problem of etymology in Swiss Romansh: *buca* 'not'. *Semasia* 5, 35–46.

Axel, Katrin (2007). *Studies on Old High German syntax: Left sentence periphery, verb placement and verb-second*. Amsterdam: John Benjamins.

Ayres-Bennett, Wendy (1994). Negative evidence: Or another look at the non-use of negative *ne* in seventeenth-century French. *French Studies* 48, 63–85.

Babby, Leonard H. (1980). *Existential sentences and negation in Russian*. Ann Arbor, Mich.: Karoma.

Babby, Leonard H. (2001). The genitive of negation: A unified analysis. In Steven Franks (ed.), *Annual Workshop on Formal Approaches to Slavic Linguistics: The Bloomington meeting 2000*, 39–55. Ann Arbor, Mich.: Michigan Slavic Publications.

Babyonyshev, Maria and Dina Brun (2002). Specificity matters: A new look at the new genitive of negation in Russian. In Jindřich Toman (ed.), *Annual Workshop on Formal Approaches to Slavic Linguistics: The second Ann Arbor meeting 2001*, 47–66. Ann Arbor, Mich.: Michigan Slavic Publications.

Bader, Christian (2006). *Parlons oromo*. Paris: L'Harmattan.

Bailyn, John F. (1995). Underlying phrase structure and 'short' verb movement in Russian. *Journal of Slavic Linguistics* 3, 13–58.

Bailyn, John F. (1997). Genitive of negation is obligatory. In Wayles Browne, Ewa Dornisch, Natasha Kondrashova, and Draga Zec (eds.), *Annual Workshop on Formal Approaches to Slavic Linguistics: The Cornell meeting 1995*, 84–114. Ann Arbor, Mich.: Michigan Slavic Publications.

Bailyn, John F. (2004). The case of Q. In Olga Arnaudova (ed.), *Annual Workshop on Formal Approaches to Slavic Linguistics: The Ottawa meeting 2003*, 1–35. Ann Arbor, Mich.: Michigan Slavic Publications.

Ball, Martin J. and Nicole Müller (eds.) (2009). *The Celtic languages*. London: Routledge.

Baptista, Marlyse (2002). *The syntax of Cape Verdean creoles: The Sotavento varieties*. Amsterdam: John Benjamins.

Barbiers, Sjef, Johan van der Auwera, Hans Bennis, Eefje Boef, Gunther De Vogelaer, and Margreet van der Ham (2008). *Syntactic atlas of the Dutch dialects*, vol. ii. Amsterdam: Amsterdam University Press.

Bartens, Raija (1996). Die positive und negative Existentiale in den finnisch-ugrischen Sprachen. *Ural-Altaische Jahrbücher* [new series] 14, 58–97.

Bartens, Raija (1999). *Mordvalaiskielten rakenne ja kehitys*. Helsinki: Suomalais-Ugrilainen Seura.

Bartens, Raija (2000). *Permiläisten kielten rakenne ja kehitys*. Helsinki: Suomalais-Ugrilainen Seura.

Basset, Louis (1989). La syntaxe de l'imaginaire: étude de modes et des négations dans l'Iliade et l'Odyssée. Lyons: Maison de l'Orient, Diffusion de Boccard.

Bayer, Josef (1990). What Bavarian negative concord reveals about the syntactic structure of German. In Joan Mascaro (ed.), *Grammar in progress*, 13–24. Dordrecht: Foris.

Bayer, Josef (2009). Nominal negative quantifiers as adjuncts. *Journal of Comparative Germanic Linguistics* 12, 5–30.

Bean, Marian (1983). *The development of word order patterns in Old English*. London: Croom Helm.

Behaghel, Otto (1918). Die Verneinung in den deutschen Sprachen. *Wissenschaftliche Beihefte zur Zeitschrift des allgemeinen deutschen Sprachvereins* 5 (38/40), 225–52.

Behaghel, Otto (1924). *Deutsche Syntax, Band II*. Heidelberg: Carl Winters Universitätsbuchhandlung.

Behaghel, Otto (1932). *Deutsche Syntax. Eine geschichtliche Darstellung*. Vol. iv: *Worstellung, Periodenbau*. Heidelberg: Winter.

Beheydt, Griet (1998). *Het gebruik en de vorm van de negatie in het zuidelijke Nederlands in een diachronisch perspectief (15e–20e eeuw)*. Licentiate's thesis, University of Leuven.

Belić, Aleksandar (1962). *Istorija srpskohrvatskog jezika*, 2nd edn. Belgrade: Naučna knjiga.

Benincà, Paola (1983). Il clitico *a* nel dialetto padovano. In Paola Benincà, Manlio Cortelazzo, Aldo Luigi Prosdocimi, Laura Vanelli, and Alberto Zamboni (eds.), *Scritti linguistici in onore di Giovan Battista Pellegrini*, 25–35. Pisa: Pacini.

Benincà, Paola (1986). Punti di sintassi comparata dei dialetti italiani settentrionali. In Günter Holtus and Kurt Ringger (eds.), *Raetia antiqua et moderna. W. Th. Elwert zum 80. Geburtstag*, 457–79. Tübingen: Niemeyer.

Benincà, Paola (1996). La struttura della frase esclamativa alla luce del dialetto padovano. In Paola Benincà, Guglielmo Cinque, Tullio De Mauro, and Nigel Vincent (eds.), *Italiano e dialetti nel tempo. Saggi di grammatica per Giulio C. Lepschy*, 23–43. Rome: Bulzoni.

Benincà, Paola and Cecilia Poletto (2005). On some descriptive generalizations in Romance. In Guglielmo Cinque and Richard S. Kayne (eds.), *The Oxford handbook of comparative syntax*, 221–58. New York: Oxford University Press.

Benincà, Paola and Laura Vanelli (2005). *Linguistica friulana*. Padua: Unipress.

Benmamoun, Elabbas (1997). Licensing of negative polarity items in Moroccan Arabic. *Natural Language and Linguistic Theory* 15, 263–87.

Benmamoun, Elabbas (2000). *The feature structure of functional categories: A comparative study of Arabic dialects*. Oxford: Oxford University Press.

Benmamoun, Elabbas (2006). Licensing configurations: The puzzle of head negative polarity items. *Linguistic Inquiry* 37, 141–7.

Bereczki, Gábor (1988). Geschichte der wolgafinnischen Sprachen. In Denis Sinor (ed.), *The Uralic languages: Description, history and foreign influence*, 314–50. Leiden: E. J. Brill.

Bergsland, Knut (1960). Lapin kieltoverbin taivutuksesta. *Virittäjä*, 327–30.

Bernini, Giuliano (1992). Forme concorrenti di negazione in italiano. In Bruno Moretti, Dario Petrini, and Sandro Bianconi (eds.), *Linee di tendenza dell'italiano contemporaneo: Atti del XXIV Congresso della Società di Linguistica Italiana*, 191–215. Rome: Bulzoni.

Bernini, Giuliano (2003). Learner varieties and language types. The case of indefinite pronouns in non-native Italian. In Anna Giacalone Ramat (ed.), *Typology and second language acquisition*, 83–124. Berlin: Mouton de Gruyter.

Bernini, Giuliano and Paolo Ramat (1996). *Negative sentences in the languages of Europe*. Berlin: Mouton de Gruyter.

Berruto, Gaetuno (1990). Note tipologiche di un non tipologo sul dialetto piemontese. In G. Gaetuno Berruto and Alberto Sobrero (eds.), *Studi di sociolinguistica e dialettologia italiana offerti a Corrado Grassi*, 3–24. Galatina: Congedo.

Besch, Werner, Anne Betten, Oskar Reichmann, and Stefan Sonderegger (eds.) (2000). *Sprachgeschichte: Ein Handbuch zur Geschichte der deutschen Sprache und ihrer Erforschung*. Berlin: Mouton de Gruyter.

Besten, Hans den (1986). Double negation and the genesis of Afrikaans. In Pieter Muysken and Neil Smith (eds.), *Substrata versus universals in creole languages. Papers from the Amsterdam creole workshop, April 1985*, 185–230. Amsterdam: John Benjamins.

Beukema, Frits and Olga Tomić (1996). Negation in English: A diachronic view. *Folia Linguistica Historica* 16, 123–35.

Bezlaj, France (1976). *Etimološki slovar slovenskega jezika*. Ljubljana: Slovenska Akademija Znanosti in Umetmosti.

Biberauer, Theresa (2008). Doubling and omission: Insights from Afrikaans negation. In Sjef Barbiers, Margreet van der Ham, Olaf Koeneman, and Marika Lekakou (eds.), *Microvariation in syntactic doubling*, 103–40. Bingley: Emerald.

Biberauer, Theresa (2009). Jespersen off course? The case of contemporary Afrikaans negation. In Elly van Gelderen (ed.), *Cyclical change*, 91–130. Amsterdam: John Benjamins.

Biberauer, Theresa (2012). Competing reinforcements: When languages opt out of Jespersen's cycle. In Ans van Kemenade and Nynke de Haas (eds.), *Historical linguistics 2009: Selected papers from the 19th International Conference on Historical Linguistics, Nijmegen, 10–14 August 2009*, 1–30. Amsterdam: John Benjamins.

Biberauer, Theresa and Ian Roberts (2011). Negative words and related expressions: A new perspective on some familiar puzzles. In Pierre Larrivée and Richard Ingham (eds.), *The evolution of negation: Beyond the Jespersen cycle*, 23–59. Berlin: Mouton de Gruyter.

Billings, Loren (1997). Negated prepositional phrases in Slavic. In Wayles Browne, Ewa Dornisch, Natasha Kondrashova, and Draga Zec (eds.), *Annual Workshop on Formal Approaches to Slavic Linguistics: The Cornell meeting 1995*, 115–34. Ann Arbor, Mich.: Michigan Slavic Publications.

Birjulin, L. A. and Victor S. Xrakovskij (2001). Imperative sentences: Theoretical problems. In Victor S. Xrakovskij (ed.), *Typology of imperative constructions*, 3–50. Munich: Lincom Europa.

Birner, Betty J. (2006). Semantic and pragmatic contributions to information status. *Acta Linguistica Hafniensia* 38, 14–32.

Bishai, Wilson (1962). Coptic grammatical influence on Egyptian Arabic. *Journal of the American Oriental Society* 82, 285–9.

Black, Jeremy, Andrew George, and Nicholas Postgate (2000). *A concise dictionary of Akkadian*. Wiesbaden: Harrassowitz.

Blanche-Benveniste, Claire and Colette Jeanjean (1987). *Le français parlé: transcription et édition*. Paris: Didier Érudition.

Błaszczak, Joanna (2001). *Investigation into the interaction between the indefinites and negation*. Berlin: Akademie-Verlag.

Błaszczak, Joanna (2002a). Towards a binding analysis of negative polarity items in Polish. In Peter Kosta and Jens Frasek (eds.), *Current approaches to formal Slavic linguistics*, 377–404. Frankfurt am Main: Peter Lang.

Błaszczak, Joanna (2002b). What do bagels and Polish *kolwiek*-pronouns have in common? In Mary Andronis, Erin Debenport, Anne Pycha, and Keiko Yoshimura (eds.), *Proceedings from the panels of the Thirty-eighth Meeting of the Chicago Linguistic Society: Volume 38-2*, 3–18. Chicago: Chicago Linguistic Society.

Błaszczak, Joanna (2008). What HAS to BE used? Existential, locative, and possessive sentences in Polish. In Andrei Antonenko, John F. Bailyn, and Christina Y. Bethin (eds.), *Annual Workshop on Formal Approaches to Slavic Linguistics: The Stony Brook meeting 2007*, 31–47. Ann Arbor, Mich.: Michigan Slavic Publications.

Bliese, Loren (1981). *A generative grammar of Afar*. Dallas, Texas: Summer Institute of Linguistics.

Boor, Helmut de and Roswitha Wisniewski (1998). *Mittelhochdeutsche Grammatik*, 10th edn. Berlin: de Gruyter.

Borg, Albert and Marie Azzopardi-Alexander (1997). *Maltese*. London: Routledge.

Borkovskij, Viktor Ivanovič (1963). Dva sintaksičeskix ètjuda. *Slavjanska Filologija* 3, 249–54.

Borkovskij, Viktor Ivanovič (1978). *Istoričeskaja grammatika russkogo jazyka: Sintaksis, Prostoe predloženie*. Moscow: Nauka.

Borkovskij, Viktor Ivanovič and Petr Savvič Kuznecov (2007 [1963]). *Istoričeskaja grammatika russkogo jazyka*, 4th edn. Moscow: Izdatel'stvo Akademii Nauk SSSR.

Borovikoff, Natalie L. (1997). Negated adjunct phrases are REALLY the genitive of negation. In Martina Lindseth and Steven Franks (eds.), *Formal Approaches to Slavic Linguistics: The Indiana meeting 1996*, 67–85. Ann Arbor, Michigan: Michigan Slavic Publications.

Borsley, Robert D. and Bob Morris Jones (2000). The syntax of Welsh negation. *Transactions of the Philological Society* 98, 15–47.

Borsley, Robert D. and Bob Morris Jones (2005). *Welsh negation and grammatical theory*. Cardiff: University of Wales Press.

Borsley, Robert D. and Mohamed Krer (2011). An HPSG approach to negation in Libyan Arabic. MS, University of Essex.

Borsley, Robert D. and Ian Roberts (eds.) (1996). *The syntax of the Celtic languages*. Cambridge: Cambridge University Press.

Borsley, Robert D. and Janig Stephens (1989). Agreement and the position of subjects in Breton. *Natural Language and Linguistic Theory* 7, 407–27.

Borsley, Robert D., Maggie Tallerman, and David Willis (2007). *The syntax of Welsh*. Cambridge: Cambridge University Press.

Boumalk, Abdallah (1996). La négation en berbère marocain. In Salem Chaker and Dominique Caubet (eds.), *La négation en berbère et en arabe maghrébin*, 35–48. Paris: L'Harmattan.

Bourciez, Édouard (1930). *Éléments de linguistique romane*. Paris: Klincksieck.

Braune, Wilhelm and Ingo Reiffenstein (2004). *Althochdeutsche Grammatik*, 15th edn., vol. i: *Laut- und Formenlehre*. Tübingen: Niemeyer.

Breitbarth, Anne (2008). The development of negation in Middle Low German. Paper presented at the Annual Meeting of the LAGB, University of Essex, Colchester.

Breitbarth, Anne (2009). A hybrid approach to Jespersen's cycle in West Germanic. *Journal of Comparative Germanic Linguistics* 12, 81–114.

Breitbarth, Anne (2013). Indefinites, negation and Jespersen's Cycle in the history of Low German. *Diachronica* 30, 171–201.

Breitbarth, Anne and Liliane Haegeman (2010). Continuity is change: The long tail of Jespersen's cycle in Flemish. In Anne Breitbarth, Christopher Lucas, Sheila Watts, and David Willis (eds.), *Continuity and change in grammar*, 61–76. Amsterdam: John Benjamins.

Breitbarth, Anne and Liliane Haegeman (forthcoming). *En* en is níet wat we dachten: A Flemish discourse particle. In Igor Yanovich and Natasha Korotkova (eds.), *Proceedings of MOSS 2*. Cambridge, Mass.: MIT Working Papers in Linguistics.

Breitbarth, Anne, Christopher Lucas, and David Willis (2013). Incipient Jespersen's cycle: The (non-)grammaticalization of new negative markers. In Jürg Fleischer and Horst J. Simon (eds.), *Comparing diachronies / Sprachwandelvergleich*, 143–64. Tübingen: Niemeyer.

Breitbarth, Anne, Christopher Lucas, and David Willis (forthcoming). *The history of negation in the languages of Europe and the Mediterranean*. Volume ii: *Patterns and processes*. Oxford: Oxford University Press.

Britain, David (2007). Grammatical variation in England. In David Britain (ed.), *Language in the British Isles*, 75–104. Cambridge: Cambridge University Press.

Brockett, A. A. (1985). *The spoken Arabic of Khābūra on the Bāṭina of Oman*. Manchester: University of Manchester.

Brown, Francis, S. R. Driver, and Charles Briggs (1999). *The Brown–Driver–Briggs Hebrew and English lexicon: With an appendix containing the Biblical Aramaic*. Peabody, Mass.: Hendrickson Publishers.

Brown, Sue (1999). *The syntax of negation in Russian: A minimalist approach*. Stanford, Calif.: CSLI.

Brown, Sue (2003). A minimalist approach to negation in Old Church Slavonic: A look at the *Codex Marianus*. In Wayles Browne, Ji-Yung Kim, Barbara H. Partee, and Robert A. Rothstein (eds.), *Annual Workshop on Formal Approaches to Slavic Linguistics: The Amherst meeting 2002*, 159–78. Ann Arbor, Mich.: Michigan Slavic Publications.

Brown, Sue and Steven Franks (1997). The syntax of pleonastic negation in Russian. In Wayles Browne, Ewa Dornisch, Natalia Kondrashova, and Draga Zec (eds.), *Annual Workshop on Formal Approaches to Slavic Linguistics: The Cornell meeting 1995*, 135–64. Ann Arbor, Mich.: Michigan Slavic Publications.

Brückner, Aleksander (1957 [1927]). *Słownik etymologiczny języka polskiego*. Warsaw: Wiedza Powszechna.

Brugnatelli, Vermondo (1987). La negazione discontinua in berbero e in arabo-magrebino. In Giuliano Bernini and Vermondo Brugnatelli (eds.), *Atti della 4a giornata di studi camito-semitici ed indoeuropei*, 53–62. Milan: Unicopli.

Brustad, Kristen (2000). *The syntax of spoken Arabic: A comparative study of Moroccan, Egyptian, Syrian and Kuwaiti dialects*. Washington, DC: Georgetown University Press.

Bubrix, D. V. (1953). *Istoričeskaja grammatika èrzjanskogo jazyka*. Saransk: Mordovskoe Knižnoe Izdatel'stvo.

Bulaxovskij, Leonid Arsen'evič (1954). *Russkij literaturnyj jazyk pervoj poloviny XIX veka*. Moscow: Gosudarstvennoe Učebno-pedagogičeskoe Izdatel'stvo Ministerstva Prosveščenija RSFSR.

Bulaxovskij, Leonid Arsen'evič (1958). *Istoričeskij kommentarij k russkomu literaturnomu jazyku*, 5th edn. Kiev: Gosudarstvennoe Učebno-pedagogičeskoe Izdatel'stvo 'Radjans'ka Škola'.

Buridant, Claude (2000). *Grammaire nouvelle de l'ancien français*. Paris: SEDES.

Büring, Daniel (1994). *Mittelfeldreport V*. In Brigitta Haftka (ed.), *Was determiniert Wortstellungsvariation? Studien zu einem Interaktionsfeld von Grammatik, Pragmatik und Sprachtypologie*, 79–96. Opladen: Westdeutscher Verlag.

Büring, Daniel and Katharina Hartmann (2001). The syntax and semantics of focus-sensitive particles in German. *Natural Language and Linguistic Theory* 19, 229–81.

Burridge, Kate (1993). *Syntactic change in Germanic: Aspects of language change in Germanic with particular reference to Middle Dutch*. Amsterdam: John Benjamins.

Buslaev, Fedor Ivanovič (1959 [1858]). *Istoričeskaja grammatika russkogo jazyka*, 5th edn. Moscow: Gosudarstvennoe Učebno-pedagogičeskoe Izdatel'stvo Ministerstva Prosveščenija RSFSR.

Butz, Beat (1981). *Morphosyntax der Mundart von Vermes (Val Terbi)*. Bern: Francke.

Bybee, Joan L. (2003). Mechanisms of change in grammaticalization: The role of frequency. In Richard D. Janda and Brian D. Joseph (eds.), *Handbook of historical linguistics*, 602–23. Oxford: Blackwell.

Bybee, Joan (2010). *Language, usage and cognition*. Cambridge: Cambridge University Press.

Cann, Ronnie, Ruth Kempson, and Lutz Marten (2005). *The dynamics of language: An introduction*. Amsterdam: Elsevier.

Catalani, Luigi (2001). *Die Negation im Mittelfranzösischen*. Frankfurt-am-Main: Peter Lang.

Caubet, Dominique (1996). La négation en arabe maghrébin. In Salem Chaker and Dominique Caubet (eds.), *La négation en berbère et en arabe maghrébin*, 79–97. Paris: L'Harmattan.

Chaker, Salem (1996). Remarques préliminaires sur la négation en berbère. In Salem Chaker and Dominique Caubet (eds.), *La négation en berbère et en arabe maghrébin*, 9–22. Paris: L'Harmattan.

Chanoch, Irene (1930). *Fremdsprachliche Einflüsse im modernen Hebräisch*. Heidelberg: Winter.

Chantraine, Pierre (1948). *Grammaire Homérique*. Paris: C. Klincksieck.

Cheshire, Jenny (1998). English negation from an interactional perspective. In Ingrid Tieken-Boon van Ostade, Gunnel Tottie, and Wim van der Wurff (eds.), *Negation in the history of English*, 29–53. Berlin: Mouton de Gruyter.

Chomsky, Noam (2001). Derivation by phase. In Michael Kenstowicz (ed.), *Ken Hale: A life in language*, 1–52. Cambridge, Mass.: MIT Press.

Christensen, Ken Ramshøj (2003). On the synchronic and diachronic status of the negative adverbial *ikke / not*. *Working Papers in Scandinavian Syntax* 72, 1–53.

Churchward, C. Maxwell (1953). *Tongan grammar*. London: Oxford University Press.

Cieślikowa, Aleksandra (1965). Partykuła *-kolwiek* w historii i dialektach języka polskiego. *Prace Językoznawcze* 15, 45–84.

Cindrić, Pavao (1955). Tkogod, kojigod ili tko god, koji god? *Jezik* 4, 32.

Cinque, Guglielmo (1976). *Mica*: Note di sintassi e pragmaticà. *Annali della Facoltà di Lettere e Filosofia dell'Università di Padova* 1, 101–12. Reprinted in Guglielmo Cinque, *Teoria linguistica e sintassi italiana*, 311–23. Bologna: Il Mulino, 1991.

Cinque, Guglielmo (1999). *Adverbs and functional heads: A cross-linguistic perspective*. New York: Oxford University Press.

Cinque, Guglielmo (2004). 'Restructuring' and functional structure. In Adriana Belletti (ed.), *Structures and beyond: The cartography of syntactic structures*, vol. iii, 132–91. Oxford: Oxford University Press.

Clackson, James (1994). *The linguistic relationship between Armenian and Greek*. Oxford: Blackwell.

Clauson, Gerard (1972). *An etymological dictionary of pre-thirteenth-century Turkish*. Oxford: Oxford University Press.

Clivio, Gianrenzo P. (1976). *Storia linguistica e dialettologia piemontese*. Turin: Ca dë Studi Piemontèis.

Collins Spanish Dictionary (2013). London: Harper Collins. Available online at <http://www.collinsdictionary.com/dictionary/english-spanish>.

Comrie, Bernard (1981). Negation and other verb categories in the Uralic languages. In Osmo Ikola (ed.), *Congressus Quintus Internationalis Fenno-Ugristarum*, vol. vi, 350–5. Turku: Suomen Kielen Seura.

Comrie, Bernard and Greville G. Corbett (eds.) (1993). *The Slavonic languages*. London: Routledge.

Cornillet, Gérard (2008). *Grammatik der bretonischen Gegenwartssprache*. Lannion: Skol Ober.

Corriente, Federico (1977). *A grammatical sketch of the Spanish Arabic dialect bundle*. Madrid: Instituto Hispano-Árabe de Cultura.

Coveney, Aidan (1996). *Variability in spoken French: A sociolinguistic study of negation and interrogation*. Exeter: Elm Bank Publications.

Cowell, Mark W. (1964). *A reference grammar of Syrian Arabic*. Washington, DC: Georgetown University Press.

Cowgill, Warren (1960). Greek *ou* and Armenian *oč*. *Language* 36, 347–50.

Croft, William (1991). The evolution of negation. *Journal of Linguistics* 27, 1–27.

Dahl, Östen (1979). Typology of sentence negation. *Linguistics* 17, 79–106.

Dahl, Östen (2001). Inflationary effects in language and elsewhere. In Joan Bybee and Paul Hopper (eds.), *Frequency and the emergence of linguistic structure*, 471–80. Amsterdam: John Benjamins.

Dal, Ingerid (1966). *Kurze deutsche Syntax auf historischer Grundlage*, 3rd edn. Tübingen: Niemeyer.

Daničić, Djuro (1880–1975). *Rječnik hrvatskoga ili srpskoga jezika*. Zagreb: Tisak Dioničke Tiskare.

Daničić, Djuro (1962 [1863–4]). *Rječnik iz književnih starina srpskih*. Graz [Belgrade]: Akademische Druck- und Verlagsanstalt [U Državnoj Štampariji].

Déprez, Viviane (2000). Parallel (a)symmetries and the internal structure of negative expressions. *Natural Language and Linguistic Theory* 18, 253–342.

Déprez, Viviane and France Martineau (2004). Micro-parametric variation and negative concord. In Julie Auger, Clancy Clements, and Barbara Vance (eds.), *Contemporary approaches to Romance linguistics*, 139–58. Amsterdam: John Benjamins

Detges, Ulrich (2003). La grammaticalisation des constructions de négation dans une perspective onomasiologique, ou: la déconstruction d'une illusion d'optique. In Peter Koch and Andreas Blank (eds.), *Kognitive romanische Onomasiologie und Semasiologie*, 213–33. Tübingen: Niemeyer.

Detges, Ulrich and Richard Waltereit (2002). Grammaticalization vs. reanalysis: A semantic-pragmatic account of functional change in grammar. *Zeitschrift für Sprachwissenschaft* 21, 151–95.

Devos, Maud and Johan van der Auwera (forthcoming). Jespersen cycles in Bantu: Double and triple negation.

Devos, Maud, Michael Kasombo Tshibanda, and Johan van der Auwera (2010). Jespersen cycles in Kanincin: Double, triple and maybe even quadruple negation. *Africana Linguistica* 16, 5–31.

Diakonoff, Igor (1996). Some reflections on the Afrasian linguistic macrofamily. *Journal of Near Eastern Studies* 55, 293–4.

Dictionary of the Irish language (2007). Edited by Gregory Toner, Maxim Fomin, Thomas Torma, and Grigory Bondarenko, electronic edn. Dublin/Belfast: Royal Irish Academy/University of Ulster. Available online at <http://www.dil.ie>.

Donhauser, Karin (1996). Negationssyntax in der deutschen Sprachgeschichte: Grammatikalisierung oder Degrammatikalisierung? In Ewald Lang and Gisela Zifonun (eds.), *Deutsch-typologisch*, 201–17. Berlin: de Gruyter.

Donhauser, Karin (1998). Negationssyntax im Althochdeutschen. Ein sprachhistorisches Rätsel und der Weg zu seiner Lösung. In Karin Donhauser and Ludwig M. Eichinger (eds.), *Deutsche Grammatik. Thema in Variationen. Festschrift für Hans-Werner Eroms*, 283–98. Heidelberg: Winter.

Dostie, Gaétane (2004). Pragmaticalisation et marqueurs discursifs: analyse sémantique et traitement lexicographique. Brussels: De Boeck/Duculot.

Dryer, Matthew S. (1988). Universals of negative position. In Michael Hammond, Edith A. Moravcsik, and Jessica Wirth (eds.), *Studies in syntactic typology*, 93–124. Amsterdam: John Benjamins.

Dryer, Matthew S. (2011). Negative morphemes. In Matthew S. Dryer and Martin Haspelmath (eds.), *The world atlas of language structures online*, <http://wals.info/chapter/112>. Munich: Max Planck Digital Library.

Durand, Olivier (2004). *L'arabo del marocco: Elementi di dialetto standard e mediano*. Rome: Università degli Studi La Sapienza.

Ebert, Robert Peter, Oskar Reichmann, Hans-Joachim Solms, and Klaus-Peter Wegera (1993). *Frühneuhochdeutsche Grammatik*. Tübingen: Niemeyer.

Eckardt, Regine (2003). The structure of change: Meaning change under reanalysis. Habilitation thesis, Humboldt Universität, Berlin.

Eckardt, Regine (2006). *Meaning change in grammaticalization*. Oxford: Oxford University Press.

Ehala, Martin (1996). Self-organisation and language change. *Diachronica* 13, 1–28.

Ehret, Christopher (1995). *Reconstructing Proto-Afro-Asiatic: Vowels, tone, consonants and vocabulary*. Berkeley, Calif.: University of California Press.

Elhalimi, Brahim (1996). La négation dans le parler arabe de mazouna (ouest algérien). In Salem Chaker and Dominique Caubet (eds.), *La négation en berbère et en arabe maghrébin*, 135–62. Paris: L'Harmattan.

Ellegård, Alvar (1953). *The auxiliary* do: *The establishment and regulation of its use in English*. Stockholm: Almqvist and Wiksell.

Emonds, Joseph (1987). The invisible category principle. *Linguistic Inquiry* 18, 613–32.

Emonds, Joseph (2000). *Lexicon and grammar: The English syntacticon*. Berlin: Mouton de Gruyter.

Erelt, Mati (2009). Typological overview of Estonian syntax. *Sprachtypologie und Universalienforschung: Language Typology and Universals* 62, 6–28.

Erman, Britt and Ulla-Britt Kotsinas (1993). Pragmaticalization: The case of *ba'* and *you know*. *Studier i Modern Språkvetenskap* 10, 76–93.

Eŕźań keľ (2000) = Cygankin, D. V., N. A. Agafonova, M. D. Imajkina, M. V. Mosin, V. P. Cypkajkina, and E. A. Abramova (eds.) (2000). *Eŕźań keľ. Morfologija*. Saransk: Krasnyj Oktjabr'.

Espinal, M. Teresa (1993). The intepretation of *no-pas* in Catalan. *Journal of Pragmatics* 19, 353–69.

Espinal, M. Teresa (2000). On the semantic status of n-words in Catalan and Spanish. *Lingua* 110, 557–80.

Evans, D. Simon (1964). *Grammar of Middle Welsh*. Dublin: Dublin Institute for Advanced Studies.

Evsev'ev, M. E. (1931). *Osnovy mordovskoj grammatiki. Eŕźań gramatika*. Moscow: Centrizdat.

Evsev'ev, M. E. (1963). *Izbrannye trudy. Tom četvertyj. Osnovy mordovskoj grammatiki*. Saransk: Mordovskoe knižnoe izdatel'stvo.

Eythórsson, Thórhallur (1995). *Functional categories, cliticization, and verb movement in the early Germanic languages*. Doctoral dissertation, Cornell University.

Eythórsson, Thórhallur (2002). Negation in C: The syntax of negated verbs in Old Norse. *Nordic Journal of Linguistics* 25, 190–224.

Faber, Alice (1988). Indefinite pronouns in early Semitic. In Yoel Arbeitman (ed.), *Fucus: A Semitic/Afrasian gathering in memory of Albert Ehrmann*, 221–38. Amsterdam: John Benjamins.

Faber, Alice (1991). The diachronic relationship between negative and interrogative markers in Semitic. In Alan S. Kaye (ed.), *Semitic studies in honor of Wolf Leslau*, 411–29. Wiesbaden: Otto Harrassowtiz.

Faber, Alice (1997). Genetic subgrouping of the Semitic languages. In Robert Hetzron (ed.), *The Semitic languages*, 3–15. London: Routledge.

Faggin, Giorgio (1997). *Grammatica friulana*. Campoformido: Ribis.

Falaus, Anamaria (2007). Le paradoxe de la double négation dans une langue à concordance négative stricte. In Franck Floricic (ed.), *La négation dans les langues romanes*, 75–97. Amsterdam: John Benjamins.

Falileyev, Alexander (2000). *Etymological glossary of Old Welsh*. Tübingen: Max Niemeyer Verlag.

Faßke, Helmut and Siegfried Michalk (1981). *Grammatik der obersorbischen Schriftsprache der Gegenwart: Morphologie*. Bautzen: VEB Domowina Verlag.

Fauconnier, Gilles (1975). Polarity and the scale principle. *Chicago Linguistic Society* 11, 188–99.

Favereau, Francis (1997). *Grammaire du breton contemporain. Yezhadur ar brezhoneg a-vremañ*. Morlaix: Skol Vreizh.

Federal'naja Služba Gosudarstvennoj Statistiki [Federal State Statistics Service] (2010). Vserossijskaja perepis'naselenija 2010 goda [All-Russian Census 2010]: Nacional'nyj sostav naselenija. Available at <http://www.perepis-2010.ru/results_of_the_census/tab5.xls>.

Fernández, Miguel Perez (1999). *An introductory grammar of Rabbinic Hebrew*. Leiden: Brill.

Fitzgibbons, Natalia (2010). Freestanding n-words in Russian: A syntactic account. *Journal of Slavic Linguistics* 18, 55–99.

Fleischman, Suzanne (1982). The future in thought and language: Diachronic evidence from Romance. Cambridge: Cambridge University Press.

Fleuriot, Léon (1964a). *Dictionnaire des gloses en vieux breton*. Paris: C. Klincksleck.

Fleuriot, Léon (1964b). *Le vieux breton: éléments d'une grammaire*. Paris: Librairie C. Klincksieck.

Fonseca-Greber, Bonnibeth Beale (2007). The emergence of emphatic 'ne' in conversational Swiss French. *French Language Studies* 17, 249–75.

Forner, Werner (1998). La 'coniugazione interrogativa' nei dialetti liguri. In Giovanni Ruffino (ed.), *Atti del XXI CILPR, Università di Palermo 18–24 settembre 1995*, 319–36. Tübingen: Niemeyer.

Foulet, Lucien (1919). Études de syntaxe française: I Quelque. *Romania* 45, 220–49.

Foulet, Lucien (1965). *Petite syntaxe de l'ancien français*. Paris: Honoré Champion.

Franks, Steven (1995). *Parameters of Slavic morphosyntax*. New York: Oxford University Press.

Franks, Steven and Katarzyna Dziwirek (1993). Negated adjunct phrases are really partitive. *Journal of Slavic Linguistics* 1, 280–305.

Frisch, Stefan (1997). The change in negation in Middle English: A NegP licensing account. *Lingua* 101, 21–64.

Fynes-Clinton, O. H. (1913). *The Welsh vocabulary of the Bangor district*. Oxford: Oxford University Press.

Gaatone, David (1971). *Étude descriptive du système de la négation en français contemporain*. Geneva: Droz.

Gardiner, Alan H. (1904). The word iwn3. *Zeitschrift für Ägyptische Sprache und Altertumskunde* 41, 130–5.

Gardiner, Alan H. (1957). *Egyptian grammar: Being an introduction to the study of hieroglyphs*. London: Oxford University Press.

Garzonio, Jacopo (2008). Diacronia e sincronia della sintassi di *punto* in Fiorentino. In Diego Pescarini and Federica Cognola (eds.), *La negazione: variazione dialettale ed evoluzione diacronica, Quaderni di Lavoro ASIt* 8, 21–42. Available online at <http://asis-cnr.unipd.it/documenti/ql8/garzonio_2008.pdf>.

Garzonio, Jacopo and Cecilia Poletto (2009). Quantifiers as negative markers in Italian dialects. In Jeroen van Cranenbroek and Johan Rooryck (eds.), *Yearbook of Linguistic Variation*, 127–52. Amsterdam: John Benjamins.

Gasca Queirazza, Giuliano (1996). Un'ipotesi sulla localizzazione dei Sermoni subalpini. *Studi Piemontesi* 25, 105–10.

Gebauer, Jan (1929). *Historická mluvnice jazyka českého: Skladba*, vol. iv. Prague: Česká Akademie Věd i Umění.

Gelderen, Elly van (2004). Economy, innovation and prescriptivism: from Spec to head and head to head. *Journal of Comparative Germanic Linguistics* 7, 59–98.

Gelderen, Elly van (2008). Negative cycles. *Linguistic Typology* 12, 195–243.

Gelderen, Elly van (ed.) (2009). *Cyclical change*. Amsterdam: John Benjamins.

Gelderen, Elly van (2011). *The linguistic cycle: Language change and the language faculty*. Oxford: Oxford University Press.

Gerken, Lou Ann (1994). A metrical template account of children's weak syllable omissions from multisyllabic words. *Journal of Child Language* 21, 565–84.

Gheno, Danilo (1994–5). Die Verneinung im Mordwinischen. *Nyelvtudományi Közlemények* 94, 133–6.

Giannakidou, Anastasia (1998). *Polarity sensitivity as (non)veridical dependency*. Amsterdam: John Benjamins.

Giannakidou, Anastasia (2000). Negative... concord? *Natural Language and Linguistic Theory* 18, 457–523.

Giannakidou, Anastasia (2001). The meaning of free choice. *Linguistics and Philosophy* 24, 659–735.

Giannakidou, Anastasia (2006). N-words and negative concord. In Martin Everaert, Henk van Riemsdijk, Rob Goedemans, and Bart Hollebrandse (eds.), *The Blackwell companion to syntax*, vol. iii, 327–91. Oxford: Blackwell.

Gildersleeve, Basil Lanneau (1880). Encroachments of *mē* on *ou* in Later Greek. *American Journal of Philology* 1, 45–57.

Givón, Talmy (1978). Negation in language: Pragmatics, function, ontology. In Peter Cole (ed.), *Pragmatics*, 69–112. New York: Academic Press.

Givón, Talmy (2001). *Syntax: An introduction*. Amsterdam: John Benjamins.

Glinert, Lewis H. (1982). Negative and non-assertive in contemporary Hebrew. *Bulletin of the School of Oriental and African Studies* 45, 434–70.

Goodwin, William Watson (1889). *Syntax of the moods and tenses of the Greek verb*. London: Macmillan.

Gragg, Gene (1997). Ge'ez. In Robert Hetzron (ed.), *The Semitic languages*, 242–60. London: Routledge.

Grammatika (1962) = Koljadenkov, M. N. and R. A. Zavodova (eds.) (1962). *Grammatika mordovskix jazykov: Fonetika i morfologija*. Saransk: Mordovskoe knižnoe izdatel'stvo.

Grammatika (1980) = Cygankin, D. V. (ed.) (1980). *Grammatika mordovskix jazykov*. Saransk: Mordovskij gosudarstvennyj universitet imeni N. P. Orgaeva.

Granda, Rosa María Medina (1999). *Polaridad negativa en occitano antiguo: Elementos de comparación con otros romances medievales*. Oviedo: Universidad de Oviedo.

Greenberg, Joseph H. (1963). *The languages of Africa*. Bloomington: Indiana University Press.

Greenberg, Joseph H. (1987). *Language in the Americas*. Stanford, Calif.: Stanford University Press.

Greszczuk, Barbara (1993). *Składniowe wykładniki negacji i ich funkcje w historii języka polskiego*. Rzeszów: Wydawnictwo Wyższej Szkoły Pedagogicznej w Rzeszowie.

Grimm, Jacob (1890). *Deutsche Grammatik*, vol. iii. Ed. G. Roethe and E. Schröder. Gütersloh: Bertelsmann.

Grimm, Jacob and Wilhelm Grimm (1877). *Deutsches Wörterbuch*, vol. x. Reprinted 1984. Munich: Deutscher Taschenbuchverlag.

Grünthal, Riho (2007). The Mordvinic languages between bush and tree: A historical reappraisal. In Jussi Ylikoski and Ante Aikio (eds.), *Sámit, sánit, sátnehámit. Riepmočála Pekka Sammallahtii miessemánu 21. beaivve*, 115–35. Helsinki: Suomalais-Ugrilainen Seura.

Guiraud, Pierre (1964). L'opposition actuel-virtuel: remarques sur l'adverbe de négation dans *Aucassin et Nicolette*. In Jean Renson, Jules Horrent, Louis Remacle, and Madeleine Tyssens (eds.), *Mélanges de linguistique romane et philologie médiévale offerts à M. Maurice Delbouille*, vol. i, 295–306. Gembloux: Duculot.

Guiraud-Weber, Marguerite (2003). Ešče raz o russkom genitive otricanija: Vzgljad so storony. *Russian Linguistics* 27, 363–84.

Güldemann, Tom (2011). Proto-Bantu and Proto-Niger-Congo: Macro-areal typology and linguistic reconstruction. In Osamu Hieda, Christa König, and Hirosi Nakagawa (eds.), *Geographical typology and linguistic areas*, 109–42. Amsterdam: John Benjamins.

Haan, Ger J. de and Fred Weerman (1984). Taaltypologie, taalverandering en mogelijke grammatica's: Het Middelnederlandse *en*. *Tijdschrift voor Nederlandse Taal- en Letterkunde* 100, 161–90.

Haeberli, Eric (2002). Inflectional morphology and the loss of V2 in English. In David Lightfoot (ed.), *Syntactic effects of morphological change*, 88–106. Oxford: Oxford University Press.

Haeberli, Eric and Richard Ingham (2007). Negation and adverbs in Middle English. *Lingua* 117, 1–25.

Haegeman, Liliane (1995). *The syntax of negation*. Cambridge: Cambridge University Press.

Haegeman, Liliane (2002). West Flemish negation and the derivation of SOV order in West Germanic. *Nordic Journal of Linguistics* 25, 154–89.

Haegeman, Liliane and Terje Lohndal (2010). Negative concord and multiple agree: A case study of West Flemish. *Linguistic Inquiry* 41, 181–211.

Haegeman, Liliane and Rafaella Zanuttini. (1991). Negative Heads and the Neg-Criterion. *Linguistic Review* 8, 233–51.

Haider, Hubert (1997). Projective Economy. On the minimal functional structure of the German clause. In Werner Abraham and Elly van Gelderen (eds.), *German: Syntactic problems—problematic syntax*, 83–103. Tübingen: Niemeyer.

Haiman, John (1980). The iconicity of grammar: Isomorphism and motivation. *Language* 56, 565–89.

Hajdú, Péter (1988). Die Samojedischen Sprachen. In Denis Sinor (ed.), *The Uralic languages: Description, history and foreign influence*, 3–40. Leiden: E. J. Brill.

Hakulinen, Lauri (1979). *Suomen kielen rakenne ja kehitys*, 4th edn. Helsinki: Otava.

Hall, Joseph (ed.) (1920). *Selections from Early Middle English, 1130–1250*. 2 vols. Oxford: Clarendon Press.

Hamari, Arja (2007). *The negation of stative relation clauses in the Mordvin languages*. Helsinki: Suomalais-Ugrilainen Seura.

Hamari, Arja (2010). The negation of stative relation clauses in the Mordvin, Mari and Permic languages. Paper presented at 11th International Congress for Fenno-Ugric Studies, Piliscsaba, Hungary.

Hamari, Arja (2011). The negation of stative relation clauses in the Mordvin, Mari and Permic languages. In Sándor Csúcs, Nóra Falk, Viktória Tóth, and Gábor Zaicz (eds.), *Congressus XI Internationalis Fenno-ugristarum, Piliscsaba, 9–14. VIII. 2010. Pars VI: Dissertationes symposiorum ad linguisticam*, 250–7. Piliscsaba: Reguly Társaság.

Hammond, Lila (2005). *Serbian: An essential grammar*. London: Routledge.

Han, Chung-hye (2001). Force, negation and imperatives. *Linguistic Review* 18, 289–325.

Hansen, Anita Berit and Isabelle Malderez (2004). Le *ne* de négation en région parisienne: une étude en temps réel. *Langage et Société* 107, 5–30.

Hansen, Björn (2004). The grammaticalization of the analytical imperatives in Russian, Polish and Serbian/Croatian. *Die Welt der Slaven* 49, 257–74.

Hansen, Maj-Britt Mosegaard (2008). *Particles at the semantics/pragmatics interface: Synchronic and diachronic issues. A study with special reference to the French phasal adverbs.* Oxford: Elsevier/Emerald.

Hansen, Maj-Britt Mosegaard (2009). The grammaticalization of negative reinforcers in Old and Middle French: A discourse-functional approach. In Maj-Britt Mosegaard Hansen and Jacqueline Visconti (eds.), *Current trends in diachronic semantics and pragmatics*, 227–51. Bingley: Emerald.

Hansen, Maj-Britt Mosegaard (2011). A pragmatic approach to historical semantics, with special reference to markers of clausal negation in Medieval French. In Kathryn Allan and Justyna Robinson (eds.), *Current methods in historical semantics*, 233-57. Berlin: De Gruyter Mouton.

Hansen, Maj-Britt Mosegaard (2012). On the evolution of temporal n-words in medieval French. *Language Sciences* 34, 76–91.

Hansen, Maj-Britt Mosegaard and Jacqueline Visconti (2009). On the diachrony of 'reinforced' negation in French and Italian. In Corinne Rossari, Claudia Ricci, and Adriana Spiridon (eds.), *Grammaticalization and pragmatics: Facts, approaches, theoretical issues*, 137–71. Bingley: Emerald.

Härd, John Evert (1973). Mittelniederdeutsch. In Hans Peter Althaus, Helmut Henne, and Herbert Ernst Wiegand (eds.), *Lexikon der Germanistischen Linguistik*, 418–21. Tübingen: Niemeyer.

Härd, John Evert (2000). Syntax des Mittelniederdeutschen. In Besch et al. (2000), 1456–63.

Harrel, Richard S. (ed.) (1966). *A dictionary of Moroccan Arabic.* Washington DC: Georgetown University Press.

Harrer-Pisarkowa, Krystyna (1959). Przypadek dopełnienia w polskim zdaniu zaprzeczonym. *Język Polski* 39, 9–32.

Harris, Martin B. (1978). *The evolution of French syntax: A comparative approach.* London: Longman.

Hartmann, Jutta M. and Nataša Milićević (2008). The syntax of existential sentences in Serbian. In Andrei Antonenko, John F. Bailyn, and Christina Y. Bethin (eds.), *Annual Workshop on Formal Approaches to Slavic Linguistics: The Stony Brook meeting 2007*, 168–84. Ann Arbor, Mich.: Michigan Slavic Publications.

Harves, Stephanie (1998). The syntax of negated prepositional phrases in Slavic. In Željko Bošković, Steven Franks, and William Snyder (eds.), *Formal Approaches to Slavic Linguistics: The Connecticut meeting 1997*, 166–86. Ann Arbor, Mich.: Michigan Slavic Publications.

Harves, Stephanie (2002). *Unaccusative syntax in Russian.* Doctoral dissertation, Princeton University.

Haspelmath, Martin (1991). Zur Grammatikalisierung von Indefinitpronomina. In Norbert Boretzky, Werner Enninger, Benedikt Jeßing, and Thomas Stolz (eds.), *Sprachwandel und seine Prinzipien*, 103–25. Bochum: Brockmeyer.

Haspelmath, Martin (1997). *Indefinite pronouns.* Oxford: Oxford University Press.

Haspelmath, Martin (1999). Why is grammaticalization irreversible? *Linguistics* 37, 1043–68.

Haspelmath, Martin (2011). Negative indefinite pronouns and predicate negation. In Matthew S. Dryer and Martin Haspelmath (eds.), *The world atlas of language structures online*. Munich: Max Planck Digital Library, feature 115A. Available online at <http://wals.info/feature/115A>. Electronic version of Haspelmath, Martin. 2005. Negative indefinite pronouns and predicate negation. In Martin Haspelmath, Matthew Dryer, David Gil, and Bernard Comrie (eds.), *World atlas of language structures*, 466–70. Oxford: Oxford University Press.

Haspelmath, Martin and Josephine Caruana (1996). Indefinite pronouns in Maltese. *Rivista di Linguistica* 8, 213–27.

Hausenblas, Karel (1958). *Vývoj předmětového genitivu v češtině*. Prague: Nakladatelství Československé Akademie Věd.

Heath, Jeffrey (2002). *Jewish and Muslim dialects of Moroccan Arabic*. London: Routledge.

Heikkilä, Kaino (1968). Itämerensuomen *-ek* johtimella vastine mordvassa? In Paavo Ravila (ed.), *Congressus Secundus Internationalis Fenno-Ugristarum. Pars I: Acta Linguistica*, 178–87. Helsinki: Suomalais-Ugrilainen Seura.

Heine, Bernd and Tania Kuteva (2005). *Language contact and grammatical change*. Cambridge: Cambridge University Press.

Helten, W. L. van (1885a). Over eenige constructies, met *en*, *ne*, *of en* en *of*. *Tijdschrift voor Nederlandsche Taal- en Letterkunde* 5, 220–33.

Helten, W. L. van (1885b). Over enkele eigenaardigheden in het gebruik van *nie*, *noit*, *nemmer*, *ne* of *en*. *Tijdschrift voor Nederlandsche Taal- en Letterkunde* 5, 235–44.

Hemon, Roparz (1975). *A historical morphology and syntax of Breton*. Dublin: Institute for Advanced Studies.

Hendrick, Randall (2011). Some Breton indefinites. In Andrew Carnie (ed.), *Formal approaches to Celtic linguistics*, 95–112. Newcastle: Cambridge Scholars.

Herburger, Elena (2001). The negative concord puzzle revisited. *Natural Language Semantics* 9, 289–333.

Herrity, Peter (2000). *Slovene: A comprehensive grammar*. London: Routledge.

Hetzron, Robert (1972). *Ethiopian Semitic: Studies in classification*. Manchester: Manchester University Press.

Hetzron, Robert (1997). Outer South Ethiopic. In Robert Hetzron (ed.), *The Semitic languages*, 535–49. London: Routledge.

Hill, Steven P. (1977). *The n-factor and Russian prepositions: Their development in 11th–20th century texts*. The Hague: Mouton.

Hoeksema, Jack (1994). On the grammaticalization of negative polarity items. In Susanne Gahl, Andy Dolbey, and Christopher Johnson (eds.), *Proceedings of the Twentieth Annual Meeting of the Berkeley Linguistics Society*, 273–82. Berkeley, Calif.: Berkeley Linguistics Society.

Hoeksema, Jack (1997). Negation and negative concord in Middle Dutch. In Danielle Forget, Paul Hirschbühler, France Martineau, and María Luisa Rivero (eds.), *Negation and polarity: Syntax and semantics*, 139–58. Amsterdam: John Benjamins.

Hoeksema, Jack (1998). On the (non)loss of the polarity sensitive Dutch *ooit*. In Richard M. Hogg and Linda van Bergen (eds.), *Historical linguistics 1995: Selected papers from the 13th International Conference on Historical Linguistics, Manchester, August 1995. Volume 2: Germanic linguistics*, 101–14. Amsterdam: John Benjamins.

Hoeksema, Jack (1999). Aantekeningen bij *ooit*, deel 2: De opkomst van niet-polair *ooit*. *Tabu* 29, 147–72.

Hoeksema, Jack (2007a). Dutch *enig*: From nonveridicality to downward entailment. In Hedde Zeijlstra and Jan-Philipp Soehn (eds.), *Proceedings of the Workshop on Negation and Polarity*, 8–15. University of Tübingen: Collaborative Research Center 441.

Hoeksema, Jack (2007b). Parasitic licensing of negative polarity items. *Journal of Comparative Germanic Literature* 10, 163–82.

Hoeksema, Jack (2009). Jespersen recycled. In Elly van Gelderen (ed.), *Cyclical change*, 15–34. Amsterdam: John Benjamins.

Holton, David, Peter Mackridge, and Irene Philippaki-Warburton (1997). *Greek: A comprehensive grammar of the modern language*. New York: Abingdon.

Honti, László (1997). Die Negation im Uralischen. *Linguistica Uralica* 33, 81–96, 161–76, 241–52.

Hopper, Paul J. (1991). On some principles of grammaticization. In Elizabeth Closs Traugott and Bernd Heine (eds.), *Approaches to grammaticalization*, vol. i, 17–35. Amsterdam: John Benjamins.

Hopper, Paul J. and Elizabeth Closs Traugott (1993). *Grammaticalization*. Cambridge: Cambridge University Press.

Horn, Laurence R. (1989). *A natural history of negation*. Chicago: University of Chicago Press.

Horrocks, Geoffrey (1997). *Greek: A history of the language and its speakers*. London: John Wiley and Sons.

Horst, Joop M. van der and Marijke J. van der Wal (1978). Negatieverschijnselen en woordvolgorde in de geschiedenis van het Nederlands. *Tijdschrift voor Nederlandse Taal- en Letterkunde* 95, 6–37.

Hoyt, Frederick (2005). Negative concord in two dialects of Arabic. MS, University of Texas at Austin.

Hoyt, Frederick (2006). An Arabic Wackernagel clitic? The morphosyntax of negation in Palestinian Arabic. In Mustafa Mughazy (ed.), *Perspectives on Arabic linguistics XX*, 105–34. Amsterdam: John Benjamins.

Hoyt, Frederick (2010). Negative concord in Levantine Arabic. Doctoral dissertation, University of Texas at Austin.

Hryščenko, A. P., H. P. Arpolenko, V. V. Nimčuk, V. M. Rusanivs'kyj, and H. X. Ščerbatjuk (1983). *Istorija ukraïns'koï movy: Syntaksys*. Kiev: Akademija Nauk Ukraïns'koï RSR, Instytut Movoznavstva im. O. O. Potebni.

Hudson, Grover (1997). Amharic and Argobba. In Robert Hetzron (ed.), *The Semitic languages*, 457–85. London: Routledge.

Iaaich, Jamal (1996). La négation en hassaniyya de tan-tan (maroc). In Salem Chaker and Dominique Caubet (eds.), *La négation en berbère et en arabe maghrébin*. Paris: L'Harmattan.

Iliescu, Maria (2011). Pour une romanistique panromane: coup d'oeil sur l'étymologie de l'italien *niente* «rien». In Anja Overbeck, Wolfgang Schweickard, and Harald Völker (eds.). *Alba pratalia araba ... Romanistische Studien zum 65. Geburtstag von Günter Holtus*, Berlin: de Gruyter.

Ingham, Richard (2000). Negation and OV order in late Middle English. *Journal of Linguistics* 36, 13–38.

Ingham, Richard (2002). Negated subjects and objects in 15th-century nonliterary English. *Language Variation and Change* 14, 291–322.

Ingham, Richard (2003). The development of Middle English expletive negative sentences. *Transactions of the Philological Society* 101, 411–52.

Ingham, Richard (2005a). NegV1 and secondary negation in Old and Middle English religious prose. *York Papers in Linguistics* (Series 2) 5, 29–49.

Ingham, Richard (2005b). The loss of Neg V → C in Middle English. *Linguistische Berichte* 202, 171–206.

Ingham, Richard (2006a). Negative concord and the loss of the negative particle *ne* in late Middle English. *Studia Anglica Posnaniensia* 42, 77–97.

Ingham, Richard (2006b). On two negative concord dialects in early English. *Language Variation and Change* 18, 241–66.

Ingham, Richard (2007). NegP and negated constituent movement in the history of English. *Transactions of the Philological Society* 105, 1–33.

Ingham, Richard (2009). Negative co-ordination in the history of English. In Anne Breitbarth, Christopher Lucas, Sheila Watts, and David Willis (eds.), *Continuity and change in grammar*, 181–200. Amsterdam: John Benjamins.

Ingham, Richard (2011). Grammar change in Anglo-Norman and Continental French: The replacement of non-affirmative indefinite *nul* by *aucun*. *Diachronica* 28, 441–67.

Israel, Michael (1998). The rhetoric of grammar: Scalar reasoning and polarity sensitivity. Doctoral dissertation, University of California, San Diego.

Israel, Michael (2001). Minimizers, maximizers and the rhetoric of scalar reasoning. *Journal of Semantics* 18, 297–331.

Itkonen, Erkki (1954). Zur Geschichte des Vokalismus der ersten Silbe im Tscheremissischen und in den permischen Sprachen. *Finnisch-Ugrische Forschungen* 31, 149–345.

Ivanova-Mirčeva, Dora (ed.) (1999). *Starobălgarski rečnik*. Sofia: Izdatelstvo Valentin Trajanov.

Ivanova-Mirčeva, Dora and Ivan Xaralampiev (1999). *Istorija na bălgarskija ezik*. Veliko Tărnovo: Faber.

Iyeiri, Yoko (2001). *Negative constructions in Middle English*. Fukuoka: Kyushu University Press.

Iyeiri, Yoko (2002a). Development of *any* from Middle English to Early Modern English: A study using the Helsinki Corpus of English Texts. In Toshio Saito, Junsaku Nakamura, and Shunji Yamazaki (eds.), *English corpus linguistics in Japan*, 211–23. Amsterdam: Rodopi.

Iyeiri, Yoko (2002b). Development of *any* in later Middle English and the decline of multiple negation. In Yoko Iyeiri and Margaret Connolly (eds.), And gladly wolde he lerne and gladly teche: *Essays on medieval English presented to Professor Matsuji Tajima on his sixtieth birthday*, 127–43. Tokyo: Kaibunsha.

Iyeiri, Yoko (ed.) (2005). *Aspects of English negation*. Amsterdam: John Benjamins.

Jaberg, Karl and Jakob Jud (1928–40). *Sprach- und Sachatlas Italiens und der Südschweiz*. Zofingen: Ringier.

Jack, George B. (1978a). Negation in later Middle English prose. *Archivum Linguisticum* 9, 58–72.

Jack, George B. (1978b). Negative adverbs in Early Middle English. *English Studies* 59, 295–309.

Jacobs, Joachim (1980). Lexical decomposition in Montague Grammar. *Theoretical Linguistics* 7, 121–36.

Jacobs, Joachim (1982). *Syntax und Semantik der Negation im Deutschen*. München: Fink.

Jacobs, Joachim (1983). *Fokus und Skalen: Zur Syntax und Semantik der Gradpartikeln im Deutschen*. Tübingen: Niemeyer.

Jacobs, Joachim (1991). Negation. In Arnim von Stechow and Dieter Wunderlich (eds.), *Semantik: Ein internationales Handbuch der zeitgenössischen Forschung*, 560–96. Berlin: Mouton de Gruyter.

Jacobsson, Bengt (1951). *Inversion in English with particular reference to the Early Modern English period*. Uppsala: Almqvist and Wiksell.

Jäger, Agnes (2005). Negation in Old High German. *Zeitschrift für Sprachwissenschaft* 24, 227–62.

Jäger, Agnes (2007). 'No' changes: On the history of German indefinite determiners in the scope of negation. In Elisabeth Stark, Elisabeth Leiss, and Werner Abraham (eds.), *Nominal determination: Typology, context constraints, and historical emergence*, 141–70. Amsterdam: John Benjamins.

Jäger, Agnes (2008). *History of German negation*. Amsterdam: John Benjamins.

Jäger, Agnes (2010). *Anything* is *nothing* is *something*: On the diachrony of polarity types of indefinites. *Natural Language and Linguistic Theory* 28, 787–822.

Jäger, Agnes and Doris Penka (2012). Development of sentential negation in German. In Peter Ackema, Rhona Alcorn, Caroline Heycock, Dany Jaspers, Jeroen Van Craenenbroeck, and Guido Vanden Wyngaerd (eds.), *Comparative Germanic syntax: The State of the art*, 199–222. Amsterdam: John Benjamins.

Janda, Richard D. and Brian D. Joseph (1999). The Modern Greek negator *mi(n)* as a morphological constellation. In G. Babiniotis (ed.), *Greek linguistics: Proceedings of the 3rd International Conference on Greek Linguistics*, 341–51. Athens: Elinika Gramata.

Janhunen, Juha (1982). On the structure of Proto-Uralic. *Finnisch-Ugrische Forschungen* 44, 23–42.

Jannaris, Antonius Nicholas (1897). *An historical Greek grammar*. London: Macmillan.

Jespersen, Otto (1917). *Negation in English and other languages*. Copenhagen: A. F. Høst og Søn.

Jouitteau, Mélanie (2005). *La syntaxe comparée du breton*. Doctoral dissertation, Université de Nantes.

Jones, Michael A. (1993). *Sardinian syntax*. London: Routledge.

Jordal, Knud (1973). Greko-russkie sintaksičeskie svjazi. *Scando-Slavica* 19, 143–64.

Joseph, Brian D. (1983). *The synchrony and diachrony of the Balkan infinitive*. Cambridge: Cambridge University Press.

Joseph, Brian D. (2001). Language contact and the development of negation in Greek and the Balkans. In *Greek linguistics '99: Proceedings of the 4th International Conference on Greek Linguistics, Nicosia, September 1999*, 346–53. Thessaloniki: University Studio Press.

Joseph, Brian D. (2002). Balkan insights into the syntax of *mē in Indo-European. In Mark R. V. Southern (ed.), *Indo-European perspectives*, 103–20. Washington, DC: Institute for the Study of Man.

Joseph, Brian D. (2005). Some ancient shared metaphors, *Acta Studia Albanica* 2, 43–6.

Kagan, Olga (2010). Genitive objects, existence and individuation. *Russian Linguistics* 34, 17–39.

Kahrel, Pieter (1996). *Aspects of negation*. Doctoral dissertation, University of Amsterdam.

Kallel, Amel (2005). *The loss of negative concord in English*. Doctoral dissertation, University of Reading.

Karlsson, Fred (1999 [1983]). *Finnish: An essential grammar*. London: Routledge.

Kayne, Richard S. (1975). *French syntax*. Cambridge, Mass.: MIT Press.

Kayne, Richard S. (1984). *Connectedness and binary branching*. Dordrecht: Foris.

Kelle, Johann (1881). *Glossar der Sprache Otfrids*. Regensburg: Manz.

Keller, Rudolf (1986). *Die deutsche Sprache und ihre historische Entwicklung*. Hamburg: Buske.

Kemenade, Ans van (1997). Negative-initial sentences in Old and Middle English. *Studia Anglica Posnaniensia* 31, 91–104.

Kemenade, Ans van (1999). Sentential negation and clause structure in Old English. In Ingrid Tieken-Boon van Ostade, Gunnel Tottie, and Wim van der Wurff (eds.), *Negation in the history of English*, 147–65. Berlin: Mouton de Gruyter.

Kemenade, Ans van (2000). Jespersen's cycle revisited: Formal properties of grammaticalization. In Susan Pintzuk, George Tsoulas, and Anthony Warner (eds.), *Diachronic syntax: Models and mechanisms*, 51–74. Oxford: Oxford University Press.

Kempson, Ruth, Wilfried Meyer-Viol, and Dov Gabbay (2001). *Dynamic syntax: The flow of language understanding*. Oxford: Blackwell.

Kenesei, István, Robert M. Vago, and Anna Fenyvesi (1998). *Hungarian*. London: Routledge.

Keresztes, László (1999). *Development of Mordvin definite conjugation*. Helsinki: Suomalais-Ugrilainen Seura.

Khalafallah, Abdelghany (1969). *A descriptive grammar of Saεi:Di colloquial Arabic*. The Hague: Mouton.

Kiparsky, Paul (1973). Elsewhere in phonology. In Stephen R. Anderson and Paul Kiparsky (eds.), *A festschrift for Morris Halle*, 93–106. New York: Holt, Rinehart, and Winston.

Kiparsky, Paul and Cleo Condoravdi (2006). Tracking Jespersen's cycle. In Mark Janse, Brian D. Joseph, and Angela Ralli (eds.), *Proceedings of the Second International Conference on Modern Greek Dialects and Linguistic Theory*, 172–97. Mytilene: Doukas.

Kiss, Katalin É. (forthcoming). A negative cycle in 12th–15th-century Hungarian. In Theresa Biberauer and George Walkden (eds.), *Syntax over time: Lexical, morphological and information-structural interactions*. Oxford: Oxford University Press.

Klein, Thomas (2000). Sozio-kulturelle Voraussetzungen und Sprachraum des Altniederdeutschen (Altsächsischen). In Besch et al. (2000), 124–7.

Klemm, Antál (1934). A mordvin tagadó és tiltó szerkezetek története. *Nyelvtudományi Közlemények* 48, 382–402.

Klenin, Emily (1978). Quantification, partitivity, and the genitive of negation in Russian. In Bernard Comrie (ed.), *Classification of grammatical categories*, 162–82. Edmonton: Linguistic Research.

Klima, Edward S. (1964). Negation in English. In Jerry A. Fodor and Jerrold J. Katz (eds.), *The structure of language*, 246–323. Englewood Cliffs, NJ: Prentice-Hall.

Kogan, Leonid (1997). Tigrinya. In Robert Hetzron (ed.), *The Semitic languages*, 424–45. London: Routledge.

König, Ekkehard (1991). *The meaning of focus particles: A comparative perspective*. London: Routledge.

Koopman, Hilda and Dominique Sportiche (1985). Theta-Theory and extraction. *GLOW Newsletter* 14, 57–8.

Koopman, Hilda and Dominique Sportiche (1991). The position of subjects. *Lingua* 85, 211–58.

Korenchy, Éva (1972). Az uráli tagadásról. *Nyelvtudományi Közlemények* 74, 155–67.

Korhonen, Mikko (1967). *Die Konjugation im Lappischen: Morphologisch-historische Untersuchung. I. Die finiten Formkategorien.* Helsinki: Suomalais-Ugrilainen Seura.

Korhonen, Mikko (1973). Zur Geschichte des negativen Präteritums und Perfekts im Ostseefinnischen und Lappischen. In *Commentationes Fenno-Ugricae in honorem Erkki Itkonen*, 174–84. Helsinki: Suomalais-Ugrilainen Seura.

Korhonen, Mikko (1981). *Johdatus lapin kielen historiaan.* Helsinki: Suomalaisen Kirjallisuuden Seura.

Kotkov, Sergej Ivanovič (ed.) (1990). *Pamjatniki južnovelikorusskogo narečija: Konec XVI–načalo XVII v.* Moscow: Nauka.

Kotkov, Sergej Ivanovič and Zinaida Danilovna Popova (1986). *Očerki po sintaksisu južnovelikorusskoj pis'mennosti XVII veka.* Moscow: Nauka.

Kozintseva, Natalia A. (2001). Imperative sentences in Armenian. In Victor S. Xrakovskij (ed.), *Typology of imperative constructions*, 245–67. Munich: Lincom Europa.

Kraft, Charles H. and Anthony Kirk-Greene (1973). *Hausa.* London: Hodder and Stoughton.

Kramer, Christina E. (1997). Negation and the grammaticalization of *have* and *want* futures in Bulgarian and Macedonian. *Canadian Slavonic Papers* 39, 407–16.

Krasovitsky, Alexander, Matthew Baerman, Dunstan Brown, and Greville G. Corbett (2011). Changing semantic factors in case selection: Russian evidence from the last two centuries. *Morphology* 21, 573–92.

Krefeld, Thomas (1997). Dame phonétique, dame sémantique et les autres: Bemerkungen zur bündnerromanischen Negation. In Günter Holtus, Johannes Kramer, and Wolfgang Schweickard (eds.), *Italica et Romanica: Festschrift für Max Pfister zum 65. Geburtstag*, vol. i, 24–9. Tübingen: Niemeyer.

Křížková, Helena (1968). K voprosu o tak nazyvaemoj dvojnoj negacii v slavjanskix jazykax. *Slavia* 37, 21–39.

Kroch, Anthony S. (1989). Reflexes of grammar in patterns of language change. *Language Variation and Change* 1, 199–244.

Kroch, Anthony and Ann Taylor (2000). The Penn–Helsinki Parsed Corpus of Middle English (PPCME2). Department of Linguistics, University of Pennsylvania.

Kryshevich, Olesya (2010). Genitive of negation in Ukrainian. MPhil. dissertation, University of Cambridge.

Kühner, Raphael and Bernhard Gerth (1898). *Ausführliche Grammatik der Griechischen Sprache.* Hanover: Hahnsche Buchhandlung.

Künnap, Ago (1993). Uurali keelte *ś-preteeritumi algupärast. In Ulla-Maija Kulonen (ed.), *Festschrift für Raija Bartens zum 25.10.1993*, 203–10. Mémoires de la Société Finno-Ougrienne. Helsinki: Suomalais-Ugrilainen Seura.

Künnap, Ago (1994). On the past tense of the negative auxiliary in the Uralic languages. In Ago Künnap (ed.), *Minor Uralic languages: Structure and development*, 83–7. Tartu: Tartu ülikool.

Künnap, Ago (2002). Some features of the Finno-Ugric verbal negation in the Volga area. In Jorma Luutonen (ed.), *Volgan alueen kielikontaktit*, 35–9. Turku: Turun yliopisto.

Kuryłowicz, Jerzy (1971). Słowiański genetivus po negacji. In Międzynarodowa Komisja Budowy Gramatycznej Języków Słowiańskich (ed.), *Sesja naukowa Międzynarodowej Komisji*

Budowy Gramatycznej Języków Słowiańskich w Krakowie w dniach 3–5 grudnia 1969 r., 11–15. Wrocław: Zakład Narodowy imienia Ossolińskich, Wydawnictwo Polskiej Akademii Nauk.

Ladusaw, William A. (1979). Polarity sensitivity as inherent scope relations. Doctoral dissertation, University of Texas, Austin.

Ladusaw, William A. (1992). Expressing negation. *Ohio State University Working Papers in Linguistics* 40, 237–59.

Ladusaw, William A. (1993). Negation, indefinites, and the Jespersen cycle. In Joshua S. Guenter, Barbara A. Kaiser, and Cheryl C. Zoll (eds.), *Proceedings of the Nineteenth Annual Meeting of the Berkeley Linguistics Society. General Session and Parasession on Semantic Typology and Semantic Universals*, 437–46. Berkeley, Calif.: Berkeley Linguistics Society.

Lafkioui, Mena (2007). *Atlas linguistique des variétés berbères du rif.* Cologne: Rüdiger Köppe.

Laka, Itziar (1990). *Negation in syntax: On the nature of functional categories and projections.* Doctoral dissertation, Massachusetts Institute of Technology.

Lallukka, Seppo (2005). Venäjän suomalais-ugrilaiset—väestönlaskentojen kertomaa. In Paula Kokkonen (ed.), *Sukukansaohjelman arki*, 28–46. Helsinki: M. A. Castrén Society, Finno-Ugrian Society and the Department of Finno-Ugrian Studies of the University of Helsinki.

Lamberti, Marcello (1991). Cushitic and its classifications. *Anthropos* 86, 552–61.

Landsman, David M. (1988). The history of some Greek negatives: Phonology, grammar and meaning. *Glossologia* 7–8, 13–31.

Lane, Edward (1863). *An Arabic–English lexicon.* London: Williams and Norgate.

Larrivée, Pierre (2011). The role of pragmatics in grammatical change: The case of French preverbal *non. Journal of Pragmatics* 43, 1987–96.

Lass, Roger (2000). Language periodisation and the concept 'middle'. In Irma Taavitsainen, Terttu Nevalainen, Päivi Pahta, and Matti Rissanen (eds.), *Placing Middle English in context*, 7–42. Berlin: Mouton.

Le Roux, Pierre (1924–63). *Atlas linguistique de la Basse-Bretagne.* Rennes: Plihon et Hommay.

Lebbe, D. (1997). Schoon Vlaams. *Taal en Tongval* 49, 158–75.

Ledgeway, Adam L. (2009). *Grammatica diacronica del napoletano.* Tübingen: Niemeyer.

Leemans, Emilienne (1966). *Syntactische kenmerken van het Gents dialect.* Licentiate's thesis, Universiteit Gent.

Lehmann, Winfred P. (1973). A structural principle of language and its implications. *Language* 49, 47–66.

Lehmann, Winfred P. (1974). *Proto-Indo-European syntax.* Austin: University of Texas Press.

Lehmann, Winfred P. (1978). Changes in the negative sentence patterns in German. In Dietrich Hartmann, Hansjürgen Linke, and Otto Ludwig (eds.), *Sprache in Gegenwart und Geschichte: Festschrift für H. M. Heinrichs*, 94–109. Cologne: Böhlau.

Leslau, Wolf (1945). Grammatical sketches in Tigré (North Ethiopic): Dialect of Mensa. *Journal of the American Oriental Society* 65, 164–203.

Leslau, Wolf (1995). *Reference grammar of Amharic.* Wiesbaden: Harrassowitz Verlag.

Leslau, Wolf (1999). *Zway Ethiopic documents: Grammar and dictionary.* Wiesbaden: Harrassowitz Verlag.

Levin, Samuel R. (1958). Negative contraction: An Old and Middle English dialect criterion. *Journal of English and Germanic Philology* 57, 492–501.

Levinson, Dmitry (2006). Polarity sensitivity in inflectional morphology. MS, Stanford University.

Lewis, Henry (1946). *Llawlyfr Cernyweg Canol.* Cardiff: Gwasg Prifysgol Cymru.

Lewis, Henry (1990). *Handbuch des Mittelkornischen.* Innsbruck: Institut für Sprachwissenschaft der Universität Innsbruck.

Lewis, Henry and Holger Pedersen (1937). *A concise comparative Celtic grammar.* Göttingen: Vandenhoeck and Ruprecht.

Lightfoot, David W. (1999). *The development of language: Acquisition, change and evolution.* Oxford: Blackwell.

Lindstedt, Jouko (2010). Mood in Bulgarian and Macedonian. In Björn Rothstein and Rolf Thieroff (eds.), *Mood in the languages of Europe*, 409–21. Amsterdam: John Benjamins.

Lipínski, Edward (1997). *Semitic languages: Outline of a comparative grammar.* Leuven: Peeters.

Liver, Ricarda (1999). *Rätoromanisch: Eine Einführung in das Bündnerromanische.* Tübingen: Narr.

Lockwood, William Burley (1968). *Historical German syntax.* Oxford: Clarendon Press.

Lockwood, William Burley (1995). *Lehrbuch der modernen jiddischen Sprache.* Hamburg: Buske.

Lomtev, Timofej Petrovič (1956). *Očerki po istoričeskomu sintaksisu russkogo jazyka.* Moscow: Izdatel'stvo Moskovskogo Universiteta.

Longobardi, Giuseppe (1988). I quantificatori. In Lorenzo Renzi (ed.), *Grande grammatica italiana di consultazione*, i. 647–96. Bologna: Il Mulino.

Lucas, Christopher (2007). Jespersen's cycle in Arabic and Berber. *Transactions of the Philological Society* 105, 398–431.

Lucas, Christopher (2009). The development of negation in Arabic and Afro-Asiatic. Doctoral thesis, University of Cambridge.

Lucas, Christopher (2010a). Negative *-š* in Palestinian (and Cairene) Arabic: Present and possible past. *Brill's Annual of Afroasiatic Languages and Linguistics* 2, 165–201.

Lucas, Christopher (2010b). A dynamic syntax account of n-words and negative concord. Paper presented at the Annual Meeting of the LAGB, University of Leeds.

Lucas, Christopher (2011). Indefinites and negative concord in Maltese: Towards a dynamic account. Paper presented at the 3rd Conference on Maltese Linguistics, University of Malta.

Lucas, Christopher and Elliott Lash (2010). Contact as catalyst: The case for Coptic influence in the development of Arabic negation. *Journal of Linguistics* 46, 379–413.

Lucas, Christopher and David Willis (2012). Never again: the multiple grammaticalization of *never* as a marker of negation in English. *English Language and Linguistics* 16, 459–85.

Ludolf, Hiob (1698). *Grammatica linguae amharicae.* Frankfurt: Prostat apud Johannem David Zunnerum. Impressit Martinus Jacquetus.

MacAulay, Donald (1992). The Celtic languages: An overview. In Donald MacAulay (ed.), *The Celtic languages*, 1–8. Cambridge: Cambridge University Press.

McKay, Kenneth L. (1986). Aspects of the imperative in Ancient Greek. *Antichthon* 20, 41–58.

Mägiste, Julius (1983). *Estnisches etymologisches Wörterbuch.* Helsinki: Finnisch-ugrische Gesellschaft.

Manzini, M. Rita and Leonardo M. Savoia (2002). Negative adverbs are neither Adv nor Neg. In Masako Hirotani (ed.), *Proceedings of NELS* 32, 327–46. Amherst, Mass.: Graduate Linguistics Student Association, University of Massachusetts. Reprinted as M. Rita Manzini and Leonardo M. Savoia (eds.) (2008). *Work notes on Romance morphosyntax / Appunti di morfosintassi romanza*, 79–97. Alessandria: Edizioni dell'Orso.

Manzini, M. Rita and Leonardo M. Savoia (2005). *I dialetti italiani e romanci. Morfosintassi generativa*, vol. iii. Alessandria: Edizioni dell'Orso.

Manzotti, Emilio and Alessandra Rigamonti (1991). La negazione. In Lorenzo Renzi, Giampaolo Salvi, and Anna Cardinaletti (eds.), *Grande grammatica italiana di consultazione*, 245–320. Bologna: Il Mulino.

Marcato, Gianna and Flavia Ursini (1998). *Dialetti veneti: Grammatica e storia*. Padua: Unipress.

Marchello-Nizia, Christiane (1997). *La langue française aux XIVe et XVe siècles*. Paris: Nathan.

Marchello-Nizia, Christiane (1999). *Le français en diachronie: douze siècles d'évolution*. Paris: Ophrys.

Marchetti, Giuseppe (1977). *Lineamenti di grammatica friulana*, 3rd edn. Udine: Società Filologica Friulana.

Martin, Robert (1972). La négation de 'virtualité' du moyen français. *Romania* 93, 20–49.

Martineau, France (1994). Movement of negative adverbs in French infinitival clauses. *Journal of French Language Studies* 4, 55–73.

Martineau, France and Raymond Mougeon (2003). A sociolinguistic study of the origins of *ne* deletion in European and Quebec French. *Language* 79, 118–52.

Martineau, France and M.-T. Vinet (2005). Microvariation in French negation markers: A historical perspective. In Montse Batllori, Maria-Lluïsa Hernanz, Carme Picallo, and Francesc Roca (eds.), *Grammaticalization and parametric variation*, 194–205. Oxford: Oxford University Press.

Martins, Ana Maria (2000). Polarity items in Romance: Underspecification and lexical change. In Susan Pintzuk, George Tsoulas, and Anthony Warner (eds.), *Diachronic syntax: Models and mechanisms*, 191–219. Oxford: Oxford University Press.

Massajoli, Pierleone and Roberto Moriani (1991). *Dizionario della cultura brigasca*, vol. i: *Lessico*. Alessandria: Edizioni dell'Orso.

Mathiassen, Terje (1996). *A short grammar of Lithuanian*. Columbus, Ohio: Slavica.

Mathiassen, Terje (1997). *A short grammar of Latvian*. Columbus, Ohio: Slavica.

Matras, Yaron (1999). The state of present-day Domari in Jerusalem. *Mediterranean Language Review* 11, 1–58.

Matras, Yaron (2007). Grammatical borrowing in Domari. In Yaron Matras and Jeanette Sakel (eds.), *Grammatical borrowing in cross-linguistic perspective*, 151–64. Berlin: Mouton de Gruyter.

May, Robert (1989). Interpreting logical form. *Linguistics and Philosophy* 12, 387–435.

Mazzon, Gabriella (2004). *A history of English negation*. London: Pearson Longman.

Meersman, Alfons de (1980). Ontkenningen in ouder Nederlands. *Studia Germanica Gandensia* 21, 5–14.

Meillet, Antoine (1912). L'évolution des formes grammaticales. *Scientia* 12, 384–400. Reprinted in Antoine Meillet (1921). *Linguistique historique et linguistique générale*, 130–48. Paris: Edouard Champion.

Meillet, Antoine and André Vaillant (1965). *Le slave commun*. Paris: Champion.

Mel'nyčuk, Oleksandr Savvyč (ed.) (1982). *Etymolohichnyj slovnyk ukrains'koj movy*. Kyiv: Naukova dumka.

Meščerskij, Nikita Aleksandrovič (1972). *Russkaja dialektologija*. Moscow: Vysšaja Škola.

Mettouchi, Amina (1996). La négation dans les langues du maghreb: synthèse. In Salem Chaker and Dominique Caubet (eds.), *La négation en berbère et en arabe maghrébin*, 177–95. Paris: L'Harmattan.

Middelnederlandsch woordenboek (1998 [1885–1941]). Edited by Eelco Verwijs, Jacob Verdam, and F. A. Stoett, 12 vols. In Instituut voor Nederlandse Lexicologie (ed.), *CD-rom Middelnederlands*. The Hague/Antwerp: Sdu Uitgevers/Standaard Uitgeverij.

Miestamo, Matti (2000). Towards a typology of standard negation. *Nordic Journal of Linguistics* 23, 65–88.

Miestamo, Matti (2005). *Standard negation: The negation of declarative verbal main clauses in a typological perspective*. Berlin: Mouton de Gruyter.

Miestamo, Matti (2007). Negation: An overview of typological research. *Language and Linguistics Compass* 1, 552–70.

Miestamo, Matti (2011). A typological perspective on negation in Finnish dialects. *Nordic Journal of Linguistics* 34, 83–104.

Mignon, Françoise (2008). Le choix du marqueur de négation dans l'expression du contraste. *Revue Romane* 43, 63–85.

Mikhailov, Nikolai (1998). *Frühslowenische Sprachdenkmäler: Die handschriftliche Periode der slowenischen Sprache (XIV. Jh. bis 1550)*. Amsterdam: Rodopi.

Miklosich, Franz (1886). *Etymologisches Wörterbuch der slavischen Sprachen*. Vienna: Braumüller.

Milićević, Nataša (2008). NPI-licensing and dependent tense in Serbian. In Marjo van Koppen and Bert Botma (eds.), *Linguistics in the Netherlands 2008*, 109–20. Amsterdam: John Benjamins.

Minkov, Vl. (1911). Genitivus negationis. *Izvestija na Seminara po Slavjanska Filologija* 3, 385–98.

Mirčev, Kiril (1932). Iztočnomakedonskoto i rodopsko *boedin*, slovensko (*n*)*obeden* i zap. slav. *žádný-żaden*. *Makedonski Pregled* 8, 9–22.

Mitchell, Bruce (1985). *Old English syntax*. Oxford: Clarendon Press.

Moerenhout, Mike and Wim van der Wurff (2005). Object–verb order in early sixteenth-century English prose: An exploratory study. *English Language and Linguistics* 9, 83–114.

Möhren, Frankwalt (1980). *Le renforcement affectif de la négation par l'expression d'une valeur minimale en ancien français*. Tübingen: Niemeyer.

Moignet, Gérard (1965). L'opposition *non/ne* en ancien français. *Travaux de linguistique et de littérature* 3, 41–65.

Moignet, Gérard (1976). *Grammaire de l'ancien français*. Paris: Klincksieck.

Mokšəń käl' (2000) = Aljamkin, N. S. (ed.) (2000). *Mokšəń käl'. Morfologija*. Saransk: Krasnyj Oktjabr'.

Molinelli, Piera (1988). *Fenomeni della negazione dal latino all'italiano*. Florence: La Nuova Italia.

Moorhouse, Alfred C. (1959). *Studies in the Greek negatives*. Cardiff: University of Wales Press.

Morgan, T. J. (1987). Simo'i'n gwbod. Sana-i'n gwbod. *Bulletin of the Board of Celtic Studies* 34, 88–93.

Morris, W. Meredith (1910). *A glossary of the Demetian dialect of north Pembrokeshire with special reference to the Gwaun Valley.* Tonypandy: Evans and Short. Repr. in Facsimile edition, Felinfach: Llanerch Publishers, 1991.

Morris-Jones, John (1913). *A Welsh grammar.* Oxford: Oxford University Press.

Morris-Jones, John (1931). *Welsh syntax: An unfinished draft.* Cardiff: University of Wales Press.

Mosin, Mixail (2005). Ètimologii èrzjanskix slov. *Linguistica Uralica* 41, 182–6.

Motylinski, Gustave-Adolphe (1904). *Le dialecte berbère de r'edamès.* Paris: Ernest Leroux.

Mouřek, Vaclav (1903). Über die Negation im Mittelhochdeutschen. *Sitzungsberichte der königlichen böhmischen Gesellschaft der Wissenschaften, Classe fur Philosophie, Geschichte und Philologie* 12, 1–30.

Muller, Claude (1991). *La négation en français: Syntaxe, sémantique et éléments de comparaison avec les autres langues romanes.* Geneva: Librairie Droz.

Musić, August (1921). Slovensko (n)obeden. *Rad Jugoslavenske Akademije Znanosti i Umjetnosti, Razredi Filologičko-historički i Filosofičko-juridički* 98, 189–220.

Mustajoki, Arto and Hannes Heino (1991). *Case selection for direct object in Russian negative clauses. Part II: Report on a statistical analysis.* Helsinki: Department of Slavonic Languages, Helsinki University.

Naro, Anthony J. (1978). A study on the origins of pidginization. *Language* 54, 314–47.

Neidle, Carol Jan (1988). *The role of case in Russian syntax.* Dordrecht: Kluwer.

Nevalainen, Terttu (2006). Negative concord as an English 'vernacular universal'. *Journal of English Linguistics* 34, 257–78.

Newman, Paul (1980). *The classification of Chadic within Afroasiatic.* Leiden: Leiden University Press.

Nocentini, Alberto (2003). La cosidetta negazione espletiva in italiano. *Archivio Glottologico Italiano* 88, 72–90.

Nyrop, Kristoffer (1930). *Grammaire historique de la langue française*, vol. vi. Copenhagen: Nordisk Forlag.

Offord, Malcolm (1976). Negation in *Bérinus*: A contribution to the study of negation in fourteenth-century French. *Zeitschrift für Romanische Philologie* 92, 313–85.

Ohkado, Masayuki (2005). On grammaticalization of negative adverbs, with special reference to Jespersen's cycle recast. In Yoko Iyeiri (ed.), *Aspects of English negation*, 39–58. Amsterdam: John Benjamins.

Otrębski, Jan (1966). Z dziejów wyrazów polskich. *Język Polski* 46, 12–17.

Ouali, Hamid (2008). Negative expressions in Moroccan Arabic and Modern Standard Arabic: NCIs or NPIs? MS, University of Wisconsin-Milwaukee.

Ouhalla, Jamal (1990). Sentential negation, Relativized Minimality and the aspectual status of auxiliaries. *Linguistic Review* 7, 183–231.

Ouhalla, Jamal (2002). The structure and logical form of negative sentences in Arabic. In Jamal Ouhalla and Ur Shlonsky (eds.), *Themes in Arabic and Hebrew syntax*, 299–320. Dordrecht: Kluwer.

Owens, Jonathan (1984). *A short reference grammar of eastern Libyan Arabic.* Wiesbaden: Otto Harrassowitz.

Owens, Jonathan (1985). *A grammar of Harar Oromo (northeastern Ethiopia).* Hamburg: Helmut Buske Verlag.

Owens, Jonathan (2006). *A linguistic history of Arabic.* Oxford: Oxford University Press.

Paasonen, Heikki (1903). *Mordwinische Lautlehre.* Helsinki: Suomalais-Ugrilainen Seura.

Paasonen, Heikki (1953). *Mordwinische Chrestomathie mit Glossar und grammatikalischem Abriss.* Helsinki: Suomalais-Ugrilainen Seura.

Paasonen, Heikki (1990–9). *H. Paasonens Mordwinisches Wörterbuch.* Edited by Kaino Heikkilä, Hans-Hermann Bartens, Aleksandr Feoktistov, Grigori Jermuschkin, and Martti Kahla. Helsinki: Suomalais-Ugrilainen Seura.

Padučeva, Elena Viktorovna (2006). Genitiv dopolnenija v otricatel'nom predloženii. *Voprosy Jazykoznanija* 2006, 21–43.

Pall, Valdek (1957). Negatsioonist mordva keeles. *Emakeele Seltsi Aastaraamat III*, 217–23.

Palmer, Frank R. (2000). *Mood and modality.* Cambridge: Cambridge University Press.

Paradisi, Umberto (1961). Testi berberi di augila (cirenaïca). *Annali dell'Istituto Orientale di Napoli* 10, 79–91.

Parry, M. Mair (1996). La negazione italo-romanza: Variazione tipologica e variazione strutturale. In Paola Benincà, Guglielmo Cinque, T. De Mauro, and Nigel Vincent (eds.), *Italiano e dialetti nel tempo: Saggi di grammatica per Giulio C. Lepschy*, 225–57. Rome: Bulzoni.

Parry, M. Mair (1997). Preverbal negation and clitic ordering, with particular reference to a group of north-west Italian dialects. *Zeitschrift für romanische Philologie* 113, 243–70.

Parry, M. Mair (2005). *Parluma 'D Còiri: Sociolinguistica e grammatica del dialetto di Cairo Montenotte.* Savona: Società Savonese di Storia Patria e Editrice Liguria.

Parry, M. Mair (2010). Prohibition and Romance: Negative imperatives in the early vernaculars of Italy. In Roberta D'Alessandro, Adam Ledgeway, and Ian Roberts (eds.), *Syntactic variation: The dialects of Italy*, 153–70. Cambridge: Cambridge University Press.

Paspati, Alexandre (1870). *Études sur les tschinghianés ou bohémiens de l'empire ottoman.* Osnabrück: Biblio.

Patkanoff, K.P. (1907–8). Some words on the dialects of the Transcaucasian gypsies. *Journal of the Gypsy Lore Society* (new series) 1, 229–57.

Patruno, Barbara and Laura Sgarioto (2004). Osservazioni sull'imperativo negativo in alcune varietà del ladino centrale. *Quaderni di lavoro ASIt* 4. Available online at <http://asit.maldura.unipd.it/documenti/ql4/patruno-sgarioto.pdf>.

Paul, Hermann (1920). *Deutsche Grammatik*, vol. iv, part iv: *Syntax*. Halle: Niemeyer.

Paul, Hermann (2007). *Mittelhochdeutsche Grammatik. Neubearb. v. Thomas Klein, Joachim Solms u. Klaus-Peter Wegera. Mit einer Syntax v. Ingeborg Schröbler, neubearb. u. hrsg. v. Heinz-Peter Prell*, 25th edn. Tübingen: Niemeyer.

Pauw, Godelieve de (1973). *Syntaktische kenmerken van het dialect van Buggenhout-Opstal.* Licentiate's dissertation, University of Ghent.

Payne, John R. (1985). Negation. In Timothy Shopen (ed.), *Language typology and syntactic description*, vol. i: *Clause structure*, 197–242. Cambridge: Cambridge University Press.

Penka, Doris (2010). *Negative indefinites.* Oxford: Oxford University Press.

Penka, Doris and Arnim von Stechow (2001). Negative Indefinita unter Modalverben. In Reimar Müller and Marga Reis (eds.), *Modalität und Modalverben im Deutschen*, 263–86. Hamburg: Buske.

Pensel, Franzjosef (1976). Die Satznegation. In Gerhard Kettmann and Joachim Schildt (eds.), *Zur Ausbildung der Norm der deutschen Literatursprache auf der syntaktischen Ebene (1470–1730). Der Einfachsatz*, 2nd edn. 287–326. Berlin: Akademie-Verlag.

Pereltsvaig, Asya (2006). Negative polarity items in Russian and the 'bagel problem'. In Sue Brown and Adam Przepiórkowski (eds.), *Negation in Slavic*, 153–78. Bloomington, Ind.: Slavica.

Perle, Friedrich (1878). Die Negation im Altfranzösischen. *Zeitschrift für Romanische Philologie* 2, 1–24.

Pesetsky, David (1982). Paths and categories. Doctoral dissertation, Massachusetts Institute of Technology.

Peškovskij, Aleksandr Matveevič (1956 [1927]). *Russkij sintaksis v naučnom osveščenii*. Moscow: Gosudarstvennoe Učebno-Pedagogičeskoe Izdatel'stvo Ministerstva Prosveščenija RSFSR.

Peters, Robert (1984). Überlegungen zu einer Karte des mittelniederdeutschen Sprachraums. *Niederdeutsches Wort* 4, 51–9.

Peters, Robert (2000a). Soziokulturelle Voraussetzungen und Sprachraum des Mittelniederdeutschen. In Besch et al. (2000), 1408–22.

Peters, Robert (2000b). Die Diagliederung des Mittelniederdeutschen. In Besch et al. (2000), 1478–90.

Philippaki-Warburton, Irene and Vassilios Spyropoulos (2004). A change of mood: The development of the Greek mood system. *Linguistics* 42, 791–817.

Pigin, M. I. (1962). Iz istorii otricatel'nyx bezličnyx predloženij v russkom jazyke. *Učenye Zapiski Petrozavodskogo Gosudarstvennogo Universiteta: Lingvističeskij Sbornik* 10, 3–42.

Pinkster, Harm (1990). *Latin syntax and semantics*. London: Routledge.

Pintzuk, Susan (2005). Arguments against a universal base: Evidence from Old English. *English Language and Linguistics* 9, 115–38.

Planta, Robert von, Florian Meicher, and Chasper Pult (eds.) (1939). *Dicziunari rumantsch Grischun*. Cuoira: Bischoflerger.

Pohl, Jacques (1968). *Ne* dans le français parlé contemporain: Les modalités de son abandon. In Antonio Quilis (ed.), *Actas del XI Congreso Internacional de Lingüística y Filología Romanicàs*, vol. iii, 1343–58. Madrid: Consejo Superior de Investiaciones Científicas.

Poletto, Cecilia (1993). *La sintassi del soggetto nei dialetti settentrionali*. Padua: Unipress.

Poletto, Cecilia (2000). *The higher functional field: Evidence from northern Italian dialects*. New York: Oxford University Press.

Poletto, Cecilia (2008). On negative doubling. In Federica Cognola and Diego Pescarini (eds.), *Quaderni di Lavoro ASIt* 8, *La negazione: Variazione dialettale ed evoluzione diacronica*, 57–84. Available online at <http://asit.maldura.unipd.it/documenti/ql8/poletto_2008.pdf>.

Poletto, Cecilia and Raffaella Zanuttini (2003). Making imperatives: Evidence from central Rhaeto-Romance. In Christina Tortora (ed.), *The syntax of Italian dialects*, 175–206. New York: Oxford University Press.

Pollock, Jean-Yves (1989). Verb movement, Universal Grammar, and the structure of IP. *Linguistic Inquiry* 20, 365–424.

Poole, Geoffrey (2009). The syntax-discourse interface and the development of n-words in Spanish. In Young-Se Kang, Sze-Wing Tang, Chul Kim, Jong-Yurl Yoon, Yong-Soon Kang, Kyoung-Ae Kim, Hyunkyung Yoo, Youngjun Jang, and Hye-Kyung Kang (eds.), *Current issues in linguistic interfaces*, 27–48. Seoul: Hankook Munhwasa.

Poole, Geoffrey (2011). Focus and the development of n-words in Spanish. In Janine Berns, Haike Jacobs, and Tobias Scheer (eds.), *Romance languages and linguistic theory 2009: Selected papers from 'Going Romance', Nice 2009*, 291–303. Amsterdam: John Benjamins.

Poppe, Erich (1995). Negation in Welsh and 'Jespersen's cycle'. *Journal of Celtic Linguistics* 4, 99–107.

Poppe, Erich (2009). Expressions of negative polarity in the Middle Welsh *Ystorya Bown de Hamtwn*. *Journal of Celtic Linguistics* 13, 117–30.

Portner, Paul and Raffaela Zanuttini (1996). The syntax and semantics of scalar negation: Evidence from Paduan. In Kusumoto, K. (ed.), *Proceedings of NELS* 26, 257–71. Amherst, Mass.: Graduate Linguistics Student Association, University of Massachusetts.

Portner, Paul and Raffaella Zanuttini (2000). The force of negation in *wh*-exclamatives and interrogatives. In Laurence R. Horn and Yasuhiko Kato (eds.), *Negation and polarity: Syntactic and semantic perspectives*, 193–231. Oxford: Oxford University Press.

Posner, Rebecca (1984). Double negatives, negative polarity and negative incorporation in Romance: A historical and comparative view. *Transactions of the Philological Society* 82, 1–26.

Posner, Rebecca (1985). Postverbal negation in non-standard French: A historical and comparative view. *Romance Philology* 39, 170–97.

Postma, Gertjan (2002). De enkelvoudige clitische negatie in het Middelnederlands en de Jespersen cyclus. *Nederlandse Taalkunde* 7, 44-82.

Postma, Gertjan and Hans J. Bennis (2006). Variatie in negatie: Het gedrag van het negatieve cliticum in het Middeldrents van rond 1400. *Taal en Tongval* 58, 148–66.

Postma, Gertjan and Wim van der Wurff (2007). How to say *no* and *don't*: Negative imperatives in Romance and Germanic. In Wim van der Wurff (ed.), *Imperative clauses in generative grammar*, 205–49. Amsterdam: Benjamins.

Potebnja, Aleksandr Afanas'evič (1968 [1899]). *Iz zapisok po russkoj grammatike*, vol. iii: *Ob izmenenii značenija i zamenax suščestvitel'nogo*. Moscow: Prosveščenie.

Prévost, Sophie and Catherine Schnedecker (2004). *Aucun(e)(s)/d'aucun(e)(s)/les aucun(e)(s)*: Evolution du français médiéval au français moderne. *SCOLIA* 18, 41–75.

Price, Glanville (1962). The negative particles *pas*, *mie* and *point* in French. *Archivum Linguisticum* 14, 14–34.

Price, Glanville (1978). L'interrogation négative sans *ne*. *Studii și Cercetări Lingvistice* 29, 599–606.

Price, Glanville (1990). The origins and syntax of *ne... goutte*. In John N. Green and Wendy Ayres-Bennett (eds.), *Variation and change in French: Essays presented to Rebecca Posner on the occasion of her sixtieth birthday*, 201–9. London: Routledge.

Price, Glanville (1993). *Pas (point)* without *ne* in interrogative clauses. *French Language Studies* 3, 191–5.

Price, Glanville (1996). Negative particles in Cornish. *Cornish Studies (2nd series)* 4, 147–51.

Progovac, Ljiljana (1994). *Negative and positive polarity: A binding approach*. Cambridge: Cambridge University Press.

Progovac, Ljiljana (2000). Coordination, c-command, and 'logophoric' n-words. In Laurence Horn and Yatsuhiko Kato (eds.), *Negation and polarity: Semantic and syntactic perspectives*, 88–114. Oxford: Oxford University Press.

Progovac, Ljiljana (2005). *A syntax of Serbian: Clausal architecture*. Bloomington: Slavica.

Provotelle, Paul (1911). *Étude sur la tamazir't, ou zénatia de qalaât es-sened*. Paris: Ernest Leroux.

Przepiórkowski, Adam (1999). Case assignment and the complement/adjunct dichotomy: A non-configurational constraint-based approach. Doctoral dissertation, University of Tübingen.

Qian, Minru (1987). *Untersuchungen zur Negation in der deutschen Gegenwartssprache: Eine mikro- und makrostrukturelle Analyse*. Heidelberg: Groos.

Quak, Arend (1981). *Die altmittel- und altniederfränkischen Psalmen und Glossen*. Amsterdam: Rodopi.

Quak, Arend and Joop M. van der Horst (2002). *Inleiding Oudnederlands*. Leuven: Universitaire Pers Leuven.

Rabhi, Allahoua (1996). De la négation en berbère: Les donnés algériennes. In Salem Chaker and Dominique Caubet (eds.), *La négation en berbère et en arabe maghrébin*, 23–34. Paris: L'Harmattan.

Ramat, Paolo (1998). Perché *veruno* significa 'nessuno'? In Paolo Ramat and Elisa Roma (eds.), *Sintassi storica: Atti del XXX Congresso della SLI*, 397–409. Rome: Bulzoni.

Ramat, Paolo (2006a). Italian negatives from a typological/areal point of view. In N. Grandi and G. Iannàccaro (eds.), *Zhì. Scritti in onore di Emanuele Banfi in occasione del suo 60° compleanno*, 355–70. Cesena: Caissa Italia.

Ramat, Paolo (2006b). Negation. In Keith Brown (ed.), *The encyclopedia of language and linguistics*, 2nd edn, 559–66. Amsterdam: Elsevier.

Rappaport, Gilbert C. (1986). On a persistent problem of Russian syntax: Sentences of the type *mne negde spat'*. *Russian Linguistics* 10, 1–31.

Räsänen, Martti (1969). *Versuch eines etymologischen Wörterbuchs der Türksprachen*. Helsinki: Suomalais-Ugrilainen Seura.

Ravila, Paavo (1929). Über eine doppelte vertretung des urfinnischwolgaischen *a der nicht-ersten silbe im mordwinischen. *Finnisch-Ugrische Forschungen* 20, 83–120.

Rédei, Károly (1970). Kiellon ilmaisemisen alkuperästä unkarissa [On the origin of the expression of negation in Hungarian]. *Virittäjä* 74, 47–51.

Rédei, Károly (1988a). Geschichte der permischen Sprachen. In Denis Sinor (ed.), *The Uralic languages: Description, history and foreign influence. Handbuch der Orientalistik*. Achte Abteilung, vol. i, 351–94. Leiden: E. J. Brill.

Rédei, Károly (ed.) (1988b). *Uralisches etymologisches Wörterbuch*, vol. i: *Uralische und finnisch-ugrische Schicht*. Wiesbaden: Otto Harrassowitz.

Reenen, Pieter van and Maaike Mulder (2000). Un corpus linguistique de 3000 chartes en moyen néerlandais du 14e siècle. In Mireille Bilger (ed.), *Corpus: Méthodologie et applications linguistiques*, 209–17. Paris: Champion et Presses Universitaires de Perpignan.

Reinhardt, Carl (1894). *Ein arabischer Dialekt gesprochen in 'Omān und Zanzibar: Nach praktischen Gesichtspunkten für das Seminar für orientalische Sprachen in Berlin*. Berlin: Spemann.

Reintges, Chris (2004). *Coptic Egyptian (Sahidic dialect).* Cologne: Rüdiger Köppe.

Richards, Melville (1948). The negativing of the verb noun in Welsh. *Études Celtiques* 4, 369–78.

Rissanen, Matti (1999). *Isn't it?* or *Is it not?* On the order of postverbal subject and negative particle in the history of English. In Ingrid Tieken-Boon van Ostade, Gunnel Tottie, and Wim van der Wurff (eds.), *Negation in the history of English*, 189–205. Berlin: Mouton de Gruyter.

Rivero, María-Luisa (1991). Long head movement and negation: Serbo-Croatian vs. Slovak and Czech. *Linguistic Review* 8, 319–51.

Rivero, María-Luisa (1994). Negation, imperatives and Wackernagel effects. *Rivista di Linguistica* 6, 39–66.

Rizzi, Luigi (1986). On the status of subject clitics in Romance. In Osvaldo Jaeggli and Carmen Silva-Corvalan (eds.), *Studies in Romance linguistics*, 391–419. Dordrecht: Foris.

Rizzi, Luigi (1996). Residual verb second and the wh-criterion. In Adriana Belletti and Luigi Rizzi (eds.), *Parameters and functional heads: Essays in comparative syntax*, 63–90. New York: Oxford University Press.

Roberts, Ian and Anna Roussou (2003). *Syntactic change: A minimalist approach to grammaticalization.* Cambridge: Cambridge University Press.

Rohlfs, Gerhard (1968). *Grammatica storica della lingua italiana e dei suoi dialetti*, vol. ii. Turin: Einaudi.

Rohlfs, Gerhard (1969). *Grammatica storica della lingua italiana e dei suoi dialetti*, vol. iii. Turin: Einaudi.

Rooth, Mats (1992). A theory of focus interpretation. *Natural Language Semantics* 1, 75–116.

Roussou, Anna (2000). On the left periphery: Modal particles and complementisers. *Journal of Greek Linguistics* 1, 65–94.

Roussou, Anna (2007). Minimalism and diachronic syntax: The development of negative expressions. In Eleni Agathopoulou, Maria Dimitrakopoulou, and Despina Papadopoulou (eds.), *Selected papers on theoretical and applied linguistics: Festschrift for Prof E. Panagopoulos*, vol. i, 11–27. Thessaloniki: Monochromia.

Rowlett, Paul (1998). *Sentential negation in French.* New York: Oxford University Press.

Rubin, Aaron (2005). *Studies in Semitic grammaticalization.* Winona Lake, Indiana: Eisenbrauns.

Ruhlen, Merritt (1994). *On the origin of languages: Studies in linguistic taxonomy.* Stanford, Calif.: Stanford University Press.

Ruijgh, Cornelis J. (1992). L'emploi le plus ancien et les emplois plus recents de la particule *κε*/*ἄν*. chez Homère. In Françoise Letoublon (ed.), *La langue et les textes en grec ancien*, 75–84. Amsterdam: Gieben.

Rusek, Jerzy (1978). Z historii zaimków nieokreślonych w języku bułgarskim. *Sprawozdania z Posiedzeń Komisji Naukowych PAN* 22, 393–4.

Russell, Paul (1995). *An introduction to the Celtic languages.* London: Longman.

Růžička, Rudolf (1994). Asymmetry and parallelism between affirmative and negative *(ne) (est') gde spat'*. *Russian Linguistics* 18, 53–72.

Rybicka-Nowacka, H. (1990). Przypadek dopełnienia w konstrukcjach zaprzeczonych we współczesnym języku polskim (norma a praktyka językowa). *Poradnik Językowy* 8, 572–7.

Saeed, John (1999). *Somali*. Amsterdam: John Benjamins.

Salo, Merja (1991). Mordvalaiset. In Johanna Laakso (ed.), *Uralilaiset kansat*, 156–84. Juva: Werner Söderström Osakeyhtiö.

Sammallahti, Pekka (2011). From Africa to the Arctic: Expansions, bottlenecks and contact in the linguistic prehistory of the Saami. In Cornelius Hasselblatt, Peter Houtzagers, and Remco van Pareren (eds.), *Language contact in times of globalization*, 197–216. Amsterdam: Rodopi.

Sandanya, Manuel Pérez (2004). La negació i la concordança negativa en catalá antic. *Estudis Romànics* 26, 65–84.

Sanders, Willy (2000). Die Textsorten des Mittelniederdeutschen. In Besch et al. (2000), 1276–82.

Sankoff, Gillian and Diane Vincent (1977). L'emploi productif de *ne* dans le français parlé à Montréal. *Le Français Moderne* 45, 243–56.

Sapir, Edward (1949 [1921]). *Language: An introduction to the study of speech*. London: Harvest.

Šapiro, Abram Borisovič (1953). *Očerki po sintaksisu russkix narodnyx govorov: Stroenie predloženija*. Moscow: Izdatel'stvo Akademii Nauk SSSR.

Savijärvi, Ilkka (1977). *Itämerensuomalaisten kielten kieltoverbi. I. Suomi*. Helsinki: Suomalaisen Kirjallisuuden Seura.

Ščemerova, V. S. (1972). K voprosu *a*-ovoj osnove glagol'nogo otricanija v mordovskix jazykax. *Sovetskoe finno-ugrovedenie* 8, 173–80.

Schaaik, Gerjan van (1994). Turkish. In Peter Kahrel and René van den Berg (eds.), *Typological studies in negation*, 35–50. Amsterdam: John Benjamins.

Schapansky, Nathalie (1994). Breton *ket*: A case study in grammaticalization. Paper presented at LSA Annual Meeting, Boston.

Schapansky, Nathalie (2000). *Negation, referentiality and boundedness in Gwenedeg Breton: A case study in markedness and asymmetry*. Munich: Lincom Europa.

Schmeller, Andreas (1872). *Bayerisches Wörterbuch*, 2nd edn. Munich: Oldenbourg.

Schmidt, Wilhelm (1993). *Geschichte der deutschen Sprache*. Revised by Helmut Langner. 6th edn. Stuttgart/Leipzig: S. Hirzel.

Schrodt, Richard (2004). *Althochdeutsche Grammatik. II. Syntax*. Tübingen: Niemeyer.

Schulze, Ursula (1997). *Das Nibelungenlied*. Stuttgart: Reclam.

Schuster-Šewc, Heinz (1978). *Historisch-etymologisches Wörterbuch der ober- und niedersorbischen Sprache*. Bautzen: VEB Domowina-Verlag.

Schuster-Šewc, Heinz (2001). *Die ältesten Drucke des Obersorbischen: Wenceslaus Warichius und Gregorius Martini*. Bautzen: Domowina-Verlag.

Schützeichel, Rudolf and Birgit Meinecke (2001). *Die älteste Überlieferung von Willirams Kommentar des Hohen Liedes: Edition, Übersetzung, Glossar*. Göttingen: Vandenhoeck and Ruprecht.

Schwegler, Armin (1983). Predicate negation and word-order change: A problem of multiple causation. *Lingua* 61, 297–334.

Schwegler, Armin (1988). Word-order changes in predicate negation strategies in Romance languages. *Diachronica* 5, 21–58.

Schwegler, Armin (1990). *Analyticity and syntheticity: A diachronic perspective with special reference to Romance languages*. Berlin: Mouton de Gruyter.

Schwegler, Armin (1991a). Negation in Palenquero: Synchrony. *Journal of Pidgin and Creole Languages* 6, 165–214.

Schwegler, Armin (1991b). Predicate negation in contemporary Brazilian Portuguese: A change in progress. *Orbis* 34, 187–214.

Schwenter, Scott A. (2005). The pragmatics of negation in Brazilian Portuguese. *Lingua* 115, 1427–56.

Schwenter, Scott A. (2006). Fine-tuning Jespersen's cycle. In Betty J. Birner and Gregory Ward (eds.), *Drawing the boundaries of meaning: Neo-Gricean studies in pragmatics and semantics in honor of Laurence R. Horn*, 327–44. Amsterdam: John Benjamins.

Searle, John R. (1975). Indirect speech acts. In Peter Cole and Jerry L. Morgan (eds.), *Speech acts*, 59–82. New York: Academic Press.

Segal, Moses Hirsch (2001). *A grammar of Mishnaic Hebrew*. Oxford: Oxford University Press.

Siatkowski, Janusz (1978). Jeszcze raz o słowiańskim *żaden* 'nikt, ani jeden'. *Slavia Orientalis* 27, 203–5.

Siller-Rungaldier, Heidi (1985). La negazione nel ladino centrale. *Revue de linguistique romane* 49, 71–85.

Simeone-Senelle, Marie-Claude (1997). The modern South Arabian languages. In Robert Hetzron (ed.), *The Semitic languages*, 378–423. London: Routledge.

Šimík, Radek (2008). On free-choice-like *wh*-based expressions in Czech. MS, University of Groningen.

Sivonen, Anja (1977). *Kiellon ilmaisemisesta ersämordvassa*. Master's dissertation. Department of Finno-Ugric Languages, University of Turku.

Skok, Petar (1971). *Etimologijski rječnik hrvatskoga ili srpskoga jezika*. Zagreb: Jugoslavenska Akademija Znanosti i Umjetnosti.

Smyth, Herbert W. (1956). *Greek grammar*. Cambridge, Mass.: Harvard University Press.

Snoj, Marko (2003). *Slovenski etimološki slovar*, 2nd edn. Ljubljana: Modrijan.

Sprinčak, Jakov Alekseevič (1960). *Očerk russkogo istoričeskogo sintaksisa (prostoe predloženie)*. Kiev: Radjans'ka Škola.

Stanislav, Ján (1967–73). *Dejiny slovenského jazyka*, 5 vols. Bratislava: Vydavatel'stvo Slovenskej Akadémie Vied.

Stecenko, Aleksej Nikitič (1972). *Istoričeskij sintaksis russkogo jazyka*. Moscow: Vysšaja Škola.

Stechow, Arnim von (1984). Comparing semantic theories of comparison. *Journal of Semantics* 3, 1–77.

Stellmacher, Dieter (1990). *Niederdeutsche Sprache: Eine Einführung*. Bern: Peter Lang.

Sten, Holger (1938). *Nægtelserne i fransk: En historisk-syntaktisk fremstilling*. Copenhagen: Nyt Nordisk Forlag.

Stoett, Frederik August (1923). *Middelnederlandsche spraakkunst: Syntaxis*. The Hague: Martinus Nijhoff.

Stussi, Alfredo (1999). Versi d'amore in volgare tra la fine del secolo XII e l'inizio del XIII. *Cultura Neolatina* 59, 1–69.

Sundquist, John D. (2007). Variable use of negation in Middle Low German. In Joseph C. Salmons and Shannon Dubenion-Smith (eds.), *Historical linguistics 2005: Selected papers from the 17th International Conference on Historical Linguistics, Madison, Wisconsin, 31 July–5 August 2005*, 149–66. Amsterdam: John Benjamins.

Sussex, Roland and Paul Cubberley (2006). *The Slavic languages*. Cambridge: Cambridge University Press.

Švedova, Natal'ja Jul'evna (ed.) (1998). *Russkij semantičeskij slovar'*. Moscow: Azbukovnik.

Swart, Henriëtte de (2006). Marking and interpretation of negation: A bi-directional Optimality Theory approach. In Rafaella Zanuttini, Héctor Campos, Elena Herburger, and Paul Portner (eds.), *Crosslinguistic research in syntax and semantics: Negation, tense and clausal architecture*, 199–218. Washington, DC: Georgetown University Press.

Swart, Henriëtte de and Ivan A. Sag (2002). Negation and negative concord in Romance. *Linguistics and Philosophy* 25, 373–417.

Szinnyei, József (1884). Volt-e a magyar nyelvnek tagadó igeje? *Magyar Nyelvőr* 13, 145–52.

Tagliavini, Carlo (1969). *Le origini delle lingue neolatine*. Bologna: Patron.

Tanaka, Shin (1994). Die Mehrfachnegation: Ein Sprachgut der Unraffinierten? In Dieter W. Halwachs and Irmgard Stütz (eds.), *Sprache, Sprechen, Handeln: Akten des 28. Linguistischen Kolloquiums, Graz 1993*, 191–8. Tübingen: Niemeyer.

Tauli, Valter (1966). *Structural tendencies of languages*, vol. ii: *Structural tendencies in Uralic languages*. The Hague: Mouton.

Tauli, Valter (1983). *Standard Estonian grammar: Syntax*. Uppsala: Acta Universitatis Upsaliensis.

Tavernier, C. (1959). Over negatief en expletief *en* in het Gents dialekt. *Taal en Tongval* 11, 245–52.

Tesnière, Lucien (1959). *Eléments de syntaxe structurale*. Paris: Klincksieck.

Teyssier, Paul (1986). La négation dans les créoles portugais. In Jean-Claude Bouvier (ed.), *Morphosyntaxe des langues romanes: Actes du XVIIe Congrès International de Linguistique et Philologie Romanes (Aix-en-Provence, 29 août–3 septembre 1983)*, vol. iv, 591–604. Aix-en-Provence: Université de Provence.

Theil, Rolf (2006). Is Omotic Afro-Asiatic? MS, University of Oslo.

Thüringisches Wörterbuch (1966–2006). Edited by Victor Michels, Hermann Hucke, and Karl Spangenberg. Sächsische Akademie der Wissenschaften zu Leipzig, Sprachwissenschaftliche Kommission. Berlin: Akademie-Verlag.

Thurneysen, Rudolf (1946). *A grammar of Old Irish*. Dublin: Dublin Institute for Advanced Studies.

Timberlake, Alan (1975). Hierarchies in the genitive of negation. *Slavic and East European Journal* 19, 125–38.

Timberlake, Alan (1977). Reanalysis and actualization in syntactic change. In Charles N. Li (ed.), *Mechanisms of syntactic change*, 141–77. Austin: University of Texas Press.

Togeby, Knud (1974). *Précis historique de l'ancien français*. Copenhagen: Akademisk Forlag.

Tomić, Olga Mišeska (1999). Negation and imperatives. *Catalan Working Papers in Linguistics* 7, 191–206.

Tonciulescu, Keren C. (2007). Indefinites and negative polarity items in Modern Hebrew. *Cahiers Linguistiques d'Ottawa* 35, 183–206.

Toso, Fiorenzo (1997). *Grammatica del genovese*. Genoa: Le Mani.

Tottie, Gunnel (1991). Lexical diffusion in syntactic change: Frequency as a determinant of linguistic conservatism in the development of negation in English. In Dieter Kastovsky (ed.), *Historical English syntax*, 439–67. Berlin: Mouton de Gruyter.

Trask, Robert L. (1997). *The history of Basque*. London: Routledge.

Tropper, Josef (2002). *Altäthiopisch: Grammatik des Ge'ez mit Übungstexten und Glossar.* Münster: Ugarit-Verlag.

Trubačev, Oleg Nikolaevič (ed.) (1974). *Ètimologičeskij slovar' slavjanskix jazykov.* Moscow: Nauka.

Trubinskij, Valentin Ivanovič (1984). *Očerki russkogo dialektnogo sintaksisa.* Leningrad: Izdatel'stvo Leningradskogo Universiteta.

Trudgill, Peter (1994). Language contact and dialect contact in linguistic change. In Ulla-Britt Kotsinas and John Helgander (eds.), *Dialektkontakt, språkkontakt och språkförändring i Norden: Föredrag från ett forskarsymposium*, 13–22. Stockholm: Institutionen för Nordiska Språk, Stockholms Universitet.

Trudgill, Peter (2004). The impact of language contact and social structure on linguistic structure: Focus on the dialects of Modern Greek. In Bernd Kortmann (ed.), *Dialectology meets typology: Dialect grammar from a cross-linguistic perspective*, 435–52. Berlin: Mouton de Gruyter.

Tryon, Darrell T. (1970). *An introduction to Maranungku (Northern Australia).* Canberra: Australian National University.

Tsangalidis, Anastasios (1999). *Will and tha: A comparative study of the category future.* Thessaloniki: University Studio Press.

Tsurska, Olena (2009). The negative cycle in Early and Modern Russian. In Elly van Gelderen (ed.), *Cyclical change*, 73–90. Amsterdam: John Benjamins.

Tubau, Susagna (2008). Negative concord in English and Romance: Syntax–morphology interface conditions on the expression of negation. Doctoral dissertation, University of Amsterdam.

Turunen, Rigina (2006). Complex morphosyntactic features of nominal predicates in Erzya. *SKY Journal of Linguistics* 19, 173–87.

Turunen, Rigina (2011). Parts of speech in non-typical function: (A)symmetrical encoding of non-verbal predicates in Erzya. *Linguistic Discovery* 9, 137–72.

Ukaji, Masatomo (1992). 'I not say': Bridge phenomenon in syntactic change. In Matti Rissanen, Ossi Ihalainen, Terttu Nevalainen, and Irma Taavitsainen (eds.), *History of Englishes: New methods and interpretations in historical linguistics*, 453–62. Berlin: Mouton de Gruyter.

Urbańczyk, Stanisław (ed.) (1953–2002). *Słownik staropolski.* Warsaw: Polska Akademia Nauk.

Vai, Massimo (1995). Alcuni aspetti della negazione in milanese da Bonvesin a oggi. In Emanuele Banfi, Giovanni Bonfadini, Patrizia Cordin, and Maria Iliescu (eds.), *Italia settentrionale: Crocevia di idiomi romanzi. Atti del convegno internazionale di studi, Trento, 21–23 ottobre 1993*, 159–69. Tübingen: Niemeyer.

Vaillant, André (1963–4 [1948]). *Manuel du vieux slave.* Paris: Institut d'Études slaves.

van der Auwera, Johan (2009). The Jespersen cycles. In Elly van Gelderen (ed.), *Cyclical change*, 35–71. Amsterdam: John Benjamins.

van der Auwera, Johan (2010a). On the diachrony of negation. In Laurence R. Horn (ed.), *The expression of negation*, 73–109. Berlin: Mouton de Gruyter.

van der Auwera, Johan (2010b). Prohibition: Constructions and markers. In Dingfang Shu and Ken Turner (eds.), *Contrasting meaning in languages of the East and West*, 443–75. Bern: Peter Lang.

van der Auwera, Johan and Lauren van Alsenoy (2011). On the diachrony of double clausal negation and negative concord. Paper presented at the 20th International Conference on Historical Linguistics, National Museum of Ethnology, Osaka, Japan.

van der Auwera, Johan and Paul Gybels (forthcoming). On negation, indefinites, and negative indefinites in Yiddish. In Björn Hansen and Marion Aptroot (eds.), *Yiddish language structures*. Berlin: Mouton de Gruyter.

van der Auwera, Johan and Ludo Lejeune (2011). The prohibitive. In Matthew S. Dryer and Martin Haspelmath (eds.), *The world atlas of language structures online*, <http://wals.info/chapter/71>. Munich: Max Planck Digital Library. Electronic version of Johan van der Auwera and Ludo Lejeune (2005). The prohibitive. In Martin Haspelmath, Matthew Dryer, David Gil, and Bernard Comrie (eds.), *World atlas of language structures*, 290–3. Oxford: Oxford University Press.

Vasmer, Max (1953). *Russisches etymologisches Wörterbuch*. Heidelberg: Winter.

Večerka, Radoslav (1958). Střídaní záporového genitivu se záporovým akusativem v staroslověnštině. In Václav Máchek (ed.), *Studie ze slovanské jazykovědy: Sborník k 70. narozeninám akademika Františka Trávníčka*, 185–200. Prague: Státní Pedagogické Nakladatelství.

Večerka, Radoslav (1989). *Altkirchenslavische (altbulgarische) Syntax I: Die lineare Satzorganisation*. Freiburg: U. W. Weiher.

Večerka, Radoslav (1993). *Altkirchenslavische (altbulgarische) Syntax II: Die innere Satzstruktur*. Freiburg: U. W. Weiher.

Večerka, Radoslav (1995). Eine Skizze zur altkirchenslavischen Syntax: Die Negation. *Byzantinoslavica* 56, 511–22.

Večerka, Radoslav (1996). *Altkirchenslavische (altbulgarische) Syntax III: Die Satztypen: Der einfache Satz*. Freiburg: U. W. Weiher.

Vennemann, Theo (1974). Topics, subjects, and word order: From SXV to SVX via TVX. In John M. Anderson and Charles Jones (eds.), *Historical linguistics: Proceedings of the First International Conference on Historical Linguistics, Edinburgh, 2nd–7th Sept. 1973*, 339–76. Amsterdam: North Holland.

Vergauts, Marleen (1971). *Syntactische kenmerken van het Geraardsbergs dialect*. Licentiate's thesis, University of Ghent.

Veselinova, Ljuba N. (2010). Standard and special negators in the Slavonic languages: Synchrony and diachrony. In Björn Hansen and Jasmina Grković-Major (eds.), *Diachronic Slavonic syntax: Gradual changes in focus*, 195–208. Munich: Otto Sagner.

Viitso, Tiit-Rein (1998). Estonian. In Daniel Abondolo (ed.), *The Uralic languages*, 115–48. London: Routledge.

Vilkuna, Maria (1998). Word order in European Uralic. In Anna Siewierska (ed.), *Constituent order in the languages of Europe*, 173–233. Berlin: Mouton de Gruyter.

Vinogradov, Viktor Vladimirovič and Natal'ja Jul'evna Švedova (1964). *Očerki po istoričeskoj grammatike russkogo literaturnogo jazyka XIX veka: Izmenenija v sisteme prostogo i osložnennogo predloženija*. Moscow: Nauka.

Visconti, Jacqueline (2009). From 'textual' to 'interpersonal': On the diachrony of the Italian particle *mica*. *Journal of Pragmatics* 41, 937–50.

Vitanova, Marijana (2002). Neopredelitelni i otricatelni mestoimenni formi, obrazuvani s *edin* v rupskite govori. *Bălgarska Reč* 8, 46–50.

Vitanova, Marijana (2005). Neopredelitelni mestoimenni formi v bălgarskite dialekti. *Naučni izsledvanija v čest na prof. dr. Bojan Bajčev*, 61–8. Veliko Tărnovo: Universitetsko izdatelstvo Sv. sv. Kiril i Metodij.

Vondrák, Václav (1928 [1906–8]). *Vergleichende slavische Grammatik*, 2nd edn. Göttingen: Vandenhoeck und Ruprecht.

Vries, Jan W. de, Roland Willemyns, and Peter Burger (1993). *Het verhaal van een taal: Negen eeuwen Nederlands.* Amsterdam: Prometheus.

Wagner, Ewald (1983). *Harari-Texte in arabischer Schrift.* Wiesbaden: Franz Steiner Verlag.

Wagner, Ewald (1997). Harari. In Robert Hetzron (ed.), *The Semitic languages*, 486–508. London: Routledge.

Wallage, Phillip (2005). Negation in Early English: Parametric variation and grammatical competition. Doctoral dissertation, University of York.

Wallage, Phillip (2008). Jespersen's cycle in Middle English: Parametric variation and grammatical competition. *Lingua* 118, 643–74.

Wallage, Phillip (2012). Identifying the role of pragmatic change in the grammaticalisation of Middle English *not*. Paper presented at New Reflections on Grammaticalisation V, University of Edinburgh.

Watanabe, Akira (2004). The genesis of negative concord: Syntax and morphology of negative doubling. *Linguistic Inquiry* 35, 559–612.

Watson, Janet C. E. (1993). *A syntax of San'ani Arabic.* Wiesbaden: Harrassowitz.

Watterson, Barbara (1988). *Coptic Egypt.* Edinburgh: Scottish Academic Press.

Webelhuth, Gert (1990). Diagnostics for structure. In Günther Grewendorf and Wolfgang Sternefeld (eds.), *Scrambling and barriers*, 41–75. Amsterdam: John Benjamins.

Wehr, Hans (1979). *A dictionary of modern written Arabic.* Wiesbaden: Harrassowitz.

Weijnen, A. (1956). *Zeventiende-eeuwse Taal*, 2nd edn. Zutphen: Thieme.

Weiß, Helmut (1998). *Syntax des Bairischen: Studien zur Grammatik einer natürlichen Sprache.* Tübingen: Niemeyer.

Wheeler, Max W. (1988). Catalan. In Martin Harris and Nigel Vincent (eds.), *The Romance languages*, 170–208. London: Routledge.

Wheelock LaBrum, Rebecca (1982). *Conditions on double negation in the history of English with comparison to similar developments in German*, Stanford University doctoral dissertation. Stanford: University Microfilms International.

Whitney, Arthur H. (1944). *Colloquial Hungarian.* London: K. Paul, Trench, Trubner & Co.

Willis, David (1998). *Syntactic change in Welsh: A study of the loss of verb-second.* Oxford: Clarendon Press.

Willis, David (2000). Verb movement in Slavonic conditionals. In Susan Pintzuk, George Tsoulas and Anthony Warner (eds.), *Diachronic syntax: Models and mechanisms*, 322–48. Oxford: Oxford University Press.

Willis, David (2006). Negation in Middle Welsh. *Studia Celtica* 40, 63–88.

Willis, David (2007). Syntactic lexicalization as a new type of degrammaticalization. *Linguistics* 45, 271–310.

Willis, David (2008). English–Welsh language contact in the realm of negation. Paper presented at Transmission and Diffusion Congress, Radboud University, Nijmegen.

Willis, David (2010). Motivating the emergence of new markers of sentential negation: The case of Welsh *ddim*. *Diachronica* 27, 110–56.

Willis, David (2011a). A minimalist approach to Jespersen's cycle in Welsh. In Dianne Jonas, John Whitman, and Andrew Garrett (eds.), *Grammatical change: Origins, nature, outcomes*, 93–119. Oxford: Oxford University Press.

Willis, David (2011b). Reconstructing last week's weather: Syntactic reconstruction and Brythonic free relatives. *Journal of Linguistics* 47, 407–46.

Willis, David (2012). Negative polarity and the quantifier cycle: Comparative perspectives from European languages. In Pierre Larrivée and Richard P. Ingham (eds.), *The evolution of negation: Beyond the Jespersen cycle*, 313–56. Berlin: Mouton de Gruyter.

Willmore, John (1901). *The spoken Arabic of Egypt*. London.

Willmott, Jo C. (2007). *The moods of Homeric Greek*. Cambridge: Cambridge University Press.

Willmott, Jo C. (2008). Dynamic optatives in Homeric Greek. *Cambridge Classical Journal* 54, 237–51.

Willmott, Jo C. (2009). Not in the mood: Modality and negation in the history of Greek. Paper presented at the 29th Annual Meeting of the Department of Linguistics, Thessaloniki. Available online at <http://jowillmott.co.uk/research/publications.html>.

Willmott, Jo C. (2010). The semantics of negative directives in Homeric Greek: A typological account. In Ignazio Putzu, Giulio Paulis, Gianfranco Nieddu, and Pierluigi Cuzzolin (eds.), *La morfologia del greco tra tipologia e diacronia: Atti del VII Incontro internazionale di linguistica greca*, 529–50. Milan: Franco Angeli.

Willmott, Jo C. (2011). Outis and what he can tell us about negation in Homeric Greek. In Paul Millett, S. P. Oakley, and R. J. E. Thompson (eds.), *Ratio et res ipsa: Classical essays presented by former pupils to James Diggle on his retirement*, 63–79. Cambridge: Cambridge Philological Society.

Winkel, J. te (1901). *Geschiedenis der nederlandsche taal*. Culemborg: Blom and Olivierse.

Winters, Margaret E. (1987). Innovations in French negation: A cognitive grammar account. *Diachronica* 4, 27–52.

Witkoś, Jacek (2008). Genitive of negation in Polish and single-cycle derivations. *Journal of Slavic Linguistics* 16, 247–87.

Wmffre, Iwan (1998). *Late Cornish*. Munich: Lincom Europa.

Woidich, Manfred (1968). *Negation und negative Sätze im Ägyptisch-Arabischen*. Doctoral dissertation, Ludwig-Maximilians-Universität, Munich.

Woidich, Manfred (2006). *Das Kairenisch-Arabische: Eine Grammatik*. Wiesbaden: Harrassowitz Verlag.

Wolf, Norbert Richard (2000). Syntax des Mittelhochdeutschen. In Besch et al. (2000), 1351–8.

Woodcock, E. C. (1959). *A new Latin syntax*. London: Methuen.

Wouden, Ton van der (1995). On the development of marked negation systems. Paper presented at the 12th International Conference on Historical Linguistics, University of Manchester. Available at <http://odur.let.rug.nl/~vdwouden/docs/ichl.pdf>.

Wouden, Ton van der (1997). *Negative contexts: Collocation, polarity and multiple negation*. London: Routledge.

Wurff, Wim van der (1999). Objects and verbs in Modern Icelandic and fifteenth-century English: A word order parallel and its causes. *Lingua* 109, 237–65.

Yvon, Henri (1948). *Pas* et *point* dans les propositions négatives. *Le Français Moderne* 16, 19–35.

Zagona, Karen (1982). *Government and proper government of verbal projections*. Doctoral dissertation, University of Washington, Seattle.

Zaliznjak, A. A. (2004). *Drevnenovgorodskij dialekt*, 2nd edn. Moscow: Jazyki Slavjanskoj Kul'tury.

Zanuttini, Raffaella (1994). Speculations on negative imperatives. *Rivista di Linguistica* 6, 67–89.

Zanuttini, Raffaella (1997). *Negation and clausal structure: A comparative study of Romance languages*. New York: Oxford University Press.

Zanuttini, Raffaella (2010). La negazione. In G. P. Salvi and Lorenzo Renzi (eds.), *Grammatica dell'italiano antico*, 569–82. Bologna: Il Mulino.

Zanuttini, Raffaella and Paul Portner (2000). The characterization of exclamative clauses in Paduan. *Language* 76, 123–32.

Zeijlstra, Hedde H. (2004). *Sentential negation and negative concord*. Doctoral dissertation, University of Amsterdam. Utrecht: LOT Publications.

Zeijlstra, Hedde H. (2006). The ban on true negative imperatives. In Olivier Bonami and Patricia Cabredo Hofherr (eds.), *Empirical issues in formal syntax and semantics 6*, 405–24. Paris: Colloque de syntaxe et sémantique à Paris. Available online at <http://www.cssp.cnrs.fr/eiss6/index_en.html>.

Zeijlstra, Hedde H. (2008). Negative concord is syntactic agreement. MS, University of Amsterdam.

Zeijlstra, Hedde H. (2009). On French negation. MS, University of Amsterdam. Available online at <http://ling.auf.net/lingBuzz/000885>.

Zeli, R. (1968). Spunti per lo studio della negazione nei dialetti del Ticino e del Moesano. *Vox Romanica* 27, 289–98.

Zenčuk, V. N. (1972). Ob odnom iz spornyx voprosov v svjazi s problemoj negacii v slavjanskix jazykax. *Slavjanskaja Filologija* 2, 54–62.

Zörner, L. (1994). I dialetti dell'Oltrepò pavese tra il lombardo, l'emiliano e il ligure. *Rivista Italiana di Dialettologia* 17, 55–98.

Zuckerman, Ghil'ad (2009). Hybridity versus revivability: Multiple causation, forms and patterns. *Journal of Language Contact* 2, 40–67.

Zwarts, Frans (1995). Nonveridical contexts. *Linguistic Analysis* 25, 286–312.

Zwarts, Frans (1998). Three types of polarity. In Fritz Hamm and Erhard Hinrichs (eds.), *Plurality and quantification*, 177–238. Dordrecht: Kluwer.

Index of languages

Index of subjects

OXFORD STUDIES IN DIACHRONIC AND HISTORICAL LINGUISTICS

PUBLISHED

1
From Latin to Romance
Morphosyntactic Typology and Change
Adam Ledgeway

2
Parameter Theory and Linguistic Change
Edited by Charlotte Galves, Sonia Cyrino, Ruth Lopes, Filomena Sandalo, and Juanito Avelar

3
Case in Semitic
Roles, Relations, and Reconstruction
Rebecca Hasselbach

4
The Boundaries of Pure Morphology
Diachronic and Synchronic Perspectives
Edited by Silvio Cruschina, Martin Maiden, and John Charles Smith

5
The History of Negation in the Languages of Europe and the Mediterranean
Volume I: *Case Studies*
Edited by David Willis, Christopher Lucas, and Anne Breitbarth

6
Constructionalization and Constructional Changes
Elizabeth Traugott and Graeme Trousdale

7
Word Order in Old Italian
Cecilia Poletto

IN PREPARATION

Diachrony and Dialects
Grammatical Change in the Dialects of Italy
Edited by Paola Benincà, Adam Ledgeway, and Nigel Vincent

Syntax over Time
Lexical, Morphological, and Information-Structural Interactions
Edited by Theresa Biberauer and George Walkden

The History of Negation in Low German
Anne Breitbarth

Pragmatic Markers from Latin to the Romance Languages
Edited by Chiara Ghezzi and Piera Molinelli

Nominal Expressions and Language Change
From Early Latin to Modern Romance
Giuliana Giusti

The Historical Dialectology of Arabic: Linguistic and Sociolinguistic Approaches
Edited by Clive Holes

A Study in Grammatical Change
The Modern Greek Weak Subject Pronoun τος and its Implications for Language Change and Structure
Brian D. Joseph

Gender from Latin to Romance
Michele Loporcaro

Vowel Quantity from Latin to Romance
Michele Loporcaro

The Syntax and Semantics of Vedic Particles
John J. Lowe

Syntactic Change and Stability
Joel Wallenberg

The History of Negation in the Languages of Europe and the Mediterranean
Volume II: *Patterns and Processes*
David Willis, Christopher Lucas, and Anne Breitbarth